PSEUDO DIFFERENTIAL OPERATORS & MARKOV PROCESSES

Contents

Preface

This is the second volume of my treatise on "Pseudo-differential operators and Markov processes", and it covers two topics: Generating Feller and sub-Markovian semigroups by starting with pseudo-differential operators with negative definite symbols, and the potential theory, more correctly parts of the potential theory, of these operators and their corresponding semigroups.

The style of the presentation changes in so far as we address the book to readers being familiar with the essential content of volume I.

Chapter 2 relies essentially on work done by my former PhD-students and myself and this chapter includes many results being presented for the first time in a monograph. I owe special thanks to W.Hoh and R.L.Schilling for allowing me to incorporate their work and ideas, partly following closely their presentations. This chapter provides large classes of pseudo-differential operators generating Feller and sub-Markovian semigroups — in a certain sense it shows the richness of the theory.

Chapter 3 is devoted to the potential theory of semigroups generated by pseudo-differential operators with negative definite symbols. In principle we provided non-trivial, new examples in Chapter 2 for various aspects of potential theory. Therefore it would be straightforward just to work out these examples. However, in order to obtain a more self-contained presentation we decided to include larger parts of "standard" potential theory. But we do not believe that we have to "re-invent the wheel", i.e. in the general theory we followed partly rather closely existing work, stressing however these aspects which are of special interest and need in handling our examples. In addition we had to extend the theory and added quite a lot of new material. Again I owe thanks to my collaborators W.Farkas, W.Hoh and R.L.Schilling for allowing me to include their (or our) work. But even more, I owe thanks to those authors from whom I borrowed for my presentation. In the "Introduction" and often in the text we make rather precise references, in fact we prefered to have the references close to the results and thus we need no additional notes.

No doubt, the following passage taken from the preface of volume I remains valid:

"One of the major difficulties in writing the book was that in different fields of mathematics different notations and normalizations were used. ... A second, much larger problem was and still is the enormous amount of work done in fields we have to deal with."

Hence it is clear that we could not cover all material related to our topic and we hope that we succeeded to give fair credits.

Preparing this volume I received support from many sides. First of all I want to thank R.L.Schilling who read the whole manuscript, suggested once again many improvements and partly, proposed new or different proofs, or even proved new results which I was allowed to include immediately. Secondly, I want to thank W.Hoh for comments to some earlier draft of parts of the manuscript, but in particular, to allow me to use for writing Chapter 2 to larger extent his "Habilitationsschrift" which unfortunately was never published separately.

Special thanks go to M.Röckner and H.Triebel for support in a critical academic year 1999/2000, as well as to my department at the University of Wales Swansea, especially A.Truman, where the major part of the book was written in the last one and a half year.

Many thanks to V.Knopova for transforming my handwritten manuscript into a TEX-file, and to B.Boettcher for proof-reading the type-written version. A.G.Tokarev is to thank for putting the TEX-file into its final form including doing all corrections, and I.M.Davies helped whenever the computer system made some difficulties — many thanks to all of them.

During the work on the book I received some financial support from DFG, contracts Ja $552_{7-1,7-2}$, and EPSRC, contract GR/R40241/01.

Last but not least many thanks to Ye Qiang and his team in Singapore who made the cooperation with the publisher a pleasure.

Swansea, April 2002

Niels Jacob

Notation

I.xyz means: page xyz in volume I

General Notation

$\mathcal{P}(G)$ power set of G
$\mathbb{N}$ natural numbers
$\mathbb{N}_0 = \mathbb{N} \cup \{0\}$
$\mathbb{N}_0^n$ set of all multiindices
$\mathbb{N}_{0,\gamma}^n = \{\beta \in \mathbb{N}_0^n; |\beta| > 0 \text{ and } 0 < |\beta| \leq \gamma\}$
$\mathbb{Z}$ integers
$\mathbb{Q}$ rational numbers
$\mathbb{R}$ real numbers
$\mathbb{R}_+ = \{x \in \mathbb{R}; x \geq 0\}$
$\mathbb{R}^n$ euclidean vector space
$Q_R = \{x \in \mathbb{R}^n; |x_i| \leq R\}$
$B_R(a) = \{x \in \mathbb{R}^n; |x - a| < R\}$
$S^{n-1} = \{x \in \mathbb{R}^n; |x| = 1\}$
ω_n volume of $B_1(0) \subset \mathbb{R}^n$
$T_x\mathbb{R}^n$ tangent space to $\mathbb{R}^n$ at x
$\mathbb{C}$ complex numbers
$S_{\theta,d_0} = \{z \in \mathbb{C}; |\arg(z - d_0)| < \theta\}$
$\mathbb{C}^n$ unitary vector space
$a \wedge b = \min(a, b)$
$a \vee b = \max(a, b)$
$\text{sign } a = \begin{cases} \frac{a}{|a|}, & a \neq 0 \\ 0, & a = 0 \end{cases}$ signum of a

$|A|$ cardinality of a finite set

$A \setminus B$ set theoretical difference of two sets

A^c complement of the set A

$diag\, A = \{(x,x); x \in A\}$

$A + B = \{x + y; x \in A \text{ and } y \in B\}$

$\lambda A = \{\lambda x; x \in A\}$, $\lambda \in \mathbb{R}$ or $\lambda \in \mathbb{C}$

$lin\, A$ linear hull of A

$conv\, A$ convex hull of A

$ext\, A$ extremal points of A

$H_1 \oplus H_2$ direct sum in a vector space

$\bar{A}$ closure of a set

$\overset{\circ}{A}$ (open) interior of a set

∂A boundary of a set

$\partial_H K$ Choquet boundary

$A_\infty = A \cup \{\infty\}$ one point compactification of A

$\mathcal{O}_\infty$ topology of A_∞

$|\alpha| = \alpha_1 + \cdots + \alpha_n$ for $\alpha \in \mathbb{N}_0^n$

$\alpha! = \alpha_1! \ldots \alpha_n!$ for $\alpha \in \mathbb{N}_0^n$

$\alpha \leq \beta$ means $\alpha_i \leq \beta_i$ for $\alpha, \beta \in \mathbb{N}_0^n$

$\binom{\alpha}{\beta} = \binom{\alpha_1}{\beta_1} \cdot \cdots \cdot \binom{\alpha_n}{\beta_n}$ for $\alpha, \beta \in \mathbb{N}_0^n$

$x^\alpha = x_1^{\alpha_1} \cdot \cdots \cdot x_n^{\alpha_n}$, $\alpha \in \mathbb{N}_0^n$ and $x \in \mathbb{R}^n$

$\partial^\alpha u = \frac{\partial^{|\alpha|} u}{\partial x_1^{\alpha_1} \ldots \partial x_n^{\alpha_n}}$

$D^\alpha u = (-i\partial)^\alpha u$

$\frac{d^+}{dt} u(x) = \lim\limits_{\substack{t \to 0 \\ t > 0}} \frac{u(x+t) - u(x)}{t}$

$\frac{d^-}{dt} u(x) = \lim\limits_{\substack{t \to 0 \\ t < 0}} \frac{u(x+t) - u(x)}{t}$

$\bar{\partial}_z u = \left(\frac{\partial}{\partial x} + i \frac{\partial}{\partial y}\right) u$, $z = x + iy$

$I^{(\alpha)}$ fractional integral

D_+^α, D_-^α fractional derivative for $\alpha \in \mathbb{R} \setminus \mathbb{N}$

$ind_\gamma(z)$ index of $z \in \mathbb{C}$ with respect to a closed path γ

$j \succeq j_0$ order relation in a net

$\lim\limits_{j \in I}$ limit with respect to a net

Functions and Distributions

χ_A characteristic function of the set A
id_A identity map in A
$f_{|A}$ restriction of f to A
f^+ positive part of f, i.e. $f^+ = f \vee 0$
f^- negative part of f, i.e. $f^- = -(f \vee 0)$
$\mathrm{Re}\, f$ real part of a function
$\mathrm{Im}\, f$ imaginary part of a function
f_∞ function extended from A to A_∞ by $f(\infty) = 0$
$(f_\nu)_{\nu\in\mathbb{N}}$ sequence of functions
$f_\nu \uparrow f$ increasing sequence of functions tending to f
$f_\nu \downarrow f$ decreasing sequence of functions tending to f
$f \circ g$ composition of functions
$f \otimes g$ tensor product of functions (or distributions)
$f * g$ convolution of functions (or distributions)
$j_\varepsilon, J_\varepsilon(u)$ Friedrichs mollifier
$Sx = -x$ reflection at the origin
$T_a x = x + a$ translation by a
$H_\lambda x = \lambda x$ homothetic map
$\pi_j(x) = x_j$ j-th coordinate projection
$B(\alpha, \beta)$ Beta-function
$J_{\frac{n}{2}}$ Bessel function of first kind and order $\frac{n}{2}$
K_1 Bessel function of third kind and index 1
$r_\nu(t)$ Rademacher functions
T^*u pullback of a function or distribution w.r.t. the diffeomorphism T
$\hat{u}$, Fu Fourier transform
$F^{-1}u$ inverse Fourier transform
$\left.\begin{array}{l}\delta_{diag}\\ \mathrm{id}_{kernel}\end{array}\right\}$ kernel of $\mathrm{id}_{C_0^\infty(G)}$
$\mathrm{supp}\, u$ support of a function or distribution
$\mathrm{ess\,sup}\, u$ essential supremum of a function
$\mathrm{sing\,supp}$ singular support of a function or distribution

Measures and Integrals

$\mathcal{B}^{(n)}$ Borel sets in $\mathbb{R}^n$
$\mathcal{B}(\mathcal{G})$ Borel sets of a topological space G

$\sigma(S)$ σ-field generated by S
$d(S)$ Dynkin system generated by S
$\lambda^{(n)}$ Lebesgue measure in $\mathbb{R}^n$
ε_a Dirac measure at $a \in \mathbb{R}^n$
ε Dirac measure at $0 \in \mathbb{R}^n$
$\mu_1 \otimes \mu_2$ product of the measures μ_1 and μ_2
$\mu_1 * \mu_2$ convolution of the measures μ_1 and μ_2
$\mu(\varphi) = \int \varphi d\mu$
$\|\mu\|$ total mass of a measure μ
$\operatorname{supp} \mu$ support of a measure μ
$(\mu_t)_{t\geq 0}$ convolution semigroup of subprobabilities
$(\mu_t^f)_{t\geq 0}$ subordinate convolution semigroup
$(\sigma_t^\alpha)_{t\geq 0}$ one–sided symmetric stable semigroup
$v.p.$ principial value of an integral
$\int_{a+}^{b} = \int_{(a,b]}$
$\int_{a}^{b+} = \int_{[a,b)}$
$\oint_\gamma$ integral over the closed path γ

Spaces of Functions, Measures and Distributions

If $X(\Omega)$ is one of the spaces in the following list, $X(\Omega, \mathbb{R})$ stands for the subspace of real–valued elements of $X(\Omega)$. In the following $G \subseteq \mathbb{R}^n$ stands always for an open set, $K \subset \mathbb{R}^N$ for a compact set.

$A(\mathbb{R}^n)$ Wiener algebra
$B(\Omega)$ Borel measurable function
$B_b(\Omega)$ bounded Borel measurable function
$B_{pq}^s(\mathbb{R}^n)$ Besov space
$B_{k,p}(\mathbb{R}^n)$ Hörmander space, see page I.207
$B_{\psi,p}^s(\mathbb{R}^n)$ see page I.207
$B_{\psi_1,\psi_2,p}^s(\mathbb{R}^n)$ see page I.212
$C(G)$ continuous functions
$C_0(G)$ continuous functions with compact support
$C_\infty(G)$ continuous functions vanishing at infinity
$C_u(G)$ uniformly continuous and bounded functions
$C_b(G)$ bounded continuous functions
$C^{0,\lambda}(G)$ Hölder continuous functions with exponent $\lambda \in (0,1]$

$C(\bar{G})) = \{u \in C(G); u \text{ extends continuously to } \bar{G}\}$
$C^m(G)$ m–times continuously differentiable functions
$C_0^m(G) = C^m(G) \cap C_0(G)$
$C_\infty^m(G) = \{u \in C^m(G); \partial^\alpha u \in C_\infty(G), |\alpha| \le m\}$
$C_u^m(G) = \{u \in C^m(G); \partial^\alpha u \in C_u(G), |\alpha| \le m\}$
$C_b^m(G) = \{u \in C^m(G); \partial^\alpha u \in C_b(G), |\alpha| \le m\}$
$C_\infty^{m,\lambda}(G) = \{u \in C^m(G); \partial^\alpha u \in C^{0,\lambda}(G), |\alpha| = m\}$
$C^m(\bar{G}) = \{u \in C^m(G); \partial^\alpha u \in C(\bar{G}), |\alpha| \le m\}$
$C^\infty(G) = \cap_{m \in \mathbb{N}} C^m(G)$
$C_0^\infty(G) = \cap_{m \in \mathbb{N}} C_0^m(G)$
$C_0^\infty(K)$ functions which are arbitrarily often differentiable and have compact support in K
$\mathcal{D}'(G)$ space of all distributions
$\mathcal{E}'(G)$ space of all distribution with compact support
$F_{pq}^s(\mathbb{R}^n)$ Triebel–Lizorkin space
$H_2^m(\mathbb{R}^n)$ classical Sobolev space of integer order
$H^{m,p}(\mathbb{R}^n)$ classical L^p-Sobolev space of integer order
$H^s(\mathbb{R}^n)$ classical Sobolev space of fractional order $s \in \mathbb{R}$
$H^{\psi,s}(\mathbb{R}^n) = \{u \in \mathcal{S}'(\mathbb{R}^n); \|u\|_{\psi,s} < \infty\}$
$H_p^{\psi,2}(\mathbb{R}^n)$ see Definition 3.3.3 on page 275
$H_p^{\psi,s}(\mathbb{R}^n)$ see (3.166) on page 279
$H_{loc}^{\psi,s}(\mathbb{R}^n) = \{u \in \mathcal{S}'(\mathbb{R}^n); \varphi u \in H^{\psi,s}(\mathbb{R}^n) \text{ for all } \varphi \in C_0^\infty(\mathbb{R}^n)\}$
$l_q = \{(a_k)_{k \in \mathbb{N}_0}; a_k \in \mathbb{C} \text{ and } \|(a_k)\|_{l_q} < \infty\}$
$L^p(\Omega, \mu)$ usual Lebesgue space over $(\Omega, \mathcal{A}, \mu)$
$L^p(G) = L^p(G, \lambda^{(n)})$
$L_{loc}^1(G) = \{u \in B(G); \chi_K u \in L^1(G) \text{ for all } K \subset G \text{ compact}\}$
$\mathcal{M}^+(\Omega)$ measures on Ω
$\mathcal{M}_b^+(\Omega)$ bounded measures on Ω
$\mathcal{M}_b^1(\Omega)$ probability measures on Ω
$\tilde{\mathcal{M}}_b(X)$ signed Radon measures on a locally compact space X with finite total mass
$\tilde{\mathcal{M}}^{\mathbb{C}}(\Omega)$ complex measure on Ω
$\mathcal{M}_b^{\mathbb{C}}(\Omega)$ bounded complex measure on Ω
$M_{p,q}$ Fourier multipliers of type (p, q)
$\mathcal{S}(\mathbb{R}^n)$ Schwartz space of tempered functions
$\mathcal{S}'(\mathbb{R}^n)$ tempered distributions
$W^{m,p}(\mathbb{R}^n)$ classical L^p-Sobolev space of integer order
$\overset{\circ}{W}{}^{m,p}(\mathbb{R}^n)$ see (2.23) on page 21

$W^{s,p}(\mathbb{R}^n)$ see (2.417) page 167

Some Families of Functions

$CN(\mathbb{R}^n)$ continuous negative definite functions
$CP(\mathbb{R}^n)$ continuous positive definite functions
$\mathcal{H}_{l.s.c.}(G)$ lower semicontinuous functions
$\mathcal{H}_{u.s.c.}(G)$ upper semicontinuous functions
$\mathcal{H}(A)$ see page I.62
$\mathcal{H}(X_0, X_1)$ see page I.71
$N(\mathbb{R}^n)$ negative definite functions
$P(\mathbb{R}^n)$ positive definite functions
$\mathcal{S}$ Stieltjes functions
$S_0(\mathbb{R}^n) = \{u \in \mathcal{S}(\mathbb{R}^n); \hat{u}(0) = \frac{\partial \hat{u}}{\partial \xi_i}(0) = 0 \text{ for } 1 \le j \le n\}$

Norms, Scalar Products and Seminorms

$|x|$ euclidean distance in $\mathbb{R}^n$
$|x|_\infty = \max\{|x_1|, \ldots, |x_n|\}$, $x \in \mathbb{R}^n$
$|x|_q = \left(\sum_{j=1}^n |x_j|^q\right)^{1/q}$
$|z|$ euclidean distance in $\mathbb{C}^n$
$\|\nabla^j u\|_{L^p}$ see (2.5) on page 16
$\|u\|_X$ norm of u in the space X
$\left.\begin{array}{l}\|u\|_{A,X,Y} = \|u\|_X + \|Au\|_Y \\ \|u\|_{A,X} = \|u\|_X + \|Au\|_X\end{array}\right\}$ graph norm with respect to the operator A
$\|A\| = \|A\|_{X,Y}$ operator norm of the operator A
$\|u\|_{X_\theta}$ see page I.72
$\|u\|_{\mathcal{H}}$ see page I.71
$\|u\|_0$, $(u, u)_0$ norm and scalar product in $L^2(\Omega, \mu)$
$\|u\|_\infty = \sup |u(x)|$ or $\operatorname{ess\,sup} |u(x)|$
$\|u\|_s$ norm in the space $H^s(\mathbb{R}^n)$
$\|u\|_{m,2}$ norm in the space $H_2^m(\mathbb{R}^n)$ and $W^{m,2}(\mathbb{R}^n)$
$\|u\|_{\psi,s}$ norm in the space $H^{\psi,s}(\mathbb{R}^n)$
$\|u\|_{\psi,s,p}$ norm in the space $B^s_{\psi,p}(\mathbb{R}^n)$
$\|u\|_{\psi_1,\psi_2,s,p}$ norm in the space $B^s_{\psi_1,\psi_2,p}(\mathbb{R}^n)$
$\|(a_k)\|_{l_q} = \left(\sum_{k=0}^\infty |a_k|^q\right)^{1/q}$

$\|(f_k)\|_{L^p(l_q)}$ see page I.227
$\|(f_k)\|_{l^q(L_p)}$ see page I.227
$p_{m_1,m_2}(u) = \sup\limits_{x\in\mathbb{R}^n} ((1+|x|^2)^{m_1/2} \sum_{|\alpha|\leq m_2} |\partial^\alpha u(x)|)$
$p_{\alpha,\beta}(u) = \sup\limits_{x\in\mathbb{R}^n} |x^\beta \partial^\alpha u(x)|$
$p_{\alpha,K}(u) = \sup\limits_{x\in K} |\partial^\alpha u(x)|$, u compact
$q_{m,K}(u) = \sum_{|\alpha|\leq m} \sup\limits_{x\in K} |\partial^\alpha u(x)|$, K compact

Notation from Functional Analysis, Operators

$\mathcal{B}_0$ base of neighbourhoods of $\mathcal{O} \in X$, X vector space
$\mathcal{B}_X = \{x + B_0, B_0 \in \mathcal{B}_0\}$
$U_{p,\varepsilon} = \{x \in X; p(x) \leq \varepsilon\}$, X vector space, p seminorm
$(X, \|\cdot\|_X)$ Banach space X with norm $\|\cdot\|_X$
X^* dual space of a topological vector space
$X \hookrightarrow Y$ continuous embedding of X into Y
$[X_0, X_1]_\theta$ complex interpolation space
$B(X,Y)$ bounded linear operators from X into Y
$B(X)$ bounded linear operators from X into itself
$s - \lim$ strong limit
$w - \lim$ weak limit
$< u, x >$ duality pairing between X^* and X
$(A, D(A))$ linear operator with domain $D(A)$
$D(A)$ domain of an operator
$R(A)$ range of an operator
$\Gamma(A)$ graph of an operator
$\mathrm{Ker}(A)$ kernel (null–space) of an operator
$\bar{A}$ closure of an operator
A^* conjugate operator or adjoint Hilbert space operator
$\rho(A)$ resolvent set of an operator
$\rho_+(A) = \rho(A) \cap (0, \infty)$
$\sigma(A)$ spectrum of an operator
$\theta(A)$ numerical range of an operator
$A_\lambda = \lambda A(\lambda - A)^{-1}$ Yosida approximation of A
$(R_\lambda)_{\lambda>0}$ resolvent of an operator
$R_\lambda = (\lambda - A)^{-1}$
$B_\lambda(u,v) = B(u,v) + \lambda(u,v)_0$ for a bilinear form B

B^{sym} symmetric part of a bilinear form B
B^{asym} antisymmetric part of a bilinear form B
K_{op} operator with kernel K
$q(x, D)$ pseudo-differential operator with symbol $q(x, \xi)$
$\psi(D)$ pseudo-differential operator with symbol $\psi(\xi)$
$(T_t)_{t\geq 0}$ one parameter semigroup of operators
$(T_t^*)_{t\geq 0}$ adjoint semigroup to $(T_t)_{t\geq 0}$
$(T_t^f)_{t\geq 0}$ subordinated semigroup
A^f generator of subordinated semigroup
$(T_t^{(\infty)})_{t\geq 0}$ semigroup on $C_\infty(\mathbb{R}^n)$
$A^{(\infty)}$ generator of $(T_t^{(\infty)})_{t\geq 0}$
$(T_t^{(p)})_{t\geq 0}$ semigroup on $L^p(\mathbb{R}^n)$, $1 < p < \infty$
$A^{(p)}$ generator of $(T_t^{(p)})_{t\geq 0}$

Notations related to Pseudo-Differential Operators

$S_{1,0}^m(\mathbb{R}^n)$ see Example 2.4.8 on page 92
$S_{\rho,\delta}^m(\mathbb{R}^n)$ see Definition 2.7.1 on page 139
$CNS_{\rho,\delta}^m(\mathbb{R}^n)$ see Definition 2.7.7 on page 140
$S_\rho^{m,\psi}(\mathbb{R}^n)$ see Definition 2.4.4 on page 89
$S_0^{m,\psi}(\mathbb{R}^n)$ see Definition 2.4.4 on page 89
$S_0^{m,m',\psi}(\mathbb{R}^n)$ see Definition 2.4.16 on page 97
$\Psi_\rho^{m,\psi}(\mathbb{R}^n)$ see Definition 2.4.10 on page 93
$\Psi_0^{m,\psi}(\mathbb{R}^n)$ see Definition 2.4.10 on page 93
Λ see Definition 2.4.3 on page 89
$\mathcal{A}$ see Definition 2.4.12 on page 94
$O_S - \int$ oscillatory integral, see Definition 2.4.12 on page 94
γ^σ see (2.361) on page 147
$\gamma(dx, dx)$ see page 147

Notations related to Potential Theory and Harmonic Analysis

Ana(G) analytic subsets of G
cap* outer capacity

cap_* inner capacity
cap capacity
$\mathrm{cap}_{r,p}$ capacity in $\mathcal{F}_{r,p}$
$\mathrm{cap}^{\psi}_{r,p}$ capacity in $H^{\psi,r}_p(\mathbb{R}^n)$
$V_r^{(p)}$ Γ-transform of $(T_t^{(p)})_{t=0}$, see Definition 3.1.17 on page 224
$\mathcal{F}_{r,p} = V_r^{(p)} L^p$ Bessel potential space of order r associated with $(T_t^{(p)})_{t=0}$
$\mathcal{F}_e$ extended Dirichlet space
$f^*(u)$ see (3.241) on page 303
$g_1(u)$ Littlewood–Paley function of order 1, see (3.294) on page 324
$g_k(u)$ Littlewood–Paley function of order k, see (3.298) on page 325
$u^*_{(p)}(x) = \sup_{t>0} |T_t^{(p)} u(x)|$
$u^*_{k,(p)}(x) = \sup_{t>0} \left|t^k \frac{\partial^k T_t^{(p)} u(x)}{\partial t^k}\right|$
$E(f)$ expectation of f
$E(f|\mathcal{F})$ conditional expectation of f w.r.t. $\mathcal{F}$

Introduction: Pseudo-Differential Operators and Markov Processes

This monograph is devoted to Markov processes with state space $\mathbb{R}^n$, in particular we will discuss jump processes. A brief, but very readable description of the development of the theory of stochastic processes until the late fifties can be found in the preface of the monograph of K.Itô and H.McKean [145]. In particular the different ways of constructing processes are pointed out: constructions using (functional) analysis (N.Wiener, A.Kolmogorov, W.Feller, K.Yosida, E.B.Dynkin) and pathwise constructions (P.Lévy, K.Itô).

Today, pathwise constructions via stochastic differential equations, for example, are in the centre of probabilists' interest and they led to the field of stochastic analysis, i.e. analysis on path spaces which is since the publication of P.Malliavin's paper [205] one of the central themes in modern probability theory. The reader is highly recommended to P.Malliavin's essays [207] to get first hand information.

However, after the publication of M.Fukushima's work [97] on Dirichlet forms and Markov processes the (functional) analytic approach to stochastic processes returned also into the centre of probabilists' interest.

In this monograph we will emphasize another point of view, namely the relation of Fourier analysis and Markov processes. This theme was first taken up by P.Lévy and in particular by S.Bochner when discussing stochastically continuous processes with stationary and independent increments, i.e. *Lévy processes.* Bochner's monograph [32] should be mentioned as the most inspiring source to these considerations. The observation is that every Lévy process $(X_t)_{t\geq 0}$ with state space $\mathbb{R}^n$ is completely determined by one and only one function $\psi : \mathbb{R}^n \to \mathbb{C}$ which is defined by the relation

$$E\left(e^{iX_t\cdot\xi}\right) = e^{-t\psi(\xi)}. \tag{0.1}$$

The function ψ, called the *characteristic exponent* of $(X_t)_{t\geq 0}$, is a continuous negative definite function and contains all information about $(X_t)_{t\geq 0}$. Some results for $(X_t)_{t\geq 0}$ are best proven by looking directly at ψ, for example the process is conservative if and only if $\psi(0) = 0$. But for pathwise considerations it is useful to take the Lévy–Khinchin representation

$$\begin{aligned}\psi(\xi) = c + ib\cdot\xi + Q(\xi) + \int_{\mathbb{R}^n\setminus\{0\}} &\left(1 - e^{-iy\cdot\xi} - \frac{iy\cdot\xi}{1+|y|^2}\right) \\ &\times \frac{1+|y|^2}{|y|^2}\mu(dy).\end{aligned} \tag{0.2}$$

Our starting point is the following observation made in [169]. Let $\left((X_t)_{t\geq 0}, P^x\right)_{x\in\mathbb{R}^n}$ be a (nice) Feller process with state space $\mathbb{R}^n$. Then the function

$$q(x,\xi) := -\lim_{t\to 0}\frac{E^x\left(e^{i(X_t-x)\cdot\xi}\right) - 1}{t} \tag{0.3}$$

completely characterizes $\left((X_t)_{t\geq 0}, P^x\right)_{x\in\mathbb{R}^n}$. In analogy to the theory of partial differential operators we will call $q(x,\xi)$ the *symbol of the process* $\left((X_t)_{t\geq 0}, P^x\right)_{x\in\mathbb{R}^n}$. Let us try to understand (0.3) heuristically from two different starting points.

First suppose that $\left((X_t)_{t\geq 0}, P^x\right)_{x\in\mathbb{R}^n}$ is given as a nice Feller process. For fixed $x \in \mathbb{R}^n$ we may consider the random variables X_t under P^x and look at their *characteristic functions*, i.e. we may consider

$$\lambda_t(x,\xi) := E^x\left(e^{i(X_t-x)\cdot\xi}\right) = e^{-ix\cdot\xi}E^x\left(e^{iX_t\cdot\xi}\right). \tag{0.4}$$

Denoting by $(\mathrm{T}_t)_{t\geq 0}$ the semigroup associated with $\left((X_t)_{t\geq 0}, P^x\right)_{x\in\mathbb{R}^n}$ we find

$$T_t u(x) = E^x(u(X_t)) = (2\pi)^{-n/2}\int_{\mathbb{R}^n} e^{ix\cdot\xi}\lambda_t(x,\xi)\hat{u}(\xi)\,d\xi, \tag{0.5}$$

which says that $(\mathrm{T}_t)_{t\geq 0}$ is a family of pseudo-differential operators and the symbol of T_t is $\lambda_t(x,\xi)$. By assumption $(\mathrm{T}_t)_{t\geq 0}$ is a Feller semigroup. Hence we may look at its generator

$$Au = \lim_{t\to 0} \frac{T_t u - u}{t}, \tag{0.6}$$

where the limit is taken in the strong sense in the space $C_\infty(\mathbb{R}^n;\mathbb{R})$. Substituting (0.5) into (0.6) we arrive at

$$Au(x) = -(2\pi)^{-n/2} \int_{\mathbb{R}^n} e^{ix\cdot\xi} q(x,\xi)\hat{u}(\xi)\, d\xi \tag{0.7}$$

with $q(x,\xi)$ as in (0.3). Thus the generator of the semigroup (and of the process) is also a pseudo-differential operator. The function $\xi \mapsto \lambda_t(x,\xi)$ must be positive definite (and continuous). Supposing for simplicity for a moment that the process is conservative we have $\lambda_t(x,0) = 1$ and from the general theory of negative definite functions it follows that $\xi \mapsto -\frac{(\lambda_t(x,\xi)-1)}{t}$ must be negative definite, which must also hold for the limit. Hence, we find that the symbol $q(x,\xi)$ of the process must be a continuous negative definite function with respect to ξ.

There is another way to understand (0.7) (or (0.3)). As generator of a Feller semigroup A has to satisfy the positive maximum principle, i.e.

$$\sup_{x\in\mathbb{R}^n} u(x) = u(x_0) \geq 0 \text{ implies } Au(x_0) \leq 0. \tag{0.8}$$

In [55] Ph.Courrège characterizes the operators satisfying the positive maximum principle. In particular he proved under the reasonable assumption that $C_0^\infty(\mathbb{R}^n;\mathbb{R}) \subset D(A)$, then for $u \in C_0^\infty(\mathbb{R}^n;\mathbb{R})$ the operator A has the representation (0.7), where $q(x,\xi)$ is a locally bounded function in x which is continuous and negative definite in ξ.

Clearly, the generator completely characterizes the semigroup, and further, the family of characteristic functions $(\lambda_t(x,\xi))_{t\geq 0}$ completely characterize the process. Now we are in the analogous situation as S.Bochner was in case of Lévy processes: The study of the process is reduced to the study of its symbol!

As in the case of Lévy processes it is sometimes more advantageous to use $q(x,\xi)$ directly, but sometimes it is better to use its Lévy–Khinchin decomposition

$$\begin{aligned} -q(x,\xi) &= c(x) + \sum_{j=1}^n b_j(x)\xi_j + \sum_{k,l=1}^n a_{kl}(x)\xi_k\xi_l \\ &\quad + \int_{\mathbb{R}^n\setminus\{0\}} \left(1 - e^{-iy\cdot\xi} - \frac{iy\cdot\xi}{1+|y|^2}\right)\frac{1+|y|^2}{|y|^2}\, N(x,dy). \end{aligned} \tag{0.9}$$

Now we may state the aim of the monograph:

Following Bochner: Construct and study Markov processes by systematically making use of their symbols.

It turns out that in doing so a lot of analysis is to be developed which is not covered by standard theories of pseudo-differential operators, of function spaces etc. This is due to the fact that the symbols under considerations do in general not belong to classical symbol classes. Thus, although stochastic processes are our aim, we have to develop our analytic tools first. We divide our presentation into three parts:

1. Fourier analysis and semigroups.
2. Generators and their potential theory.
3. Markov processes and applications.

For each part we will give a separate introduction.

Part II

Generators and Their Potential Theory

Chapter 1

Introduction

In this volume we discuss two major topics: Constructing a Feller semigroup or an L^p-sub-Markovian semigroup by starting with a pseudo-differential operator $-q(x,D)$, $\xi \mapsto q(x,\xi)$ being a continuous negative definite function, and in a further chapter we study the potential theory of these generators and the corresponding semigroups. In fact, some important part of this potential theory, namely balayage theory, is postponed to volume III since then will have the probabilistic counterpart to our disposal.

Our discussion relies on volume I: Fourier Analysis and Semigroups, references to this volumes are preceded by I in the enumeration, i.e. Theorem I.4.1.33 refers to Theorem 4.1.33 in volume I, and (I.3.136) refers to formula (3.136) in volume I.

When designing Chapter 2 we did not intend to enclose proofs for results on second order elliptic differential operators. This part of the theory is to our opinion by now well known and established. But since in our discussion on subordination we partly depend on these results for differential operators acting on function defined on $\mathbb{R}^n$, we wanted to summarize these results and to give precise references. Many books on one-parameter semigroups do mention these results for bounded domains (with Dirichlet or Neumann conditions imposed on the boundary) without giving a proof, a few times the results for $\mathbb{R}^n$ are mentioned, unfortunately, sometimes with giving an incorrect reference. For the case $p = 2$ the situation is slightly better. On the other hand it is also difficult to find a monograph on elliptic partial differential operators proving the needed a priori estimates in L^p, $p \neq 2$, for $\mathbb{R}^n$. In [202] A.Lunardi gives very precise statements and estimates needed in $L^p(\mathbb{R}^n)$, especially in Chapter 3, but she does not provide proofs and we could not see why the papers [2], [3] of S.Agmon, A.Douglis and L.Nirenberg are the precise reference for the case $\mathbb{R}^n$. They are the best references for a bounded domain when boundary conditions are involved. To our feeling, F.Browder's paper [37] is the better reference. Therefore, in Section 2.1 we started with some preparatory material often taken from D.Edmunds and D.Evans' monograph [80]. Then we followed Browder's work to prove some of the basic a priori estimates needed to treat a second order elliptic differential operator in $L^p(\mathbb{R}^n)$. We prove all estimates in order to adapt an existence proof for weak L^p-solution using the Hahn–Banach theorem. This proof follows closely the presentation of M.Schechter [245], the original source is of course L.Hörmander [135]. However, we do not provide the proof of the elliptic regularity result. This can be found in Browder's paper [37]. The techniques needed are either similar to those already used to derive Theorem 2.1.32, the basic lower bound (estimate (2.44)), or they are a type of mollifier argument, and such arguments are discussed in later sections in different, but similar situations. We hope that this approach is considered to be a reasonable compromise for this section: to demonstrate how the basic techniques work but not to do all details. Thus, in the end of Section 2.1 we know that large classes of second order elliptic differential operators extend to generators of Feller and L^p-sub-Markovian semigroups.

Second order elliptic differential operators as generators of semigroups are often discussed and used as examples in semigroup theory. Degenerate second order elliptic operators, especially Hörmander type operators $\sum_{j=1}^m X_j^2$ are very seldom mentioned in monographs on semigroups. However the enormous amount of work devoted to these operators, in particular in the realm of "heat kernel bounds, anisotropic metrics and sub-Riemannian geometry" which is related to our interests, convinced us to include a short section on semigroups generated by these operators. Again, to our great surprise we could not find any hint in the literature where these semigroups had been constructed using the Hille–Yosida (–Ray) theorem. We owe P.Malliavin [208] a confirmation of this observation as well as the hint that existence results using stochastic differential equations are available, compare P.Malliavin [206], or N.Ikeda and S.Watanabe [143]. Clearly, in many cases other techniques such as constructing a fundamental kernel or quadratic form techniques yield the existence of corresponding semigroups. In Section 2.2 we just "play" a bit with these operators to extend the classes of examples which we may use to illustrate subordination in the sense of Bochner or the theory of Dirichlet forms. Since this direction is more on the border of our investigations we have given only a few references and comments.

In Section 2.3 we start to investigate pseudo-differential operators with negative definite symbols where we concentrate ourselves to the case where the quadratic form part of the symbol (the diffusion part of the process) vanishes. From Chapter 4 in volume I of this treatise it is clear why we have to concentrate on these operators. Many of the results given in our work are taken from or are slight modifications of parts earlier papers of the author and his students, especially we owe much to the papers [121]–[128] of W.Hoh, compare also [152]–[156]. However, there is no "smooth" existence theory so far, i.e. still we only know results for some (large) classes, a unifying approach which will yield a "general" existence result is still missing.

Section 2.3 — Section 2.6 discuss in detail the construction of Feller and sub-Markovian semigroups with a pseudo-differential operator as pregenerator using the Hille–Yosida theorem. As early as in 1965/66 Ph.Courrège in [55] recognized the fact that large classes of operators satisfying the positive maximum principle must be pseudo-differential operators $-q(x, D)$ with a symbol $q(x, \xi)$ being a continuous negative definite function with respect to ξ for each x fixed. Following W.Hoh we call such operators now *pseudo-differential operators with negative definite symbols.* It seems that Ph.Courrège's observation had been unnoticed for a long time. To our best knowledge the first who really started to consider pseudo-differential operator theory to construct and study Markov processes (and the corresponding semigroups) had been T.Komatsu in [183], see also [158]. He considered operators which belong to some classical symbol classes as considered by the pioneers of the theory, see A.P.Calderón [40], A.P.Calderón and A.Zygmund [42], J.J.Kohn and L.Nirenberg [179] or L.Hörmander [137] and [138], and which are classical elliptic pseudo-differential operators of order α, $0 < \alpha < 2$, with symbol $q(x, \xi)$ where $\xi \mapsto q(x, \xi)$ is a continuous negative definite function satisfying some additional conditions. However, still his work used much the representation of these pseudo-differential operators as integro-differential operators, or Lévy-type operators. (For Lévy-type operators with a dominating second order differential operator a lot of results are known, we refer to the book of M.Garroni and J.Menaldi [104], the work of R.Mikulevicius and H.Pragarauskas or B.Rozovskii [215], [216] and [217], D.W.Stroock [257] and the references given there).

As we know from volume I the transfer from the pseudo-differential operator representation to the integro-differential operator representation uses the Lévy–Khinchin formula. In the pseudo-differential operator representation conditions are posed on the symbol $q(x,\xi)$, often in terms of estimates relating $q(x,\xi)$ to a fixed continuous negative definite function, whereas in the Lévy-type operator representation conditions on the Lévy-kernel (and the diffusion as well as drift coefficients) are posed. In general these condition do not transform easily into each other. Thus both pictures are complementary and we should make use of both representation. This monograph of course emphases the pseudo-differential operator point of view, for the other point of view we refer especially to the work of J.Jacod, [168] and the joint monograph [169] with A.Shiryaev and the references given therein.

The idea that pseudo-differential operators with non-classical symbols might be helpful for constructing Dirichlet forms had been discussed first in [149], see also the closely related papers [150] and [151], [152]. The method which later turned out to be rather successful was proposed in [153] with non-trivial examples provided in [154] and [155].

An important technical improvement was obtained by W.Hoh [122] who pointed out that Peetre's inequality holds for $\sqrt{\psi}$ if ψ is a (real-valued) continuous negative definite function. In fact, also W.Hoh's papers [121]–[125] on the martingale problem for pseudo-differential operators with continuous negative definite symbols contain not only deep results for constructing processes which we partly will discuss in volume III, but they contain a lot of fine auxiliary results to handle the operators we are interested in.

After these general remarks we may name the paper [155] as the principle source for Section 2.3 with some improvements taken from W.Hoh [127] and [128]. This section states basic properties of pseudo-differential operators with negative definite symbols. The results in Section 2.4 are essentially due to W.Hoh [126]–[128], to my opinion they are still the finest in the whole field of constructing Markov processes starting with a pseudo-differential operator. They combine deep inside in the structure of continuous negative definite functions and symbolic calculi for pseudo-differential operators. The estimates in Section 2.5 are essentially taken from W.Hoh [127], [128] but have partly their origin in [155]. These three sections, 2.3–2.5, are rather technical. In Section 2.6 the method proposed in [153] is used in combination with the results of Section 2.3–2.5 to construct Feller and L^2-sub-Markovian semigroups by starting with a pseudo-differential operator with negative definite symbol. Theorem 2.6.4 and 2.6.6. are taken from [153], Theorem 2.6.9 is due to W.Hoh [127]. The L^p-results are taken from the joint paper [86] with W.Farkas and R.Schilling extending work in [156].

Since Section 2.7 is in some sense a survey of the constructions of Markov processes generated by pseudo-differential operators different of our approach, we do not need to give a detailed description here, all is included there.

Section 2.8 discusses perturbation results. First we just collect the "obvious" applications of abstract perturbation results as given in volume I, in particular see Theorem 2.8.1, 2.8.4, and 2.8.5. Then we start to combine abstract perturbation results with Hoh's symbolic calculus relying on W.Hoh [127] and [128]. The final results, Theorem 2.8.13 and Theorem 2.8.15, give so far the largest class of pseudo-differential operators generating Markov processes with "general" negative definite symbols, i.e. symbols which are not smooth with respect to ξ and which are comparable to a generic continuous negative definite function ψ_1 (satisfying a minimal growth condition at infinity) without special structure assumptions. In the rest of Section 2.8 we discuss some non-local perturbations of second order elliptic differential operators. These results seem to be new, but in a certain sense they are straightforward to obtain.

Subordination in the sense of Bochner is (according to our considerations in volume I) a possibility to obtain from a given generator of a Feller or sub-Markovian semigroup a new one. In Section 2.9 we do not spent time on these straightforward applications, but in the center of our interest is to study subordinate semigroups by using results of the semigroups we started with. Moreover, we are concerned with the relation of a symbolic calculus to an operator calculus. For the latter see the discussion in volume I or the paper [247] of R.L.Schilling. For a fractional power calculus see the recent monograph [209] of C.Martinez Carracedo and M.Sanz Alix. Obviously one may subordinate second order elliptic differential operators generating Feller or sub-Markovian semigroups. In case where these operators have smooth coefficients and are given in divergence form, Aronson's estimates [7] are valid comparing the fundamental solutions of these operators with the fundamental solutions of multiples of the Laplacians. This leads directly to estimates for semigroups, both for the semigroups generated by the differential operators and all corresponding subordinate semigroups. Therefore we decided to include a proof of Aronson's estimates following the paper of E.B.Fabes and D.W.Stroock [84]. (Note that quite recently R.Bass [18] proposed a simplified proof for the upper estimate, results for non-divergent operators are given by M.R.Faradzheva [85]). Once having Aronson's estimates at our disposal, as mentioned above, the corresponding subordinate semigroups are easily to handle. More interesting is now the question whether the generator of a sub-Markovian semigroup has a symbol comparable to $\xi \mapsto f(|\xi|^2)$ when f is the Bernstein function used for the subordination. This leads to the second topic in Section 2.9. When working within a symbolic calculus, either W.Hoh's calculus or the Weyl calculus proposed by F.Baldus, an extension of the symbolic calculus to subordination is possible, see for example Theorem 2.9.18. It should be noted that these results had been much influenced by the joint papers [163] and [164] with R.Schilling. Their results are discusses at the end of Section 2.9 and they give first ideas what might be achieved in a situation where a symbolic calculus is not at our disposal. In particular some comparison results for the norms of semigroups are derived. Also we do not refer explicitly to J.Bertoin's work [27] and [28], and K.-I.Sato's monograph [244], they are of course very useful references when studying subordinate semigroups.

The final section in Chapter 2 has as starting point the observation that for every continuous negative definite function $\psi(\xi)$ or every negative definite symbol $q(x,\xi)$, for a function $\alpha(x)$, $0 < \alpha(x) \leq 2$, negative symbols are given by $(x,\xi) \mapsto \psi(\xi)^{\alpha(x)}$ and $(x,\xi) \mapsto q(x,\xi)^{\alpha(x)}$, respectively. In case of $\psi(\xi) = |\xi|^2$ the corresponding processes are called *stable-like processes.* Important contributions to the analysis of stable-like processes are due to A.Negoro and M.Tsuchiya, see [222] and [223] and T.Komatsu [187]. Following a joint paper [162] with H.-G.Leopold, K.Kikuchi and A.Negoro in [173], [174] develops a more sophisticated theory of pseudo-differential operators related to stable-like processes. Again, W.Hoh's calculus will lead to the most general results, see Theorem 2.10.1 and his paper [129].

Chapter 3 is a chapter on potential theory. It is easy to define, for example, group theory, as the theory of groups. But a precise definition of potential theory is rather complicated and we do not try to give a definition here. In our presentation the semigroups and quadratic forms are in the center, other approaches could start with resolvents, see C.Dellacherie and P.-A.Meyer [63], [64], or with Markov processes, see for example the monograph of R.Blumenthal and R.Getoor [31], with potentials, see N.S.Landkof's monograph [190], and of course with differential operators as it is done in any classical text. Our main purpose is to pass from analytic objects, semigroups and their generators as well as corresponding Dirichlet forms, to probabilistic objects, namely Markov processes, and we will provide of course whenever it is possible (and non-trivial) examples related to pseudo-differential operators with negative definite symbols.

Section 3.1 is devoted to capacities and abstract Bessel potential spaces associated with a sub-Markovian semigroup. In a first part we systematically developed the theory of Choquet capacities and our presentation uses the works of G.Choquet [49]–[51], often R.Bass [17] and M.Fukushima, Y.Oshima,

M.Takeda [102], and sometimes C.Dellacherie and P.-A.Meyer [64]–[67]. Of some help were occasionally the lecture notes [39] of R.Burckel, see also the survey [93] by J.Frehse.

From Definition 3.1.17 on, where we introduce the gamma transform $V_r^{(p)}$ of an L^p-sub-Markovian semigroup $(T_t^{(p)})_{t\geq 0}$, we turn to capacities associated with these semigroups. Our first result proves, following a joint paper with W.Farkas and R.Schilling [86], that $V_r^{(p)}$ is the inverse operator to $(\mathrm{id}-A^{(p)})^{r/2}$, $A^{(p)}$ being the generator of $(T_t^{(p)})_{t\geq 0}$. For $p=2$ and symmetric semigroups this is of course known for a long time. Next we introduce the associated Bessel potential spaces $\mathcal{F}_{r,p} = V_r^{(p)}L^p = D((\mathrm{id}-A^{(p)})^{r/2}) = D((-A^{(p)})^{r/2})$, and establish many of their properties — often following the paper [86], but being much influenced by M.Fukushima's work [100] and that of M.Fukushima, Y.Oshima and M.Takeda [102]. Of particular interest is the family of capacities $\mathrm{cap}_{(r,p)}$ associated with the spaces $\mathcal{F}_{r,p}$, and a lot of work is done in order to prove that $\mathrm{cap}_{r,p}$ is (often) an outer Choquet capacity. For this a discussion of (r,p)-exceptional sets, (r,p)-equilibrium potentials,, (r,p)-quasi-continuous functions and modifications is needed as well as the notion of regularity and the role of the truncation property.

Section 3.2 investigates now rather systematically the spaces $\mathcal{F}_{r,p}$ and the properties of the generating semigroup and its gamma-transform in these spaces. Once again, besides [86] the works of M.Fukushima and coauthors [101], [100] and [102] are of central importance to our presentation. A new aspect in our work is the careful discussion of the role of analyticity of the involved semigroups. By Stein's result, see volume I, we know that symmetric sub-Markovian semigroups are analytic and the observation in Theorem 3.2.1 is that "nice" properties of $T_t^{(p)}$, $t>0$, can be deduced from good embeddings for $D((-A^{(p)})^k)$. Next we investigate in the general situation when $T_t^{(p)}$, $t>0$, allows certain kernel representations. For this we follow partly some more recent joint work with R.L.Schilling [166].

In a next step we consider the spaces $\mathcal{F}_{r,p}$ as building blocks for a theory of generalized functions. In particular, by determining their dual spaces, see Theorem 3.2.15. With these tools we return to the capacities $\mathrm{cap}_{r,p}$ and try to find a characterization of (r,p)-equilibrium potentials analogous to that used in the theory of Dirichlet forms. This leads to a non-linear potential theory involving monotone operators in the sense of F.Browder and G.Minty. We discuss parts of this (in general non-linear) theory in detail following E.Zeidler [289] and turn to the application mentioned above. The final results are Proposition 3.2.31, Corollary 3.2.34 and Lemma 3.2.36. These results are taken from a forthcoming paper [133] with W.Hoh. This paper is a companion of the joint paper [166] with R.Schilling where the corresponding non-linear potential theory á la V.Maz'ya and V.Havin [211], see also the monograph [1] of D.Adams and L.Hedberg, is developed.

We indicated already the need of embedding results for the spaces $\mathcal{F}_{r,p}$. When working with pseudo-differential operators having negative definite symbols comparable to a fixed continuous negative definite function ψ it is reasonable to introduce function spaces associated with ψ and try to characterize $\mathcal{F}_{r,p}$ in terms of these function spaces. This is done in Section 3.3 following essentially the joint work with W.Farkas and R.Schilling [87]. We denote these function spaces by $H_p^{\psi,s}(\mathbb{R}^n)$ and call them ψ-Bessel potential spaces. In particular we identify th space $H_p^{\psi,2}$ with the domain of the L^p-generator of the L^p-sub-Markovian semigroup associated with ψ by $(T_t u)^\wedge(\xi) = e^{-t\psi(\xi)}\hat{u}(\xi)$. In addition we prove various embedding results, as well as interpolation theorems and we characterize the dual space of $H_p^{\psi,s}$. Finally we use capacities in $H_p^{\psi,s}$ to get quasi-continuous modifications of functions belonging to these spaces.

Already in volume I, Theorem I.4.2.12, we proved a major result due to E.M.Stein [255], namely the analyticity of symmetric sub-Markovian semigroups. This result has enormous consequences for the analysis of these semigroups as well as in the study of the corresponding stochastic processes.

In [255] E.M.Stein provided many more deep and important results for sub-Markovian semigroups, best known in harmonic analysis, we refer only to the M.Cowling [56] and [57], but to our big surprise, a lot of eminent scholars working on semigroup theory (or in the theory of stochastic processes) do not know these results, or do not attribute them correctly. Therefore we devoted Section 3.4 to Stein's work.

Maybe the most important result (besides Theorem I.4.2.12) for our purposes is Corollary 3.4.40, stating that imaginary powers of generators of symmetric L^p-sub-Markovian semigroups are bounded in L^p.

Thus Stein's theory deserves a natural place in our considerations. In addition, the fact that martingales are used to obtain these results makes it very attractive to include these considerations in a treatise which links analysis and stochastic processes. This applies even more since we do not discuss the role of martingale theory in function spaces within our treatise, compare for this the monograph [199] of R.Long.

However, we made the same experience as D.Stroock in writing [259]: It is hard to find a presentation of Stein's work being much different from his presentation! Thus it does not appeal to us to be reasonable to reproduce all parts of [255] in detail here, especially since [255] is easily accessible. Therefore we explain some results rather detailed, especially when "martingale proofs" are involved, but we always will follow the presentation of E.M.Stein. However, other parts, especially the rather involved derivation of estimates for (higher order) Littlewood–Paley functions will only be quoted.

We hope that this is an acceptable compromise in, on the one hand side, familiarizing the reader with Stein's work and, on the other hand, not just reproducing parts of his monograph in a modified form.

The notion of global properties of symmetric L^2-sub-Markovian semigroups or Dirichlet forms was introduced by M.Fukushima, Y.Oshima and M.Takeda in [102]. Section 3.5 is devoted to study such global properties for in general non-symmetric L^p-sub-Markovian semigroups. Since the dual semigroup $(T_t^{(p)*})_{t\geq 0}$ of a given L^p-sub-Markovian semigroup $(T_t^{(p)})_{t\geq 0}$ will be of importance in many situations, we start the section with a result taken from a joint paper [166] with R.Schilling giving necessary and sufficient condition that $(T_t^{(p)*})_{t\geq 0}$ is sub-Markovian too. (The reader should also consult the paper [193], [196] of Y.LeJan where non-symmetric Dirichlet forms had been studied first, as will as the book of Z.-M.Ma and M.Röckner [204].) Next we turn to the notion of invariant sets, more precisely we introduce invariant and strongly invariant sets with respect to a semigroup, both notions coincide for symmetric semigroups. We investigate some of the properties of these sets and give certain characterizations including Y.Oshima's result for symmetric Dirichlet forms. Naturally we turn our interest to the irreducibility problem for semigroups and provide some results taken from E.B.Davies [60].

A major topic is the discussion of transience and recurrence, and along with this we introduce the Green operator associated with a semigroup. As an important tool we need some results from ergodic theory. In our presentation we follow closely M.Fukushima, Y.Oshima and M.Takeda [102], a good reference is also U.Krengel's monograph [188]. In fact larger parts starting from Proposition 3.5.21 to Remark 3.5.32 are an adaptation of [102] to the non-symmetric situation. When turning to extended Dirichlet spaces, starting with Definition 3.4.33, and transient Dirichlet forms we come even closer to the presentation in [102] since most statements reduce to a statements on the symmetric part of a given non-symmetric Dirichlet form. In handling examples, mainly translation invariant Dirichlet forms entirely characterized by a continuous negative definite function ψ, we provide more new material due to the fact that a non-symmetric theory is at our disposal. Further we discuss extended Dirichlet spaces also from the point of view of function spaces, touching shortly the theory of homogeneous function spaces, see H.Triebel [268]. We end the discussion with stating necessary and sufficient criteria for the recurrence, transience and conservativeness of Dirichlet forms and give comparison results. Again we use the monograph [102] of M.Fukushima, Y.Oshima and M.Takeda as an important source, but we provide some new examples related to pseudo-differential operators with negative definite symbols.

Finally we treat K.Yosida's theory of abstract potential operators which we first discuss in a general setting and then, following the presentation of Chr.Berg and G.Forst [26] we investigate the corresponding theory for convolution semigroups. In particular we discuss the deep results of F.Hirsch and M.Itô on transient convolution semigroups.

We tried to avoid in Section 3.5 to provide "trivial" examples by which we mean to take a pseudo-differential operator discussed in Chapter 2 adding if necessary some additional condition and give a straightforward application of results from Chapter 3. Much more could be achieved if we would have L^p-estimates for some of the operators considered in Chapter 2 obtained independently from L^2- and L^∞-bounds by interpolation. However the lack of homogeneity of a generic negative definite function ψ prevents us in the moment to get such bounds — new techniques are needed. A first step in this direction is taken in the recent paper [88] by W.Farkas and H.-G.Leopold.

Our final section shortly indicates to what extend Nash-type and Sobolev-type inequalities are helpful in our setting. But as described in the introduction to Section 3.6, due to the missing of a clear relation to geometry, in the moment the impact of this theory is limited when dealing with non-local generators which are in the center of our interest.

A final remark: Often we write $q(x,\xi) \in S_\rho^{m,\psi}(\mathbb{R}^n)$ etc. when we mean that the symbol q, $(x,\xi) \mapsto q(x,\xi)$, belongs to the class $S_\rho^{m,\psi}(\mathbb{R}^n)$. Clearly this is an abuse of notation, but it is rather helpful.

Chapter 2

Generators of Feller and Sub-Markovian Semigroups

This chapter is devoted to the construction of Feller and sub-Markovian semigroups starting with a pseudo-differential operator. Although we are mainly interested in non-local generators, i.e. pseudo-differential operators with non-polynomial symbols, the first two sections threat differential operators, namely second order elliptic operators and Hörmander type operators, i.e. sums of squares of vector fields. Parts of the results from these section will be used later when discussing subordination. Sections 2.3–2.6 give the central construction of Feller and sub-Markovian semigroups starting with a pseudo-differential operator with negative definite symbol, in Section 2.7 we briefly describe related results. In Section 2.8 we handle perturbation of given semigroups and Section 2.9 deals with subordination in the sense of Bochner. The final section treats pseudo-differential operators of variable order of differentiation as generators of Feller of sub-Markovian semigroups.

2.1 Second Order Elliptic Differential Operators as Generators of Feller and Sub-Markovian Semigroups

In this and the following section we prove that certain second order differential operators are generators of Feller and sub-Markovian semigroups. Although the aim of this monograph is to handle mainly non-local generators, we do discuss these results in greater detail for several reasons. First of all these operators we handle give rise to the best known class of stochastic processes, namely to diffusion processes. Secondly, we will encounter several methods and techniques needed later on, and finally, perturbating and subordinating the semigroups generated by these operators will lead us to semigroups with non-local generators, see Section 2.8 and Section 2.9.

We start with elliptic differential operators of second order as generators. For this we need some preparatory materials on *Sobolev spaces* which goes beyond our considerations in Section I.3.10 and Section I.3.11.

Definition 2.1.1. Let $G \subset \mathbb{R}^n$ be an open set, $m \in \mathbb{N}$ and $1 \leq p < \infty$. The space $W^{m,p}(G)$ consists of all $u \in L^p(G)$ with distributional derivative $\partial^\alpha u \in L^p(G)$ for all $\alpha \in \mathbb{N}_0^n$, $|\alpha| \leq m$. On $W^{m,p}(G)$ we consider the norm

$$\|u\|_{W^{m,p}} = \Big(\sum_{|\alpha| \leq m} \|\partial^\alpha u\|_{L^p}^p \Big)^{\frac{1}{p}}. \tag{2.1}$$

Clearly, (2.1) is equivalent to the norm $\sum_{|\alpha|\leq m} \|\partial^\alpha u\|_{L^p}$.

Lemma 2.1.2. *The space $W^{m,p}(G)$ is a Banach space and $W^{m,2}(G)$ is a Hilbert space.*

Proof: Let $(u_j)_{j\in\mathbb{N}}$ be a Cauchy sequence in $W^{m,p}(G)$. It follows that $(\partial^\alpha u_j)_{j\in\mathbb{N}}$, $|\alpha| \leq m$, is a Cauchy sequence in $L^p(G)$, hence converging in $L^p(G)$ to some $u^{(\alpha)}$. We claim that $\partial^\alpha u^{(0)} = u^{(\alpha)}$. In fact, for $\varphi \in C_0^\infty(G)$ we find with $\frac{1}{p} + \frac{1}{p'} = 1$ that

$$|\langle \partial^\alpha u_j - u^{(\alpha)}, \varphi\rangle| = \Big| \int_G \varphi(\partial^\alpha u_j - u^{(\alpha)}) dx \Big| \leq \|\partial^\alpha u_j - u^{(\alpha)}\|_{L^p} \|\varphi\|_{L^{p'}},$$

implying that $\partial^\alpha u_j \to u^{(\alpha)}$ in $D'(G)$, and we find further

$$\begin{aligned} \langle u^{(\alpha)}, \varphi\rangle &= \int_G u^{(\alpha)} \varphi \, dx = \lim_{j\to\infty} \int_G \varphi \partial^\alpha u_j \, dx \\ &= \lim_{j\to\infty} (-1)^{|\alpha|} \int_G u_j \partial^\alpha \varphi dx = \langle u^{(0)}, \partial^\alpha \varphi\rangle, \end{aligned}$$

i.e. $\partial^\alpha u^{(0)} = u^{(\alpha)}$ in $D'(G)$ (and therefore also in $L^p(G)$). □

In [267], Section 2.3, or in [268], p.88, the following results are proved which we quote for the readers convenience.

Theorem 2.1.3. *For $m \in \mathbb{N}$ and $1 < p < \infty$ we have*

$$W^{m,p}(\mathbb{R}^n) = H_p^m(\mathbb{R}^n) \tag{2.2}$$

and

$$W^{m,p}(\mathbb{R}^n) = F_{p,2}^m(\mathbb{R}^n). \tag{2.3}$$

This theorem tells in particular that $C_0^\infty(\mathbb{R}^n)$ is dense in $W^{m,p}(\mathbb{R}^n)$, $1 < p < \infty$, a result which does also hold for $p = 1$ as we will see soon. More generally we have the famous result of N.Meyers and J.Serrin [214]:

Theorem 2.1.4. *For $m \in \mathbb{N}$ and $1 \leq p < \infty$ the subspace $C^\infty(G) \bigcap W^{m,p}(G)$ is dense in $W^{m,p}(G)$ for any open set $G \subset \mathbb{R}^n$.*

Proof: Let $(G_j)_{j\in\mathbb{N}}$ be a covering of G by open subsets having the property that $\overline{G_j}$ is compact and $\overline{G}_j \subset G_{j+1}$ with $G_1 = \varnothing$. Further assume that $G_{j+2}\backslash\overline{G}_j$ is also an open covering of G. Take a partition of unity $(\varphi_j)_{j\in\mathbb{N}}$ subordinated to the latter open covering of G. For $u \in W^{m,p}(G)$ the function $\varphi_j u$ belongs to $W^{m,p}(G)$ and $\operatorname{supp}(\varphi_j u) \subset G_{j+2}\backslash\overline{G_j}$.

Fix $\varepsilon > 0$. For $\eta_j > 0$ small enough the *Friedrichs mollification* $J_{\eta_j}(\varphi_j u)$ has support in $G_{j+3}\backslash\overline{G}_{j-1}$ and $\|J_{\eta_j}(\varphi_j u) - u\varphi_j\|_{W^{m,p}} < \frac{\varepsilon}{2^j}$, which follows immediately from Proposition I.2.3.17. Given $K \subset G$ compact. Then only a finite number of the functions $J_{\eta_j}(\varphi_j u)$ are non-zero on K, and $v := \sum_{j=1}^\infty J_{\eta_j}(\varphi_j u) \in C^\infty(G)$. Moreover we have

$$\partial^\alpha v(x) - \partial^\alpha u(x) = \sum_{j=1}^\infty \partial^\alpha (J_{\eta_j}(\varphi_j u) - u\varphi_j)(x).$$

But there are at most four non-vanishing terms in this series, which yields with some constant

$$|\partial^\alpha v(x) - \partial^\alpha u(x)|^p \leq c \sum_{j=1}^\infty |\partial^\alpha (J_{\eta_j}(\varphi_j u - u\varphi_j)(x)|^p,$$

which gives

$$\|v - u\|_{W^{m,p}} \leq c^{1/p} \sum_{j=1}^\infty \frac{\varepsilon}{2^j} = c^{1/p}\varepsilon,$$

and the theorem is proved. □

In the proof of Theorem 2.1.4 we used the fact that for $u \in W^{m,p}(G)$ and $\varphi \in C_0^\infty(G)$ the function φu is also an element of $W^{m,p}(G)$, which is obvious by the estimate

$$\|\partial^\alpha(\varphi u)\|_{L^p} \le \sum_{\beta\le\alpha} \binom{\alpha}{\beta} \|\partial^{\alpha-\beta}\varphi\|_{L^\infty} \|\partial^\beta u\|_{L^p}.$$

In fact this estimate implies $u\varphi \in W^{m,p}(G)$ for $u \in W^{m,p}(G)$ and $\varphi \in C_b^m(G)$. Now take $\psi \in C_0^\infty(\mathbb{R}^n)$ such that $0 \le \psi \le 1$, $\psi\big|_{B_1(0)} = 1$ and $\operatorname{supp}\psi \subset B_2(0)$. For $j \in \mathbb{N}$ we put $\psi_j(x) := \psi\left(\frac{x}{j}\right)$. Using the estimate just stated, it follows easily that $\|\psi_j u - u\|_{W^{m,p}} \to 0$ as $j \to \infty$ for $u \in W^{m,p}(\mathbb{R}^n)$. Moreover, if $(u_j)_{j\in\mathbb{N}}$ is a sequence in $C^\infty(\mathbb{R}^n) \bigcap W^{m,p}(\mathbb{R}^n)$ such that $\|u_j - u\|_{W^{m,p}} \to 0$ as $j \to \infty$, we find

$$\begin{aligned}\|\psi_j u_j - u\|_{W^{m,p}} &\le \|\psi_j(u_j - u)\|_{W^{m,p}} + \|\psi_j u - u\|_{W^{m,p}} \\ &\le C\|u_j - u\|_{W^{m,p}} + \|\psi_j u - u\|_{W^{m,p}},\end{aligned}$$

i.e. $\psi_j u_j \to u$ in $W^{m,p}(\mathbb{R}^n)$. Since $\psi_j u_j \in C_0^\infty(\mathbb{R}^n)$, we find that $C_0^\infty(\mathbb{R}^n)$ is dense in $W^{m,p}(\mathbb{R}^n)$ for $m \in \mathbb{N}$ and $1 \le p < \infty$. Moreover, we have the following *Leibniz rule*:

Lemma 2.1.5. *Let $u \in W^{m,p}(G)$ and $v \in W^{m,p'}(G)$, $1 < p < \infty$, $\frac{1}{p} + \frac{1}{p'} = 1$. Then we have for all $\alpha \in \mathbb{N}_0^n$, $|\alpha| \le m$, that $\partial^\alpha(u \cdot v) \in L^1(G)$ and*

$$\partial^\alpha(u\cdot v) = \sum_{\beta\le\alpha} \binom{\alpha}{\beta} \partial^{\alpha-\beta}u\,\partial^\beta v. \tag{2.4}$$

Proof: For $u \in W^{m,p}(G) \bigcap C^\infty(G)$ and $v \in W^{m,p'}(G) \bigcap C^\infty(G)$ we have by Hölder's inequality

$$\|\partial^\alpha(u\cdot v)\|_{L^1} \le \sum_{\beta\le\alpha} \binom{\alpha}{\beta} \|\partial^{\alpha-\beta}u\|_{L^p} \|\partial^\beta v\|_{L^{p'}}.$$

The general result follows now from Theorem 2.1.4. □

It is convenient to introduce the notation:

$$\|\nabla^j u\|_{L^p} := \Big(\sum_{|\alpha|=j} \|\partial^\alpha u\|_{L^p}^p\Big)^{\frac{1}{p}}. \tag{2.5}$$

The next theorem and its corollary give important *intermediate estimate* for the norm $\|.\|_{W^{m,p}}$. It is due to E.Gagliardo [103] and L.Nirenberg [224]. We will follow the proof given in the monograph [80] of D.E.Edmunds and W.D.Evans, but refer also to the new proof due to V.Maz'ya and T.Shaposhnikova [212].

Theorem 2.1.6 (Gagliardo–Nirenberg). *Let $j, m \in \mathbb{N}_0$ with $0 \le j \le m$, let $q, r \in [1, \infty]$ and define p by*

$$\frac{1}{p} := \frac{1}{q} + \frac{j}{m}\left(\frac{1}{r} - \frac{1}{q}\right). \tag{2.6}$$

Then there exists a constant $c = c(j, m, q, r, n)$ such that for all $u \in C_0^m(\mathbb{R}^n)$ we have

$$\|\nabla^j u\|_{L^p} \le c\|\nabla^m u\|_{L^r}^{\frac{j}{m}} \|u\|_{L^q}^{\frac{m-j}{m}}. \tag{2.7}$$

Proof: We consider first the case $j = 1$, $m = 2$ and $n = 1$, which gives $\frac{2}{p} = \frac{1}{q} + \frac{1}{r}$. Moreover we assume for a moment that $1 \le q < \infty$ and $1 < r < \infty$. For $a, b \in \mathbb{R}, a < b$, we put $l = \frac{1}{4}(b-a)$. Take $x_1 \in (a, a+l)$ and $x_2 \in (a+3l, b)$. The mean-value theorem yields the existence of $\xi \in (x_1, x_2)$ such that

$$u(x_2) - u(x_1) = (x_2 - x_1)u'(\xi). \tag{2.8}$$

Moreover, for $x \in (a, b)$ we have

$$u'(x) = u'(\xi) + \int_\xi^x u''(y)dy. \tag{2.9}$$

From (2.8) and (2.9) we get

$$|u'(x)| \le \frac{|u(x_1)| + |u(x_2)|}{2l} + \int_a^b |u''(y)|dy. \tag{2.10}$$

Integrating (2.10) with respect to x_1 over $(a, a+l)$ and with respect to x_2 over $(a + 3l, b)$ we obtain

$$l^2|u'(x)| \le \frac{1}{2}\int_a^b |u(y)|dy + l^2 \int_a^b |u''(y)|dy. \tag{2.11}$$

Applying Hölder's inequality in the form $(A + B)^p \le 2^{p-1}(A^p + B^p)$ to (2.11) gives

$$l^{2p}|u'(x)|^p \le 2^{p-1}\Big(2^{-p}\Big(\int_a^b |u(y)|dy\Big)^p + l^{2p}\Big(\int_a^b |u''(y)|dy\Big)^p\Big),$$

thus we have

$$2|u'(x)|^p \leq 2^{-2p}\Big(\int_a^b |u(y)|dy\Big)^p + 2^p\Big(\int_a^b |u''(y)|dy\Big)^p,$$

and therefore, once again by using Hölder's inequality, we get

$$\begin{aligned}\int_a^b |u'(x)|^p dx \\ \leq c^p l^{1+p-\frac{p}{r}}\Big(\int_a^b |u''(y)|^r dy\Big)^{\frac{p}{r}} + c^p l^{-(1+p-\frac{p}{r})}\Big(\int_a^b |u(y)|^q dy\Big)^{\frac{p}{q}}.\end{aligned} \tag{2.12}$$

Suppose now that $u\big|_{(0,\infty)}$ is not identically zero and let $L > 0$ be an arbitrary real number. Take $k \in \mathbb{N}$, $k > 1$, and let us apply (2.12) with $a = 0$ and $b = \frac{L}{k}$. Now, if the first term on the right-hand side of (2.12) is greater than the second term, we set $I_1 := [0, \frac{L}{k}]$, if not, we consider the interval of the form $[0, \delta + \frac{L}{k}]$, $\delta > 0$, and increase δ until the two terms are equal. We call this δ now δ_1 and set in this case $I_1 := [0, \delta_1 + \frac{L}{k}]$. Thus we find

$$\int_{I_1} |u'(x)|^p dx \leq \begin{cases} 2c^p\left(\dfrac{L}{k}\right)^{1+p-\frac{p}{r}}\Big(\displaystyle\int_0^L |u''(y)|^r dy\Big)^{\frac{p}{r}}, \text{ if } I_1 = \left[0, \dfrac{L}{k}\right] \\ 2c^p\Big(\displaystyle\int_{I_1} |u''(y)|^r dy\Big)^{\frac{p}{2r}}\Big(\displaystyle\int_{I_1} |u''(y)|^q dy\Big)^{\frac{p}{2q}}, \\ \qquad\qquad \text{if } I_1 = \left[0, \delta_1 + \dfrac{L}{k}\right]. \end{cases}$$

For $l(I_1) \geq L$, $l(I_1)$ denoting the length of I_1, it follows that

$$\int_0^L |u'(x)|^p dx \leq 2c^p\Big(\int_0^\infty |u''(y)|^r dy\Big)^{\frac{p}{2r}}\Big(\int_0^\infty |u(y)|^q dy\Big)^{\frac{p}{2q}} \tag{2.13}$$

If $l(I_1) < L$ we may repeat the process starting at the right-hand end-point of I_1 to construct an interval I_2. After at most k steps we obtain intervals $I_1, \ldots, I_s$ until $[0, L]$ is covered and the last interval I_s has the property that $l(I_1) + \cdots + l(I_s) \geq l(I_1) + \cdots + l(I_{s-1})$. Therefore we get

$$\begin{aligned}\int_0^L |u'(x)|^p dx &\leq \sum_{j=1}^s \int_{I_j} |u'(x)|^p dx \\ &\leq 2kc^p\left(\frac{L}{k}\right)^{(1+p-\frac{p}{r})}\Big(\int_0^L |u''(y)|^r dy\Big)^{\frac{p}{r}}\end{aligned}$$

$$+ 2c^p \sum_{j=1}^{s} \Big(\int_{I_j} |u''(y)|^r dy \Big)^{\frac{p}{2r}} \Big(\int_{I_j} |u(y)|^q dy \Big)^{\frac{p}{2q}}$$

$$\le 2kc^p \left(\frac{L}{k}\right)^{(1+p-\frac{p}{2r})} \Big(\int_0^L |u''(y)|^r dy \Big)^{\frac{p}{r}}$$

$$+ 2c^p \Big(\int_0^\infty |u''(y)|^r dy \Big)^{\frac{p}{2r}} \Big(\int_0^\infty |u(y)|^q dy \Big)^{\frac{p}{2q}}, \qquad (2.14)$$

where we used in the the last step Hölder's inequality and the fact that $\frac{p}{2r} + \frac{p}{2q} = 1$. Since $r > 1$ it follows from (2.14) for $k \to \infty$ that (2.13) holds also in this case. Thus we have for all $u \in C_0^\infty(\mathbb{R})$ such that u'' is not identically zero on $(0, \infty)$ the estimate (2.13). Since u has compact support, it follows that (2.13) holds also when u'' is identically zero in $(0, \infty)$. Clearly, (2.13) implies (2.7) in our special case. Next we establish (2.7) for $j = 1$ and $m = 2$ and r, q as before but for arbitrary $n \in \mathbb{N}$. For $1 \le i \le n$ we know already

$$\int_{\mathbb{R}} \Big| \frac{\partial}{\partial x_j} u(x) \Big|^p dx_j \le c^p \Big(\int_{\mathbb{R}} \Big| \frac{\partial^2}{\partial x_j^2} u(x) \Big|^r dx_j \Big)^{\frac{p}{2r}} \Big(\int_{\mathbb{R}} |u(x)|^q dx_j \Big)^{\frac{p}{2q}}.$$

Integrating with respect to the other $n - 1$ variables and using a further time Hölder's inequality, we get with $d\tilde{x} = dx_1 ... dx_{j-1} dx_{j+1} ... dx_n$

$$\int_{\mathbb{R}^n} \Big| \frac{\partial u}{\partial x_j} \Big|^p dx \le c^p \int_{\mathbb{R}^{n-1}} \Big\{ \Big(\int_{\mathbb{R}} \Big| \frac{\partial^2 u}{\partial x_j^2} \Big|^r dx_j \Big)^{\frac{p}{2r}} \Big(\int_{\mathbb{R}} |u|^q dx_j \Big)^{\frac{p}{2q}} \Big\} d\tilde{x}$$

$$\le c^p \Big(\int_{\mathbb{R}^n} \Big| \frac{\partial^2 u}{\partial x_j^2} \Big|^r dx \Big)^{\frac{p}{2r}} \Big(\int_{\mathbb{R}^n} |u|^q dx \Big)^{\frac{p}{2q}}, \qquad (2.15)$$

note that $\frac{p}{2r} + \frac{p}{2q} = 1$. Thus (2.7) is proved for $j = 1$ and $m = 2$ with $1 \le q < \infty$ and $1 < r < \infty$. The extreme cases $q = \infty$ and $r = 1$ or $r = \infty$ are obtained from (2.15) as limit case. Finally we want to get rid of the restriction $j = 1$ and $m = 2$. The case $j = 0$ and $m \in \mathbb{N}$ is obvious. First we claim that (2.7) holds for $j = m - 1$. For $m = 1$ nothing is to prove. Suppose that (2.7) holds for $m = k$ and $j = k - 1$. Take $u \in C_0^{k+1}(\mathbb{R}^n)$ and $\alpha \in \mathbb{N}_0^n$ such that $|\alpha| = k - 1$. We apply (2.7) with $j = 1$ and $m = 2$ to $\partial^\alpha u$. Defining p and $\tilde{q}$ by

$$\frac{1}{p(k+1)} = \frac{1}{q} + \frac{k(\frac{1}{r} - \frac{1}{q})}{k+1} \quad \text{and} \quad \frac{1}{\tilde{q}} = \frac{2}{q(k+1)} + \frac{k+1}{r(k+1)}$$

we obtain

$$\|\nabla \partial^\alpha u\|_{L^{p(k+1)}} \le c \|\nabla^2 \partial^\alpha u\|_{L^r}^{\frac{1}{2}} \|\partial^\alpha u\|_{L^{\tilde{q}}}^{\frac{1}{2}},$$

implying (recall (2.5)) that

$$\|\nabla^k u\|_{L^{p(k+1)}} \leq c\|\nabla^{k+1}u\|_{L^r}^{\frac{1}{2}}\|\nabla^{k-1}u\|_{L^{\tilde{q}}}^{\frac{1}{2}}.$$

Using the induction hypothesis we find further with $\frac{1}{k}+\frac{1}{k'}=1$ that

$$\|\nabla^k u\|_{L^{p(k+1)}} \leq c\|\nabla^{k+1}u\|_{L^r}^{\frac{1}{2}}\|\nabla^k u\|_{L^{p(k+1)}}^{\frac{1}{2k'}}\|u\|_{L^q}^{\frac{1}{2k}}.$$

Thus we have (2.7) with $m=k+1$ and $j=k$, i.e. (2.7) holds with $j=m-1$ for all $m\in\mathbb{N}$. Now the proof is completed by a backward induction. We assume that (2.7) holds for some j, $2\leq j\leq m-1$, and define $\tilde{p}(j)$ by $\frac{1}{\tilde{p}(j)}=\frac{1}{q}+\frac{(j-1)}{m}\left(\frac{1}{r}-\frac{1}{q}\right)$. For $u\in C_0^m(\mathbb{R}^n)$ we find by the arguments just used

$$\|\nabla^{j-1}u\|_{L^{\tilde{p}(j-1)}} \leq c\|\nabla^j u\|_{L^{\tilde{p}(j)}}^{(j-1)/j}\|u\|_{L^q}^{\frac{1}{j}}$$

and by induction hypothesis we get

$$\|\nabla^{j-1}u\|_{L^{\tilde{p}(j-1)}} \leq c\|\nabla^m u\|_{L^r}^{(j-1)/m}\|u\|_{L^q}^{\frac{(m-j+1)}{m}}$$

which is (2.7) with j replaced by $j-1$, and the theorem is proved. □

Corollary 2.1.7. *In the situation of Theorem* 2.1.6 *we have for any* $\varepsilon>0$ *and* $q=r=p$

$$\|\nabla^j u\|_{L^p} \leq \varepsilon\|\nabla^m u\|_{L^p}+c(\varepsilon)\|u\|_{L^p}. \tag{2.16}$$

Proof: First observe that $q=r$ implies $p=r$ and (2.7) reads as

$$\|\nabla^j u\|_{L^p} \leq c\|\nabla^m u\|_{L^p}^{\frac{j}{m}}\|u\|_{L^p}^{\frac{m-j}{m}}.$$

Next we note that with $s=\frac{m}{j}$ we have $s'=\frac{m}{m-j}$, $\frac{1}{s}+\frac{1}{s'}=1$, and estimate (I.2.4) yields for $\eta>0$

$$\begin{aligned} c\|\nabla^m u\|_{L^r}^{\frac{1}{s}}\|u\|_{L^r}^{\frac{1}{s'}} &\leq \frac{\eta}{s}\left(\|\nabla^m u\|_{L^r}^{\frac{1}{s}}\right)^s+\frac{1}{\eta^{s/s'}s'}\left(c\|u\|_{L^r}^{\frac{1}{s'}}\right)^{s'} \\ &= \eta\frac{j}{m}\|\nabla^m u\|_{L^r}+\frac{m-j}{\eta^{j/m-j}m}c^{m/m-j}\|u\|_{L^r}. \end{aligned}$$

Taking $\eta=\frac{m}{j}\varepsilon$ the corollary follows. □

The density of $C_0^\infty(\mathbb{R}^n)$ in $W^{m,p}(\mathbb{R}^n)$, $1\leq p<\infty$, gives immediately

Corollary 2.1.8. *A. The estimate* (2.16) *holds for all* $u \in W^{m,p}(\mathbb{R}^n)$, $1 \leq p < \infty$, *and* $m \in \mathbb{N}$. *Moreover, for* $\alpha \in \mathbb{N}_0^n$ *and* $|\alpha| < m$ *we have*

$$\|\partial^\alpha u\|_{L^p} \leq \varepsilon \|\nabla^m u\|_{L^p} + c(\varepsilon)\|u\|_{L^p}. \tag{2.17}$$

B. On $W^{m,p}(\mathbb{R}^n)$ *an equivalent norm is given by*

$$\|\nabla^m u\|_{L^p} + \|u\|_{L^p}. \tag{2.18}$$

Moreover, we have the following important commutator estimate. Recall that for two operators on a linear space their *commutator* is defined by

$$[A, B] = AB - BA. \tag{2.19}$$

In particular we have for a function φ and a multiindex $\alpha \in \mathbb{N}_0^n$ for suitable functions u

$$[\varphi, \partial^\alpha]u = \varphi \partial^\alpha u - \partial^\alpha(\varphi u). \tag{2.20}$$

Corollary 2.1.9. *Let* $\varphi \in C_b^\infty(\mathbb{R}^n)$ *and* $u \in W^{m,p}(\mathbb{R}^n)$. *Further let* $\alpha \in \mathbb{N}_0^n$, $|\alpha| \leq m$, *and* $\varepsilon > 0$ *be given. Then we have for* $0 \leq k \leq m - |\alpha|$

$$\|[\varphi, \partial^\alpha]u\|_{W^{k,p}} \leq c(\varphi)\|u\|_{W^{|\alpha|+k-1,p}}, \tag{2.21}$$

implying

$$\|[\varphi, \partial^\alpha]u\|_{W^{k,p}} \leq \varepsilon \|u\|_{W^{|\alpha|+k,p}} + \tilde{c}(\varphi, \varepsilon)\|u\|_{L^p}. \tag{2.22}$$

Proof: By Corollary 2.1.8.A it is obvious that (2.21) gives (2.22). Now, by the definition of $[\varphi, \partial^\alpha]u$ we have

$$[\varphi, \partial^\alpha]u = \varphi \partial^\alpha u - \partial^\alpha(\varphi u) = \sum_{\beta \leq \alpha, \beta \neq \alpha} \binom{\alpha}{\beta} \partial^{\alpha-\beta}\varphi \partial^\beta u,$$

which implies (2.21). □

Now let $G \subset \mathbb{R}^n$ be any open set. By definition we have

$$\overset{\circ}{W}{}^{m,p}(G) := \overline{C_0^\infty(G)}^{\|\ \|_{W^{m,p}}}. \tag{2.23}$$

Obviously, $\overset{\circ}{W}{}^{m,p}(G) \subset W^{m,p}(G)$ and it is a Banach space. Note that for a bounded set $G \subset \mathbb{R}^n$ we always have $\overset{\circ}{W}{}^{m,p}(G) \neq W^{m,p}(G)$. This follows from the fact that the function $x \mapsto 1$ belongs for any bounded set to $W^{m,p}(G)$, but it is not possible to approximate this function in the norm $\|\cdot\|_{W^{m,p}}$, $m \geq 1$, by test functions. In fact, in this situation we have $W^{m,p}(G) \subset W^{1,1}(G)$ and for $\varphi \in C_0^\infty(G)$ we find

$$\|1-\varphi\|_{W^{1,1}} = \int_G |1-\varphi(x)|dx + \int_G \sum_{j=1}^n \left|\frac{\partial}{\partial x_j}\varphi(x)\right|dx.$$

Moreover, since

$$\varphi(x) = \int_{-\infty}^{x_1} \frac{\partial}{\partial y_1}\varphi(y_1, x_2, \ldots, x_n)dy_1,$$

we find, see also Theorem 2.1.10,

$$\|\varphi\|_{L^1} \leq c(G)\int_G \left|\frac{\partial\varphi(x)}{\partial x_1}\right|dx, \tag{2.24}$$

which yields

$$\begin{aligned}\|1-\varphi\|_{W^{1,1}} &\geq \int_G 1dx - \int_G |\varphi(x)|dx + \int_G \sum_{j=1}^n \left|\frac{\partial\varphi(x)}{\partial x_j}\right|dx \\ &\geq \lambda^{(n)}(G) - (c(G)-1)\int_G \sum_{j=1}^n \left|\frac{\partial\varphi(x)}{\partial x_j}\right|dx \\ &\geq \lambda^{(n)}(G) - (c(G)-1)\|1-\varphi\|_{W^{1,1}},\end{aligned}$$

thus we have

$$\|1-\varphi\|_{W^{1,1}} \geq \frac{\lambda^{(n)}(G)}{c(G)}$$

which tells us that we can not approximate $x \mapsto 1$ by test functions in the norm $\|\cdot\|_{W^{1,1}}$.

We want to give a more general version of (2.24).

Theorem 2.1.10 (Poincaré's inequality). *Let $G \subset \mathbb{R}^n$ be a bounded open set and suppose that $1 \leq p < \infty$. Then, for all $u \in \overset{\circ}{W}{}^{1,p}(G)$*

$$\|u\|_{L^p} \leq c(G)\|\nabla u\|_{W^{1,p}} \tag{2.25}$$

holds, where $c(G)$ is a constant depending only on G.

Proof: It is enough to prove (2.25) for all $u \in C_0^\infty(G)$ which we consider as usual as subspace of $C_0^\infty(\mathbb{R}^n)$. Since G is bounded we may find $l > 0$ such that G lies between the two hyperplanes $x_1 = a$ and $x_1 = a + l$. Without loss of generality we may choose $a = 0$. For $x \in G$ we find with $\frac{1}{p} + \frac{1}{p'} = 1$ that

$$\begin{aligned} |u(x)| &= \left| \int_{-\infty}^{x_1} \frac{\partial}{\partial y_1} u(y_1, x_2, \ldots, x_n) dy_1 \right| \\ &\leq \left(\int_0^l \left| \frac{\partial}{\partial y_1} u(y_1, x_2, \ldots, x_n) \right|^p dy_1 \right)^{1/p} l^{1/p'}. \end{aligned}$$

Thus we get

$$\int_0^l |u(x_1, \ldots, x_n)|^p dx_1 \leq l^p \int_0^l \left| \frac{\partial}{\partial y_1} u(y_1, x_2, \ldots, x_n) \right| dy_1,$$

which yields

$$\int_G |u|^p dx \leq l^p \int_G |\nabla u|^p dx. \quad \square$$

There are several generalizations of Poincaré's inequality which we will not discuss here, but we refer to the monograph [80] of D.E.Edmunds and W.D.Evans, the book [81] of Yu.Egorov and V.Kondratiev, or the monograph [106] of D.Gilbarg and N.S.Trudinger.

We need a further embedding theorem for the space $W^{m,p}(\mathbb{R}^n)$, namely Sobolev's embedding theorem. Our presentation is oriented at the treatment of the subject by Chr.Sogge [253], and we need the following

Lemma 2.1.11. *For $s > 0$ define*

$$K_s(x) := (2\pi)^{-n/2} \int_{\mathbb{R}^n} e^{ix\cdot\xi} (1+|\xi|^2)^{-s/2} d\xi. \tag{2.26}$$

It follows that K_s is a function and if $0 < s < n$ we have for every $N \in \mathbb{N}$ the estimate

$$|K_s(x)| \leq c_N |x|^{-n+s} (1+|x|)^{-N}. \tag{2.27}$$

Proof: First note that since $s > 0$ the function $\xi \mapsto (1+|\xi|^2)^{-s/2}$ is bounded and measurable, thus (2.26) make sense at least in $S'(\mathbb{R}^n)$. Suppose first that

$|x| \geq 1$. For $m \in \mathbb{N}$ observe that

$$\begin{aligned}(2\pi)^{n/2} K_s(x) &= \int_{\mathbb{R}^n} |x|^{-2m} |x|^{2m} e^{ix\cdot\xi} (1+|\xi|^2)^{-s/2} d\xi \\ &= |x|^{-2m} \int_{\mathbb{R}^n} \left((-\Delta_\xi)^m e^{ix\cdot\xi}\right) (1+|\xi|^2)^{-s/2} d\xi \\ &= |x|^{-2m} \int_{\mathbb{R}^n} e^{ix\cdot\xi} (-\Delta)^m (1+|\xi|^2)^{-s/2} d\xi.\end{aligned}$$

Since

$$|(-\Delta)^m (1+|\xi|^2)^{-s/2}| \leq c_m (1+|\xi|^2)^{-\frac{s}{2}-m} \tag{2.28}$$

is an integrable function for $m > \frac{n}{2}$, it follows that (2.27) holds for $|x| \geq 1$. Next we prove (2.27) for $|x| \leq 1$. For this take $\psi \in C_0^\infty(\mathbb{R}^n)$ such that $0 \leq \psi \leq 1$, supp $\psi \subset \overline{B_{r_0}(0)}$ and $\psi\Big|_{B_{r_0/2}(0)} = 1$. For $R > 0$ we find

$$\begin{aligned}&\left|(2\pi)^{-n/2} \int_{\mathbb{R}^n} e^{ix\cdot\xi} \psi\left(\frac{\xi}{R}\right) (1+|\xi|^2)^{-s/2} d\xi\right| \\ &\qquad \leq c \int_{|\xi| \leq r_0 R} (1+|\xi|^2)^{-s} d\xi \leq \tilde{c} R^{n-s}.\end{aligned}$$

On the other hand, for $m > \frac{n}{2}$ we get

$$\begin{aligned}&\left|(2\pi)^{-n/2} \int_{\mathbb{R}^n} e^{ix\cdot\xi} \left(1 - \psi\left(\frac{\xi}{R}\right)\right) (1+|\xi|^2)^{-s/2} d\xi\right| \\ &= \left|(2\pi)^{-n/2} \int_{\mathbb{R}^n} |x|^{-2m} \left((-\Delta_\xi)^m e^{ix\cdot\xi}\right) \left(1 - \psi\left(\frac{\xi}{R}\right)\right) (1+|\xi|^2)^{-s/2} d\xi\right| \\ &\leq c \int_{|\xi| > \frac{r_0}{2} R} |x|^{-2m} |\xi|^{-s-2m} d\xi \leq \tilde{c'} |x|^{-2m} R^{n-s-2m},\end{aligned}$$

where we used once again (2.28). Finally, taking $R = |x|^{-1}$, we obtain (2.27) for all x, which also implies that K_s is a function. (Compare also [256], p. 132.) □

Theorem 2.1.12 (Sobolev's embedding theorem). *Let $m - k > \frac{n}{p}$, $p \geq 1$. Then $W^{m,p}(\mathbb{R}^n)$ is continuously embedded into the space $C_\infty^k(\mathbb{R}^n)$. More precisely, each element of $W^{m,p}(\mathbb{R}^n)$ has a representation by an element of $C_\infty^k(\mathbb{R}^n)$ and we have the estimate*

$$\|u\|_{C_\infty^k} \leq c\|u\|_{W^{m,p}}. \tag{2.29}$$

Proof: We prove (2.29) first for $k = 0$. Since $C_0^\infty(\mathbb{R}^n)$ is dense in $W^{m,p}(\mathbb{R}^n)$, it is sufficient to show (2.29) for all $u \in C_0^\infty(\mathbb{R}^n)$. Defining $v := (\mathrm{id} - \Delta)^{m/2} u$, it is sufficient to prove

$$\|K_m * v\|_\infty \leq c\|v\|_{L^p}. \tag{2.30}$$

In fact we have

$$\begin{aligned} K_m * v &= F^{-1}(F(K_m * v)) \\ &= (2\pi)^{n/2} F^{-1}(\widehat{K}_m \widehat{v}) \\ &= (2\pi)^{n/2} F^{-1}\left((1+|\cdot|^2)^{-m/2}(1+|\cdot|^2)^{m/2}\widehat{u}\right) \\ &= (2\pi)^{n/2} u, \end{aligned}$$

and further we have

$$\|v\|_{L^p} = \|(\mathrm{id} - \Delta)^{m/2} u\|_{L^p} \leq c\|u\|_{W^{m,p}}. \tag{2.31}$$

Now, if $m > \frac{n}{p}$ it follows that $-np' + mp' > -1$ and (2.27) implies that $K_m \in L^{p'}(\mathbb{R}^n)$. From Lemma I.2.3.15, Young's inequality, we deduce finally (2.30) and therefore (2.29) for $k = 0$. The general case follows by induction. □

Note that (2.31) used in the proof is trivial for $m = 2l$, $l \in \mathbb{N}$, or $p = 2$. For general $m \in \mathbb{N}$ and $p \geq 1$ this is just the content of (2.2) and is proved using some Fourier multiplier results, we will discuss this result in a more general context in a later section. Note that (2.3) combined with Theorem I.3.11.12 yields also

$$\|u\|_{L^{np/(n-p)}} \leq c\|u\|_{W^{1,p}} \tag{2.32}$$

for $p < n$ and all $u \in W^{1,p}(\mathbb{R}^n)$. As noted after Corollary I.3.11.13, it is possible to show that for $m - k > \frac{n}{p}$ elements of $W^{m,p}(\mathbb{R}^n) = H_p^m(\mathbb{R}^n)$ do not only belong to $C_\infty^k(\mathbb{R}^n)$, but their k^{th} partial derivatives are also Hölder continuous with some Hölder exponent $\gamma \in (0, 1]$ such that $(m - k - \gamma) > \frac{n}{p}$. Since we need this embedding result explicitly for the case $m = 1$, we will give a proof which is essentially borrowed from W.Ziemer [290].

Theorem 2.1.13. *If $u \subset W^{1,p}(\mathbb{R}^n)$, $p > n$, then we have with $\gamma = 1 - \frac{n}{p}$ that $u \in C^{0,\gamma}(\mathbb{R}^n)$ and*

$$\frac{|u(x) - u(y)|}{|x - y|^\gamma} \leq c\|u\|_{W^{1,p}}. \tag{2.33}$$

Proof: Since $C_0^\infty(\mathbb{R}^n)$ is dense in $W^{1,p}(\mathbb{R}^n)$ it is sufficient to prove (2.33) for $u \in C_0^\infty(\mathbb{R}^n)$. Let $x \in \mathbb{R}^n$ and $y \in \mathbb{R}^n$ fixed. Put $y - x = r\, e_\vartheta$, $r \in \mathbb{R}_+$, and $e_\vartheta \in S^{n-1}$. It follows that

$$\begin{aligned} u(y) - u(x) &= \int_0^1 (y-x) \cdot \operatorname{grad} u(x + t(y-x))dt \\ &= \int_0^1 r\, e_\vartheta \cdot \operatorname{grad} u(x + tr\, e_\vartheta)dt \\ &= \int_0^1 r\, e_\vartheta \cdot \operatorname{grad} u(x + tr\, e_\vartheta)\chi_{\overline{B_1(0)}}(x + tr\, e_\vartheta)dt. \end{aligned}$$

Integrating over S^{n-1} we find with $\omega_n = \lambda^{(n-1)}(S^{n-1})$

$$\begin{aligned} &\omega_n(u(x) - u(y)) \\ &= \int_{S^{n-1}} \int_0^1 r\, e_\vartheta \cdot \operatorname{grad} u(x + tr\, e_\vartheta)t^{n-1} \frac{\chi_{\overline{B_1(0)}}(x + tr\, e_\vartheta)}{t^{n-1}} dS(\vartheta)dt \\ &= c_1 \int_{\overline{B_r(x)}} (\operatorname{grad} u)(z) \frac{\chi_{\overline{B_1(x)}}(z)}{|x-z|^{n-1}} dz. \end{aligned}$$

Using Hölder's inequality we get

$$\begin{aligned} \left|u(y) - u(x)\right| &\leq c_2 \int_{B_r(x)} |\operatorname{grad} u(z)| \frac{\chi_{\overline{B_r(x)}}(z)}{|x-z|^{n-1}} dz \\ &\leq c_3 \Big(\int_{\mathbb{R}^n} |\operatorname{grad} u(z)|^p dz \Big)^{\frac{1}{p}} \Big(\int_{\overline{B_r(x)}} \Big(\frac{dz}{|x-z|^{n-1}} \Big)^{p'} \Big)^{\frac{1}{p'}} \\ &\leq c_4 \|u\|_{W^{1,p}} \Big(\int_{\overline{B_r(x)}} \frac{1}{|x-z|^{np'-p'}} dz \Big)^{\frac{1}{p'}}. \end{aligned}$$

Since

$$\Big(\int_{\overline{B_r(x)}} \frac{1}{|x-z|^{np'-p'}} dz \Big)^{\frac{1}{p'}} = c_5 r^{1-\frac{n}{p}} = c_5 |x-y|^{1-\frac{n}{p}}$$

we finally arrive at

$$\frac{|u(x) - u(y)|}{|x-y|^{1-\frac{n}{p}}} \leq c\|u\|_{W^{1,p}}$$

proving the theorem. □

In order to construct sub-Markovian semigroups and Dirichlet form we will discuss now some special properties of the space $W^{1,p}(\mathbb{R}^n)$. We refer to D.Gilbarg and N.S.Trudinger [106] or to D.Kinderlehrer and G.Stampacchia [175] as our main reference. First we prove the *chain rule* for functions in $W^{1,p}(\mathbb{R}^n)$.

Lemma 2.1.14. *Let $f \in C^1(\mathbb{R})$ with $f' \in C_b(\mathbb{R})$ and $u \in W^{1,p}(\mathbb{R}^n)$. Then $f \circ u \in W^{1,p}(\mathbb{R}^n)$ and we have for $1 \le j \le n$ the chain rule $\frac{\partial}{\partial x_j}(f \circ u) = f'(u)\frac{\partial u}{\partial x_j}$.*

Proof: We know that $u \in W^{1,p}(\mathbb{R}^n)$ may be approximated in the norm $\|\cdot\|_{W^{1,p}(\mathbb{R}^n)}$ by a sequence $(u_m)_{m\in\mathbb{N}}$ of functions $u_m \in C_0^\infty(\mathbb{R}^n)$. Without loss of generality we may assume that $u_m \to u$ almost everywhere. Since f' is bounded, we have for all functions $v, \omega \in W^{1,p}(\mathbb{R}^n)$

$$|f(v(x)) - f(\omega(x))| \le \|f'\|_\infty |v(x) - \omega(x)|.$$

Thus we find

$$\|f(u_m) - f(u)\|_{L^p} \le \|f'\|_\infty \|u_m - u\|_{L^p},$$

and

$$\begin{aligned}
&\left\|f'(u_m)\frac{\partial}{\partial x_j}u_m - f'(u)\frac{\partial}{\partial x_j}u\right\|_{L^p} \\
&\quad\le \left\|f'(u_m)\Big(\frac{\partial}{\partial x_j}u_m - \frac{\partial}{\partial x_j}u\Big)\right\|_{L^p} + \left\|(f'(u_m) - f'(u))\frac{\partial}{\partial x_j}u\right\|_{L^p} \\
&\quad\le \|f'\|_\infty \left\|\frac{\partial}{\partial x_j}(u_m - u)\right\|_{L^p} \\
&\qquad + \Big(\int_{\mathbb{R}^n} |f'(u_m(x)) - f'(u(x))|^p \Big|\frac{\partial u(x)}{\partial x_j}\Big|^p dx\Big)^{\frac{1}{p}}.
\end{aligned}$$

By our assumption we have $u_m \to u$ in $L^p(\mathbb{R}^n)$ and almost everywhere as well as $\frac{\partial}{\partial x_j}u_m \to \frac{\partial}{\partial x_j}u$ in $L^p(\mathbb{R}^n)$. Moreover, since f' is continuous we have $f'(u_m(x)) \to f'(u(x))$ almost everywhere and

$$|f'(u_m(x)) - f'(u(x))| \cdot \left|\frac{\partial u}{\partial x_j}(x)\right| \le 2\|f'\|_\infty \left|\frac{\partial u}{\partial x_j}(x)\right|,$$

implying that $f(u_m) \to f(u)$ and $f'(u_m)\frac{\partial u_m}{\partial x_j} \to f'(u)\frac{\partial u}{\partial x_j}$ in $L^p(\mathbb{R}^n)$. Thus we find $\frac{\partial}{\partial x_j}f(u_m) \to f'(u)\frac{\partial u}{\partial x_j}$, which proves the lemma. □

Corollary 2.1.15. *Let $u \in W^{1,p}(\mathbb{R}^n;\mathbb{R})$. Then u^+, u^- and $|u|$ are also elements in $W^{1,p}(\mathbb{R}^n)$ and we have for $1 \le j \le n$*

$$\frac{\partial}{\partial x_j}u^+ = \begin{cases} \frac{\partial}{\partial x_j}u, & \text{if } u > 0 \\ 0, & \text{if } u \le 0 \end{cases}, \tag{2.34}$$

$$\frac{\partial}{\partial x_j}u^- = \begin{cases} 0, & \text{if } u \ge 0 \\ -\frac{\partial}{\partial x_j}u, & \text{if } u < 0 \end{cases}, \tag{2.35}$$

and

$$\frac{\partial}{\partial x_j}|u| = \begin{cases} \frac{\partial}{\partial x_j}u, & \text{if } u > 0 \\ 0 & \text{if } u = 0 \\ -\frac{\partial}{\partial x_j}u, & \text{if } u < 0 \end{cases}. \tag{2.36}$$

Recall that derivatives of $W^{1,p}$-functions are distributional derivatives belonging to $L^p(\mathbb{R}^n)$.

Proof: For $\varepsilon > 0$ consider the function

$$f_\varepsilon(t) := \begin{cases} (t^2+\varepsilon^2)^{\frac{1}{2}} - \varepsilon, & \text{if } t > 0 \\ 0, & \text{if } t \le 0. \end{cases}$$

Applying Lemma 2.1.14 we find for any $\varphi \in C_0^\infty(\mathbb{R}^n)$

$$\int_{\mathbb{R}^n} \frac{\partial f_\varepsilon(u)}{\partial x_j}\varphi\, dx = -\int_{u>0} \varphi u \frac{\partial u}{\partial x_j}(u^2+\varepsilon^2)^{-\frac{1}{2}}\, dx$$

and for $\varepsilon \to 0$ we obtain

$$\int_{\mathbb{R}^n} u^+ \frac{\partial}{\partial x_j}\varphi\, dx = -\int_{u>0} \varphi u \frac{\partial u}{\partial x_j}\, dx$$

which gives (2.34). Since $u^- = (-u)^+$ and $|u| = u^+ + u^-$ we also get (2.35) and (2.36). □

Corollary 2.1.16. *Let $u \in W^{1,p}(\mathbb{R}^n)$. Then we have* $\operatorname{grad} u = 0$ *on any open set where u is constant.*

Proof: First note that if u is constant on a set with infinite measure, this constant must be zero. Now, if $u = c$ on a bounded open set G, we may multiply u by a test function ψ being equal to 1 in a neighbourhood of G. Taking another test function which is equal to c on the support of ψ, we find by subtracting that we may assume that $c = 0$ in any case. But now the corollary follows from $\operatorname{grad} u = \operatorname{grad} u^+ - \operatorname{grad} u^-$ and Corollary 2.1.15. □

Theorem 2.1.17. *Let $f : \mathbb{R} \to \mathbb{R}$ be a continuous function with a continuous and bounded first order derivative except of finite numbers of points where the graph of f may have corners. The set of these corners is denoted by E. Further let $u \in W^{1,p}(\mathbb{R}^n, \mathbb{R})$. Then we have $f \circ u \in W^{1,p}(\mathbb{R}^n, \mathbb{R})$ and*

$$\frac{\partial}{\partial x_j}(f \circ u) = \begin{cases} f'(u(x))\frac{\partial u(x)}{\partial x_j}, & \text{if } (x, u(x)) \notin E \\ 0, & \text{if } (x, u(x)) \in E \end{cases} \tag{2.37}$$

Proof: Clearly we may assume that E consists only of the origin. Take $f_1, f_2 \in C_b^1(\mathbb{R}^n)$ such that $f_1(0) = f_2(0)$ and $f_1(u) = f(u)$ for $u \geq 0$ and $f_2(u) = f(u)$ for $u \leq 0$. Since now $f(u) = f_1(u^+) + f_2(-u^-)$, the result follows from Lemma 2.1.14 and Corollary 2.1.15. □

Corollary 2.1.18. *A. Let $u \in W^{1,p}(\mathbb{R}^n, \mathbb{R})$. Then the function $(0 \vee u) \wedge 1$ belongs to $W^{1,p}(\mathbb{R}^n, \mathbb{R})$ and we have*

$$\frac{\partial}{\partial x_j}((0 \vee u) \wedge 1) = \begin{cases} \frac{\partial u}{\partial x_j}, & \text{if } 0 < u < 1 \\ 0, & \text{otherwise} \end{cases}. \tag{2.38}$$

B. Moreover, for $u \in W^{1,p}(\mathbb{R}^n, \mathbb{R})$ the function $(u - 1)^+$ is an element of $W^{1,p}(\mathbb{R}^n, \mathbb{R})$, and

$$\frac{\partial}{\partial x_j}(u-1)^+ = \begin{cases} \frac{\partial u}{\partial x_j}, & \text{if } u > 1 \\ 0, & \text{if } u \leq 0 \end{cases}. \tag{2.39}$$

holds.

Proof: A. The function $t \mapsto f(t) = \frac{t+|t|+2-|t+|t|-2|}{4}$ fulfills the requirements of Theorem 2.1.17, but $f(t) = (0 \vee t) \wedge 1$.

B. Since $t \mapsto g(t) := \frac{t-1+|t-1|}{2}$ satisfies the assumptions of Theorem 2.1.17 and $(t-1)^+ = g(t)$, we may apply once more Theorem 2.1.17. □

In Section I.4.6 we introduce the concept of a Dirichlet operator in the space $L^p(\mathbb{R}^n, \mathbb{R})$. Our next aim is to determine classes of second order differential operators which might have a closure being a Dirichlet operator. For this let us consider the operator

$$L(x,D) = \sum_{k,l=1}^{n} a_{kl}(x)\frac{\partial^2}{\partial x_k\,\partial x_l} + \sum_{j=1}^{n} b_j(x)\frac{\partial}{\partial x_j} + c(x) \tag{2.40}$$

with coefficients $a_{kl}, b_j, c \in C_b(\mathbb{R}^n, \mathbb{R})$.

Lemma 2.1.19. *Let $L(x,D)$ be as in (2.40) with $a_{kl}, b_j, c \in C_b(\mathbb{R}^n, \mathbb{R})$. Then $L(x,D)$ maps $W^{2,p}(\mathbb{R}^n, \mathbb{R})$ continuously to $L^p(\mathbb{R}^n, \mathbb{R})$, $1 \le p < \infty$, and it maps $C^2_\infty(\mathbb{R}^n, \mathbb{R})$ continuously to $C_\infty(\mathbb{R}^n, \mathbb{R})$. Moreover we have the estimates*

$$\|L(x,D)u\|_{L^p} \le c\|u\|_{W^{2,p}} \tag{2.41}$$

and

$$\|L(x,D)u\|_\infty \le c\|u\|_{C^2_\infty}. \tag{2.42}$$

Proof: It is sufficient to prove (2.41) and (2.42) for all $u \in C_0^\infty(\mathbb{R}^n, \mathbb{R})$. For $1 \le p < \infty$ we find

$$\begin{aligned}
&\|L(x,D)u\|_{L^p}\\
&\le \sum_{k,l=1}^{n} \|a_{kl}\|_\infty \left\|\frac{\partial^2 u}{\partial x_k\,\partial x_l}\right\|_{L^p} + \sum_{j=1}^{n}\|b_j\|_\infty\left\|\frac{\partial u}{\partial x_j}\right\|_{L^p} + \|c\|_\infty\|u\|_{L^p}\\
&\le c\|u\|_{W^{2,p}},
\end{aligned}$$

and (2.42) follows analogously. □

Now, let us consider $L(x,D)$ as operator on $L^p(\mathbb{R}^n, \mathbb{R})$, $1 < p < \infty$, with domain $W^{2,p}(\mathbb{R}^n, \mathbb{R})$. We are longing for conditions on the coefficients to prove

$$\int_{\mathbb{R}^n} (L(x,D)u)((u-1)^+)^{p-1}dx \le 0$$

for all $u \in W^{2,p}(\mathbb{R}^n, \mathbb{R})$. First we observe that for $c(x) \le 0$ for all $x \in \mathbb{R}^n$, we get

$$\int_{\mathbb{R}^n} c(x)u(x)((u(x)-1)^+)^{p-1}dx = \int_{u>1} c(x)u(x)(u(x)-1)^{p-1}dx \le 0.$$

Next we find by Corollary 2.1.18.B that

$$\begin{aligned}
\frac{\partial u}{\partial x_j}((u-1)^+)^{p-1} &= \chi_{\{u>1\}}\frac{\partial(u-1)}{\partial x_j}(u-1)^{p-1}\\
&= \chi_{\{u>1\}}\frac{1}{p}\frac{\partial(u-1)^p}{\partial x_j},
\end{aligned}$$

and therefore we have for $b_j \in C_b^1(\mathbb{R}^n, \mathbb{R})$ and $u \in W^{2,p}(\mathbb{R}^n, \mathbb{R})$

$$\int_{\mathbb{R}^n} b_j(x)\frac{\partial u(x)}{\partial x_j}(u(x)-1)^+)^{p-1}dx = -\int_{u>1}\frac{1}{p}(u(x)-1)^p\frac{\partial}{\partial x_j}b_j(x)\,dx.$$

Thus, if $\sum_{j=1}^{n}\frac{\partial b_j}{\partial x_j} \geq 0$, we have for $u \in W^{2,p}(\mathbb{R}^n, \mathbb{R})$

$$\int_{\mathbb{R}^n} b_j(x)\frac{\partial u(x)}{\partial x_j}((u(x)-1)^+)^{p-1}dx \leq 0.$$

Finally we get for $a_{kl} \in C_b^2(\mathbb{R}^n, \mathbb{R})$

$$\begin{aligned}
&\int_{\mathbb{R}^n}\sum_{k,l=1}^{n} a_{kl}(x)\frac{\partial^2 u(x)}{\partial x_k\,\partial x_l}((u-1)^+(x))^{p-1}dx \\
&\quad= \int_{\mathbb{R}^n}\sum_{k,l=1}^{n}\frac{\partial}{\partial x_k}\left(a_{kl}(x)\frac{\partial u(x)}{\partial x_l}\right)((u-1)^+(x))^{p-1}\,dx \\
&\quad- \int_{\mathbb{R}^n}\sum_{k,l=1}^{n}\frac{\partial a_{kl}(x)}{\partial x_k}\frac{\partial u(x)}{\partial x_l}((u-1)^+(x))^{p-1}\,dx \\
&\quad= \int_{\mathbb{R}^n}\chi_{\{u>1\}}(x)(p-1)((u-1)(x))^{p-2} \\
&\quad\times\sum_{k,l=1}^{n} a_{kl}(x)\frac{\partial(u-1)(x)}{\partial x_l}\frac{\partial(u-1)(x)}{\partial x_k}\,dx \\
&\quad- \frac{1}{p}\int_{\mathbb{R}^n}\chi_{\{u>1\}}(x)\frac{\partial a_{kl}(x)}{\partial x_k}\frac{\partial((u-1)(x))^p}{\partial x_k}\,dx,
\end{aligned}$$

where we used once again Corollary 2.1.18.B. Now it follows

Theorem 2.1.20. *Suppose that $L(x,D)$ has coefficients $a_{kl} \in C_b^2(\mathbb{R}^n, \mathbb{R})$, $b_j \in C_b^1(\mathbb{R}^n, \mathbb{R})$ and $c \in C_b(\mathbb{R}^n, \mathbb{R})$ such that $c \leq 0$, $a_{kl} = a_{lk}$ and $\sum_{k,l=1}^n a_{kl}(x)\xi_k\xi_l \geq 0$ for all $x \in \mathbb{R}^n$ and $\xi \in \mathbb{R}^n$. If in addition we assume that $\sum_{j=1}^n \frac{\partial}{\partial x_j}\left(b_j(x) - \sum_{k=1}^n \frac{\partial a_{kj}(x)}{\partial x_k}\right) \geq 0$, then we have for all $u \in W^{2,p}(\mathbb{R}^n, \mathbb{R})$, $1 < p < \infty$,*

$$\int_{\mathbb{R}^n}(L(x,D)u(x))\left((u-1)^+(x)\right)^{p-1}dx \leq 0. \tag{2.43}$$

Note that we need only the regularity assumptions $a_{kl} \in C_b^1(\mathbb{R}^n; \mathbb{R})$ and $b_j - \sum_{k=1}^n \frac{\partial a_{kj}}{\partial x_k} \in C_b^1(\mathbb{R}^n; \mathbb{R})$. This leads immediately to

Corollary 2.1.21. *Suppose that $L(x,D)$ is given by*

$$L(x,D)u(x) = \sum_{k,l=1}^{n} \frac{\partial}{\partial x_k}\left(a_{kl}(x)\frac{\partial u(x)}{\partial x_l}\right) + \sum_{j=1}^{n} b_j(x)\frac{\partial u(x)}{\partial x_j} + c(x)u(x),$$

with $a_{kl} = a_{lk} \in C_b^1(\mathbb{R}^n;\mathbb{R})$, $b_j \in C_b^1(\mathbb{R}^n;\mathbb{R})$ and $c \in C_b(\mathbb{R}^n;\mathbb{R})$ such that $\sum_{k,l=1}^{n} a_{kl}(x)\xi_k\xi_l \geq 0$ for al $x \in \mathbb{R}^n$ and $\xi \in \mathbb{R}^n$, and $c(x) \leq 0$. Then (2.43) holds provided that $\sum_{j=1}^{n} \frac{\partial b_j(x)}{\partial x_k} \geq 0$.

For $p = 2$ we may weaken the sign condition on

$$\sum_{j=1}^{n} \frac{\partial}{\partial x_j}\left(b_j(x) - \sum_{k=1}^{n} \frac{\partial a_{kj}(x)}{\partial x_k}\right)$$

in the following way. Consider the form

$$B(u,v) = \sum_{k,l=1}^{n} \int_{\mathbb{R}^n} a_{kl}(x)\frac{\partial u(x)}{\partial x_l}\frac{\partial v(x)}{\partial x_k}\,dx$$
$$+ \sum_{j=1}^{n} \int_{\mathbb{R}^n} \left(b_j(x) - \sum_{k=1}^{n} \frac{\partial a_{kj}(x)}{\partial x_k}\right)\frac{\partial u(x)}{\partial x_j}\,v(x)\,dx + \int_{\mathbb{R}^n} c(x)u(x)v(x)\,dx.$$

Suppose that B satisfies the sector condition, i.e. (I.4.333) in Definition I.4.7.12. If we add to $h(x) := c(x) - \sum_{j=1}^{n} \frac{\partial}{\partial x_j}\left(b_j(x) - \sum_{k=1}^{n} \frac{\partial a_{kj}(x)}{\partial x_k}\right)$ a function $x \mapsto d(x)$ such that $h(x) + d(x) \geq 0$, we find by Theorem I.4.7.20 that the contraction condition (I.4.347) is fulfilled, compare Example I.4.7.35. Thus $B(\cdot,\cdot) + (d(\cdot)\cdot,\cdot)_0$ extends to a semi-Dirichlet form with domain $H^1(\mathbb{R}^n, \mathbb{R}) = W^{1,2}(\mathbb{R}^n, \mathbb{R})$ and $L(x,D) - d(x)\cdot \mathrm{id}$ extends to a Dirichlet operator. In order to identify $(L(x,D), W^{2,p}(\mathbb{R}^n, \mathbb{R}))$ as a Dirichlet operator on $L^p(\mathbb{R}^n, \mathbb{R})$, $1 < p < \infty$, we have to prove closedness. Thus we have to show that $\|\cdot\|_{W^{2,p}}$ is equivalent to the graph norm $\|L(x,D)u\|_{L^p} + \|u\|_{L^p}$. Taking into account Lemma 2.1.19 it remains to prove

$$\|u\|_{W^{2,p}} \leq c\left(\|L(x,D)u\|_{L^p} + \|u\|_{L^p}\right). \tag{2.44}$$

This leads to the concept of ellipticity.

Definition 2.1.22. Let $L(x,D)$ be given by (2.40) with coefficients $a_{kl} = a_{lk} \in C_b(\mathbb{R}^n;\mathbb{R})$. We call $L(x,D)$ *uniformly elliptic* if for some $\lambda_0 > 0$

$$\lambda_0|\xi|^2 \leq \sum_{k,l=1}^{n} a_{kl}(x)\xi_k\xi_l \tag{2.45}$$

holds for all $x \in \mathbb{R}^n$ and $\xi \in \mathbb{R}^n$.

We will prove (2.44) for uniformly elliptic differential operator of second order. Our presentation follows mainly the paper [37] of F.E.Browder, we refer also to the papers [2], [3] of S.Agmon, A.Douglis and L.Nirenberg, as well as to the work [252] of Chr.Simader. Let us start with some preparatory material. Given $\varepsilon > 0$ and let $(\psi_k)_{k\in\mathbb{Z}^n}$ be the *partition of unity* from the second part of Lemma I.2.1.3.B. The family $(\psi_k^l)_{k\in\mathbb{Z}^n}$, $l > 0$, defined by

$$\psi_k^l(x) := \psi\Big(\frac{x-k}{l}\Big) \tag{2.46}$$

fulfills

$$\operatorname{diam}(\operatorname{supp}\psi_k^l) \le (1+\varepsilon)l\cdot n^{1/2}, \tag{2.47}$$

$$|\partial^\alpha \psi_k^l(x)| \le \Big(\frac{\varepsilon}{l}\Big)^{-|\alpha|} M, \quad |\alpha| \le m \quad \text{and } M = M(n,m) \tag{2.48}$$

and

$$\sum_{k\in\mathbb{Z}^n} (\psi_k^l(x))^2 = 1. \tag{2.49}$$

Moreover, there exists $N = N(\varepsilon, n, l)$ such that the intersection of $N+1$ of the supports of the functions ψ_k^l is empty. In fact, we may choose N independent of l.

Lemma 2.1.23. *Given $u \in W^{m,p}(\mathbb{R}^n)$ and for $\varepsilon > 0$, $l > 0$ let $(\psi_k^l)_{k\in\mathbb{Z}^n}$ be given by (2.46). Then we have for all $j \in \mathbb{N}_0$, $j \le m$,*

$$\|u\|_{W^{j,p}}^p \le c_j \sum_{k\in\mathbb{Z}^n} \|(\psi_k^l)^2 u\|_{W^{j,p}}^p < \infty \tag{2.50}$$

and

$$\sum_{k\in\mathbb{Z}^n} \|(\psi_k^l)^2 u\|_{W^{j,p}}^p \le \tilde{c}_j \sum_{j_1\le j} \Big(\frac{\varepsilon}{l}\Big)^{-(j-j_1)p} \|u\|_{W^{j_1,p}}^p < \infty. \tag{2.51}$$

Proof: We prove first (2.51). For this observe that

$$\begin{aligned}\sum_{|k|\le K} \|(\psi_k^l)^2 u\|_{W^{j,p}}^p &= \sum_{|k|\le K}\sum_{|\alpha|\le j} \int_{\mathbb{R}^n} \Big|\partial^\alpha((\psi_k^l)^2 u)\Big|^p dx \\ &\le c_0 \sum_{|k|\le K}\sum_{|\alpha|\le j}\sum_{|\beta|+|\gamma|+|\delta|=|\alpha|} \int_{\mathbb{R}^n} |\partial^\beta\psi_k^l|^p\,|\partial^\gamma\psi_k^l|^p\,|\partial^\delta u|^p\,dx,\end{aligned}$$

where we used

$$\Big| \sum_{|\beta|+|\gamma|+|\delta|=|\alpha|} c_{\beta,\gamma,\delta}(\partial^\beta \psi_k^l)(\partial^\gamma \psi_k^l)(\partial^\delta u)\Big|^p \\ \leq c_0 \sum_{|\beta|+|\gamma|+|\delta|=|\alpha|} |\partial^\beta \psi_k^l|^p |\partial^\gamma \psi_k^l|^p |\partial^\delta u|^p.$$

Since $\operatorname{supp} \partial^\beta \psi_k^l \subset \operatorname{supp} \psi_k^l$, it follows from the conditions on the supports that

$$\sum_{|k|\leq K} |\partial^\beta \psi_k^l||\partial^\gamma \psi_k^l| \leq (N+1)M^2 \left(\frac{\varepsilon}{l}\right)^{-|\beta|-|\gamma|},$$

implying that

$$c_0 \sum_{|\alpha|\leq j} \sum_{|\beta|+|\gamma|+|\delta|=|\alpha|} \int_{\mathbb{R}^n} \sum_{|k|\leq K} |\partial^\beta \psi_k^l|^p |\partial^\gamma \psi_k^l|^p |\partial^\delta u|^p \, dx \\ \leq c_0(N+1)M^2 \sum_{j_1=0}^{j} \left(\frac{\varepsilon}{l}\right)^{-(j-j_1)p} \int_{\mathbb{R}^n} \sum_{|\delta|\leq j_1} |\partial^\delta u|^p \, dx.$$

In particular we have that

$$\sum_{|k|\leq K} \|(\psi_k^l)^2 u\|_{W1,p}^p$$

has a finite bound independent of K, hence as $K \to \infty$ it converges to a finite limit and (2.51) is proved. Now (2.50) follows easily since

$$\int_{\mathbb{R}^n} |\partial^\alpha u|^p \, dx = \int_{\mathbb{R}^n} |\partial^\alpha \sum_{k\in\mathbb{Z}^n} (\psi_k^l)^2 u|^p \, dx,$$

and for each $x \in \mathbb{R}^n$ at most $N+1$ terms of the integral on the right-hand side are different from zero. Hence the integral is less than

$$2^{(N+1)(p-1)} \sum_{k\in\mathbb{Z}^n} \int_{\mathbb{R}^n} |\partial^\alpha((\psi_k^l)^2 u)|^p \, dx \leq 2^{(N+1)(p-1)} \sum_{k\in\mathbb{Z}^n} \|(\psi_k^l)^2 u\|_{W^{j,p}}^p.$$

Adding up over all $\alpha \in \mathbb{N}_0^n$, $|\alpha| \leq m$, gives (2.50). □

We will prove (2.44) for uniformly elliptic operator in several steps. First we observe

Proposition 2.1.24. *Suppose that for the second order differential operator* $L_{pr}(x,D) = \sum_{|\alpha|=2} a_\alpha(x) D^\alpha$ *we have the estimate*

$$\|u\|_{W^{2,p}} \le c_p(\|L_{pr}(x,D)u\|_{L^p} + \|u\|_{L^p}) \tag{2.52}$$

for all $u \in \overset{\circ}{W}{}^{2,p}(G)$, $G \subset \mathbb{R}^n$ *being an open set. Suppose further that* $b_j \in C_b(G)$, $1 \le j \le n$, *and* $c \in C_b(G)$ *are given functions. Then the differential operator*

$$L(x,D) := L_{pr}(x,D) + \sum_{j=1}^n b_j(x) \frac{\partial}{\partial x_j} + c(x)$$

satisfies with a suitable constant $\tilde{c}_p > 0$

$$\|u\|_{W^{2,p}} \le \tilde{c}_p \left(\left\| L(x,D)u \right\|_{L^p} + \|u\|_{L^p} \right) \tag{2.53}$$

for all $u \in \overset{\circ}{W}{}^{2,p}(G)$.

Proof: For $u \in C_0^\infty(G)$ (or $u \in \overset{\circ}{W}{}^{2,p}(G)$) we find with $\eta > 0$

$$\begin{aligned} \|L_{pr}(x,D)u\|_{L^p} &\le \|L(x,D)u\|_{L^p} + \left\| \sum_{j=1}^n b_j(\cdot) \frac{\partial u}{\partial x_j} \right\|_{L^p} + \|c(\cdot)u\|_{L^p} \\ &\le \|L(x,D)u\|_{L^p} + \gamma_0 \|u\|_{W^{1,p}} \\ &\le \|L(x,D)u\|_{L^p} + \eta \|u\|_{W^{2,p}} + \gamma_1(\eta) \|u\|_{L^p}, \end{aligned}$$

where we used Corollary 2.1.8. But now the proposition follows immediately from (2.52). □

Theorem 2.1.25. *Let* $L(D)u = \sum_{|\alpha|=2} a_\alpha D^\alpha u$ *be a uniformly elliptic operator with constant coefficients* $a_\alpha \in \mathbb{R}$. *Then we have for all* $u \in W^{2,p}(\mathbb{R}^n)$

$$\Big(\sum_{|\alpha|=2} \|D^\alpha u\|_{L^p}^p \Big)^{1/p} \le c \|L(D)u\|_{L^p}. \tag{2.54}$$

Proof: It is sufficient to prove (2.54) for all $\varphi \in C_0^\infty(\mathbb{R}^n)$. For $\varphi \in C_0^\infty(\mathbb{R}^n)$ we find

$$(L(D)\varphi)^\wedge(\xi) = L(\xi)\hat{\varphi}(\xi) = \sum_{|\alpha|=2} a_\alpha \xi^\alpha \hat{\varphi}(\xi).$$

Since $\xi^\alpha\hat{\varphi}(\xi) = (D^\alpha\varphi)^\wedge(\xi)$ we find for $|\alpha| = 2$

$$(D^\alpha\varphi)^\wedge(\xi) = \frac{\xi^\alpha}{L(\xi)}(L(D)\varphi)^\wedge(\xi).$$

The polynomial $\xi \mapsto L(\xi)$ is homogeneous of degree 2 and by assumption uniformly elliptic. Thus for $\xi \neq 0$ we find

$$\left|D^\beta\left(\frac{\xi^\alpha}{L(\xi)}\right)\right| \leq c_0|\xi|^{-|\beta|}$$

for some $c_0 = c_0(n, \lambda_0)$, λ_0 being the ellipticity constant and $|\alpha| = 2$. Next we define

$$m(\xi) := \begin{cases} \frac{\xi^\alpha}{L(\xi)}, & \xi \neq 0 \\ 0, & \xi = 0 \end{cases},$$

thus $|m(\xi)| \leq K$ for all $\xi \in \mathbb{R}^n$. Since $L(D)\varphi \in C_0^\infty(\mathbb{R}^n)$, we find that

$$\left|(L(D)\varphi)^\wedge(\xi)\right| \leq (1+|\xi|^2)^{-(n+1)/2}$$

and the Fourier inversion formula gives

$$\begin{aligned} D^\alpha\varphi(x) &= (2\pi)^{-n/2}\int_{\mathbb{R}^n} e^{ix\cdot\xi}\,\frac{\xi^\alpha}{L(\xi)}(L(D)\varphi)^\wedge(\xi)\,d\xi \\ &= (2\pi)^{-n/2}\int_{\mathbb{R}^n} e^{ix\cdot\xi}\,m(\xi)\,(L(D)\varphi)^\wedge(\xi)\,d\xi. \end{aligned}$$

Thus by the *Michlin–Hörmander multiplier theorem*, Theorem I.3.12.3 (in the form of Corollary I.3.12.4) we have

$$\|D^\alpha\varphi\|_{L^p} = \|F^{-1}(m(\xi)\,(L(D)\varphi)^\wedge)\|_{L^p} \leq \|L(D)\varphi\|_{L^p},$$

which implies (2.54). □

Remark 2.1.26. A. Clearly (2.54) implies (2.44) in case of constant coefficients. We only have to take into account (2.17).

B. Since D^α and $L(D)$ are local operators, we find with constants $\tilde{c}_p$ and c'_p independent of $G \subset \mathbb{R}^n$, G open, the estimates

$$\left(\sum_{|\alpha|=2}\|D^\alpha\varphi\|_{L^p}^p\right)^{1/p} \leq \tilde{c}_p\|L(D)u\|_{L^p} \tag{2.55}$$

and

$$\|\varphi\|_{W^{2,p}} \le c_p'(\|L(D)\varphi\|_{L^p} + \|\varphi\|_{L^p}) \tag{2.56}$$

for all $\varphi \in C_0^\infty(G)$ or equivalently for all $u \in \overset{\circ}{W}{}^{2,p}(G)$.

Lemma 2.1.27. *Let $L(x, D) = \sum_{|\alpha|=2} a_\alpha(x) D^\alpha$ be a uniformly elliptic operator with coefficients $a_\alpha \in C_b(\mathbb{R}^n, \mathbb{R})$ and let $G \subset \mathbb{R}^n$ be an open set. Further suppose that for some $x_0 \in G$ we have with c_p' from (2.56)*

$$(c_p')^p \sup_{x\in G} \sum_{|\alpha|=2} |a_\alpha(x) - a_\alpha(x_0)|^p \le \left(\frac{1}{2}\right)^p \eta^p \tag{2.57}$$

with some $\eta \le 2^{-N(p-1)/p}$, where $N = \frac{n(n+1)}{2}$ is the cardinality of the set $\{\alpha \in \mathbb{N}_0^n; \quad |\alpha| = 2\}$. Then we have for all $u \in \overset{\circ}{W}{}^{2,p}(G)$

$$\|u\|_{W^{2,p}} \le 2c_p'(\|L(x, D)u\|_{L^p} + \|u\|_{L^p}). \tag{2.58}$$

Proof: It is sufficient to prove that (2.58) holds for all $u \in C_0^\infty(G)$. Moreover, by (2.56) it follows that

$$\begin{aligned}\|u\|_{W^{2,p}} &\le c_p'(\|L(x_0, D)u\|_{L^p} + \|u\|_{L^p}) \\ &\le c_p'\big(\|L(x_0, D)u\|_{L^p} + \|(L(x, D) - L(x_0, D))u\|_{L^p} + \|u\|_{L^p}\big)\end{aligned}$$

and it is suffices to prove that (2.57) implies (2.58) which will follow from

$$c_p'\|(L(x, D) - L(x_0, D))u\|_{L^p} \le \frac{1}{2}\|u\|_{W^{2,p}}. \tag{2.59}$$

But for $u \in C_0^\infty(G)$ we have

$$\begin{aligned}&(c_p')^p\|(L(x, D) - L(x_0, D))u\|_{L^p}^p \\ &= (c_p')^p \int_G \Big|\sum_{|\alpha|=2} (a_\alpha(x) - a_\alpha(x_0)) D^\alpha u(x)\Big|^p dx \\ &\le (c_p')^p \, 2^{N(p-1)} \Big(\sup_{x\in G} \sum_{|\alpha|=2} |a_\alpha(x) - a_\alpha(x_0)|^p\Big) \sum_{|\alpha|=2} \int_G |D^\alpha u(x)|^p \, dx \\ &\le \left(\frac{1}{2}\right)^p \|u\|_{W^{2,p}}^p,\end{aligned}$$

where we used in the final step (2.57). □

To proceed further, we need the following commutator estimate:

Lemma 2.1.28. *Let* $L(x,D) = \sum_{|\alpha|=2} a_\alpha(x) D^\alpha$ *be a differential operator with coefficients* $a_\alpha \in C_b(\mathbb{R}^n)$. *For any* $\varphi \in C_b^\infty(\mathbb{R}^n)$ *we have for all* $u \in W^{1,p}(\mathbb{R}^n)$

$$\left\| [\varphi, L(x,D)]u \right\|_{L^p} \leq c\|u\|_{W^{1,p}} \tag{2.60}$$

Proof: Observe that

$$\begin{aligned}\left[\varphi, \sum_{|\alpha|=2} a_\alpha(x) D^\alpha\right] u &= \sum_{|\alpha|=2} [\varphi, a_\alpha(x) D^\alpha] u \\ &= \sum_{|\alpha|=2} a_\alpha(x)[\varphi, D^\alpha] u,\end{aligned}$$

and the lemma follows from Corollary 2.1.9.

Theorem 2.1.29. *Let* $L(x,D) = \sum_{|\alpha|=2} a_\alpha(x) D^\alpha$ *be a uniformly elliptic differential operator with coefficients* $a_\alpha \in C_b(\mathbb{R}^n, \mathbb{R})$. *Further let* $G \subset \mathbb{R}^n$ *be a bounded open set. Then we have for all* $u \in \overset{\circ}{W}{}^{2,p}(G)$

$$\|u\|_{W^{2,p}} \leq \lambda_p \left(\left\| L(x,D)u \right\|_{L^p} + \|u\|_{L^p} \right), \tag{2.61}$$

where $\lambda_p = \lambda_p(G, a_\alpha)$.

Proof: Once again, it is sufficient to prove (2.61) for all $u \in C_0^\infty(G)$. On $\overline{G} + \overline{B_1(0)}$ the coefficients a_α are uniformly continuous and therefore we may find a finite open covering $\left(U_j\right)_{j=1,\cdots,M}$ of $\overline{G}$ such that

$$(c_p')^p \max_{1\leq j\leq M} \sup_{x,y\in G\cap U_j} \sum_{|\alpha|=2} |a_\alpha(x) - a_\alpha(y)|^p \leq \left(\frac{1}{2}\right)^p 2^{-N(p-1)} \tag{2.62}$$

where c_p' is as in (2.56) and as before we take $N = \frac{n(n+1)}{2}$. We know by Lemma I.2.1.3.B that there exists a partition of unity subordinated to $(U_j)_{j=1,\dots,M}$, i.e. there are functions $\varphi_j \in C_0^\infty(\mathbb{R}^n)$ such that $0 \leq \varphi_j \leq 1$, $\operatorname{supp}\varphi_j \subset U_j$, and $\sum_{j=1}^M \varphi_j(x) = 1$ for all $x \in G$. From Lemma 2.1.27 we know that (2.62) implies

$$\|\varphi_j u\|_{W^{2,p}(G\cap U_j)} \leq 2c_p'(\|L(x,D)(\varphi_j u)\|_{L^p(G\cap U_j)} + \|\varphi_j u\|_{L^p(G\cap U_j)})$$

which leads to

$$\|\varphi_j u\|_{W^{2,p}(G\cap U_j)} \leq 2c_p'(\|L(x,D)(\varphi_j u)\|_{L^p(G)} + \|\varphi_j u\|_{L^p(G)}). \tag{2.63}$$

Now it follows for $u \in C_0^\infty(G)$ by Lemma 2.1.28 that

$$\begin{aligned}
\|u\|^p_{W^{2,p}(G)} &= \sum_{|\alpha|\le 2}\int_G |D^\alpha u|^p\,dx = \sum_{|\alpha|\le 2}\int_G \Big(\sum_{j=1}^M \varphi_j\Big)^p |D^\alpha u|^p\,dx \\
&\le \sum_{|\alpha|\le 2} 2^{M(p-1)} \int_G \sum_{j=1}^M |\varphi_j D^\alpha u|^p\,dx \\
&= 2^{M(p-1)} \sum_{|\alpha|\le 2}\int_G \sum_{j=1}^M \big|D^\alpha(\varphi_j\, u) + [\varphi_j,\, D^\alpha]u\big|^p\,dx \\
&\le 2^{(M+2)(p-1)} \sum_{|\alpha|\le 2}\int_G \Big(\sum_{j=1}^M |D^\alpha(\varphi_j\, u)|^p + \sum_{j=1}^M \big|[\varphi_j,\, D^\alpha]u\big|^p\Big)\,dx \\
&= 2^{(M+2)(p-1)}\Big(\sum_{j=1}^M \|\varphi_j\, u\|^p_{W^{2,p}(G\bigcap U_j)} + \sum_{j=1}^M\sum_{|\alpha|\le 2}\int_{U_j\cap G} \big|[\varphi_j,\, D^\alpha]u\big|^p\,dx\Big) \\
&\le c\Big(\sum_{j=1}^M \|\varphi_j u\|^p_{W^{2,p}(G\bigcap U_j)} + \|u\|^p_{W^{2,p}(G)}\Big).
\end{aligned} \tag{2.64}$$

Further we find using (2.63) and Lemma 2.1.28

$$\begin{aligned}
\|\varphi_j\, u\|^p_{W^{2,p}(G\bigcap U_j)} &\le 2\,c_p'\Big(\big\|L(x,D)(\varphi_j u)\big\|_{L^p(G)} + \|u\|_{L^p(G)}\Big) \\
&\le 2\,c_p'\Big(\big\|\varphi_j L(x,D)u\big\|_{L^p(G)} + \big\|[\varphi_j, L(x,D)]u\big\|_{L^p(G)} + \|u\|_{L^p(G)}\Big) \\
&\le 2\,c_p'\big\|L(x,D)u\big\|_{L^p(G)} + \tilde c\|u\|_{W^{1,p}(G)} + 2\,c_p'\|u\|_{L^p(G)} \\
&\le 2\,c_p'\big\|L(x,D)u\big\|_{L^p(G)} + \eta\|u\|_{W^{1,p}(G)} + \tilde c(\eta,p)\|u\|_{L^p(G)},
\end{aligned}$$

where we used Corollary 2.1.8 for $\eta > 0$ arbitrarily given. Thus we find

$$\begin{aligned}
&\sum_{j=1}^M \|\varphi_j\, u\|^p_{W^{2,p}(G\bigcap U_j)} + \|u\|^p_{W^{2,p}(G)} \\
&\qquad \le \gamma\Big(\big\|L(x,D)u\big\|^p_{L^p(G)} + 2\eta^p\|u\|^p_{W^{2,p}(G)} + \|u\|^p_{L^p(G)}\Big).
\end{aligned}$$

Taking $2\gamma\eta^p = \frac{1}{2}$ we finally arrive at

$$\|u\|^p_{W^{2,p}(G)} \le 2\gamma\Big(\big\|L(x,D)u\big\|^p_{L^p(G)} + \|u\|^p_{L^p(G)}\Big),$$

which of course implies (2.61). □

It was crucial in the proof of Theorem 2.1.29 that we could work with a finite covering of G. If the coefficients a_α are Lipshitz continuous, i.e.

$$\max_{|\alpha|=2}|a_\alpha(x)-a_\alpha(y)| \leq \kappa|x-y| \tag{2.65}$$

if follows that M in the proof of Theorem 2.1.29 depends only on diam(G) and κ. Thus we have

Corollary 2.1.30. *Suppose that $L(x,D) = \sum_{|\alpha|=2} a_\alpha(x)D^\alpha$ is a uniformly elliptic differential operator with coefficients $a_\alpha \in C_b(\mathbb{R}^n, \mathbb{R})$ satisfying* (2.65). *Then there exists a constant $\mu = \mu(p,\kappa,R)$ such that*

$$\|u\|_{W^{2,p}(G)} \leq \mu(\|L(x,D)u\|_{L^p(G)}+\|u\|^p_{L^p(G)}) \tag{2.66}$$

holds for all $u \in \overset{\circ}{W}{}^{2,p}(G)$ with diam$(G) \leq R$.

Now we will use the partition of unity considered in Lemma 2.1.23 and Corollary 2.1.30 to prove (2.44) for $G = \mathbb{R}^n$ and $L(x,D) = \sum_{|\alpha|=2} a_\alpha(x)D^\alpha$ with Lipschitz continuous coefficients.

Theorem 2.1.31. *Let $L(x,D) = \sum_{|\alpha|=2} a_\alpha(x)D^\alpha$ is a uniformly elliptic differential operator with coefficients $a_\alpha \in C_b(\mathbb{R}^n, \mathbb{R})$ satisfying* (2.65). *Then we have for all $u \in W^{2,p}(\mathbb{R}^n)$*

$$\|u\|_{W^{2,p}(G)} \leq c_0(\|L(x,D)u\|_{L^p(G)}+\|u\|^p_{L^p(G)}). \tag{2.67}$$

Proof: We may take the partition of unity from Lemma 2.1.23 with $\varepsilon = \frac{1}{2}$ and we denote it by $(\psi_k)_{k\in\mathbb{Z}^n}$. Further we put $Q_k := \operatorname{supp}\psi_k$. Note that diam$(Q_k) \leq \frac{3}{2}n^{1/2}$. From Corollary 2.1.30 it follows now for all $k \in \mathbb{Z}^n$ and $u \in W^{2,p}(\mathbb{R}^n)$ that

$$\|u\psi_k^2\|_{W^{2,p}(G)} \leq \mu(\|L(x,D)(u\psi_k^2)\|_{L^p}+\|u\|_{L^p}),$$

where μ depends only on p, $L(x,D)$, and n. Now using Lemma 2.1.23 and intermediate estimates for the norm $\|\cdot\|_{W^{1,p}}$ we find with $\eta > 0$ sufficiently small using the support properties of ψ_k

$$\begin{aligned}
\|u\|^p_{W^{2,p}(G)} &\le c \sum_{k\in\mathbb{Z}^n} \|\psi_k^2 u\|^p_{W^{2,p}} \\
&\le c'\Big(\sum_{k\in\mathbb{Z}^n} \|L(x,D)(\psi_k^2 u)\|^p_{L^p} + \sum_{k\in\mathbb{Z}^n} \|\psi_k^2 u\|^p_{L^p}\Big) \\
&\le c''\Big(\sum_{k\in\mathbb{Z}^n} \|\psi_k^2 L(x,D)u\|^p_{L^p} + \|u\|^p_{W^{1,p}(G)} + \|u\|^p_{L^p}\Big) \\
&\le \tilde{c}(\|L(x,D)u\|^p_{L^p} + \eta\|u\|^p_{W^{2,p}(G)} + c(\eta)\|u\|^p_{L^p})
\end{aligned}$$

which yields (2.67). □

Combining Theorem 2.1.31 with Proposition 2.1.24 we finally arrive at

Theorem 2.1.32. *Let* $L(x,D) = \sum\limits_{k,l=1}^{n} a_{kl}(x)\frac{\partial^2}{\partial x_k \partial x_l} + \sum_{j=1}^n b_j(x)\frac{\partial}{\partial x_j} + c(x)$ *be a uniformly elliptic operator with coefficients* $a_{kl} = a_{lk}$, a_{kl}, b_j, $c \in C_b(\mathbb{R}^n, \mathbb{R})$, *such that*

$$\max_{1\le k,l\le n} |a_{kl}(x) - a_{kl}(y)| \le \kappa |x-y|.$$

Then (2.44) *holds for all* $u \in W^{2,p}(\mathbb{R}^n)$.

Our next aim is to use (2.44) in order to prove that for some $\lambda > 0$ the equation

$$(\lambda - L(x,D))u = f \tag{2.68}$$

has for all $f \in L^p(\mathbb{R}^n)$ a unique solution in $W^{2,p}(\mathbb{R}^n)$. We will first introduce the concept of a weak solution to (2.68), prove the existence of a unique weak solution, and then we will discuss why weak solutions are already solutions to (2.68).

Let $L(x,D) = \sum\limits_{|\alpha|\le 2} a_\alpha(x)\partial^\alpha$ be a differential operator with coefficients $a_\alpha \in C_b^{|\alpha|}(\mathbb{R}^n, \mathbb{R})$. The operator $L^*(x,D)$ is defined by the relation

$$\int_{\mathbb{R}^n} (L(x,D)\varphi)\psi \, dx = \int_{\mathbb{R}^n} \varphi(L^*(x,D)\psi)\, dx \tag{2.69}$$

for all $\varphi, \psi \in C_0^\infty(\mathbb{R}^n, \mathbb{R})$, i.e.

$$\begin{aligned} L^*(x, D)\psi(x) &= \sum_{|\alpha|\leq 2} (-1)^{|\alpha|}\partial(a_\alpha(x)\psi(x)) \\ &= \sum_{|\alpha|\leq 2} a_\alpha^*(x)\partial^\alpha\psi(x). \end{aligned} \tag{2.70}$$

Note that if $L(x, D)$ is elliptic with principal part

$$L_{pr}(x, D) = \sum_{|\alpha|=2} a_\alpha(x)\partial^\alpha,$$

then $L^*(x, D)$ is also elliptic with the same principal part, in this case we have $a_\alpha = a_{kl} = a_{lk}$ for $\alpha = \varepsilon_k + \varepsilon_l$ by our definition of ellipticity.

Definition 2.1.33. A. Let $L(x, D) = \sum_{|\alpha|\leq 2} a_\alpha(x)\partial^\alpha$ be a second order (elliptic) differential operator with coefficients $a_\alpha \in C_b^{|\alpha|}(\mathbb{R}^n, \mathbb{R})$. Given $f \in L^p(\mathbb{R}^n)$, $1 \leq p < \infty$. We call $u \in W^{2,p}(\mathbb{R}^n)$ an *L^p-solution* of (2.68) if this equality holds in the sence of the space $L^p(\mathbb{R}^n)$.

B. We call $u \in L^p(\mathbb{R}^n)$ a *weak L^p-solution* of (2.68) if

$$\int_{\mathbb{R}^n} u\, L^*(x, D)\varphi\, dx = \int_{\mathbb{R}^n} f\varphi\, dx \tag{2.71}$$

holds for all $\varphi \in C_0^\infty(\mathbb{R}^n)$ (or equivalently, for all $\varphi \in W^{2,p'}(\mathbb{R}^n)$ for $\frac{1}{p} + \frac{1}{p'} = 1$, $p > 1$).

Corollary 2.1.34. *Every L^p-solution of* (2.68) *is also a weak L^p-solution of* (2.68).

To establish the existence of a weak L^p-solution for $\lambda - L(x, D)$ we need a further estimate.

Theorem 2.1.35. *Let $L(x, D)$ be as in Theorem* 2.1.32. *Then there exists $R \geq 0$ and $\hat{c}_p \geq 0$ such that*

$$\lambda\|u\|_{L^p} \leq \hat{c}_p\|(\lambda - L(x, D))u\|_{L^p} \tag{2.72}$$

holds for all $\lambda \geq R$ and $u \in W^{2,p}(\mathbb{R}^n)$.

Proof: We consider on $W^{2,p}(\mathbb{R}^{n+1})$ the operator $L(x,D)+\frac{\partial^2}{\partial y^2}$. Since this operator as a differential operator acting on functions defined on $\mathbb{R}^{n+1}$ fulfills the assumptions of Theorem 2.1.32 we have for all $v \in W^{2,p}(\mathbb{R}^{n+1})$

$$\|v\|_{W^{2,p}(\mathbb{R}^{n+1})} \le c_p\Big(\Big\|L(x,D)v+\frac{\partial^2 v}{\partial y^2}\Big\|_{L^p(\mathbb{R}^{n+1})}+\|v\|_{L^p(\mathbb{R}^{n+1})}\Big). \tag{2.73}$$

We choose $\zeta \in C^\infty(\mathbb{R})$ such that $0 \le \zeta \le 1$, $\zeta\big|_{[-\frac{1}{2},\frac{1}{2}]} = 1$ and $\zeta\big|_{[-1,1]^c} = 0$. Further let $r > 0$ and $u \in W^{2,p}(\mathbb{R}^n)$. The function $v(x,y) := \zeta(y)e^{iry}u(x)$ belong to $W^{2,p}(\mathbb{R}^{n+1})$. First we observe that

$$\frac{\partial^2}{\partial y^2}v(x,y) = u(x)\Big(e^{iry}\zeta''(y) + 2ire^{iry}\zeta'(y) - r^2e^{iry}\zeta(y)\Big),$$

which implies by our assumptions on ζ

$$\|v\|^p_{L^p(\mathbb{R}^{n+1})} = \int_{\mathbb{R}^n}\int_{\mathbb{R}} |\zeta(y)e^{iry}|^p\,|u(x)|^p\,dy\,dx \le 2\|u\|^p_{L^p(\mathbb{R}^n)},$$

and

$$\begin{aligned}\|v\|^p_{W^{2,p}(\mathbb{R}^n\times(-\frac{1}{2},\frac{1}{2}))} &= \int_{\mathbb{R}^n}\int_{-\frac{1}{2}}^{\frac{1}{2}}\sum_{|\alpha|\le 2}\big|D^\alpha(ue^{iry}\zeta(y))\big|^p\,dy\,dx \\ &= \int_{\mathbb{R}^n}\int\Big((1+r^p+r^{2p})|u|^p + (1+2r^p) \\ &\qquad \times\sum_{j=1}^n\Big|\frac{\partial u}{\partial x_j}\Big|^p + \sum_{k,l=1}^n\Big|\frac{\partial^2 u}{\partial x_k\partial x_l}\Big|^p\Big)\,dx \\ &\ge r^{2p}\|u\|^p_{L^p(\mathbb{R}^n)} + r^p\|\nabla u\|^p_{L^p(\mathbb{R}^n)} + \|\nabla^2 u\|^p_{L^p(\mathbb{R}^n)}.\end{aligned}$$

In addition we find

$$\begin{aligned}&\Big\|L(x,D)v+\frac{\partial^2 v}{\partial y^2}\Big\|_{L^p(\mathbb{R}^{n+1})} \\ &= \Big\|\zeta e^{ir\cdot}L(x,D)u - e^{ir\cdot}u(r^2\zeta - 2ir\zeta' - \zeta'')\Big\|_{L^p(\mathbb{R}^{n+1})} \\ &\le \|\zeta(L(x,D)u - r^2u)\|_{L^p(\mathbb{R}^{n+1})} + \|(2ir\zeta' + \zeta'')u\|_{L^p(\mathbb{R}^{n+1})} \\ &\le 2\|L(x,D)u - r^2u\|_{L^p(\mathbb{R}^n)} + 2r\|\zeta'\|_{L^p(\mathbb{R}^n)}\|u\|_{L^p(\mathbb{R}^n)} \\ &\quad + \|\zeta''\|_{L^p(\mathbb{R}^n)}\|u\|_{L^p(\mathbb{R}^n)}.\end{aligned}$$

Using (2.73) we get now

$$\begin{aligned} r^2\|u\|_{L^p(\mathbb{R}^n)} + r\|\nabla u\|_{L^p(\mathbb{R}^n)} + \|\nabla^2 u\|_{L^p(\mathbb{R}^n)} &\leq 3\|v\|_{W^{2,p}(\mathbb{R}^{n+1})} \\ &\leq 3c_p\Big(\Big\|L(x,D)v + \frac{\partial^2 v}{\partial y^2}\Big\|_{L^p(\mathbb{R}^{n+1})} + \|v\|_{L^p(\mathbb{R}^{n+1})}\Big) \\ &\leq \tilde{c}_p(\|L(x,D)u - r^2 u\|_{L^p(\mathbb{R}^n)} + (1+r)\|u\|_{L^p(\mathbb{R}^n)}). \end{aligned}$$

With $\lambda = \sqrt{r}$ and r such large that $\tilde{c}_p(1+r) \leq \frac{r^2}{2}$ we find

$$\begin{aligned} &\lambda\|u\|_{L^p(\mathbb{R}^n)} + \lambda^{1/2}\|\nabla u\|_{L^p(\mathbb{R}^n)} + \|\nabla^2 u\|_{L^p(\mathbb{R}^n)} \\ &\qquad \leq 2\tilde{c}_p\|\lambda u - L(x,D)u\|_{L^p(\mathbb{R}^n)}, \end{aligned} \tag{2.74}$$

which gives in particular for λ sufficiently large (2.72). □

Corollary 2.1.36. *Let $L(x,D)$ be as in Theorem 2.1.32 and λ sufficiently large. Then the equation (2.68) admits at most one L^p-solution.*

Proof: Let $u, v \in W^{2,p}(\mathbb{R}^n)$ be two solutions of (2.68) for some $f \in L^p(\mathbb{R}^n)$. By Theorem 2.1.35 we find

$$\begin{aligned} \lambda\|u-v\|_{W^{2,p}(\mathbb{R}^n)} &\leq \hat{c}_p\|\lambda(u-v) - L(x,D)(u-v)\|_{L^p(\mathbb{R}^n)} \\ &= \hat{c}_p\|\lambda u - L(x,D)u - (\lambda v - L(x,D)v)\|_{L^p(\mathbb{R}^n)} \\ &= \hat{c}_p\|f-f\|_{L^p(\mathbb{R}^n)} = 0. \quad \square \end{aligned}$$

Corollary 2.1.37. *Let $L(x,D)$ be as in Theorem 2.1.32 with coefficients $a_{kl} = a_{lk} \in C_b^2(\mathbb{R}^n, \mathbb{R})$, $b_j \in C_b^1(\mathbb{R}^n, \mathbb{R})$ and $c \in C_b(\mathbb{R}^n, \mathbb{R})$. Then Theorem 2.1.32 and Theorem 2.1.35 also apply to $L^*(x,D)$.*

Next we give a necessary condition for the existence of a weak L^p-solution to (2.68).

Proposition 2.1.38. *A necessary condition for (2.68) to have a weak L^p-solution is that*

$$\Big|\int_{\mathbb{R}^n} f\varphi\,dx\Big| \leq c\|(L^*(x,D)-\lambda)\varphi\|_{L^{p'}} \tag{2.75}$$

is valid for all $\varphi \in C_0^\infty(\mathbb{R}^n)$ (or equivalently for all $\varphi \in W^{2,p'}(\mathbb{R}^n)$).

Proof: By definition of a weak L^p-solution we have

$$\begin{aligned}\left|\int_{\mathbb{R}^n} f\varphi\,dx\right| &= \left|\int_{\mathbb{R}^n} u(L^*(x,D)-\lambda)\varphi\,dx\right| \\ &\leq \|u\|_{L^p}\big\|(L^*(x,D)-\lambda)\varphi\big\|_{L^{p'}}. \quad \square\end{aligned}$$

We prove now that (2.75) is a sufficient criterion for the existence of a weak L^p-solution to (2.68).

Theorem 2.1.39. *Let $L(x,D)=\sum_{|\alpha|\leq 2} a_\alpha(x)\partial^\alpha$ be a second order differential operator with coefficients $a_\alpha \in C_b^{|\alpha|}(\mathbb{R}^n;\mathbb{R})$ and let $L^*(x,D)$ be defined by* (2.69). *Moreover, suppose that for some $f \in L^p(\mathbb{R}^n)$ we have* (2.75). *Then the equation $\lambda u - L(x,D)u = f$ admits a weak L^p-solution.*

Proof: Consider the set

$$W_\lambda := \{w \in L^{p'}(\mathbb{R}^n); \text{there exists } \varphi \in C_0^\infty(\mathbb{R}^n) \text{ such that } (L^*(x,D)-\lambda)\,\varphi = w\}.$$

Clearly W_λ is a linear subset of $L^{p'}(\mathbb{R}^n)$ and since $L^*(x,D)$ maps $W^{2,p'}(\mathbb{R}^n)$ continuously into $L^{p'}(\mathbb{R}^n)$, W_λ is a non-empty set, in particular we have $(L^*(x,D)-\lambda)\,(C_0^\infty(\mathbb{R}^n)) \subset W_\lambda$. On W_λ we define the linear functional

$$F_\lambda : W_\lambda \to \mathbb{C}, \quad w \to F_\lambda w := \int_{\mathbb{R}^n} \varphi f\,dx,$$

where $\varphi \in C_0^\infty(\mathbb{R}^n)$ is a solution of the equation $(L^*(x,D)-\lambda)\,\varphi = w$. First note that F_λ is independent of φ. Indeed, if ψ is a further solution of $(L^*(x,D)-\lambda)\,\psi = w$, we find

$$\begin{aligned}\left|\int_{\mathbb{R}^n} f(\varphi-\psi)\,dx\right| &\leq c\,\|(L^*(x,D)-\lambda)\,(\varphi-\psi)\|_{L^p(\mathbb{R}^n)} \\ &= c\|w-w\|_{L^p(\mathbb{R}^n)} = 0.\end{aligned}$$

In addition we have

$$|F_\lambda w| = \left|\int_{\mathbb{R}^n} f\varphi\,dx\right| \leq c\,\|(L^*(x,D)-\lambda)\varphi\|_{L^p(\mathbb{R}^n)} = c\|w\|_{L^p(\mathbb{R}^n)}.$$

Hence F_λ is a continuous linear functional on W_λ. By the Hahn–Banach Theorem, Theorem I.2.7.4, we may extend F_λ to a continuous linear functional on $L^{p'}(\mathbb{R}^n)$ which we denote once again by F_λ. For this reason we find some $u \in L^p(\mathbb{R}^n)$ such that

$$F_\lambda w = \int_{\mathbb{R}^n} uw\, dx$$

holds for all $w \in L^{p'}(\mathbb{R}^n)$. For $\varphi \in C_0^\infty(\mathbb{R}^n)$ we know that $L^*(x,D)\varphi - \lambda\varphi$ belongs to W_λ. Thus we get

$$\int_{\mathbb{R}^n} u(L^*(x,D)\varphi - \lambda\varphi)\, dx = \int_{\mathbb{R}^n} f\varphi\, dx,$$

which shows that $u \in L^p(\mathbb{R}^n)$ is a weak L^p-solution of the equation $L(x,D)u - \lambda u = f$. □

Corollary 2.1.40. *Suppose that $L(x,D)$ satisfies the assumptions of Theorem* 2.1.32 *with coefficients $a_{kl} = a_{lk} \in C_b^2(\mathbb{R}^n;\mathbb{R})$, $b_j \in C_b^1(\mathbb{R}^n;\mathbb{R})$ and $c \in C_b(\mathbb{R}^n;\mathbb{R})$. In particular, $L(x,D)$ is uniformly elliptic. Then there exists $\lambda_0 \geq 0$ such that for $\lambda \geq \lambda_0$ equation* (2.68) *has for all $f \in L^p(\mathbb{R}^n)$ a unique weak L^p-solution, $1 < p < \infty$.*

Proof: According to Corollary 2.1.37 we have for some $\lambda_0 \geq 0$ the estimate

$$\|\varphi\|_{W^{2,p'}} \leq c\|(L^*(x,D) - \lambda)\varphi\|_{L^p}$$

for all $\varphi \in W^{2,p'}(\mathbb{R}^n)$ and $\lambda \geq \lambda_0$. Given $f \in L^p(\mathbb{R}^n)$ we find for $\lambda \geq \lambda_0$

$$\begin{aligned}\left|\int_{\mathbb{R}^n} f\varphi\, dx\right| &\leq \|f\|_{L^p}\|\varphi\|_{L^{p'}} \leq \|f\|_{L^p}\|\varphi\|_{W^{2,p'}} \\ &\leq c\|f\|_{L^p}\|(L^*(x,D) - \lambda)\varphi\|_{L^{p'}},\end{aligned}$$

and the corollary follows. □

To proceed further we need the following regularity result which we will not prove, for a proof we refer to F.Browder [37].

Theorem 2.1.41. *Let $L(x,D)$ be a uniformly elliptic differential operator with coefficients $a_{kl} = a_{lk} \in C_b^{2+m}(\mathbb{R}^n;\mathbb{R})$, $b_j \in C_b^{m+1}(\mathbb{R}^n;\mathbb{R})$ and $c \in C_b^m(\mathbb{R}^n;\mathbb{R})$ for some $m \in \mathbb{N}_0$. Further let $u \in L^p(\mathbb{R})$, $1 < p < \infty$, a (unique) weak L^p-solution to the equation $L(x,D)u - \lambda u = f$, $\lambda \geq \lambda_0$. If $f \in W^{m,p}(\mathbb{R}^n;\mathbb{R})$, then $u \in W^{m+2,p}(\mathbb{R}^n;\mathbb{R})$ and the estimate*

$$\|u\|_{W^{m+2,p}} \leq c(\|f\|_{W^{m,p}} + \|u\|_{L^p}) \tag{2.76}$$

holds.

Taking Theorem 2.1.41 for granted we have

Theorem 2.1.42. *Let $L(x, D)$ be a uniformly elliptic differential operator with coefficients $a_{kl} = a_{lk} \in C_b^2(\mathbb{R}^n; \mathbb{R})$, $b_j \in C_b^1(\mathbb{R}^n; \mathbb{R})$ and $c \in C_b(\mathbb{R}^n; \mathbb{R})$. Further let $\lambda \geq \lambda_0$ sufficiently large. Then the operator $L(x, D) - \lambda$ and $L^*(x, D) - \lambda$ are continuous and bijective operators from $W^{2,p}(\mathbb{R}^n; \mathbb{R})$ to $L^p(\mathbb{R}^n; \mathbb{R})$ with continuous inverse. If in addition $a_{kl}, b_j, c \in C_b^m(\mathbb{R}^n; \mathbb{R})$, $m \in \mathbb{N}_0$, then $L(x, D)$ and $L^*(x, D)$ map $W^{m+2,p}(\mathbb{R}^n; \mathbb{R})$ continuously into $W^{m,p}(\mathbb{R}^n; \mathbb{R})$.*

Proof: It remains only to show the last statement which follows immediately from Leibniz' rule by observing that for $|\alpha| \leq m$ and $d \in C_b^m(\mathbb{R}^n; \mathbb{R})$

$$\|\partial^\alpha(d\partial^\beta u)\|_{L^p} \leq \sum_{\gamma \leq \alpha} \binom{\alpha}{\gamma} \|\partial^\gamma d\|_\infty \|\partial^{\beta+\alpha-\gamma} u\|_{L^p}. \quad \square$$

Next we will prove that $L(x, D)$ extends to a generator of a Feller semigroup $\left(T_t^{(\infty)}\right)_{t\geq 0}$ as well as to a generator of an L^p-sub-Markovian semigroup $\left(T_t^{(p)}\right)_{t\geq 0}$ for $1 < p < \infty$. First we turn our interest to the Feller semigroup. For this let $L(x, D)$ be a uniformly elliptic differential operator of second order with appropriate coefficients. From Theorem 2.1.41 we know that for $\lambda \geq \lambda_0$ the equation

$$L(x, D)u - \lambda u = f$$

has for all $f \in W^{1,p}(\mathbb{R}^n; \mathbb{R})$, $1 < p < \infty$, a unique solution $u \in W^{3,p}(\mathbb{R}^n; \mathbb{R})$. Further, by Theorem 2.1.42 we have for all $u \in W^{3,p}(\mathbb{R}^n; \mathbb{R})$ the estimate

$$\|L(x, D)u\|_{W^{1,p}} \leq c\|u\|_{W^{3,p}}.$$

Let us suppose that $p > n$. Sobolev's embedding theorem, Theorem 2.1.12, yields now that $(L(x, D), W^{3,p}(\mathbb{R}^n; \mathbb{R}))$, is a densely defined operator on $C_\infty(\mathbb{R}^n; \mathbb{R})$ and it is clear that $L(x, D)$ satisfied on $W^{3,p}(\mathbb{R}^n; \mathbb{R}) \subset C_\infty^2(\mathbb{R}^n; \mathbb{R})$ also the positive maximum principle provided that $c(x) \leq 0$. Thus we see that $(L(x, D), W^{3,p}(\mathbb{R}^n; \mathbb{R}))$ satisfies all requirements of the Hille–Yosida–Ray theorem, Theorem I.4.5.3.

Theorem 2.1.43. *Let $L(x, D)$ be a uniformly elliptic differential operator with coefficients $a_{kl} = a_{lk} \in C_b^3(\mathbb{R}^n; \mathbb{R})$, $b_j \in C^2(\mathbb{R}^n; \mathbb{R})$ and $c \in C_b^1(\mathbb{R}^n; \mathbb{R})$. Further assume that $c(x) \leq 0$ for all $x \in \mathbb{R}^n$. Then the operator $(L(x, D), W^{3,p}(\mathbb{R}^n; \mathbb{R}))$, $p > n$, extends to a generator of a Feller semigroup.*

In case of L^p-sub-Markovian semigroups we find

Theorem 2.1.44. *Let $L(x,D)$ be a uniformly elliptic differential operator with coefficients $a_{kl} = a_{lk} \in C_b^2(\mathbb{R}^n;\mathbb{R})$, $b_j \in C_b^1(\mathbb{R}^n;\mathbb{R})$ and $c \in C_b(\mathbb{R}^n;\mathbb{R})$. In addition suppose that $c(x) \leq 0$ and*

$$\sum_{j=1}^{n} \frac{\partial}{\partial x_j}\left(b_j(x) - \sum_{k=1}^{n} \frac{\partial}{\partial x_k} a_{kj}(x)\right) \geq 0$$

Then $(L(x,D), W^{2,p}(\mathbb{R}^n;\mathbb{R}))$ is a generator of an L^p-sub-Markovian semigroup for $1 < p < \infty$.

Proof: First note that Lemma 2.1.19 and Theorem 2.1.31 implies that $(L(x,D), W^{2,p}(\mathbb{R}^n;\mathbb{R}))$ is a closed operator in $L^p(\mathbb{R}^n;\mathbb{R})$, $1 < p < \infty$. Moreover, Theorem 2.1.20 yields that the operator $L^p(\mathbb{R}^n;\mathbb{R})$ is a Dirichlet operator, in particular, it is dissipative according to Proposition I.4.6.12. Thus it remains to prove that for some $\lambda > 0$ we have $R(L(x,D) - \lambda) = L^p(\mathbb{R}^n;\mathbb{R})$, which is however the content of Theorem 2.1.44. □

2.2 Some Second Order Differential Operators with Non-Negative Characteristic Form as Generators of Sub-Markovian Semigroups

In this section we want briefly discuss some second order differential operators with non-negative characteristic form which are not elliptic but still have an extension to a generators of a sub-Markovian semigroup. We restrict our attention mainly to the Hilbert space case, i.e. to L^2-sub-Markovian semigroups, and we will use the result from Section I.4.7, i.e. the theory of Dirichlet forms, to construct these semigroups.

However, since in this monograph we are mainly concerned with pseudo-differential operators which are not differential operators, we do not prove all results in this section.

We start with symmetric operators in divergence form. Let

$$L(x,D) = \sum_{k,l=1}^{n} \frac{\partial}{\partial x_k}\left(a_{kl}(\cdot)\frac{\partial}{\partial x_l}\right) \tag{2.77}$$

have symmetric coefficients $a_{kl} = a_{lk} \in C_b^1(\mathbb{R}^n; \mathbb{R})$ and assume

$$\sum_{k,l=1}^{n} a_{kl}(x)\xi_k\xi_l \geq 0 \tag{2.78}$$

for all $x \in \mathbb{R}^n$. The operator is considered first as an operator on $L^2(\mathbb{R}^n; \mathbb{R})$ with domain $C_0^\infty(\mathbb{R}^n; \mathbb{R})$. On $C_0^\infty(\mathbb{R}^n; \mathbb{R})$ we associate with $L(x, D)$ the following form

$$B(u, v) := \int_{\mathbb{R}^n} \sum_{k,l=1}^{n} a_{kl}(x) \frac{\partial u(x)}{\partial x_l} \frac{\partial v(x)}{\partial x_k}\, dx \tag{2.79}$$

which is symmetric and non-negative, i.e

$$B(u, v) = B(v, u) \qquad \text{and} \qquad B(u, u) \geq 0$$

for all $u, v \in C_0^\infty(\mathbb{R}^n; \mathbb{R})$. Moreover we have the estimates

$$\|L(x, D)u\|_0 \leq c_1 \|u\|_{W^{2,2}}, \quad u \in W^{2,2}(\mathbb{R}^n; \mathbb{R}), \tag{2.80}$$

and

$$|B(u, v)| \leq c_2 \|u\|_{W^{1,2}} \|v\|_{W^{1,2}}, \quad u, v \in W^{1,2}(\mathbb{R}^n; \mathbb{R}), \tag{2.81}$$

implying that $L(x, D)$ extends to a continuous operator from $W^{2,2}(\mathbb{R}^n; \mathbb{R})$ to $L^2(\mathbb{R}^n; \mathbb{R})$ and B extends to a continuous bilinear form $B : W^{1,2}(\mathbb{R}^n; \mathbb{R}) \times W^{1,2}(\mathbb{R}^n; \mathbb{R}) \to \mathbb{R}$. By Corollary 2.1.21 it follows that

$$\int_{\mathbb{R}^n} L(x, D)u(x)(u-1)^+(x)\, dx \leq 0 \tag{2.82}$$

for all $u \in W^{2,2}(\mathbb{R}^n; \mathbb{R})$. Thus $L(x, D)$ is a candidate for a Dirichlet operator in $L^2(\mathbb{R}^n; \mathbb{R})$, see Definition I.4.6.7. In fact, Remark I.4.6.13.A yields that $(L(x, D), C_0^\infty(\mathbb{R}^n; \mathbb{R}))$ is closable and (2.82) still holds for the closure. Note that also the operator $(L(x, D), W^{2,2}(\mathbb{R}^n; \mathbb{R}))$ is closable and and its closure coincides with that of $(L(x, D), C_0^\infty(\mathbb{R}^n; \mathbb{R}))$. It is convenient to denote this closure of $L(x, D)$ by $(-A, D(A))$. Moreover, the symmetry of B implies estimate (I.4.324), i.e.

$$|B(u, v)| \leq c(B_1(u, u))^{1/2}(B_1(v, v))^{1/2}, \tag{2.83}$$

where as usual $B_1(u,v)$ stands for $B(u,v)+(u,v)_0$. In addition, since $B(u,u) \geq 0$, we find

$$\|u\|_0^2 \leq B_1(u,u), \tag{2.84}$$

i.e. estimate (I.4.323) holds. Corollary I.4.7.14 implies that $(B, C_0^\infty(\mathbb{R}^n;\mathbb{R}))$ and $(B, W^{1,2}(\mathbb{R}^n;\mathbb{R}))$ are closable and have the same closure. Let us denote this closure again by B, its domain by $D(B)$. Thus we have

$$B(u,v) = (-Au, v)_0 \tag{2.85}$$

for all $u \in D(A)$ and $v \in D(B)$, and we have $D(A) \subset D(B)$. Now combining Theorem I.4.7.10 with Theorem I.4.7.23 we obtain

Theorem 2.2.1. *Let $L(x,D) = \sum_{k,l=1}^n \frac{\partial}{\partial x_k}\left(a_{kl}(\cdot)\frac{\partial}{\partial x_l}\right)$ be a second order differential operator with coefficients $a_{kl} = a_{lk} \in C_b^1(\mathbb{R}^n;\mathbb{R})$ such that $\sum_{k,l=1}^n a_{kl}\xi_k\xi_l \geq 0$ holds for all $x \in \mathbb{R}^n$ and $\xi \in \mathbb{R}^n$. Consider $L(x,D)$ as operator on $L^2(\mathbb{R}^n;\mathbb{R})$ with domain $C_0^\infty(\mathbb{R}^n;\mathbb{R})$ (or $W^{2,2}(\mathbb{R}^n;\mathbb{R})$). Then it is closable, its closure is a selfadjoint Dirichlet operator which is the generator of a symmetric sub-Markovian semigroup $(T_t^{(2)})_{t\geq 0}$ on $L^2(\mathbb{R}^n;\mathbb{R})$ and therefore it generates a symmetric Dirichlet form $(B, D(B))$.*

Remark 2.2.2. In case of an elliptic operator we may characterize the domains $D(A)$ and $D(B)$, namely $D(A) = W^{2,2}(\mathbb{R}^n;\mathbb{R})$ and $D(B) = W^{1,2}(\mathbb{R}^n;\mathbb{R})$. The first result follows from the estimate given in Section 2.1, especially from (2.39) combined with Theorem 2.1.32. Whereas the second result follows from (2.81) together with the observation that

$$\sum_{k,l=1}^n \int_{\mathbb{R}^n} a_{kl}(x)\frac{\partial u(x)}{\partial x_l}\frac{\partial u(x)}{\partial x_k}\,dx \geq \lambda_0 \int_{\mathbb{R}^n} |\nabla u(x)|^2\,dx,$$

where $\lambda_0 > 0$ is the ellipticity constant, i.e.

$$\sum_{k,l=1}^n a_{kl}(x)\xi_k\xi_l \geq \lambda_0|\xi|^2.$$

However, for a general operator $L(x,D)$ as in Theorem 2.2.1 we do not know explicitly $D(A)$ or $D(B)$. We only know the two inclusions

$$W^{2,2}(\mathbb{R}^n;\mathbb{R}) \subset D(A) \subset L^2(\mathbb{R}^n;\mathbb{R}) \tag{2.86}$$

and

$$W^{1,2}(\mathbb{R}^n;\mathbb{R}) \subset D(B) \subset L^2(\mathbb{R}^n;\mathbb{R}). \tag{2.87}$$

Later in this section we will discuss the question to improve (2.86) or (2.87) in order to obtain inclusions of the type $D(A) \subset H^{2s}(\mathbb{R}^n;\mathbb{R})$ or $D(B) \subset H^s(\mathbb{R}^n;\mathbb{R})$, respectively, for some $s > 0$. In a next step we want to add a drift term to the operator $L(x,D)$, i.e. a first order differential operator. Thus we consider the operator

$$L(x,D) = \sum_{k,l=1}^{n} \frac{\partial}{\partial x_k}\left(a_{kl}(\cdot)\frac{\partial}{\partial x_l}\right) + \sum_{j=1}^{n} b_j(\cdot)\frac{\partial}{\partial x_j}, \tag{2.88}$$

where we assume that $\sum_{k,l=1}^{n} \frac{\partial}{\partial x_k}\left(a_{kl}(\cdot)\frac{\partial}{\partial x_l}\right)$ satisfies the assumptions of Theorem 2.2.1. In order that $L(x,D)$ extends to a Dirichlet operator, we assume according to Corollary 2.1.21 that $b_j \in C_b^1(\mathbb{R}^n;\mathbb{R})$ and that

$$\sum_{j=1}^{n} \frac{\partial b_j(x)}{\partial x_j} \geq 0, \tag{2.89}$$

holds for all $x \in \mathbb{R}^n$. The linear form associated with $L(x,D)$ is now given on $C_0^\infty(\mathbb{R}^n;\mathbb{R})$ by

$$E(u,v) = \sum_{k,l=1}^{n} \int_{\mathbb{R}^n} a_{kl}(x)\frac{\partial u(x)}{\partial x_l}\frac{\partial v(x)}{\partial x_k}\,dx + \sum_{j=1}^{n}\int_{\mathbb{R}^n} b_j(x)\frac{\partial u(x)}{\partial x_j}v(x)\,dx. \tag{2.90}$$

In order to apply results from Section I.4.7 we need the sector condition to hold for E. Denoting by B the symmetric part of the form E, i.e.

$$B(u,v) := E^{sym}(u,v) = \sum_{k,l=1}^{n} \int_{\mathbb{R}^n} a_{kl}(x)\frac{\partial u(x)}{\partial x_l}\frac{\partial v(x)}{\partial x_k}\,dx,$$

we are longing for the estimate

$$\left|\sum_{j=1}^{n}\int_{\mathbb{R}^n} b_j(x)\frac{\partial u(x)}{\partial x_j}v(x)\,dx\right| \leq c(B_1(u,u)^{1/2}(B_1(v,v)^{1/2} \tag{2.91}$$

for all $u,v \in C_0^\infty(\mathbb{R}^n;\mathbb{R})$ (or $W^{1,2}(\mathbb{R}^n;\mathbb{R})$). Clearly, for an elliptic operator (2.91) follows easily since

$$\left|\sum_{j=1}^{n}\int_{\mathbb{R}^n} b_j(x)\frac{\partial u(x)}{\partial x_j}v(x)\,dx\right| \leq \sum_{j=1}^{n} \|b_j\|_\infty \left\|\frac{\partial u}{\partial x_j}\right\|_0 \|v\|_0, \tag{2.92}$$

and $\|\frac{\partial u}{\partial x_j}\|_0 \leq \|u\|_{W^{1,2}}$, $\|v\|_0 \leq \|v\|_{W^{1,2}}$. In case of non-elliptic operator we have to examine the non-symmetric part of E, i.e. the linear form

$$\sum_{j=1}^{n} \int_{\mathbb{R}^n} b_j(x) \frac{\partial u(x)}{\partial x_j} v(x)\, dx, \tag{2.93}$$

more carefully. For this we recall Lemma I.4.7.34 where we proved with $\psi_j(\xi) = |\xi_j|$

$$\begin{aligned} &\Big| \sum_{j=1}^{n} \int_{\mathbb{R}^n} b_j(x) \frac{\partial u(x)}{\partial x_j} v(x)\, dx \Big| \\ &\qquad \leq \sum_{j=1}^{n} \|b_j\|_\infty^{1/2} \left(\|b_j\|_\infty^2 + \left\| \frac{\partial b_j}{\partial x_j} \right\|_\infty^2 \right)^{1/4} \|u\|_{\psi_j,1/2} \|v\|_{\psi_j,1/2}. \end{aligned} \tag{2.94}$$

Now let

$$L(x, D) = \sum_{k,l=1}^{n} \frac{\partial}{\partial x_k} \left(a_{kl}(\cdot) \frac{\partial}{\partial x_l} \right)$$

be a linear differential operator as in Theorem 2.2.1. In addition we assume for some $\lambda_0 > 0$ that

$$B(u, u) \geq \lambda_0 \sum_{j \in J} \left\| \frac{\partial u}{\partial x_j} \right\|_0^2, \tag{2.95}$$

where $J \subset \{1, \ldots, n\}$. We set

$$\psi_J(\xi) := 1 + \sum_{j \in J} |\xi_j|^2 \tag{2.96}$$

and find

$$B_1(u, u) \geq (\lambda_0 \wedge 1) \|u\|_{\psi_J,1}. \tag{2.97}$$

Since

$$\frac{1}{|J|} \sum_{j \in J} \psi_j^2(\xi) \leq \psi_J(\xi) = 1 + \sum_{j \in J} \psi_j^2(\xi)$$

we see immediately that

$$\|u\|_{\psi_j,1/2} \leq \|u\|_{\psi_J,1}. \tag{2.98}$$

Thus we obtain

Theorem 2.2.3. *Suppose that $L(x,D)$ is as in Theorem 2.2.1 and satisfies (2.97). In addition let $b_j \in C_b^1(\mathbb{R}^n;\mathbb{R})$, $j \in J$, be given functions such that $\sum_{j\in J} \frac{\partial b_j}{\partial x_j}(x) \geq 0$ for all $x \in \mathbb{R}^n$. Then the bilinear form*

$$E(u,v) = \sum_{k,l=1}^{n} \int_{\mathbb{R}^n} a_{kl}(x) \frac{\partial u(x)}{\partial x_l} \frac{\partial v(x)}{\partial x_k}\, dx + \sum_{j\in J} \int_{\mathbb{R}^n} b_j(x) \frac{\partial u(x)}{\partial x_j} v(x)\, dx \tag{2.99}$$

originally defined on $C_0^\infty(\mathbb{R}^n;\mathbb{R})$ is closable and its closure is a semi-Dirichlet form. Thus the operator $L(x,D) + \sum_{j\in J} b_j(x)\frac{\partial u(x)}{\partial x_j}$ has a closure which is a Dirichlet operator and it generates a sub-Markovian semigroup on $L^2(\mathbb{R}^n;\mathbb{R})$.

Proof: It remains only to prove that E satisfies the sector condition. But from (2.94) we deduce using (2.97)

$$\begin{aligned}\Big|\sum_{j\in J} \int_{\mathbb{R}^n} b_j(x) \frac{\partial u(x)}{\partial x_j} v(x)\, dx\Big| &\leq c \sum_{j\in J} \|u\|_{\psi_j,1/2} \|v\|_{\psi_j,1/2} \\ &\leq c' \|u\|_{\psi_J,1} \|v\|_{\psi_J,1} \leq \tilde{c}(B_1(u,u))^{1/2}(B_1(v,v))^{1/2},\end{aligned}$$

and the theorem is proved. □

Example 2.2.4. Consider a matrix $(a_{kl})_{k,l=1,\dots,n}$ with $a_{kl} = a_{lk} \in C_b^1(\mathbb{R}^n;\mathbb{R})$ having the property that for some $0 < m < n$ it follows

$$\sum_{k,l=1}^{m} a_{kl}(x)\xi_k\xi_l \geq \lambda_0 \sum_{j=1}^{m} \xi_j^2$$

for all $x \in \mathbb{R}^n$ and all $\xi \in \mathbb{R}^m$. Moreover we assume that $a_{kl} = 0$ if $k > m$ and $l < m$, which implies by symmetry that $a_{kl} = 0$ for $k < m$ and $l > m$, and further we require that

$$\sum_{k,l=m+1}^{n} a_{kl}(x)\xi_k\xi_l \geq 0$$

for all $x \in \mathbb{R}^n$ and all $\xi \in \mathbb{R}^{n-m}$. In this case we have for the operator $\sum_{k,l=1}^{n} \frac{\partial}{\partial x_k}\left(a_{kl}(x)\frac{\partial}{\partial x_l}\right)$ estimate (2.95) with $J = \{1,\dots,m\}$.

In Theorem 2.2.3 we face the situation that the drift acts only in directions where $L(x,D)$ is elliptic. In order to treat more general example we need

Definition 2.2.5. Let $L(x,D) = \sum_{k,l=1}^{n} \frac{\partial}{\partial x_k}\left(a_{kl}(x)\frac{\partial}{\partial x_l}\right)$ be a second order differential operator with non-negative characteristic form, i.e. $a_{kl} = a_{lk}$ and $\sum_{k,l=1}^{n} a_{kl}(x)\xi_k\xi_l \geq 0$ for all $x \in \mathbb{R}^n$ and all $\xi \in \mathbb{R}^n$. Further assume $a_{kl} \in C_b^1(\mathbb{R}^n;\mathbb{R})$.

A. We say that $L(x,D)$ satisfies a *sub-elliptic estimate in the strict sense* if for some $\varepsilon \in (0,1]$ and all compact sets $K \subset \mathbb{R}^n$ the estimate

$$\|u\|_\varepsilon^2 \leq c_K\left((L(x,D)u,u)_0 + \|u\|_0^2\right) \tag{2.100}$$

holds for all $u \in C_0^\infty(\mathbb{R}^n)$, $\operatorname{supp} u \subset K$.

B. We call $L(x,D)$ to satisfy a *uniformly sub-elliptic estimate* if for some $\varepsilon \in (0,1]$

$$\|u\|_\varepsilon^2 \leq c\left((L(x,D)u,u)_0 + \|u\|_0^2\right) \tag{2.101}$$

holds for all $u \in C_0^\infty(\mathbb{R}^n)$.

Remark 2.2.6. A. If $L(x,D)$ is an operator satisfying a sub-elliptic estimate in the strict sense, then $L(x,D)$ is hypo-elliptic provided that the coefficients belong to $C^\infty(\mathbb{R}^n)$. Moreover, in this case (2.100) is equivalent to

$$\|u\|_{2\varepsilon+s} \leq c_K(\|L(x,D)u\|_s + \|u\|_0) \tag{2.102}$$

for all $u \in C_0^\infty(\mathbb{R}^n)$, $\operatorname{supp} u \subset K$, and $s \geq 0$. Operators satisfying (2.102) are called *sub-elliptic operators* and we will make some remarks to their analysis later on.

B. The notion of uniformly sub-ellipticity seems not to be present in the literature, but see Example 2.2.15 which is related to *Hörmander-type operator*. Since we have to consider certain anisotropic pseudo-differential operators $q(x,D)$ satisfying with some $0 < \varepsilon_1 < \varepsilon_2 \leq 1$ the estimates

$$\|u\|_{\varepsilon_1}^2 \leq (q(x,D)u,u)_0 + \|u\|_0^2 \leq c_1\|u\|_{\varepsilon_2}^2 \tag{2.103}$$

for all $u \in C_0^\infty(\mathbb{R}^n)$ we are naturally led to this definition.

Proposition 2.2.7. *Let $L(x,D)$ given by (2.77) be an operator satisfying a uniformly sub-elliptic estimate for some $\varepsilon \geq \frac{1}{2}$. Moreover let $b_j \in C_b^1(\mathbb{R}^n;\mathbb{R})$, $1 \leq j \leq n$, be given functions such that $\sum_{j=1}^{n} \frac{\partial b_j(x)}{\partial x_j} \geq 0$ holds. Then a semi-Dirichlet form is given by the closure of*

$$E(u,v) := \sum_{k,l=1}^{n} \int_{\mathbb{R}^n} a_{kl}(x)\frac{\partial u(x)}{\partial x_l}\frac{\partial v(x)}{\partial x_k}\,dx - \sum_{j=1}^{n}\int_{\mathbb{R}^n} b_j(x)\frac{\partial u(x)}{\partial x_j}v(x)\,dx$$

originally defined on $C_0^\infty(\mathbb{R}^n;\mathbb{R})$.

Proof: It remains only to prove that E fulfills the sector condition, i.e.

$$\Big|\sum_{j=1}^{n}\int_{\mathbb{R}^n} b_j(x)\frac{\partial u(x)}{\partial x_j}v(x)\,dx\Big| \leq c(B_1(u,u))^{1/2}(B_1(v,v))^{1/2},$$

where

$$B_1(u,u) := \sum_{k,l=1}^{n}\int_{\mathbb{R}^n} a_{kl}(x)\frac{\partial u(x)}{\partial x_l}\frac{\partial v(x)}{\partial x_k}\,dx + \int_{\mathbb{R}^n}|u(x)|^2\,dx.$$

Using Lemma I.4.7.34 we obtain for $u, v \in C_0^\infty(\mathbb{R}^n;\mathbb{R})$

$$\begin{aligned}\Big|\sum_{j=1}^{n}\int_{\mathbb{R}^n} b_j(x)\frac{\partial u(x)}{\partial x_j}v(x)\,dx\Big| &\leq c\|u\|_{1/2}\|v\|_{1/2}\\ &\leq \|u\|_\varepsilon\|v\|_\varepsilon \leq \tilde{c}(B_1(u,u))^{1/2}(B_1(v,v))^{1/2},\end{aligned}$$

proving the proposition. □

More important (see below) is the following result:

Proposition 2.2.8. *Let $L(x,D)$ satisfy a sub-elliptic estimate in the strict sense with (2.100) to hold with some $\varepsilon \geq \frac{1}{2}$. In addition let $b_j \in C^1(\mathbb{R}^n;\mathbb{R})$, $1 \leq j \leq n$, be given functions such that for a fixed compact set $K \subset \mathbb{R}^n$ we have $\operatorname{supp} b_j \subset K$ for $1 \leq j \leq n$. Then there exists at least one function $c \in C_b(\mathbb{R}^n;\mathbb{R})$ such that the operator*

$$L(x,D) + \sum_{j=1}^{n} b_j(x)\frac{\partial}{\partial x_j} - c(x) \tag{2.104}$$

extends from $C_0^\infty(\mathbb{R}^n;\mathbb{R})$ to a Dirichlet operator.

Proof: Denoting by B the linear form associating with $L(x,D)$, i.e.

$$B(u,u) := \sum_{k,l=1}^{n}\int_{\mathbb{R}^n} a_{kl}(x)\frac{\partial u(x)}{\partial x_l}\frac{\partial v(x)}{\partial x_k}\,dx,$$

we prove first

$$\Big|\sum_{j=1}^{n}\int_{\mathbb{R}^n} b_j(x)\frac{\partial u(x)}{\partial x_j}v(x)\,dx\Big| \leq c(B_1(u,u))^{1/2}(B_1(v,v))^{1/2} \tag{2.105}$$

for all $u, v \in C_0^\infty(\mathbb{R}^n; \mathbb{R})$. For this let $\psi \in C_0^\infty(\mathbb{R}^n; \mathbb{R})$ such that $\psi|_K = 1$ and $0 \le \psi(x) \le 1$. Due to the support properties of the functions b_j we find using Lemma I.4.7.34

$$\begin{aligned}\Big|\sum_{j=1}^n \int_{\mathbb{R}^n} b_j(x) \frac{\partial u(x)}{\partial x_j} v(x)\, dx\Big| &= \Big|\sum_{j=1}^n \int_{\mathbb{R}^n} b_j(x) \frac{\partial (\psi u)(x)}{\partial x_j} (\psi v)(x)\, dx\Big| \\ &\le c_b \|\psi u\|_{1/2} \|\psi v\|_{1/2} \le c_b \|\psi u\|_\varepsilon \|\psi v\|_\varepsilon \\ &\le c_K c_b (B_1(\psi u, \psi u))^{1/2} (B_1(\psi v, \psi v))^{1/2}.\end{aligned}$$

Moreover we have

$$\begin{aligned}B_1(\psi u, \psi u) &= \sum_{k,l=1}^n \int_{\mathbb{R}^n} a_{kl}(x) \frac{\partial(\psi u)(x)}{\partial x_l} \frac{\partial (\psi u)(x)}{\partial x_k}\, dx \\ &+ \int_{\mathbb{R}^n} \psi^2(x) u^2(x)\, dx \\ &= \sum_{k,l=1}^n \int_{\mathbb{R}^n} a_{kl}(x) \frac{\partial u(x)}{\partial x_l} \frac{\partial u(x)}{\partial x_k}\, dx \\ &- \sum_{k,l=1}^n \int_{\mathbb{R}^n} a_{kl}(x)(1-\psi^2(x)) \frac{\partial u(x)}{\partial x_l} \frac{\partial u(x)}{\partial x_k}\, dx \\ &- \sum_{k,l=1}^n \int_{\mathbb{R}^n} \left(a_{kl}(x)\psi(x) \frac{\partial^2 \psi(x)}{\partial x_l \partial x_k} + \frac{\partial a_{kl}(x)}{\partial x_l} \psi(x) \frac{\partial \psi(x)}{\partial x_k} - \psi(x)^2 \right) |u(x)|^2 dx.\end{aligned}$$

Since

$$\sum_{k,l=1}^n a_{kl}(x) \xi_k \xi_l \ge 0$$

for all $x \in \mathbb{R}^n$ we find with $\xi = (1-\psi^2)^{1/2} \nabla u$ that

$$\int_{\mathbb{R}^n} \sum_{k,l=1}^n a_{kl}(x)(1-\psi^2) \frac{\partial u(x)}{\partial x_l} \frac{\partial u(x)}{\partial x_k}\, dx \ge 0,$$

implying that

$$B_1(\psi u, \psi u) \le B(u,u) + c_A \int_{\mathbb{R}^n} |u(x)|^2\, dx,$$

where

$$c_A := \sup_{x\in\mathbb{R}^n} \left| a_{kl}(x)\psi(x)\frac{\partial^2\psi(x)}{\partial x_l \partial x_k}\frac{\partial a_{kl}(x)}{\partial x_l}\psi(x)\frac{\partial\psi(x)}{\partial x_k} - \psi(x)^2 \right|.$$

Thus we obtain (2.105) which yields that for any function $c \in C_b(\mathbb{R}^n, \mathbb{R})$, $c(x) \geq 0$, the bilinear form

$$\begin{aligned} E(u,v) &:= (-L(x,D)u - \sum_{k,l=1}^n b_j \frac{\partial u}{\partial x_j} + c(\cdot)u, v)_0 \\ &= B(u,v) - \sum_{k,l=1}^n \int_{\mathbb{R}^n} b_j(x)\frac{\partial u(x)}{\partial x_j} v(x)\,dx + \int_{\mathbb{R}^n} c(x)u(x)v(x)\,dx \end{aligned}$$

satisfies the sector condition. Moreover, from

$$\begin{aligned} &-\sum_{k,l=1}^n \int_{\mathbb{R}^n} b_j(x)\frac{\partial u(x)}{\partial x_j} u(x)\,dx + \int_{\mathbb{R}^n} c(x)|u(x)|^2\,dx \\ &\qquad = \int_{\mathbb{R}^n} \Big(c(x) + \frac{1}{2}\sum_{k,l=1}^n \frac{\partial b_j(x)}{\partial x_j}\Big)|u(x)|^2\,dx, \end{aligned}$$

we deduce that for $c(x) + \frac{1}{2}\sum_{k,l=1}^n \frac{\partial b_j(x)}{\partial x_j} \geq 0$, it follows that $B(u,u) \geq 0$ for all $u \in C_0^\infty(\mathbb{R}^n; \mathbb{R})$. A careful inspection of the proof of the Theorem I.4.7.20 shows that the contraction property holds for E if we prove that

$$E((0 \vee u) \wedge 1, u - (0 \vee u) \wedge 1) \geq 0$$

holds for all (suitable) functions u. However, for the bilinear form $B(u,v) + \int_{\mathbb{R}^n} c(x)u(x)v(x)\,dx$ this follows since $L(x,D) - c(x)$ is a Dirichlet operator, while for $1 \leq j \leq n$ we have

$$\int_{\mathbb{R}^n} b_j(x)\frac{\partial}{\partial x_j}((0 \vee u) \wedge 1)(u - (0 \vee u) \wedge 1)\,dx = 0$$

which proves finally the proposition. □

We now want to discuss briefly a class of second order partial differential operator which satisfy certain sub-elliptic estimates. The class we are interested in is the class of the so-called *Hörmander-type operators*, often also called the *sum of squares of vector fields*. Our aim is not a complete discussion of these operators — this would be beyond the scope of our book, but we want to consider some aspects related to the fact that these operators occur as generators of semigroups.

For $1 \leq j \leq n$ and $1 \leq l \leq m$, $m \in \mathbb{N}$, let $a_j^l : \mathbb{R}^n \to \mathbb{R}$ be given C^∞-functions and consider the vector fields

$$X_l := \sum_{j=1}^n a_j^l(x) \frac{\partial}{\partial x_j}. \tag{2.106}$$

As usual we obtain by $[X_l, X_k]$ the *commutator field*, i.e.

$$[X_l, X_k] = X_l X_k - X_k X_l = \sum_{j=1}^n c_j^{lk}(x) \frac{\partial}{\partial x_j}, \tag{2.107}$$

where

$$c_j^{lk}(x) = \sum_{j=1}^n \left(a_i^l(x) \frac{\partial a_j^k(x)}{\partial x_i} - a_i^k(x) \frac{\partial a_j^l(x)}{\partial x_i} \right).$$

The formal L^2-adjoint of X_l considered as a differential operator is given by

$$X_l^* = -X_l - \sum_{j=1}^n \frac{\partial a_j^l(x)}{\partial x_j}. \tag{2.108}$$

The operators we are interested in are given by

$$L^{(1)}(x, D) = \sum_{l=1}^m X_l^2 \tag{2.109}$$

or

$$L^{(2)}(x, D) = -\sum_{l=1}^m X_l^* X_l. \tag{2.110}$$

A straightforward calculation shows that for the principal part of (2.109) and (2.110) we have

$$L_{pr}^{(\nu)} = \sum_{i,j=1}^n \left(\sum_{l=1}^m a_j^l(x) a_i^l(x) \right) \frac{\partial^2}{\partial x_j \partial x_i}, \quad \nu = 1, 2, \tag{2.111}$$

and since for $x \in \mathbb{R}^n$ and $\xi \in \mathbb{R}^n$

$$\sum_{i,j=1}^{n} \sum_{l-1}^{m} a_j^l(x) a_i^l(x) \xi_j \xi_i = \sum_{l=1}^{m} \Big(\sum_{j=1}^{n} a_j^l(x) \xi_j \Big)^2 \geq 0,$$

the operators $L^{(1)}(x, D)$ and $L^{(2)}(x, D)$ have non-negative characteristic form. Moreover, in case of the operator $L^{(2)}(x, D)$ we find for $u, v \in C_0^\infty(\mathbb{R}^n; \mathbb{R})$

$$\begin{aligned}(L^{(2)}(x, D)u, v)_0 &= -\int_{\mathbb{R}^n} \Big(\sum_{l=1}^{m} X_l^* X_l u(x) \Big) v(x)\, dx \\ &= \sum_{l=1}^{m} \int_{\mathbb{R}^n} \int_{\mathbb{R}^n} X_l(u)(x) X_l(v)(x)\, dx,\end{aligned}$$

thus

$$B^{(2)}(u, v) := \sum_{l=1}^{m} \int_{\mathbb{R}^n} (X_l u)(x)(X_l v)(x)\, dx \tag{2.112}$$

is a pre-Dirichlet form and by an analogous argument which led to Theorem 2.2.1 we conclude that the closure of $B^{(2)}$ is a symmetric Dirichlet form on $L^2(\mathbb{R}^n; \mathbb{R})$, hence the operator $-L^2(x, D) = \sum_{l=1}^{m} X_l^* X_l$ extends to a generator of a symmetric sub-Markovian semigroup on $L^2(\mathbb{R}^n; \mathbb{R})$. Since we may write

$$L^{(1)}(x, D) = -\sum_{l=1}^{m} X_l^* X_l + \sum_{l=1}^{m} \sum_{j=1}^{n} \frac{\partial a_j^l(x)}{\partial x_j} X_l,$$

the difference of $L^{(1)}(x, D)$ and $L^{(2)}(x, D)$ is a first order drift term. For $a_j^l \in C_b^1(\mathbb{R}^n; \mathbb{R})$, $1 \leq l \leq m$, $1 \leq j \leq n$, we find

$$\begin{aligned}&\Big| \int_{\mathbb{R}^n} \sum_{l=1}^{m} \sum_{j=1}^{n} \frac{\partial a_j^l(x)}{\partial x_j} X_l(u)(x) v(x)\, dx \Big| \\ &\leq \max_{l \leq l \leq m} \Big\| \sum_{j=1}^{n} \frac{\partial a_j^l}{\partial x_j} \Big\|_\infty \sum_{l=1}^{m} \|X_l u\|_0 \|v\|_0,\end{aligned}$$

whereas for $\lambda \geq 0$

$$B_\lambda^{(2)}(u, u) = B^{(2)}(u, u) + \lambda(u, u)_0 = \sum_{l=1}^{m} \|X_l u\|_0^2 + \lambda \|u\|_0^2.$$

Thus we find

$$\begin{aligned}
&\Big|\sum_{l=1}^{m}\sum_{j=1}^{n}\int_{\mathbb{R}^n}\frac{\partial a_j^l(x)}{\partial x_j}X_l(u)(x)v(x)\,dx\Big| \\
&\quad\leq c\sum_{l=1}^{m}\|X_l(u)\|_0\|v\|_0 \leq \frac{c}{2}\sum_{l=1}^{m}\|X_l u\|_0^2+\frac{m^2c}{2}\|v\|_0^2 \\
&\quad\leq \tilde{c}(B_1^{(2)}(u,u))^{1/2}(B_1^{(2)}(v,v))^{1/2},
\end{aligned}$$

implying that

$$\begin{aligned}
B^{(1)}(u,v) :&= (-L^{(1)}(x,D)u,v)_0 \\
&= B^{(2)}(u,v)+\sum_{l=1}^{m}\sum_{j=1}^{n}\int_{\mathbb{R}^n}\frac{\partial a_j^l(x)}{\partial x_j}X_l(u)(x)v(x)\,dx
\end{aligned} \tag{2.113}$$

extends from $C_0^\infty(\mathbb{R}^n;\mathbb{R})$ to a semi-Dirichlet form on $L^2(\mathbb{R}^n;\mathbb{R})$. Hence $L^{(1)}(x,D)$ extends from $C_0^\infty(\mathbb{R}^n;\mathbb{R})$ to a generator of a sub-Markovian semigroup on $L^2(\mathbb{R}^n;\mathbb{R})$.

It is not easy to characterize the domains $D(B^{(2)})$ and $D(L^{(2)}(x,D))$ or $D(L^{(1)}(x,D))$, respectively, in terms of (classical) function spaces. Denoting by X the system $X_1,\ldots,X_l$, one may of course introduce Sobolev-type spaces $H_X^1(\mathbb{R}^n)$ as completion of $C_0^\infty(\mathbb{R}^n)$ with respect to the scalar product $B_1^{(2)}$. In order to get some reasonable scale of function spaces further assumptions on X are necessary.

In fact, such assumptions are also necessary to get sub-elliptic estimates for $L^{(1)}(x,D)$ or $L^{(2)}(x,D)$, respectively.

Definition 2.2.9. Let X be a system of vector fields $X_1,\ldots,X_m$ on $\mathbb{R}^n$ with C^∞-coefficients.

A. We say that the system satisfies the *Hörmander condition* if for any point $x\in\mathbb{R}^n$ the vector fields $X_1,\ldots,X_m$ together with a finite number of iterated commutators of them do span the tangent space at x.

B. We say that the system X satisfies a *uniform Hörmander condition* if there exists an integer k independent of $x\in\mathbb{R}^n$ such that for all $x\in\mathbb{R}^n$ the fields $X_1,\ldots,X_m$ together with all iterated commutators of length less or equal to k do span the tangent space $T_x\mathbb{R}^n$.

Clearly, by an *iterated commutator* of the system X we mean a vector field of the the type

$$[X_{l_1},[X_{l_2},[\ldots[X_{l_{j-1}},X_{l_j}]\ldots]]],\quad l_i\in\{1,\ldots,m\} \tag{2.114}$$

The operator $-\Delta_K$ is the *Kohn-Laplacian* (on the Heisenberg group), sometimes it is called the *sub-Laplacian.* The commutator relations for X_j and Y_j are

$$[X_j, X_l] = [Y_j, Y_l] = 0, \quad 1 \le j, l \le n,$$

$$[X_j, Y_l] = 0 \quad \text{for } j \ne l,$$

and

$$[X_j, Y_j] = -4\frac{\partial}{\partial t}.$$

Thus $X_1, \dots, X_n, Y_1, \dots Y_n$ satisfy on $\mathbb{R}^{2n+1}$ a uniform Hörmander condition with $k = 2$. Subelliptic estimates for the Kohn-Laplace operator are proved using pseudo-differential operators, by J.J.Kohn [178].

Note that a uniform Hörmander condition leads in general not to uniform sub-elliptic estimates in the sense of Definition 2.2.5 since the operator $L(x, D)$ may have unbounded coefficients.

The following example gives an operator in $\mathbb{R}^2$ satisfying a uniform sub-elliptic estimate.

Example 2.2.15. Let $a : \mathbb{R} \to \mathbb{R}$ be a C^∞-function such that $0 \le a(t) \le 1$, $a|_{[-1,1]} = 0$, $a|_{[-2,2]^c} = 1$. Further let $b : \mathbb{R} \to \mathbb{R}$ be a C^∞-function such that $b|_{[-1,1]} = 1$, $b|_{[-2,2]^c} = 0$ and $0 \le b(t) \le 1$. We consider the vector fields

$$X_1 = \frac{\partial}{\partial x_1} \quad \text{and } X_2 = \rho(x_1)\frac{\partial}{\partial x_2},$$

with $\rho(x_1) = a(x_1) + x_1 b(x_1)$, and the associated operator

$$L(x, D) = X_1^2 + X_2^2 = \frac{\partial^2}{\partial x_1^2} + \rho^2(x_1)\frac{\partial^2}{\partial x_2^2}.$$

Since

$$[X_1, X_2] = \left(\frac{\partial a}{\partial x_1}(x_1) + b(x_1) + x_1\frac{\partial b}{\partial x_1}(x_1)\right)\frac{\partial}{\partial x_2},$$

it follows that X_1 and X_2 satisfy a uniform Hörmander condition with $k = 2$ provided either $\frac{\partial a}{\partial x_1}(x_1) \cdot \frac{\partial b}{\partial x_1}(x_1) \ne 0$ for $x \in [1, 2] \cup [-2, -1]$ or $b(x_1) \ne 0$ for $x \in [1, 2] \cup [-2, -1]$.

Moreover, for all $u \in C_0^\infty(\mathbb{R}^2;\mathbb{R})$, $\operatorname{supp} u \subset K$, $K \subset \mathbb{R}^2$ compact we have

$$\|u\|_{1/2}^2 \le c_K(B(u,u)+\|u\|_0^2), \tag{2.121}$$

where

$$B(u,v) = \int_{\mathbb{R}}\int_{\mathbb{R}} \Big(\frac{\partial u(x)}{\partial x_1}\frac{\partial v(x)}{\partial x_1} + \rho^2(x_1)\frac{\partial u(x)}{\partial x_2}\frac{\partial v(x)}{\partial x_2}\Big)\, dx_1 dx_2. \tag{2.122}$$

In addition, for $u \in C_0^\infty(\mathbb{R}^2;\mathbb{R})$ and $\operatorname{supp} u \subset Q^c$, where $Q = [-2,2]\times[-2,2]$ we have since $\rho^2(x_1) = 1$ for $|x_1| \ge 2$ that

$$\begin{aligned}\|u\|_1^2 &= \int_{\mathbb{R}}\int_{\mathbb{R}} \Big(\Big(\frac{\partial u}{\partial x_1}(x)\Big)^2 + \Big(\frac{\partial u}{\partial x_2}(x)\Big)^2\Big)\, dx_1 dx_2 + \|u\|_0^2 \\ &\le B(u,u) + \|u\|_0^2.\end{aligned}$$

Now, take $\chi \in C^\infty(\mathbb{R}^2)$ such that $\chi|_Q = 1$, $0 \le \chi \le 1$, and $\operatorname{supp}\chi \subset [-4,4]\times[-4,4]$. For any $C_0^\infty(\mathbb{R}^n;\mathbb{R})$ we have

$$u = \chi u + (1-\chi)u.$$

Since $\operatorname{supp}\chi u \subset [-4,4]\times[-4,4]$ and $\operatorname{supp}(1-\chi)u \subset Q^c$ we find combining (2.121) and (2.122)

$$\begin{aligned}\frac{1}{2}\|u\|_{1/2}^2 &\le \|\chi u\|_{1/2}^2 + \|(1-\chi)u\|_{1/2}^2 \\ &\le \|\chi u\|_{1/2}^2 + \|(1-\chi)u\|_1^2 \\ &\le c(B(\chi u,\chi u) + \|\chi u\|_0^2) \\ &\quad + B((1-\chi)u,(1-\chi)u) + \|(1-\chi)u\|_0^2.\end{aligned}$$

A straightforward computation using the properties of ρ and χ leads now to

$$\|u\|_{1/2}^2 \le \tilde{c}(B(u,u)+\|u\|_0^2)$$

for all $u \in C_0^\infty(\mathbb{R}^2)$, since χ was arbitrary.

Let X be a system of vector fields $X_1,\dots,X_m$ satisfying a uniform Hörmander condition. In this case it is possible to introduce anisotropic Sobolev spaces $W_X^{k,p}(\mathbb{R}^2;\mathbb{R})$ defined by

$$\begin{aligned}&W_X^{k,p}(\mathbb{R}^2;\mathbb{R}) \\ &\quad := \{u \in L^p(\mathbb{R}^2;\mathbb{R});\ X^\alpha u \in L^p(\mathbb{R}^2;\mathbb{R}) \text{ for } \alpha = (\alpha_1,\dots,\alpha_s),\ |\alpha| \le k\},\end{aligned} \tag{2.123}$$

where the norm $\|\cdot\|_{W_X^{k,p}}$ is given by

$$\|u\|_{W_X^{k,p}} = \Big(\sum_{|\alpha|\le k} \|X^\alpha u\|_{L^p}^p\Big)^{1/p} \tag{2.124}$$

and

$$X^\alpha u = X_{j_{\alpha_1}} \cdots X_{j_{\alpha_s}} u, \quad j_{\alpha_i} \in \{1,\dots,m\}.$$

These spaces where introduced by L.Hörmander [142], see also the paper [47] of V.M.Chernikov and S.K.Vodop'yanov as well as C.-J. Xu [282], [283] who related these spaces using a type of Sobolev embedding theorem to spaces of Hölder continuous functions when the Hölder continuity is defined by an intrinsic metric obtained by the system. This metric was introduced in [220] by A.Nagel, E.M.Stein and S.Wainger, see also the paper [90] of C.Fefferman and D.H.Phong. In the paper [105] U.Gianazza and V.Vespri used these spaces to obtain analytic semigroups generated by Hörmander type operators on $[W_X^{1,p}(\mathbb{R}^n)]^*$. Note that $W_X^{1,2}(\mathbb{R}^n;\mathbb{R})$ is a Dirichlet space under the scalar product $B_1^{(2)}$, $B_1^{(2)}$ associated with $L^{(2)}(x,D)$. Hence also $(B^{(2)}, W_X^{1,2}(\mathbb{R}^n;\mathbb{R}))$ is a Dirichlet space.

Furthermore, fix a system X of vector fields $X_1,\dots X_m$ satisfying a (uniform) Hörmander condition and let $(a_{kl}(\cdot))_{k,l=1,\dots,m}$ be a strictly positive defined matrix of $C_b^\infty(\mathbb{R}^n;\mathbb{R})$-functions, i.e. $a_{kl} = a_{lk}$ and

$$\sum_{k,l=1}^m a_{kl}(x)\xi_k\xi_l \ge \lambda_0|\xi|^2, \quad \lambda_0 > 0.$$

Consider the operator

$$-\sum_{k,l=1}^m X_k^*(a_{kl}(\cdot)X_l). \tag{2.125}$$

Using the paper [242] of L.Saloff-Coste and D.W.Stroock, it is also possible to construct symmetric sub-Markovian semigroups on $L^{(2)}(\mathbb{R}^n;\mathbb{R})$ generated by an extension of the operator (2.125). Further we can find that

$$\sum_{k,l=1}^m \int_{\mathbb{R}^n} a_{kl}(x)X_l(u)(x)X_k(u)(x)\,dx \tag{2.126}$$

is a symmetric Dirichlet form with domain $W_X^{1,2}(\mathbb{R}^n;\mathbb{R})$. In fact, if we are not interested in the hypoellipticity of the operator (2.125) we may reduce the regularity assumptions in (2.126) to $L^\infty(\mathbb{R}^n;\mathbb{R})$-coefficients and still obtain a symmetric Dirichlet form on $W_X^{1,2}(\mathbb{R}^n;\mathbb{R})$.

2.3 Some Properties of Pseudo-Differential Operators with Negative Definite Symbols

In some sense, starting with this section we come to the central theme of this monograph, namely to study these pseudo-differential operators which may extend to a generator of a Feller semigroup or a sub-Markovian semigroup. In this section we collect some first results. The class of pseudo-differential operators we are interested in is determined by Courrège's result, see Section I.4.5. Thus we will consider operator

$$q(x,D)u(x) = (2\pi)^{-n/2}\int_{\mathbb{R}^n} e^{ix\cdot\xi} q(x,\xi)\hat{u}(\xi)\,d\xi, \tag{2.127}$$

where we assume that $q:\mathbb{R}^n\times\mathbb{R}^n\to\mathbb{C}$ is a locally bounded function such that for every $x\in\mathbb{R}^n$ the function $q(x,\cdot):\mathbb{R}^n\to\mathbb{C}$ is a negative definite and continuous. For reference sake we give

Definition 2.3.1. We call a function $q:\mathbb{R}^n\times\mathbb{R}^n\to\mathbb{C}$ a *(continuous) negative definite symbol* if q is locally bounded (continuous) and for each $x\in\mathbb{R}^n$ the function $q(x,\cdot):\mathbb{R}^n\to\mathbb{C}$ is negative definite and continuous.

Note that for a negative definite symbol q for every compact set $K\subset\mathbb{R}^n$ there exists a constant c_K such that

$$|q(x,\xi)| \le c_K(1+|\xi|^2) \tag{2.128}$$

holds for all $x\in K$ and $\xi\in\mathbb{R}^n$. In fact we may take

$$c_K = 2\sup_{\substack{x\in K\\ |\eta|\le 1}} |q(x,\eta)|.$$

The pseudo-differential operator corresponding to a negative definite symbol is called a *pseudo-differential operator with negative definite symbol.*

From Theorem I.4.5.7 we deduce that a pseudo-differential operator with negative definite symbol maps $C_0^\infty(\mathbb{R}^n)$ into $B_b(\mathbb{R}^n)$. Using the Lévy–Khinchin formula we see that a pseudo-differential operator with negative definite symbol has also the representation

$$\begin{aligned} q(x,D)u(x) = &-\sum_{k,l=1}^{n} a_{kl}(x)\frac{\partial^2 u(x)}{\partial x_k \partial x_l} + \sum_{j=1}^{n} d_j(x)\frac{\partial u(x)}{\partial x_j} + c(x)u(x) \\ &- \int_{\mathbb{R}^n\setminus\{0\}} \Big(u(x-y) - u(x) + \sum_{j=1}^{n} \frac{y_j}{1+|y|^2}\frac{\partial u(x)}{\partial x_j}\Big) \\ &\times \frac{1+|y|^2}{|y|^2} N(x,dy) \end{aligned} \tag{2.129}$$

with a_{kl}, d_j, c and $N(x,dy)$ as in Section I.4.5. In particular $-q(x,D)$ satisfies the positive maximum principle on $C_0^\infty(\mathbb{R}^n;\mathbb{R})$ and in addition, $q(x,D)$ maps $C_0^\infty(\mathbb{R}^n;\mathbb{R})$ into $B_b(\mathbb{R}^n;\mathbb{R})$, i.e. it maps real-valued functions onto real-valued functions. The latter fact follows already from $q(x,\xi) = \overline{q(x,-\xi)}$ which holds for every negative definite symbol, see (I.3.122).

We are interested in proving estimates for the operator $q(x,D)$ in Sobolev spaces related to continuous negative definite functions, more precisely, in the Hilbert spaces $H^{\psi,s}(\mathbb{R}^n)$ introduced in Section I.3.10 for a fixed continuous negative definite function ψ and $s \in \mathbb{R}$. Note that the two different representations of $q(x,D)$, i.e. (2.127) and (2.129), give rise to different conditions on $q(x,D)$ in order to prove estimates since different techniques have to be used. In this and the following sections we will not make use of (2.129), but later, in Section 2.8 we will combine both representations to get more general results. In some estimates we need a more precise control on constants. For this we note

$$\int_{\mathbb{R}^n} (1+|x|^2)^{-k/2}\,dx = \frac{\pi^{n/2}\Gamma(\frac{k-n}{2})}{\Gamma(\frac{k}{2})} =: \tilde{c}_{n,k}, \quad k>n, \tag{2.130}$$

and $\tilde{\gamma}_{m,n}$ is the smallest constant such that

$$(1+|\xi|^2)^{m/2} \le \tilde{\gamma}_{m,n} \sum_{|\alpha|\le m} |\xi^\alpha| \tag{2.131}$$

holds. An upper bound for $\tilde{\gamma}_{m,n}$ is $(n+1)^{m/2}$. Finally, for a continuous negative definite function $\psi : \mathbb{R}^n \to \mathbb{R}$ the constant c_ψ is defined as the smallest constant such that

$$\psi(\xi) \le c_\psi(1+|\xi|^2) \tag{2.132}$$

holds.

For a (negative definite) symbol $q(x,\xi)$ we set

$$\hat{q}(\eta,\xi) := (2\pi)^{-n/2} \int_{\mathbb{R}^n} e^{-ix\cdot\eta} q(x,\xi)\,dx = F_{x\to\eta}(q(x,\xi))(\eta), \tag{2.133}$$

whenever this Fourier transform exists.

Lemma 2.3.2. *Let $q : \mathbb{R}^n \times \mathbb{R} \to \mathbb{C}$ be a measurable function and further let $\psi : \mathbb{R}^n \to \mathbb{R}$ be a continuous negative definite function. In addition assume that $q(\cdot,\xi) : \mathbb{R}^n \to \mathbb{C}$ is m-times continuously differentiable.*

A. If for every multiindex $\alpha \in \mathbb{N}_0^n$, $|\alpha| \le m$, the estimate

$$|\partial_x^\alpha q(x,\xi)| \le c_\alpha(1+\psi(\xi)) \tag{2.134}$$

holds we find for all $\varphi \in C_0^\infty(\mathbb{R}^n)$

$$|(\varphi q)^\wedge(\eta,\xi)| \le c_\varphi(1+|\eta|^2)^{-m/2}(1+\psi(\xi)). \tag{2.135}$$

B. Suppose that there are functions $\varphi_\alpha \in L^1(\mathbb{R}^n)$, $|\alpha| \le m$, such that

$$|\partial_x^\alpha q(x,\xi)| \le \varphi_\alpha(x)(1+\psi(\xi)) \tag{2.136}$$

holds. Then we have for all $k \in \mathbb{N}_0$, $|k| \le m$, the estimate

$$|\hat{q}(\eta,\xi)| \le \tilde{\gamma}_{k,n} \sum_{|\alpha|\le k} \|\varphi_\alpha\|_{L^1}(1+|\eta|^2)^{-k/2}(1+\psi(\xi)). \tag{2.137}$$

Proof: A. For $\beta \in \mathbb{N}_0^n$, $|\beta| \le m$, we have

$$\begin{aligned}
\left|\eta^\beta \int_{\mathbb{R}^n} e^{-ix\cdot\eta}\varphi(x)q(x,\xi)\,dx\right| &= \left|\int_{\mathbb{R}^n} \left(\partial_x^\beta(e^{-ix\cdot\eta})\right)\varphi(x)q(x,\xi)\,dx\right| \\
&= \left|\int_{\mathbb{R}^n} e^{-ix\cdot\eta}\partial_x^\beta(\varphi(x)q(x,\xi))\,dx\right| \\
&\le \sum_{\gamma\le\beta}\binom{\beta}{\gamma} c_\gamma(1+\psi(\xi))\|\partial^{\beta-\gamma}\varphi\|_{L^1} \le c_{\beta,\varphi}(1+\psi(\xi)),
\end{aligned}$$

which leads to

$$(1+|\eta|^2)^{m/2}|(\varphi q)^\wedge(\eta,\xi)| \le (2\pi)^{-n/2}\tilde{c}_\varphi(1+\psi(\xi)),$$

i.e. (2.135) is proved.

B. We may use the calculation of part A to obtain

$$\left|\eta^\beta \int_{\mathbb{R}^n} e^{-ix\cdot\eta} q(x,\xi)\,dx\right| = \left|\int_{\mathbb{R}^n} e^{-ix\cdot\eta}\partial_x^\beta q(x,\xi)\,dx\right| \leq \|\varphi_\beta\|_{L^1}(1+\psi(\xi)),$$

which gives for $k \leq m$

$$(1+|\eta|^2)^{k/2}|\hat{q}(\eta,\xi)| \leq \tilde{\gamma}_{k,n} \sum_{|\beta|\leq k} \|\varphi_\beta\|_{L^1}(1+\psi(\xi)),$$

and (2.137) is proved. □

To prove a first consequence of Lemma 2.3.2 we need

Lemma 2.3.3. *Let $k \in L^1(\mathbb{R}^n)$. Then we have for all $u, v \in L^2(\mathbb{R}^n)$*

$$\left|\int_{\mathbb{R}^n}\int_{\mathbb{R}^n} k(\xi-\eta)u(\eta)v(\xi)\,d\eta d\xi\right| \leq \|k\|_{L^1}\|u\|_0\|v\|_0. \tag{2.138}$$

Proof: Using the Cauchy–Schwarz inequality we find

$$\begin{aligned}
&\left|\int_{\mathbb{R}^n}\int_{\mathbb{R}^n} k(\xi-\eta)u(\eta)v(\xi)\,d\eta d\xi\right| \\
&\leq \int_{\mathbb{R}^n}\left(\left(\int_{\mathbb{R}^n}|k(\xi-\eta)|\,d\eta\right)^{1/2}\left(\int_{\mathbb{R}^n}|k(\xi-\eta)||u(\eta)|^2\,d\eta\right)^{1/2}|v(\xi)|\right)d\xi \\
&\leq \|k\|_{L^1}^{1/2}\|v\|_0\left(\int_{\mathbb{R}^n}\int_{\mathbb{R}^n}|k(\xi-\eta)||u(\eta)|^2\,d\eta d\xi\right)^{1/2} \\
&\leq \|k\|_{L^1}\|v\|_0\|u\|_0,
\end{aligned}$$

where in the last step Young's inequality, Lemma I.2.3.15, was used. □

Now we may show

Proposition 2.3.4. *Suppose that q is a function satisfying the assumptions of Lemma 2.3.2.A for $m \geq n+1$. Then the operator $q(x,D)$ maps $H^{\psi,2}(\mathbb{R}^n)$ into the space $L^2_{loc}(\mathbb{R}^n)$, and we have for every test function $\varphi \in C_0^\infty(\mathbb{R}^n)$ the estimate*

$$\|\varphi q(x,D)u\|_0 \leq \tilde{c}_\varphi\|u\|_{\psi,2}. \tag{2.139}$$

Proof: We will prove (2.139) for all $\varphi \in C_0^\infty(\mathbb{R}^n)$ and all $u \in S(\mathbb{R}^n)$. Note that for $u, v \in S(\mathbb{R}^n)$ we have by (2.135) and Lemma 2.3.3

$$\begin{aligned}|(\varphi q(x,D)u,v)_0| &= \left|\int_{\mathbb{R}^n} (\varphi q)^\wedge(\xi-\eta,\eta)\hat{u}(\eta)\overline{\hat{v}(\xi)}\,d\eta d\xi\right| \\ &\leq c_\varphi \int_{\mathbb{R}^n}\int_{\mathbb{R}^n} (1+|\xi-\eta|^2)^{-(n+1)/2}(1+\psi(\eta))|\hat{u}(\eta)||\hat{v}(\xi)|\,d\eta d\xi \\ &\leq c_\varphi \tilde{c}_{n,n+1}\|1+\psi(\cdot)\hat{u}\|_0\|v\|_0 \\ &\leq \tilde{c}_\varphi\|u\|_{\psi,2}\|v\|_0,\end{aligned}$$

or

$$\sup_{\substack{v\in L^2(\mathbb{R}^n)\\ v\neq 0}} \frac{|(\varphi q(x,D)u,v)_0|}{\|v\|_0} \leq \tilde{c}_\varphi\|u\|_{\psi,2},$$

which implies the proposition. □

The proof of Proposition 2.3.4 shows already a principal problem when estimating a pseudo-differential operator with negative definite symbol in some of the spaces $H^{\psi,s}(\mathbb{R}^n)$. We need also a control on the Fourier transform of the function $x \mapsto q(x,\xi)$, i.e. we need a control on $\hat{q}(\eta,\xi)$. This problem is well known in the theory of pseudo-differential operators. For example in order to improve (2.139) to a global estimate , i.e. to

$$\|q(x,D)u\|_0 \leq c\|u\|_{\psi,2}$$

additional assumptions on $q(x,D)$ are necessary. We follow in this section the idea presented in the monograph [225] of O.A.Oleinik and E.V.Radkevič, and consider symbols $q(x,\xi)$

$$q(x,\xi) = q_1(\xi) + q_2(x,\xi). \tag{2.140}$$

Of course we have still to assume that $q(x,\cdot) : \mathbb{R}^n \to \mathbb{C}$ is for all $x \in \mathbb{R}^n$ a continuous negative definite function. In addition we assume that $q_1 : \mathbb{R}^n \to \mathbb{C}$ is also a continuous negative definite function. The operators $q_1(D)$ and $q_2(x,D)$ will be estimated separately. Note that decomposition (2.140) may arise when freezing the coefficients, i.e.

$$q(x,\xi) = q(x_0,\xi) + (q(x,\xi) - q(x_0,\xi)).$$

Let us pose some general assumptions on q_1 and q_2.

Assumption 2.3.5. We assume that the function $q : \mathbb{R}^n \times \mathbb{R}^n \to \mathbb{C}$ is a continuous negative definite symbol having the decomposition (2.140) into a continuous negative definite function $q_1 : \mathbb{R}^n \to \mathbb{C}$ and a continuous function $q_2 : \mathbb{R}^n \times \mathbb{R}^n \to \mathbb{C}$. Further let $\psi : \mathbb{R}^n \to \mathbb{R}$ be a fixed continuous negative definite (reference) function.

A.1. The function q_1 is continuous and negative definite and satisfies with $\gamma_0 > 0$ and $\gamma_1, \gamma_2 \geq 0$

$$\gamma_0\psi(\xi) \leq \operatorname{Re} q_1(\xi) \leq \gamma_1\psi(\xi) \quad \text{for all } |\xi| \geq 1, \tag{2.141}$$

and

$$|\operatorname{Im} q_1(\xi)| \leq \gamma_2 \operatorname{Re} q_1(\xi) \quad \text{for all } \xi \in \mathbb{R}^n. \tag{2.142}$$

(Note that (2.141) and (2.142) implies

$$1+\operatorname{Re} q_1(\xi) \leq 1+|q_1(\xi)| \leq \tau_1(1+\psi(\xi)) \tag{2.143}$$

for all $\xi \in \mathbb{R}^n$ with some $\tau_1 > 0$).

A.2.m. For $m \in \mathbb{N}_0$ the function $x \mapsto q_2(x, \xi)$ belongs to $C^m(\mathbb{R}^n)$ and we have the estimate

$$|\partial_x^\alpha q_2(x,\xi)| \leq \varphi_\alpha(x)(1+\psi(\xi)) \tag{2.144}$$

for all $\alpha \in \mathbb{N}_0^n$, $|\alpha| \leq m$, with function $\varphi_\alpha \in L^1(\mathbb{R}^n)$.

Later we will add a smallness assumption for $q_2(x, \xi)$. We start with estimates for the operator $q_1(D)$.

Proposition 2.3.6. *We assume A.1. For any $s \in \mathbb{R}$ the operator $q_1(D)$ maps $H^{\psi,s}(\mathbb{R}^n)$ continuously into the space $H^{\psi,s-2}(\mathbb{R}^n)$. Further we have the estimates*

$$\|q_1(D)u\|_{\psi,s-2} \leq \tau_1\|u\|_{\psi,s} \tag{2.145}$$

and

$$\|q_1(D)u\|_{\psi,s-2} \geq \gamma_0\|u\|_{\psi,s} - \lambda_0\|u\|_{\psi,s-2}, \tag{2.146}$$

with a suitable constant λ_0.

Proof: It is sufficient to prove (2.145) and (2.146) for all $u \in S(\mathbb{R}^n)$. For $s \in \mathbb{R}$ we find using (2.143)

$$\begin{aligned}\|q_1(D)u\|_{\psi,s-2}^2 &= \int_{\mathbb{R}^n} (1+\psi(\xi))^{s-2}|q_1(\xi)\hat{u}(\xi)|^2\,d\xi \\ &\leq \tau_1^2 \int_{\mathbb{R}^n} (1+\psi(\xi))^{s-2}(1+\psi(\xi))^2|\hat{u}(\xi)|^2\,d\xi \\ &= \tau_1^2\|u\|_{\psi,s}^2,\end{aligned}$$

proving (2.145). To prove (2.146) observe that

$$\begin{aligned}\|q_1(D)u\|_{\psi,s-2}^2 &= \int_{B_1^c(0)} (1+\psi(\xi))^{s-2}|q_1(\xi)|^2|\hat{u}(\xi)|^2\,d\xi \\ &\quad + \int_{B_1(0)} (1+\psi(\xi))^{s-2}|q_1(\xi)|^2|\hat{u}(\xi)|^2\,d\xi \\ &\geq \int_{B_1^c(0)} (1+\psi(\xi))^{s-2}(\operatorname{Re} q_1(\xi))^2|\hat{u}(\xi)|^2\,d\xi \\ &\geq \gamma_0^2 \int_{B_1^c(0)} (1+\psi(\xi))^{s-2}(\psi(\xi))^2|\hat{u}(\xi)|^2\,d\xi \\ &\geq \gamma_0^2 \int_{B_1^c(0)} (1+\psi(\xi))^{s}|\hat{u}(\xi)|^2\,d\xi \\ &\quad - \gamma_0^2 \int_{B_1^c(0)} (1+\psi(\xi))^{s-2}|\hat{u}(\xi)|^2\,d\xi \\ &= \gamma_0^2\|u\|_{\psi,s}^2 - \gamma_0^2 \int_{B_1^c(0)} (1+\psi(\xi))^{s-2}|\hat{u}(\xi)|^2\,d\xi \\ &\quad - \gamma_0^2 \int_{B_1(0)} (1+\psi(\xi))^{s}|\hat{u}(\xi)|^2\,d\xi \\ &\geq \gamma_0^2\|u\|_{\psi,s}^2 - \gamma_0^2 \sup_{\xi\in B_1(0)} (1+\psi(\xi))^2\|u\|_{\psi,s-2}^2,\end{aligned}$$

which implies (2.146). □

Corollary 2.3.7. *Suppose that* $\lim_{|\xi|\to\infty} \psi(\xi) = \infty$. *Then we have in the situation of Proposition* 2.3.6 *for every* $\varepsilon > 0$ *and* $s \geq 2$

$$\|q_1(D)u\|_{\psi,s-2} \geq (\gamma_0-\varepsilon)\|u\|_{\psi,s} - \lambda_\varepsilon\|u\|_0 \tag{2.147}$$

for $\lambda_\varepsilon \geq 0$.

Proof: Since $\lim_{|\xi|\to\infty} \psi(\xi) = \infty$ we find for $\varepsilon > 0$ a suitable constant $\rho_\varepsilon \geq 0$ such that

$$(1+\psi(\xi))^{s-2} \leq \frac{\varepsilon^2}{\lambda_0^2}(1+\psi(\xi))^s$$

holds for all $\xi \in \mathbb{R}^n$, $|\xi| \geq \rho_\varepsilon$. Note that the case $\lambda_0 = 0$ is trivial. It follows that

$$\begin{aligned}
\lambda_0^2\|u\|^2_{\psi,s-2} &= \lambda_0^2 \int_{\mathbb{R}^n} (1+\psi(\xi))^{s-2} |\hat{u}(\xi)|^2\, d\xi \\
&= \lambda_0^2 \int_{B^c_{\rho_\varepsilon}(0)} (1+\psi(\xi))^{s-2} |\hat{u}(\xi)|^2\, d\xi \\
&\quad + \lambda_0^2 \int_{B_{\rho_\varepsilon}(0)} (1+\psi(\xi))^{s-2} |\hat{u}(\xi)|^2\, d\xi \\
&\leq \varepsilon^2 \int_{B^c_{\rho_\varepsilon}(0)} (1+\psi(\xi))^{s} |\hat{u}(\xi)|^2\, d\xi \\
&\quad + \lambda_0^2 \left(\sup_{x\in B_{\rho_\varepsilon}(0)} (1+\psi(\xi))^{s-2} \right) \int_{B^c_{\rho_\varepsilon}(0)} |\hat{u}(\xi)|^2\, d\xi \\
&\leq \varepsilon^2\|u\|^2_{\psi,s} + \lambda_\varepsilon^2\|u\|_0^2.
\end{aligned}$$

Hence (2.146) implies

$$\|q_1(D)u\|_{\psi,s-2} \geq \gamma_0\|u\|_{\psi,s} - \lambda_0\|u\|_{\psi,s-2} \geq (\gamma_0-\varepsilon)\|u\|_{\psi,s} - \lambda_\varepsilon\|u\|_0. \quad \square$$

Remark 2.3.8. It is possible to get a smaller value for λ_ε, but we do not need it later on.

Next we want to estimate the operator $q_2(x, D)$ assuming that $q_2(x,\xi)$ fulfills A.2.m with a suitable large m. We start with the following commutator estimate:

Theorem 2.3.9. *Let $s \geq \frac{1}{2}$ and assume that $q_2(x,\xi)$ satisfies A.2.m with $m > n + 2s$. Then we have*

$$\|[(1+\psi(\xi))^s, q_2(x,D)]u\|_0 \leq \kappa_{n,m,s,\psi} \sum_{|\alpha|\leq m} \|\varphi_\alpha\|_{L^1}\|u\|_{\psi,2s+1} \tag{2.148}$$

for all $u \in H^{\psi,2s+1}(\mathbb{R}^n)$, where

$$\kappa_{n,m,s,\psi} = (2\pi)^{-n/2}2^{2s+3}s\, c_\psi \tilde{\gamma}_{m,n}\tilde{c}_{n,m-2s} \tag{2.149}$$

with $\tilde{c}_{n,m-2s}$, $\tilde{\gamma}_{m,n}$ and c_ψ as in (2.130)–(2.132).

Proof: For $s \geq \frac{1}{2}$ we have using Lemma I.3.6.21 and Peetre's inequality, Lemma I.3.6.23,

$$\begin{aligned}
|\,(1+\psi(\xi))^s - (1+\psi(\eta))^s\,| &= 2s\Big|\int_{(1+\psi(\eta))^{1/2}}^{(1+\psi(\xi))^{1/2}} t^{2s-1}\,dt\Big| \\
&\leq 2s\left((1+\psi(\xi))^{\frac{2s-1}{2}} + (1+\psi(\eta))^{\frac{2s-1}{2}}\right)\left|(1+\psi(\xi))^{1/2} - (1+\psi(\eta))^{1/2}\right| \\
&\leq 2^{2s+3} s c_\psi \left(1+|\xi-\eta|^2\right)^s (1+\psi(\eta))^{\frac{2s-1}{2}},
\end{aligned} \tag{2.150}$$

where we use also estimate (2.132). Further, by Lemma 2.3.2.B we have

$$|\hat{q}_2(\xi-\eta,\eta)| \leq \tilde{\gamma}_{m,n} \sum_{|\alpha|\leq m} \|\varphi_\alpha\|_{L^1} \left(1+|\xi-\eta|^2\right)^{-m/2} (1+\psi(\eta))\,. \tag{2.151}$$

Now, for $u, v \in C_0^\infty(\mathbb{R}^n)$ we find using (2.150) and (2.151)

$$\begin{aligned}
&|(([1+\psi(D))^s, q_2(x,D)]u, v)_0| \\
&\quad = (2\pi)^{-n/2}\Big|\int_{\mathbb{R}^n}\int_{\mathbb{R}^n} \hat{q}_2(\xi-\eta,\eta)\Big((1+\psi(\xi))^s \\
&\qquad - (1+\psi(\eta))^s\Big)\hat{u}(\eta)\hat{v}(\xi)\,d\eta d\xi\Big| \\
&\quad \leq (2\pi)^{-n/2} 2^{2s+3} s c_\psi \tilde{\gamma}_{m,n} \sum_{|\alpha|\leq m} \|\varphi_\alpha\|_{L^1} \\
&\qquad \times \int_{\mathbb{R}^n}\int_{\mathbb{R}^n} (1+|\xi-\eta|^2)^{-\frac{m}{2}+s}(1+\psi(\eta))^{\frac{2s+1}{2}} |\hat{u}(\eta)||\hat{v}(\xi)|\,d\eta d\xi \\
&\quad \leq \kappa_{n,m,s,\psi} \sum_{|\alpha|\leq m} \|\varphi_\alpha\|_{L^1} \|u\|_{\psi,2s+1} \|v\|_0,
\end{aligned}$$

which implies (2.148) and (2.149). □

Corollary 2.3.10. *Let $s \geq 0$, $t \geq \frac{1}{2}$ and assume that $q_2(x,\xi)$ satisfies A.2.m with $m > n + 2s + 2t$. Then we have with a suitable constant $\tilde{c}$*

$$\|[(1+\psi(D))^s, q_2(x,D)]u\|_{\psi,2t} \leq \tilde{c} \sum_{|\alpha|\leq m} \|\varphi_\alpha\|_{L^1} \|u\|_{\psi,2s+2t+1}. \tag{2.152}$$

Proof: Since

$$\|[(1+\psi(D))^s, q_2(x,D)]u\|_{\psi,2t} = \|(1+\psi(D))^t[(1+\psi(D))^s, q_2(x,D)]u\|_0$$

and

$$\begin{aligned}&(1+\psi(D))^t\left[(1+\psi(D))^s, q_2(x,D)\right]u\\&=\left[(1+\psi(D))^{t+s}, q_2(x,D)\right]u-\left[(1+\psi(D))^t, q_2(x,D)\right](1+\psi(D))^s u\end{aligned}$$

if follows from Theorem 2.3.9 that

$$\begin{aligned}&\|[(1+\psi(D))^s, q_2(x,D)]u\|_{\psi,2t}\\&\quad\le\|[(1+\psi(D))^{t+s}, q_2(x,D)]u\|_0\\&\qquad+\|[(1+\psi(D))^t, q_2(x,D)](1+\psi(D))^s u\|_0\\&\quad\le c\sum_{|\alpha|\le m}\|\varphi_\alpha\|_{L^1}\|u\|_{\psi,2s+2t+1}+c'\sum_{|\alpha|\le m}\|\varphi_\alpha\|_{L^1}\|(1+\psi(D))^s u\|_{\psi,2t}\\&\quad\le\tilde{c}\sum_{|\alpha|\le m}\|\varphi_\alpha\|_{L^1}\|u\|_{\psi,2s+2t+1}.\quad\square\end{aligned}$$

Now we want to estimate $q_2(x,D)$.

Theorem 2.3.11. *Let $s\ge 1$ and suppose that $q_2(x,\xi)$ satisfies A.2.m with $m\ge n+[s]+1$. Then we have for all $t\ge 0$ and $u\in H^{\psi,t+2}(\mathbb{R}^n)$*

$$\|q_2(x,D)u\|_{\psi,t}\le\tilde{c}_{n,m,t,\psi}\|u\|_{\psi,t+2},\tag{2.153}$$

and for $t=s$ we have

$$\tilde{c}_{n,m,s,\psi}=\left(\tilde{\gamma}_{n+1,n}\tilde{c}_{n,n+1}+\kappa_{n,m,\frac{s}{2},\psi}\right)\sum_{|\alpha|\le m}\|\varphi_\alpha\|_{L^1},\tag{2.154}$$

where $\kappa_{n,m,\frac{s}{2},\psi}$ is defined as in (2.149).

Proof: First note that for $u,v\in S(\mathbb{R}^n)$ we have

$$\int_{\mathbb{R}^n}q_2(x,D)u(x)v(x)\,dx=\int_{\mathbb{R}^n}\int_{\mathbb{R}^n}\hat{q}_2(\xi-\eta,\eta)\hat{u}(\eta)\overline{\hat{v}(\xi)}\,d\eta d\xi$$

and applying Lemma 2.3.2.B yields

$$\begin{aligned}|(q_2(x,D)u,v)_0|&\le\tilde{\gamma}_{n+1,n}\sum_{|\alpha|\le n+1}\|\varphi_\alpha\|_{L^1}\int_{\mathbb{R}^n}\int_{\mathbb{R}^n}(1+|\xi-\eta|^2)^{-\frac{(n+1)}{2}}\\&\qquad\times(1+\psi(\eta))|\hat{u}(\eta)||\hat{v}(\xi)|\,d\eta d\xi\\&\le\tilde{\gamma}_{n+1,n}\tilde{c}_{n,n+1}\sum_{|\alpha|\le n+1}\|\varphi_\alpha\|_{L^1}\|u\|_{\psi,2}\|v\|_0,\end{aligned}$$

which implies

$$\|q_2(x,D)u\|_0 \le \tilde{\gamma}_{n+1,n}\tilde{c}_{n,n+1} \sum_{|\alpha|\le n+1} \|\varphi_\alpha\|_{L^1}\|u\|_{\psi,2}, \tag{2.155}$$

i.e. (2.153) for $t=0$. Next let $s\ge 1$ and observe that

$$\begin{aligned}\|q_2(x,D)u\|_{\psi,s} \le \|q_2(x,D)(1+\psi(D))^{s/2}u\|_0 \\ + \|[(1+\psi(D))^{s/2}, q_2(x,D)]u\|_0.\end{aligned}$$

Since $(1+\psi(D))^{r/2}$ maps $H^{\psi,t}(\mathbb{R}^n)$ bijectively and bicontinuously onto $H^{\psi,t-r}(\mathbb{R}^n)$ we find for $u\in H^{\psi,s+2}(\mathbb{R}^n)$ that $(1+\psi(D))^{s/2}u =: v \in H^{\psi,2}(\mathbb{R}^n)$ and therefore

$$\|q_2(x,D)\,(1+\psi(D))^{s/2}\,u\|_0 \le \tilde{\gamma}_{n+1,n}\tilde{c}_{n,n+1} \sum_{|\alpha|\le n+1} \|\varphi_\alpha\|_{L^1}\|u\|_{\psi,s+2}. \tag{2.156}$$

On the other hand we may apply Theorem 2.3.9 with s replaced by $s/2$ to get

$$\|[(1+\psi(D))^{s/2}, q_2(x,D)]u\|_0 \le \kappa_{n,m,\frac{s}{2},\psi} \sum_{|\alpha|\le m} \|\varphi_\alpha\|_{L^1}\|u\|_{\psi,s+1}, \tag{2.157}$$

which leads together with (2.156) to

$$\|q_2(x,D)u\|_{\psi,s} \le \left(\tilde{\gamma}_{n+1,n}\tilde{c}_{n,n+1} + \kappa_{n,m,\frac{s}{2},\psi}\right) \sum_{|\alpha|\le m} \|\varphi_\alpha\|_{L^1}\|u\|_{\psi,s+2}$$

and the theorem proved for $s\ge 1$ and $m > [s]+n+2$. For $0<s<1$ the result follows by interpolation, see Theorem I.3.10.16 and Theorem I.2.8.7. □

Remark 2.3.12. Note that our proof of Theorem 2.3.11 and (I.2.215) yields the estimate

$$\|q_2(x,D)u\|_{\psi,t} \le \tilde{c}^*_{n,t,\psi,m} \sum_{|\alpha|\le m} \|\varphi_\alpha\|_{L^1}\|u\|_{\psi,t+2} \tag{2.158}$$

for all $t\ge 0$ with m suitably large.

In the following theorem the relation of the constants in (2.147) with $s-2$ replaced by s and the constant $c_{n,m,s,\psi}$ from (2.154) will become important. For simplicity let us restate (2.147) as

$$\|q_1(D)u\|_{\psi,s} \ge \eta\gamma_0\|u\|_{\psi,s+2} - \gamma_{\eta,s}\|u\|_0 \tag{2.159}$$

which holds for all s and η, $0 < \eta < 1$. Further recall the estimate (2.153) for $s \geq 1$ and $m \geq n + [s] + 1$:

$$\|q_2(x,D)u\|_{\psi,s} \leq \tilde{c}_{n,m,s,\psi} \sum_{|\alpha|\leq m} \|\varphi_\alpha\|_{L^1} \|u\|_{\psi,s+2}, \tag{2.160}$$

where $\tilde{c}_{n,m,s,\psi} = \left(\tilde{\gamma}_{n+1,n}\tilde{c}_{n,n+1} + \kappa_{n,m,\frac{s}{2},\psi}\right)$. Combining (2.159) and (2.160) we get

$$\begin{aligned} \|q(x,D)u\|_{\psi,s} &\geq \|q_1(D)u\|_{\psi,s} - \|q_2(x,D)u\|_{\psi,s} \\ &\geq \eta\gamma_0\|u\|_{\psi,s+2} - \gamma_{\eta,s}\|u\|_0 - \tilde{c}_{n,m,s,\psi} \sum_{|\alpha|\leq m} \|\varphi_\alpha\|_{L^1}\|u\|_{\psi,s+2} \\ &= \Big(\eta\gamma_0 - \tilde{c}_{n,m,s,\psi} \sum_{|\alpha|\leq m} \|\varphi_\alpha\|_{L^1}\Big)\|u\|_{\psi,s+2} - \gamma_{\eta,s}\|u\|_0. \end{aligned}$$

Thus we have proved

Theorem 2.3.13. *Suppose that $q(x,\xi) = q_1(\xi) + q_2(x,\xi)$ satisfies A.1 and A.2.m with $m > n + [s] + 1$ for some $s \geq 1$. Further assume that for some $\eta \in (0,1)$ we have*

$$\frac{1}{4}\tilde{c}_{n,m,s,\psi} \sum_{|\alpha|\leq m} \|\varphi_\alpha\|_{L^1} < \gamma_0. \tag{2.161}$$

Then we have for all $u \in H^{\psi,s}(\mathbb{R}^n)$ the lower estimate

$$\|q(x,D)u\|_{\psi,s} \geq \delta_0\|u\|_{\psi,s+2} - \gamma_{\eta,s}\|u\|_0 \tag{2.162}$$

with

$$\delta_0 := \eta\gamma_0 - \tilde{c}_{n,m,s,\psi} \sum_{|\alpha|\leq m} \|\varphi_\alpha\|_{L^1} > 0. \tag{2.163}$$

Remark 2.3.14. Later on, we see that often

$$(q(x,D)u,u)_0 \geq 0 \tag{2.164}$$

holds for all $u \in H^{\psi,2}(\mathbb{R}^n;\mathbb{R})$. Taking in (2.162) $s = 2$ we find then for all $\lambda \geq 0$

$$\begin{aligned} \|q(x,D)u + \lambda u\|_0^2 &= \|q(x,D)u\|_0^2 + 2\lambda(q(x,D)u,u)_0 + \lambda^2\|u\|_0^2 \\ &\geq \|q(x,D)u\|_0^2 + \lambda^2\|u\|_0^2 \end{aligned}$$

or

$$\begin{aligned}\|q(x,D)u+\lambda u\|_0 &\geq \|q(x,D)u\|_0+\lambda\|u\|_0\\ &\geq \delta_0\|u\|_{\psi,2}-\gamma_{\eta,2}\|u\|_0+\lambda\|u\|_0.\end{aligned}$$

Thus for $\lambda \geq \gamma_{\eta,2}$ we have under the assumption (2.164)

$$\|q(x,D)u+\lambda u\|_0 \geq \delta_0\|u\|_{\psi,2}. \tag{2.165}$$

In order to prove regularity results for solutions of the equation $q(x,D)u = f$ we have to have a closer look to the Friedrichs mollifier.

Proposition 2.3.15. *Let J_ε be defined as in (I.2.78). For any $s \geq 0$ and $u \in H^{\psi,s}(\mathbb{R}^n)$ we have*

$$J_\varepsilon(u) \in \bigcap_{t\geq 0} H^{\psi,t}(\mathbb{R}^n) \cap C^\infty(\mathbb{R}^n) \tag{2.166}$$

and

$$\|J_\varepsilon(u)\|_{\psi,s} \leq \|u\|_{\psi,s} \tag{2.167}$$

as well as

$$\lim_{\varepsilon\to 0}\|J_\varepsilon(u)-u\|_{\psi,s}=0. \tag{2.168}$$

In addition, if for $\varepsilon \in (0,\rho)$, $\rho > 0$, we have for some $u \in L^2(\mathbb{R}^n)$

$$\|J_\varepsilon(u)\|_{\psi,s} \leq c_{u,s} \tag{2.169}$$

with a constant independent of ε, it follows that $u \in H^{\psi,s}(\mathbb{R}^n)$.

Proof: For $t \geq 0$ we find

$$\begin{aligned}\|J_\varepsilon(u)\|^2_{\psi,t} &= \int_{\mathbb{R}^n}(1+\psi(\xi))^t|(j_\varepsilon * u)^\wedge(\xi)|^2\,d\xi\\ &= (2\pi)^n\int_{\mathbb{R}^n}(1+\psi(\xi))^t|\hat{j}_\varepsilon(\xi)|^2|\hat{u}(\xi)|^2\,d\xi.\end{aligned}$$

Since $j_\varepsilon \in C_0^\infty(\mathbb{R}^n)$, it follows that $\hat{j}_\varepsilon \in S(\mathbb{R}^n)$. Hence there is a constant $c_{t,s,\varepsilon}$ such that

$$(1+\psi(\xi))^t\,|\hat{j}_\varepsilon(\xi)| \leq c_{t,s,\varepsilon}\,(1+\psi(\xi))^s,$$

which implies $\|J_\varepsilon(u)\|^2_{\psi,t} \le c_{t,s,\varepsilon}\|u\|^2_{\psi,s}$, hence we have $J_\varepsilon(u) \in \bigcap_{t\ge 0} H^{\psi,t}(\mathbb{R}^n)$, and (2.166) follows with Proposition I.2.2.17. Moreover, the Lemma of Riemann–Lebesgue, Theorem I.3.2.1, implies

$$|\hat{j}_\varepsilon(\xi)| = |\hat{j}(\varepsilon\xi)| \le (2\pi)^{-n/2}\|j\|_{L^1} = (2\pi)^{-n/2},$$

which yields

$$\begin{aligned}\|J_\varepsilon(u)\|^2_{\psi,t} &= (2\pi)^n \int_{\mathbb{R}^n} (1+\psi(\xi))^s |\hat{j}_\varepsilon(\xi)|^2 |\hat{u}(\xi)|^2\, d\xi \\ &\le \int_{\mathbb{R}^n} (1+\psi(\xi))^s |\hat{u}(\xi)|^2\, d\xi = \|u\|^2_{\psi,s}.\end{aligned}$$

In order to show (2.168), observe that

$$\begin{aligned}\|J_\varepsilon(u) - u\|^2_{\psi,t} &= \int_{\mathbb{R}^n} (1+\psi(\xi))^s |(j_\varepsilon * u)^\wedge(\xi) - \hat{u}(\xi)|^2\, d\xi \\ &= (2\pi)^n \int_{\mathbb{R}^n} (1+\psi(\xi))^s |\hat{j}_\varepsilon(\xi) - (2\pi)^{-n/2}|^2 |\hat{u}(\xi)|^2\, d\xi.\end{aligned}$$

Since $\hat{j}_\varepsilon(\xi) = \hat{j}(\varepsilon\xi) \to \hat{j}(0) = (2\pi)^{-n/2}$ and $|\hat{j}_\varepsilon(\xi)| \le (2\pi)^{-n/2}$ for all $\xi \in \mathbb{R}^n$, the dominated convergence theorem implies

$$\lim_{\varepsilon\to 0} \|J_\varepsilon(u) - u\|^2_{\psi,t} = 0.$$

It remains to prove that (2.169) implies $u \in H^{\psi,s}(\mathbb{R}^n)$. From (2.169) it follows that $\left(J_{1/n}(u)\right)_{n>\frac{1}{\rho}}$ converges weakly in $H^{\psi,s}(\mathbb{R}^n)$ to some element $v \in H^{\psi,s}(\mathbb{R}^n)$. By the linearity of the continuous embedding of $H^{\psi,s}(\mathbb{R}^n)$ into $L^2(\mathbb{R}^n)$ it follows that $\left(J_{1/n}(u)\right)_{n>\frac{1}{\rho}}$ converges also weakly in $L^2(\mathbb{R}^n)$ to v. But by Proposition I.2.3.17 we know that $\left(J_{1/n}(u)\right)_{n>\frac{1}{\rho}}$ converges strongly in $L^2(\mathbb{R}^n)$ to u, hence $u = v$ and $u \in H^{\psi,s}(\mathbb{R}^n)$. □

Theorem 2.3.16. *Suppose that $q_2(x,\xi)$ satisfies A.2.m. Further let $s \ge 0$ such that $|s-1|+n+1 < m$. Then we have for all $\varepsilon \in (0,1]$ and all $u \in H^{\psi,s+1}(\mathbb{R}^n)$*

$$\|[J_\varepsilon, q_2(x,D)]u\|_{\psi,s} \le c\|u\|_{\psi,s+1} \tag{2.170}$$

with a constant c independent of $\varepsilon \in (0,1]$.

Proof: First observe that

$$\begin{aligned}&([J_\varepsilon, q_2(x,D)]u)^\wedge(\xi)\\&= (2\pi)^{-n/2}\int_{\mathbb{R}^n} \hat{q}_2(\xi-\eta,\eta)\left(\hat{j}(\varepsilon\xi)-\hat{j}(\varepsilon\eta)\right)\hat{u}(\eta)\,d\eta.\end{aligned}\tag{2.171}$$

Further we claim

$$|\hat{j}(\varepsilon\xi)-\hat{j}(\varepsilon\eta)|\left(1+|\xi|^2\right)^{1/2} \le c^*\left(1+|\xi-\eta|^2\right)^{1/2}\tag{2.172}$$

with c^* independent of $\varepsilon \in (0,1]$. In fact, since $|\hat{j}(\xi)| \le (2\pi)^{-n/2}$ for all $\xi \in \mathbb{R}^n$, we have for $|\xi-\eta| \ge \frac{|\xi|}{2}$

$$\begin{aligned}&|\hat{j}(\varepsilon\xi)-\hat{j}(\varepsilon\eta)|(1+|\xi|^2)^{1/2}\\&\qquad\le 2(2\pi)^{-n/2}\left(1+|\xi|^2\right)^{1/2} \le c'\left(1+|\xi-\eta|^2\right)^{1/2}.\end{aligned}$$

On the other hand, for $|\xi-\eta| \le \frac{|\xi|}{2}$, since $\hat{j} \in S(\mathbb{R}^n)$ and therefore $|\nabla\hat{j}(\xi)| \le \tilde{c}\left(1+|\xi|^2\right)^{-1/2}$, the mean-value theorem yields for $\varepsilon \le \sqrt{2}$

$$\begin{aligned}|\hat{j}(\varepsilon\xi)-\hat{j}(\varepsilon\eta)|\left(1+|\xi|^2\right)^{1/2} &\le |\varepsilon\xi-\varepsilon\eta|\tilde{c}\left(1+\left|\frac{\varepsilon\xi}{2}\right|\right)^{-1/2}\left(1+|\xi|^2\right)^{1/2}\\&\le c''\left(1+|\xi-\eta|^2\right)^{1/2}\left(\frac{1}{\varepsilon^2}+\frac{|\xi|^2}{2}\right)^{-1/2}\left(1+|\xi|^2\right)^{1/2},\end{aligned}$$

and (2.172) is proved. Now we get using (2.170), Lemma 2.3.2.B, (2.172), Peetre's inequality for continuous negative definite functions, and the estimate $((1+\psi(\xi)) \le c_\psi\left(1+|\xi|^2\right)$

$$\begin{aligned}&|\left(1+\psi(\xi)\right)^{s/2}([J_\varepsilon, q_2(x,D)]u)^\wedge(\xi)|\\&=(2\pi)^{-n/2}\left|\int_{\mathbb{R}^n}\hat{q}_2(\xi-\eta,\eta)\left(\hat{j}(\varepsilon\xi)-\hat{j}(\varepsilon\eta)\right)\left(1+\psi(\xi)\right)^{s/2}\hat{u}(\eta)\,d\eta\right|\\&\le c'''\int_{\mathbb{R}^n}\left(1+|\xi-\eta|^2\right)^{-m/2}\left(1+\psi(\eta)\right)\left(1+|\xi-\eta|^2\right)^{1/2}\\&\qquad\times\frac{\left(1+\psi(\xi)\right)^{1/2}}{\left(1+|\xi|^2\right)^{1/2}}\left(1+\psi(\xi)\right)^{\frac{s-1}{2}}|\hat{u}(\xi)|\,d\xi\\&\le c^{**}\int_{\mathbb{R}^n}\left(1+|\xi-\eta|^2\right)^{-\frac{m+1}{2}}\left(1+\psi(\xi-\eta)\right)^{\frac{|s-1|}{2}}\left(1+\psi(\eta)\right)^{\frac{s+1}{2}}|\hat{u}(\eta)|\,d\eta\\&\le\bar{c}\int_{\mathbb{R}^n}\left(1+|\xi-\eta|^2\right)^{-\frac{m+1+|s-1|}{2}}\left(1+\psi(\eta)\right)^{\frac{s+1}{2}}|\hat{u}(\eta)|\,d\eta,\end{aligned}$$

with $\bar{c}$ independent of $\varepsilon \in (0,1]$. Now, we finally get by Young's inequality

$$\begin{aligned}\|[J_\varepsilon, q_2(x,D)]u\|_{\psi,s} &\leq \bar{c}\|(1+|\cdot|^2)^{-\frac{m+1+|s-1|}{2}} * (1+\psi(\cdot))^{\frac{s+1}{2}}|\hat{u}(\cdot)|\|_0 \\ &\leq c\|u\|_{\psi,s+1},\end{aligned}$$

which is independent of $\varepsilon \in (0,1]$. □

Remark 2.3.17. Theorem 2.3.16 was originally proved in [155], we followed here the more polished proof given by W.Hoh in [128].

Corollary 2.3.18. *Suppose that $q_1(\xi)$ satisfies A.1 and that $q_2(x,\xi)$ satisfies A.2.m. Further let $s \geq 0$ such that $|s-1|+n+1 < m$. Then for all $u \in H^{\psi,s+1}(\mathbb{R}^n)$*

$$\|[J_\varepsilon, q(x,D)]u\|_{\psi,s} \leq c\|u\|_{\psi,s+1} \tag{2.173}$$

holds with a constant c independent of $\varepsilon \in (0,1]$.

Proof: Recall that $q(x,D) = q_1(D) + q_2(x,D)$. In order to see (2.173) we only have to note that $[J_\varepsilon, q_1(D)] = 0$ since both J_ε and $q_1(D)$ are translation invariant operators. The rest follows from Theorem 2.3.16. □

Combining the estimates for $q(x,D)$ we can derive the following regularity theorem.

Theorem 2.3.19. *Suppose that $q(x,\xi) = q_1(\xi) + q_2(x,\xi)$ fulfills A.1, A.2.m with $m \geq [s]+n+2$ for some $s \geq 0$ and (2.161). Further suppose that for some $\lambda \in \mathbb{R}$ and $f \in H^{\psi,s}(\mathbb{R}^n)$ we have a solution $u \in H^{\psi,s+1}(\mathbb{R}^n)$ to the equation*

$$q_\lambda(x,D)u = q(x,D)u + \lambda u = f. \tag{2.174}$$

Then it follows that $u \in H^{\psi,s+1}(\mathbb{R}^n)$.

Proof: Using Theorem 2.3.13 we find

$$\begin{aligned}\delta_0\|J_\varepsilon(u)\|_{\psi,s+2} &- \gamma_{\eta,s}\|J_\varepsilon(u)\|_0 - \|\lambda J_\varepsilon(u)\|_{\psi,s} \\ &\leq \|q(x,D)J_\varepsilon(u)\|_{\psi,s} - \|\lambda J_\varepsilon(u)\|_{\psi,s} \\ &\leq \|q(x,D)J_\varepsilon(u) + \lambda J_\varepsilon(u)\|_{\psi,s} \\ &\leq \|J_\varepsilon(q(x,D)u + \lambda u)\|_{\psi,s} + \|[J_\varepsilon, q_2(x,D)](u)\|_{\psi,s} \\ &= \|J_\varepsilon(f)\|_{\psi,s} + \|[J_\varepsilon, q_2(x,D)](u)\|_{\psi,s}.\end{aligned}$$

Thus we get by Theorem 2.3.16

$$\begin{aligned}\delta_0\|J_\varepsilon(u)\|_{\psi,s+2} &\leq \|J_\varepsilon(f)\|_{\psi,s} + \|[J_\varepsilon, q_2(x,D)](u)\|_{\psi,s} \\ &\quad + |\lambda|\|J_\varepsilon(u)\|_{\psi,s} + \gamma_{\eta,s}\|J_\varepsilon(u)\|_{\psi,s} \\ &\leq \|f\|_{\psi,s} + \tilde{c}\|u\|_{\psi,s} + c\|u\|_{\psi,s+1},\end{aligned}$$

with $\tilde{c}$ independent of $\varepsilon \in (0,1]$, and the theorem follows from Proposition 2.3.15. □

Let us introduce the sesquilinear form

$$B(u,v) := (q(x,D)u, v)_0 \tag{2.175}$$

which is associated to $q(x,D)$ and defined on $C_0^\infty(\mathbb{R}^n)$. Clearly we have the decomposition

$$B(u,v) = B^{q_1}(u,v) + B^{q_2}(u,v), \tag{2.176}$$

where $B^{q_1}(u,v) := (q_1(D)u, v)_0$ and $B^{q_2}(u,v) := (q_2(x,D)u, v)_0$.

Proposition 2.3.20. *Suppose that q_1 satisfies A.1. The sesquilinear form B^{q_1} has a continuous extension onto $H^{\psi,1}(\mathbb{R}^n)$ and the following estimates holds:*

$$|B^{q_1}(u,v)| \leq \tau_1\|u\|_{\psi,1}\|v\|_{\psi,1} \tag{2.177}$$

with τ_1 as in (2.143). *Furthermore we have with some $\tilde{\lambda}_0 \geq 0$*

$$|B^{q_1}(u,u)| \geq \operatorname{Re} B^{q_1}(u,u) \geq \gamma_0\|u\|^2_{\psi,1} - \tilde{\lambda}_0\|u\|_0^2. \tag{2.178}$$

Proof: It is sufficient to prove (2.177) and (2.178) for all $u, v \in S(\mathbb{R}^n)$. From (2.143) we deduce

$$\begin{aligned}|B^{q_1}(u,v)| &= \Big|\int_{\mathbb{R}^n} q_1(\xi)\hat{u}(\xi)\overline{\hat{v}(\xi)}\,d\xi\Big| \\ &\leq \tau_1 \int_{\mathbb{R}^n} (1+\psi(\xi))|\hat{u}(\xi)||\hat{v}(\xi)|\,d\xi \\ &\leq \tau_1\|u\|_{\psi,1}\|v\|_{\psi,1},\end{aligned}$$

which proves (2.177). To see (2.178) observe that

$$\begin{aligned}|B(u,u)| \geq \operatorname{Re} B(u,u) &= \int_{\mathbb{R}^n} \operatorname{Re} q_1(\xi)|\hat{u}(\xi)|^2\,d\xi \\ &\geq \gamma_0 \int_{B_1^c(0)} \psi(\xi)|\hat{u}(\xi)|^2\,d\xi + \int_{B_1(0)} \operatorname{Re} q_1(\xi)|\hat{u}(\xi)|^2\,d\xi \\ &\geq \gamma_0\|u\|^2_{\psi,1} - \gamma_0\|u\|_0^2 - \sup_{|\xi|\leq 1} |\operatorname{Re} q_1(\xi) - \gamma_0\psi(\xi)|\;\|u\|_0^2,\end{aligned}$$

and (2.178) is shown. □

Remark 2.3.21. Since $q_1(D)$ maps real-valued functions onto real-valued functions, it follows that for $u \in H^{\psi,1}(\mathbb{R}^n;\mathbb{R})$ we have

$$B^{q_1}(u,u) \geq \gamma_0 \|u\|^2_{\psi,1} - \tilde{\lambda}_0 \|u\|^2_0. \tag{2.179}$$

Next we will estimate B^{q_2}.

Proposition 2.3.22. *Suppose that A.2.m holds for $m \geq n+2$. Then for all $u, v \in H^{\psi,1}(\mathbb{R}^n;\mathbb{R})$ the estimate*

$$|B^{q_2}(u,v)| \leq \kappa_2 \sum_{|\alpha| \leq n+2} \|\varphi_\alpha\|_{L^1} \|u\|_{\psi,1} \|v\|_{\psi,1} \tag{2.180}$$

Proof: As in the proof of Theorem 2.3.11 we find for $u, v \in S(\mathbb{R}^n)$

$$\begin{aligned}
|B^{q_2}(u,v)| &= \left| \int_{\mathbb{R}^n} q_2(x,D)u(x)\overline{v(x)}\,dx \right| \\
&\leq \tilde{\gamma}_{n+2,n} \sum_{|\alpha| \leq n+2} \|\varphi_\alpha\|_{L^1} \\
&\quad \times \int_{\mathbb{R}^n}\int_{\mathbb{R}^n} (1+|\xi-\eta|^2)^{-\frac{(n+2)}{2}} (1+\psi(\eta))|\hat{v}(\eta)||\hat{u}(\xi)|\,d\eta d\xi \\
&\leq \tilde{\gamma}_{n+2,n} \sum_{|\alpha| \leq n+2} \|\varphi_\alpha\|_{L^1} \int_{\mathbb{R}^n}\int_{\mathbb{R}^n} (1+|\xi-\eta|^2)^{-\frac{(n+2)}{2}} \left(\frac{1+\psi(\eta)}{1+\psi(\xi)}\right)^{1/2} \\
&\quad \times (1+\psi(\eta))^{1/2} |\hat{v}(\eta)| (1+\psi(\xi))^{1/2} |\hat{u}(\xi)|\,d\eta d\xi \\
&\leq \sqrt{2}\tilde{\gamma}_{n+2,n} (1 \vee c_\psi)^{1/2} \sum_{|\alpha| \leq n+2} \|\varphi_\alpha\|_{L^1} \\
&\quad \int_{\mathbb{R}^n}\int_{\mathbb{R}^n} (1+|\xi-\eta|^2)^{-\frac{(n+2)}{2}} (1+\psi(\eta))^{1/2} |\hat{v}(\eta)||\hat{u}(\xi)|\,d\eta d\xi \\
&\leq \kappa_2 \sum_{|\alpha| \leq n+2} \|\varphi_\alpha\|_{L^1} \|v\|_{\psi,1} \|u\|_{\psi,1}
\end{aligned}$$

with

$$\kappa_2 = \sqrt{2}(1 \vee c_\psi)^{1/2} \tilde{\gamma}_{n+2,n} \tilde{c}_{n,n+1}. \quad \square \tag{2.181}$$

Combining Proposition 2.3.20 and Proposition 2.3.22 we obtain

Theorem 2.3.23. *Suppose that $q_1(\xi)$ and $q_2(x,\xi)$ satisfy A.1 and A.2.m with $m \geq n+2$. Then B is continuous on $H^{\psi,1}(\mathbb{R}^n)$, i.e. for all $u, v \in H^{\psi,1}(\mathbb{R}^n)$ we have*

$$|B(u,v)| \leq c\|u\|_{\psi,1}\|v\|_{\psi,1}. \tag{2.182}$$

Further we have

Theorem 2.3.24. *Suppose that $q_1(\xi)$ and $q_2(x,\xi)$ satisfy A.1 and A.2.m with $m \geq n+2$. Assume further with κ_2 from* (2.181) *that*

$$\delta_1 := \gamma_0 - \kappa_2 \sum_{|\alpha|\leq n+2} \|\varphi_\alpha\|_{L^1} > 0. \tag{2.183}$$

Then we have for all $u \in H^{\psi,1}(\mathbb{R}^n)$

$$|B(u,u)| \geq \operatorname{Re} B(u,u) \geq \delta_1\|u\|^2_{\psi,1} - \tilde{\lambda}_0\|u\|^2_0, \tag{2.184}$$

where $\tilde{\lambda}_0$ is taken from (2.178).

Proof: Using Proposition 2.3.20 and Proposition 2.3.22 we get for $u \in H^{\psi,1}(\mathbb{R}^n)$

$$\begin{aligned}|B(u,u)| &\geq \operatorname{Re} B(u,u) \geq B^{q_1}(u,u) - |B^{q_2}(u,u)| \\ &\geq \gamma_0\|u\|^2_{\psi,1} - \tilde{\lambda}_0\|u\|^2_0 - \kappa_2 \sum_{|\alpha|\leq n+2} \|\varphi_\alpha\|_{L^1}\|u\|^2_{\psi,1} \\ &= \delta_1\|u\|^2_{\psi,1} - \tilde{\lambda}_0\|u\|^2_0. \quad \square\end{aligned}$$

Corollary 2.3.25. *In the situation of Theorem* 2.3.24 *we have for all $\lambda \geq \tilde{\lambda}_0$*

$$|B_\lambda(u,u)| \geq \operatorname{Re} B_\lambda(u,u) = \operatorname{Re} B(u,u) + \lambda\|u\|^2_{\psi,1} \geq \delta_1\|u\|^2_{\psi,1}. \tag{2.185}$$

Let us suppose that $q_1(\xi)$ and $q_2(x,\xi)$ satisfy A.1 and A.2.m , $m \geq n+2$. Further let $f \in L^2(\mathbb{R}^n)$.

Definition 2.3.26. We call $u \in H^{\psi,1}(\mathbb{R}^n)$ a *variational solution* to the equation

$$q_\lambda(x,D)u = q(x,D)u + \lambda u = f, \quad \lambda \in \mathbb{R}, \tag{2.186}$$

if

$$B_\lambda(u,\varphi) = B(u,\varphi) + \lambda(u,\varphi)_0 = (\varphi, f)_0 = \overline{(f,\varphi)_0} \tag{2.187}$$

holds for all $\varphi \in C_0^\infty(\mathbb{R}^n)$, or equivalently for all $\varphi \in H^{\psi,1}(\mathbb{R}^n)$.

Note that a variational *solution* of (2.186) which belongs to $H^{\psi,2}(\mathbb{R}^n)$ satisfies (2.186) already in the *strong sense*, i.e. $q_\lambda(x,D)u \in L^2(\mathbb{R}^n)$ and (2.186) is an equality in $L^2(\mathbb{R}^n)$. Using the Lax–Milgram theorem, Theorem I.2.7.41, we deduce

Theorem 2.3.27. *Let $q_1(\xi)$ and $q_2(x,\xi)$ satisfy A.1 and A.2.m , $m \geq n+2$, and take $\tilde{\lambda}_0$ from* (2.179). *Moreover assume* (2.183), *i.e.* $\delta_1 > 0$. *Then for every $\lambda \geq \tilde{\lambda}_0$ there exists a unique variational solution $u \in H^{\psi,1}(\mathbb{R}^n)$ to equation* (2.186) *for all $f \in L^2(\mathbb{R}^n)$.*

Proof: First note that by

$$|(\varphi, f)_0| \leq \|f\|_0 \|\varphi\|_0 \leq \|f\| \|\varphi\|_{\psi,1} \tag{2.188}$$

every $f \in L^2(\mathbb{R}^n)$ defines a continuous linear functional on $H^{\psi,1}(\mathbb{R}^n)$. Moreover, Theorem 2.3.23 and Theorem 2.3.24 imply

$$|B_\lambda(u,v)| \leq \|u\|_{\psi,1}\|v\|_{\psi,1}$$

and

$$|B_\lambda(u,u)| \geq \operatorname{Re} B_\lambda(u,u) \geq \delta_1 \|u\|^2_{\psi,1}, \quad \delta_1 > 0.$$

Thus Theorem I.2.7.41 gives the existence of a unique element $u \in H^{\psi,1}(\mathbb{R}^n)$ such that

$$B_\lambda(u,\varphi) = (\varphi, f)_0$$

holds for $\varphi \in C_0^\infty(\mathbb{R}^n)$, which proves the theorem. □

We want to prove that the unique variational solution constructed in Theorem 2.3.27 has more regularity properties.

Theorem 2.3.28. *Let $q(x,D) = q_1(D) + q_2(x,D)$, $u \in H^{\psi,1}(\mathbb{R}^n)$ and $f \in L^2(\mathbb{R}^n)$ be as in Theorem* 2.3.27 *and assume* (2.141) *with $s = 0$. Then u belongs to $H^{\psi,2}(\mathbb{R}^n)$.*

Proof: Denote by J_ε, $\varepsilon \in (0,1]$, the Friedrichs mollifier and let $(u_k)_{k\in\mathbb{N}}$ be a sequence in $C_0^\infty(\mathbb{R}^n)$ converging to u in $H^{\psi,1}(\mathbb{R}^n)$. It follows that

$$\begin{aligned} B_\lambda(J_\varepsilon(u_k),\varphi) &= (q(x,D)J_\varepsilon(u_k) + \lambda J_\varepsilon(u_k), \varphi)_0 \\ &= (J_\varepsilon((q(x,D)+\lambda)u_k), \varphi)_0 - ([J_\varepsilon, q(x,D)]u_k, \varphi)_0 \\ &= B_\lambda(u_k, J_\varepsilon(\varphi)) - ([J_\varepsilon, q_2(x,D)]u_k, \varphi)_0. \end{aligned}$$

From Corollary 2.3.18 we obtain (for k large)

$$\|[J_\varepsilon, q_2(x,D)]u_k\|_0 \le c\|u_k\|_{\psi,1} \le \tilde{c}\|u\|_{\psi,1},$$

implying that $[J_\varepsilon, q_2(x,D)]u_k \to \omega_\varepsilon$ in $L^2(\mathbb{R}^n)$ for some $\omega_\varepsilon \in L^2(\mathbb{R}^n)$ and $\|\omega_\varepsilon\|_0 \le \tilde{c}$ for $\varepsilon \in (0,1]$ with $\tilde{c}$ independent of ε. Thus for $k \to \infty$ we obtain

$$\begin{aligned} B_\lambda(J_\varepsilon(u),\varphi) &= B_\lambda(u, J_\varepsilon(\varphi)) - (\omega_\varepsilon,\varphi)_0 \\ &= (J_\varepsilon(\varphi), f)_0 - (\omega_\varepsilon,\varphi)_0 = (\varphi, J_\varepsilon(f))_0 - (\omega_\varepsilon,\varphi)_0. \end{aligned}$$

It follows that

$$\|q_\lambda(x,D)J_\varepsilon(u)\|_0 \le \|J_\varepsilon(f)\|_0 + \|\omega_\varepsilon\|_0 \le \|f\|_0 + \tilde{c}$$

or

$$\|q_\lambda(x,D)J_\varepsilon(u))\|_0 \le |\lambda|\|u\|_0 + \|f\|_0 + \tilde{c},$$

which implies by Theorem 2.3.13 that

$$\|J_\varepsilon(u)\|_{\psi,2} \le c$$

for all $\varepsilon \in (0,1]$ with c independent of $\varepsilon \in (0,1]$. Thus we have $u \in H^{\psi,2}(\mathbb{R}^n)$ by Proposition 2.3.15. □

Remark 2.3.29. Theorem 2.3.28 was proved in [155], but we borrowed the more polished version from W.Hoh [128].

Finally we have by Theorem 2.3.19:

Theorem 2.3.30. *Suppose that* $q_1(\xi)$ *and* $q_2(x,\xi)$ *satisfy A.1 and A.2.m,* $m \ge [t] + n + 2$, $t \ge 0$. *Moreover assume* (2.161) *as well as* (2.183) *(with t instead of s). If* $\lambda \ge \tilde{\lambda}_0$, $\tilde{\lambda}_0$ *taken from* (2.178), *and* $f \in H^{\psi,t}(\mathbb{R}^n)$, *then there exists a unique variational solution* $u \in H^{\psi,t+2}(\mathbb{R}^n)$ *to* (2.186).

Note that Theorem 2.3.30 tells that every variational solution to (2.186) is already a strong solution.

Finally in this section we will prove a commutator estimate for the operator $q_1(D)$ and elements of $B^s_{\psi,1}(\mathbb{R}^n)$.

Theorem 2.3.31. *Suppose that* q_1 *fulfills A.1 and let* $a \in B^{s+1}_{\psi,1}(\mathbb{R}^n)$. *Then the commutator*

$$[a, q_1(D)]u = aq_1(D)u - q_1(D)(au) \tag{2.189}$$

is defined on $H^{\psi,s}(\mathbb{R}^n)$, $s \geq 1$, *and maps the space* $H^{\psi,s}(\mathbb{R}^n)$ *continuously into* $H^{\psi,s-1}(\mathbb{R}^n)$. *Moreover we have the estimate*

$$\|[a, q_1(D)]u\|_{\psi,s-1} \leq c_{n,s,\psi}\|a\|_{\psi,s+1,1}\|u\|_{\psi,s}. \tag{2.190}$$

Proof: Of course it is sufficient to prove (2.190) for all $u \in S(\mathbb{R}^n)$. First note that

$$\begin{aligned}([a, q_1(D)]u)^\wedge(\xi) &= (aq_1(D)u)^\wedge(\xi) - (q_1(D)(au))^\wedge(\xi) \\ &= (2\pi)^{-n/2}\int_{\mathbb{R}^n} \hat{a}(\xi-\eta)(q_1(\eta) - q_1(\xi))\hat{u}(\eta)\,d\eta.\end{aligned}$$

Further, we have by Lemma I.3.6.21 and the generalization of Peetre's inequality to continuous negative definite function, Lemma I.3.6.23,

$$\begin{aligned}|q_1(\eta) - q_1(\xi)| &= |q_1^{1/2}(\eta) - q_1^{1/2}(\xi)||q_1^{1/2}(\eta) + q_1^{1/2}(\xi)| \\ &\leq q_1^{1/2}(\xi-\eta)\left(2^{1/2}\left(1 + q_1^{1/2}(\xi-\eta)\right) + 1\right)\left(1 + q_1^{1/2}(\eta)\right) \\ &\leq (1 + q_1(\xi-\eta))\,(1 + q_1(\eta))^{1/2}.\end{aligned}$$

Thus it follows that

$$\begin{aligned}&(1+\psi(\xi))^{\frac{s-1}{2}}|\,([a, q_1(D)]u)^\wedge(\xi)| \\ &\quad\leq c_{s,n,\psi}\int_{\mathbb{R}^n} |\hat{u}(\xi-\eta)|\,(1+\psi(\xi-\eta))^{\frac{s+1}{2}}\,(1+\psi(\eta))^{s/2}\,|\hat{u}(\eta)|\,d\eta.\end{aligned}$$

Now, $\hat{a}(\cdot)\,(1+\psi(\cdot))^{\frac{s+1}{2}} \in L^1(\mathbb{R}^n)$ by our assumptions and $(1+\psi(\cdot))^{s/2}\,|\hat{u}(\cdot)| \in L^1(\mathbb{R}^n)$ since $u \in S(\mathbb{R}^n)$ (or $H^{\psi,s}(\mathbb{R}^n)$). Hence, Young's inequality implies

$$\begin{aligned}\|[a, q_1(D)]u\|_{\psi,s-1} &= \|(1+\psi(\cdot))^{\frac{s-1}{2}}\,|([a, q_1(D)]u)^\wedge|\|_0 \\ &\leq c_{s,n,\psi}\|\hat{a}(\cdot)\,(1+\psi(\cdot))^{\frac{s+1}{2}}\,\|_{L^1}\|\,(1+\psi(\cdot))^{s/2}\,|\hat{u}(\cdot)|\|_0 \\ &= c_{s,n,\psi}\|a\|_{\psi,s+1,1}\|u\|_{\psi,s},\end{aligned}$$

and the theorem is proved. □

2.4 Hoh's Symbolic Calculus for Pseudo-Differential Operators with Negative Definite Symbols

In the last section we solved the equation $q_\lambda(x, D)u = f$, $\lambda \in \mathbb{R}$ sufficiently large, for a pseudo-differential operator with a negative definite symbol $q(x, \xi)$

having the decomposition $q(x,\xi) = q_1(\xi) + q_2(x,\xi)$ with $q_2(x,\xi)$ being a small perturbation of $q_1(\xi)$. So far it was not possible to get rid of this smallness condition since the application of suitable partition of unity as in Section 2.1 fails due to the *non-locality* of the operators $q_1(D)$ and $q_2(x,D)$. In particular, for $\varphi_1, \varphi_2 \in C_0^\infty(\mathbb{R}^n)$ such that $\operatorname{supp}\varphi_1 \cap \operatorname{supp}\varphi_2 = \varnothing$ it does not follow (in general) that $(\varphi_1 q_1(D)\varphi_2)(x) = 0$ or $(\varphi_1 q_2(x,D)\varphi_2)(x) = 0$ for all $x \in \mathbb{R}^n$.

Another way of solving the equation $q_\lambda(x,D)u = f$ is to construct a parametrix for $q(x,D)$ or $q_\lambda(x,D)$. In fact this was the central idea when introducing (elliptic) pseudo-differential operators, compare [179] or [137]. However, typical continuous negative definite functions do not fulfill typical conditions posed on 'classical' symbols. It was W.Hoh, see [126], [127] and [128] who introduced for large classes of negative definite symbols a corresponding symbolic calculus which enables us to construct parametrices for operators corresponding to 'elliptic' symbols. In this section we present this symbolic calculus following closely the papers W.Hoh [127] and [128]. We start with recalling Theorem I.3.7.13:

Let $\psi : \mathbb{R}^n \to \mathbb{R}$ be a continuous negative definite function with Lévy–Khinchin representation

$$\psi(\xi) = c + Q(\xi) + \int_{\mathbb{R}^n\setminus\{0\}} (1 - \cos x\cdot\xi)\,\nu(dx). \tag{2.191}$$

Suppose that for $2 \le l \le m$ all absolute moments of the measure ν exist, i.e.

$$M_l := \int_{\mathbb{R}^n\setminus\{0\}} |x|^l\,\nu(dx) < \infty, \quad 2 \le l \le m. \tag{2.192}$$

Then ψ is of class $C^m(\mathbb{R}^n;\mathbb{R})$ and for $\alpha \in \mathbb{N}_0^n$, $|\alpha| \le m$, we have the estimate

$$|\partial_\xi^\alpha \psi(\xi)| \le c_{|\alpha|} \begin{cases} \psi(\xi), & \alpha = 0 \\ \psi^{1/2}(\xi), & |\alpha| = 1\,, \\ 1, & |\alpha| \ge 2 \end{cases} \tag{2.193}$$

with $c_0 = 1$, $c_1 = (2M_2)^{1/2} + 2\lambda^{1/2}$, $c_2 = M_2 + 2\lambda$ and $c_l = M_l$, $2 \le l \le m$, where λ is the maximal eigenvalue of the quadratic form given in (2.191).

Remark 2.4.1. Note that for a measure ν with compact support in $\mathbb{R}^n\setminus\{0\}$, (2.192) holds for all $k \in \mathbb{N}$, $k \ge 2$.

We will often need the subadditive function

$$\begin{aligned} \rho &: \mathbb{N}_0 \to \mathbb{N}_0 \\ k &\mapsto \rho(k) := k \wedge 2. \end{aligned} \tag{2.194}$$

Lemma 2.4.2. *Suppose that the continuous negative definite function $\psi : \mathbb{R}^n \to \mathbb{R}$ satisfies* (2.192) *for all $k \in \mathbb{N}$. For all $m \in \mathbb{R}$ and all $\alpha \in \mathbb{N}_0^n$ we have*

$$|\partial_\xi^\alpha (1+\psi(\xi))^{m/2}| \le c_{|\alpha|}(1+\psi(\xi))^{\frac{m-\rho(|\alpha|)}{2}}. \tag{2.195}$$

Proof: From (2.193) it follows that

$$|\partial_\xi^\alpha (1+\psi(\xi))| \le c_{|\alpha|}(1+\psi(\xi))^{\frac{2-\rho(|\alpha|)}{2}} \tag{2.196}$$

holds for all $\alpha \in \mathbb{N}_0^n$ which yields

$$\frac{|\partial_\xi^\alpha (1+\psi(\xi))|}{(1+\psi(\xi))} \le c_{|\alpha|}(1+\psi(\xi))^{\frac{-\rho(|\alpha|)}{2}}. \tag{2.197}$$

Using formula (I.2.25) we get

$$\begin{aligned} &\partial_\xi^\alpha (1+\psi(\xi))^{m/2} \\ &\qquad = (1+\psi(\xi))^{m/2} \sum_{\beta_1+\cdots+\beta_{|\alpha|}=\alpha} c_{\beta_1,\ldots,\beta_{|\alpha|},m} \prod_{j=1}^{|\alpha|} \frac{\partial_\xi^{\beta_j}(1+\psi(\xi))}{(1+\psi(\xi))}, \end{aligned}$$

which gives

$$\begin{aligned} &|\partial_\xi^\alpha (1+\psi(\xi))^{m/2}| \\ &\qquad \le c_{|\alpha|}\,(1+\psi(\xi))^{m/2} \sum_{\beta_1+\cdots+\beta_{|\alpha|}=\alpha} \prod_{j=1}^{|\alpha|} (1+\psi(\xi))^{-\rho(|\beta_j|)} \\ &\qquad = c_{|\alpha|}\,(1+\psi(\xi))^{m/2} \sum_{\beta_1+\cdots+\beta_{|\alpha|}=\alpha} (1+\psi(\xi))^{-\sum_{j=1}^{|\alpha|}\rho(|\beta_j|)} \\ &\qquad \le \tilde{c}_{|\alpha|}\,(1+\psi(\xi))^{-\frac{m-\rho(|\alpha|)}{2}}, \end{aligned}$$

where we used the subadditivity of ρ. $\square$

Definition 2.4.3. We say that a continuous negative definite function $\psi : \mathbb{R}^n \to \mathbb{R}$ belongs to the *class* Λ if it satisfies (2.196) for all $\alpha \in \mathbb{N}_0^n$.

Definition 2.4.4. A. Let $m \in \mathbb{R}$ and $\psi \in \Lambda$. We call a C^∞-function $q : \mathbb{R}^n \times \mathbb{R}^n \to \mathbb{C}$ a *symbol* in the *class* $S_\rho^{m,\psi}(\mathbb{R}^n)$ if for all $\alpha, \beta \in \mathbb{N}_0^n$ there are constants $c_{\alpha,\beta} \ge 0$ such that

$$|\partial_\xi^\alpha \partial_x^\beta q(x,\xi)| \le c_{\alpha\beta}\,(1+\psi(\xi))^{\frac{m-\rho(|\alpha|)}{2}} \tag{2.198}$$

holds for all $x \in \mathbb{R}^n$ and $\xi \in \mathbb{R}^n$. We call $m \in \mathbb{R}$ the order of the symbol $q(x, \xi)$.

B. Let $\psi \in \Lambda$ and suppose that for a C^∞- function $q : \mathbb{R}^n \times \mathbb{R}^n \to \mathbb{C}$ the estimate

$$|\partial_\xi^\alpha \partial_x^\beta q(x, \xi)| \leq \tilde{c}_{\alpha\beta} \left(1 + \psi(\xi)\right)^{m/2} \tag{2.199}$$

holds for all $\alpha, \beta \in \mathbb{N}_0^n$ and $x \in \mathbb{R}^n$ and $\xi \in \mathbb{R}^n$. In this case we call q a *symbol* of the *class* $S_0^{m,\psi}(\mathbb{R}^n)$.

Clearly we have $S_\rho^{m,\psi}(\mathbb{R}^n) \subset S_0^{m,\psi}(\mathbb{R}^n)$. First of all we want to give examples for symbols in the class $S_\rho^{m,\psi}(\mathbb{R}^n)$.

Example 2.4.5. From Lemma 2.4.2 it follows immediately that for $\psi \in \Lambda$ the function $\xi \mapsto (1 + \psi(\xi))^{m/2}$ belongs to $S_\rho^{m,\psi}(\mathbb{R}^n)$.

Example 2.4.6. Let ν be a symmetric measure on $\mathbb{R}^n \backslash \{0\}$, i.e. $\nu(B) = \nu(-B)$, such that the function

$$\tilde{\psi}(\xi) := \int_{\mathbb{R}^n \backslash \{0\}} (1 - \cos(y \cdot \xi)) \; \nu(dy) \tag{2.200}$$

is a continuous negative definite function. Clearly the function

$$\psi(\xi) := \int\limits_{0<|y|\leq 1} (1 - \cos(y \cdot \xi)) \; \nu(dy) \tag{2.201}$$

is continuous and negative definite too and the measure $\chi_{\overline{B_1(0)}\backslash\{0\}}(y)\nu(dy)$ has all absolute moments of order greater or equal to 2. Further let $f : \mathbb{R}^n \times \mathbb{R}^n \to \mathbb{R}$, $f(x, y) \geq 0$, be a bounded measurable function such that for all $\alpha \in \mathbb{N}_0^n$ the functions $(x, y) \mapsto \partial_x^\alpha f(x, y)$ are bounded. We define the kernel

$$\mu(x, dy) := \chi_{\overline{B_1(0)}\backslash\{0\}}(y) f(x, y) \nu(dy) \tag{2.202}$$

and consider the symbol

$$\begin{aligned} q(x, \xi) :&= \int_{\mathbb{R}^n \backslash \{0\}} (1 - \cos(y \cdot \xi)) \; \mu(x, dy) \\ &= \int_{0<|y|\leq 1} (1 - \cos(y \cdot \xi)) \, f(x, y) \, \nu(dy). \end{aligned} \tag{2.203}$$

First of all it follows that $\xi \mapsto q(x,\xi)$ is a continuous negative definite function. Furthermore, for $\beta \in \mathbb{N}_0^n$ we find

$$
\begin{aligned}
|\partial_x^\beta q(x,\xi)| &\le \left| \int_{0<|y|\le 1} (1-\cos(y\cdot\xi))\, \partial_x^\beta f(x,y)\, \nu(dy) \right| \\
&\le \|\partial_x^\beta f\|_\infty \int_{0<|y|\le 1} (1-\cos(y\cdot\xi))\, \nu(dy) = c_\beta\,(1+\psi(\xi))\,.
\end{aligned}
$$

In addition, for $\alpha \in \mathbb{N}_0^n$ we get

$$
\begin{aligned}
|\partial_\xi^\alpha \partial_x^\beta q(x,\xi)| &\le \left| \partial_\xi^\alpha \partial_x^\beta \int_{0<|y|\le 1} (1-\cos(y\cdot\xi))\, f(x,y)\, \nu(dy) \right| \\
&\le \|\partial_x^\beta f\|_\infty \Big(\int_{0<|y|\le 1} |\partial_\xi^\alpha (1-\cos(y\cdot\xi))|\, \nu(dy) \Big) \\
&= \|\partial_x^\beta f\|_\infty \int_{0<|y|\le 1} |y^\alpha \cos^{(\alpha)}(y\cdot\xi)|\, \nu(dy) \\
&\le c_\beta c_\alpha\,(1+\psi(\xi))^{\frac{2-\rho(|\alpha|)}{2}}\,,
\end{aligned}
$$

where we argued as in the proof of Theorem I.3.7.13. Thus we have proved $q \in S_\rho^{2,\psi}(\mathbb{R}^n)$ with ψ defined by (2.201).

Example 2.4.7. Let $f : (0,\infty) \to \mathbb{R}$ be a Bernstein function satisfying $f(s) \le cs^r$ for $s \ge 1$ and some $r \in (0,1]$. Further let $\psi \in \Lambda$ and $q \in S_\rho^{m,\psi}(\mathbb{R}^n)$, $m \ge 2$, be a symbol satisfying with $\gamma_0 > 0$

$$
q(x,\xi) \ge \gamma_0\,(1+\psi(\xi))^{m/2}\,. \tag{2.204}
$$

Then we have $f \circ q \in S_\rho^{m\cdot r,\psi}(\mathbb{R}^n)$. Indeed , for any function $g \in C^\infty(\mathbb{R}^{2n})$ and any $\gamma \in \mathbb{N}_0^{2n}$ we have, compare (I.2.25),

$$
\partial^\gamma (f\circ g) = \sum_{j=1}^{|\gamma|} f^{(j)}\circ g \sum_{\substack{\gamma_1+\cdots+\gamma_j=\gamma \\ \gamma_1,\ldots,\gamma_j \in \mathbb{N}_0^n}} c_{\gamma_1,\ldots,\gamma_j} \prod_{l=1}^{j} \partial^{\gamma_l} g. \tag{2.205}
$$

Now, for $\alpha,\beta \in \mathbb{N}_0^n$ let $\gamma = (\alpha,\beta) \in \mathbb{N}_0^{2n}$ and put $g = q(x,\xi)$. In addition recall Lemma I.3.9.34.D which says that for the Bernstein function f the estimates

$$
|f^{(j)}(s)| \le \frac{j!}{s^j} f(s), \quad j \in \mathbb{N}_0,
$$

hold. Thus (2.205) yields

$$
\begin{aligned}
|\partial_\xi^\alpha \partial_x^\beta (f\circ q)(x,\xi)| &\le c\sum_{j=1}^{|\gamma|} |f^{(j)}(q(x,\xi))| \sum_{\substack{\alpha_1+\cdots+\alpha_j=\alpha\\ \beta_1+\cdots+\beta_j=\beta}} \prod_{l=1}^{j} \left|\partial_\xi^{\alpha_l}\partial_x^{\beta_l} q(x,\xi)\right| \\
&\le c\sum_{j=1}^{|\gamma|} j! f(q(x,\xi)) \sum_{\substack{\alpha_1+\cdots+\alpha_j=\alpha\\ \beta_1+\cdots+\beta_j=\beta}} \prod_{l=1}^{j} \left|\frac{\partial_\xi^{\alpha_l}\partial_x^{\beta_l} q(x,\xi)}{q(x,\xi)}\right| \\
&\le cf(1+q(x,\xi))\sum_{j=1}^{|\gamma|} \sum_{\substack{\alpha_1+\cdots+\alpha_j=\alpha\\ \beta_1+\cdots+\beta_j=\beta}} \prod_{l=1}^{j} (1+\psi(\xi))^{-\frac{1}{2}\rho(|\alpha_l|)} \\
&\le \tilde{c}_{\alpha\beta}(1+q(x,\xi))^r\,(1+\psi(\xi))^{-\rho(|\alpha|)/2},
\end{aligned}
$$

where we used again the subadditivity of ρ and also the monotonicity of f. Hence we get

$$
\begin{aligned}
|\partial_\xi^\alpha \partial_x^\beta (f\circ q)(x,\xi)| &\le c_{\alpha,\beta}\,(1+\psi(\xi))^{\frac{m}{2}\cdot r}\,(1+\psi(\xi))^{-\frac{\rho(|\alpha|)}{2}} \\
&= c_{\alpha,\beta}\,(1+\psi(\xi))^{\frac{m\cdot r-\rho(|\alpha|)}{2}},
\end{aligned}
$$

i.e. $f\circ q \in S_\rho^{m\cdot r,\psi}(\mathbb{R}^n)$.

Example 2.4.8. Obviously the continuous negative definite function $\xi \mapsto |\xi|^2$, $\xi\in\mathbb{R}^n$, belongs to the class Λ. Now by definition a *classical symbol* $q \in S_{1,0}^m(\mathbb{R}^n)$, $m\in\mathbb{R}$, is an arbitrarily often differentiable function $q:\mathbb{R}^n\times\mathbb{R}^n\to\mathbb{C}$ such that

$$
|\partial_\xi^\alpha \partial_x^\beta q(x,\xi)| \le c_{\alpha,\beta}\left(1+|\xi|^2\right)^{\frac{m-|\alpha|}{2}} \tag{2.206}
$$

holds for all $\alpha,\beta\in\mathbb{N}_0^n$ and $x\in\mathbb{R}^n$, $\xi\in\mathbb{R}^n$. Since for all $m\in\mathbb{R}$

$$
\left(1+|\xi|^2\right)^{\frac{m-|\alpha|}{2}} \le \left(1+|\xi|^2\right)^{\frac{m-\rho(|\alpha|)}{2}},
$$

it follows that $S_{1,0}^m(\mathbb{R}^n)\subset S_\rho^{m,|\cdot|^2}(\mathbb{R}^n)$.

Lemma 2.4.9. *Let $\psi\in\Lambda$.*

A. The sets $S_\rho^{m,\psi}(\mathbb{R}^n)$ and $S_0^{m,\psi}(\mathbb{R}^n)$ are vector spaces.

B. For $q_1\in S_\rho^{m_1,\psi}(\mathbb{R}^n)$ and $q_2\in S_\rho^{m_2,\psi}(\mathbb{R}^n)$ it follows that $q_1\cdot q_2\in S_\rho^{m_1+m_2,\psi}(\mathbb{R}^n)$. Thus $\bigcup_{m\in\mathbb{R}} S_\rho^{m,\psi}(\mathbb{R}^n)$ is a graded algebra (which respects the order of the symbols).

Proof: Part A is trivial. In order to see Part B, we apply Leibniz' rule to find

$$
\begin{aligned}
|\partial_\xi^\alpha \partial_x^\beta (q_1 \cdot q_2)(x,\xi)| &\leq \sum_{\substack{\alpha'+\alpha''=\alpha \\ \beta'+\beta''=\beta}} |\partial_\xi^{\alpha'} \partial_x^{\beta'} q_1(x,\xi)||\partial_\xi^{\alpha''} \partial_x^{\beta''} q_2(x,\xi)| \\
&\leq \tilde{c} \sum_{\substack{\alpha'+\alpha''=\alpha \\ \beta'+\beta''=\beta}} (1+\psi(\xi))^{\frac{m_1-\rho(|\alpha'|)}{2}} (1+\psi(\xi))^{\frac{m_2-\rho(|\alpha''|)}{2}},
\end{aligned}
$$

and the subadditivity of ρ gives

$$
|\partial_\xi^\alpha \partial_x^\beta (q_1 \cdot q_2)(x,\xi)| \leq c' \, (1+\psi(\xi))^{\frac{m_2+m_2-\rho(|\alpha|)}{2}}, \tag{2.207}
$$

which proves the lemma. □

Note that Part B of Lemma 2.4.9 holds of course also for $S_0^{m_j,\psi}(\mathbb{R}^n)$, $j=1,2$.

Next we introduce pseudo-differential operators associated with the symbol classes $S_\rho^{m,\psi}(\mathbb{R}^n)$ and $S_0^{m_1,\psi}(\mathbb{R}^n)$, respectively.

Definition 2.4.10. For $q \in S_\rho^{m,\psi}(\mathbb{R}^n)$ or $q \in S_0^{m,\psi}(\mathbb{R}^n)$ we define on $S(\mathbb{R}^n)$ the *pseudo-differential operator* $q(x,D)$ by

$$
q(x,D)u(x) := (2\pi)^{-n/2} \int_{\mathbb{R}^n} e^{ix\cdot\xi} q(x,\xi)\hat{u}(\xi)\, d\xi. \tag{2.208}
$$

The classes of these operators are denoted by $\Psi_\rho^{m,\psi}(\mathbb{R}^n)$ and $\Psi_0^{m,\psi}(\mathbb{R}^n)$, respectively. Obviously we have the inclusion $\Psi_\rho^{m,\psi}(\mathbb{R}^n) \subset \Psi_0^{m,\psi}(\mathbb{R}^n)$.

Since for fixed $x \in \mathbb{R}^n$ the function $\xi \mapsto q(x,\xi)$ is polynomially bounded, it follows that $q(x,D)u(x)$ is well defined for $u \in S(\mathbb{R}^n)$.

Theorem 2.4.11. *Let $q \in S_0^{m_1,\psi}(\mathbb{R}^n)$. Then $q(x,D)$ maps $S(\mathbb{R}^n)$ continuously into itself. In particular $q(x,D) \in \Psi_\rho^{m,\psi}(\mathbb{R}^n)$ is continuous on $S(\mathbb{R}^n)$.*

Proof: For $\alpha, \beta \in \mathbb{N}_0^n$ take $N > |\beta| + m + n$. It follows that

$$
\begin{aligned}
&\left|\partial_x^\beta \left(x^\alpha (2\pi)^{(-n/2)} \int_{\mathbb{R}^n} e^{ix\cdot\xi} q(x,\xi)\hat{u}(\xi)\, d\xi\right)\right| \\
&\quad \leq (2\pi)^{(-n/2)} \left|\partial_x^\beta \int_{\mathbb{R}^n} e^{ix\cdot\xi} D_\xi^\alpha \left(q(x,\xi)\hat{u}(\xi)\right) d\xi\right|
\end{aligned}
$$

$$= (2\pi)^{(-n/2)} \Big| \int_{\mathbb{R}^n} \sum_{\substack{\alpha'+\alpha''=\alpha \\ \beta'+\beta''=\beta}} \binom{\alpha}{\alpha'} \binom{\beta}{\beta'} (i\xi)^{\beta'} e^{ix\cdot\xi}$$

$$\times \partial_x^{\beta''} D_\xi^{\alpha'} q(x,\xi) D_\xi^{\alpha''} \hat{u}(\xi)\, d\xi \Big|$$

$$\leq c \int_{\mathbb{R}^n} \left(1+|\xi|^2\right)^{|\beta|/2} (1+\psi(\xi))^{m/2} \sum_{|\gamma|\leq|\alpha|} |\partial_\xi^\gamma \hat{u}(\xi)|\, d\xi$$

$$\leq c' \int_{\mathbb{R}^n} \left(1+|\xi|^2\right)^{\frac{|\beta|+m-N}{2}} d\xi \sup_{\xi\in\mathbb{R}^n} \Big(\left(1+|\xi|^2\right)^{N/2} \sum_{|\gamma|\leq|\alpha|} |\partial_\xi^\gamma \hat{u}(\xi)| \Big),$$

where we used the estimate $(1+\psi(\xi)) \leq c_\psi \left(1+|\xi|^2\right)$. With the notation of Section I.2.6 we arrive at

$$p_{\alpha,\beta}(q(x,D)u) \leq c\, p_{N,|\alpha|}(\hat{u}). \tag{2.209}$$

Since the Fourier transform maps $S(\mathbb{R}^n)$ continuously into itself, (2.209) implies the theorem. □

To handle elements of $\Psi_0^{m,\psi}(\mathbb{R}^n)$it is convenient to introduce double symbols and oscillatory integrals.

Definition 2.4.12. A. The *class* $\mathcal{A}$ consists of all functions $a \in C^\infty(\mathbb{R}^n \times \mathbb{R}^n)$ such that for all $\alpha, \beta \in \mathbb{N}_0^n$ and all $\eta \in \mathbb{R}^n$, $y \in \mathbb{R}^n$ the estimate

$$|\partial_\eta^\alpha \partial_y^\beta a(\eta,y)| \leq c_{\alpha\beta}(1+|\eta|^2)^{\frac{m+\delta|\beta|}{2}}(1+|y|^2)^{\tau/2} \tag{2.210}$$

holds for suitable $m \in \mathbb{R}$, $\delta \in [0,1)$ and $\tau \geq 0$.

B. For $a \in \mathcal{A}$ and $\chi \in S(\mathbb{R}^n \times \mathbb{R}^n)$ such that $\chi(0,0) = 1$ we define the *oscillatory integral* by

$$O_s - \int_{\mathbb{R}^n} \int_{\mathbb{R}^n} e^{-iy\cdot\eta} a(\eta,y)\, dy d\eta := \lim_{\varepsilon\to 0} \int_{\mathbb{R}^n} \int_{\mathbb{R}^n} e^{-iy\cdot\eta} \chi(\varepsilon\eta, \varepsilon y) a(\eta,y)\, dy d\eta. \tag{2.211}$$

Lemma 2.4.13. *Let* $\chi \in S(\mathbb{R}^n)$ *such that* $\chi(0) = 1$. *Then we have uniformly on compact sets*

$$\chi(\varepsilon x) \to 1 \quad \text{as } \varepsilon \to 0, \tag{2.212}$$

and for all $\alpha \in \mathbb{N}_0^n \setminus \{0\}$ *we have uniformly on* $\mathbb{R}^n$

$$\partial_x^\alpha \chi(\varepsilon x) \to 0 \quad \text{as } \varepsilon \to 0. \tag{2.213}$$

In addition, with a constant c_α independent of $\varepsilon \in (0,1)$ we have for $0 \leq \sigma \leq |\alpha|$

$$|\partial_x^\alpha \chi(\varepsilon x)| \leq c_\alpha \varepsilon^\sigma (1+|x|^2)^{-\frac{(|\alpha|-\sigma)}{2}}. \tag{2.214}$$

Proof: Since $\partial_x^\alpha \chi(\varepsilon x) = \varepsilon^{|\alpha|} (\partial_y \chi(y))|_{y=\varepsilon x}$, (2.212) and (2.213) follow immediately. Now, write

$$|\partial_x^\alpha \chi(\varepsilon x)| = \varepsilon^\sigma \left(\varepsilon^{(|\alpha|-\sigma)} |\partial_y^\alpha \chi(y)|_{y=\varepsilon x}| \right),$$

which implies (2.214) for $|x| \leq 1$. But for $|x| > 1$ we find for $0 \leq \sigma \leq |\alpha|$

$$\begin{aligned} \varepsilon^{(|\alpha|-\sigma)} |\partial_y^\alpha \chi(y)|_{y=\varepsilon x}| &= |y|^{(|\alpha|-\sigma)} |\partial_y^\alpha \chi(y)|_{y=\varepsilon x}| |x|^{-(|\alpha|-\sigma)} \\ &\leq c_\alpha \left(1+|x|^2\right)^{-\frac{(|\alpha|-\sigma)}{2}}, \end{aligned}$$

and the lemma is proved. □

Theorem 2.4.14. *For $a \in \mathcal{A}$ the oscillatory integral* (2.211) *exists and is independent of the choice of χ.*

Proof: By our assumptions there exists $m \in \mathbb{R}$, $\delta \in [0,1)$ and $\tau \geq 0$ such that

$$|\partial_\eta^\alpha \partial_y^\beta a(\eta, y)| \leq c_{\alpha\beta} (1+|\eta|^2)^{\frac{m+\delta|\beta|}{2}} (1+|y|^2)^{\tau/2}$$

holds for all $\alpha, \beta \in \mathbb{N}_0^n$ and $y, \eta \in \mathbb{R}^n$. We choose $l, l' \in \mathbb{N}$ such that

$$-2l(1-\delta)+m < -n \quad \text{and } -2l' + \tau < -n. \tag{2.215}$$

It follows that

$$(\eta, y) \mapsto \left| \left(1+|y|^2\right)^{-l'} (1-\Delta_\eta)^{l'} \left\{ \left(1+|\eta|^2\right)^{-l} (1-\Delta_y)^l a(\eta, y) \right\} \right|$$

is an element of $L^1(\mathbb{R}^n \times \mathbb{R}^n)$. Since

$$\left(1+|\eta|^2\right)^{-l} (1-\Delta_y)^l e^{-iy\cdot\eta} = e^{-iy\cdot\eta}$$

and

$$\left(1+|y|^2\right)^{-l'} (1-\Delta_\eta)^{l'} e^{-iy\cdot\eta} = e^{-iy\cdot\eta},$$

we find by integrating by parts

$$\begin{aligned} I_\varepsilon &:= \int_{\mathbb{R}^n}\int_{\mathbb{R}^n} e^{-iy\cdot\eta}\chi(\varepsilon\eta,\varepsilon y)a(\eta,y)\,dyd\eta \\ &= \int_{\mathbb{R}^n}\int_{\mathbb{R}^n} e^{-iy\cdot\eta}\left(1+|\eta|^2\right)^{-l}(1-\Delta_y)^l\left(\chi(\varepsilon\eta,\varepsilon y)a(\eta,y)\right)dyd\eta \\ &= \int_{\mathbb{R}^n}\int_{\mathbb{R}^n} e^{-iy\cdot\eta}\left(1+|y|^2\right)^{-l'} \\ &\quad\times(1-\Delta_\eta)^{l'}\Big(\left(1+|\eta|^2\right)^{-l}(1-\Delta_y)^l\left(\chi(\varepsilon\eta,\varepsilon y)a(\eta,y)\right)\Big)\,dyd\eta. \end{aligned}$$

On the other hand, Lemma 2.4.13 implies with $c_{\alpha,\beta}$ independent of $0<\varepsilon<1$

$$|\partial_\eta^\alpha\partial_y^\beta\left(\chi(\varepsilon\eta,\varepsilon y)a(\eta,y)\right)| \le c_{\alpha,\beta}|a|_{|\alpha+\beta|}\left(1+|\eta|^2\right)^{\frac{m+\delta|\beta|}{2}}\left(1+|y|^2\right)^{\tau/2}, \tag{2.216}$$

where

$$|a|_k := \max_{|\alpha+\beta|\le k}\ \sup_{\eta,y\in\mathbb{R}^n}\{|\partial_\eta^\alpha\partial_y^\beta a(\eta,y)|\left(1+|\eta|^2\right)^{-\frac{m+\delta|\beta|}{2}}\left(1+|y|^2\right)^{-\frac{\tau}{2}}\}.$$

Since

$$|\partial_\eta^\alpha(1+|\eta|^2)^{s/2}| \le c_{s,\alpha}(1+|\eta|^2)^{\frac{s-|\alpha|}{2}} \tag{2.217}$$

we obtain

$$\begin{aligned} &|\partial_\eta^\alpha\{\left(1+|\eta|^2\right)^{-l}(1-\Delta_y)^l\left(\chi(\varepsilon\eta,\varepsilon y)a(\eta,y)\right)\}| \\ &\qquad\le c_{l,\alpha}|a|_{2l+|\alpha|}\left(1+|\eta|^2\right)^{\frac{m-2l(1-\delta)}{2}}\left(1+|y|^2\right)^{\tau/2}, \end{aligned}$$

which yields with $l_0=2(l+l')$ and $c_{l,l'}$ independent of $0<\varepsilon<1$ the estimate

$$\begin{aligned} &|\left(1+|y|^2\right)^{-l'}(1-\Delta_\eta)^{l'}\{\left(1+|\eta|^2\right)^{-l}(1-\Delta_y)^l\left(\chi(\varepsilon\eta,\varepsilon y)a(\eta,y)\right)\}| \\ &\qquad\le c_{l,l'}|a|_{l_0}\left(1+|\eta|^2\right)^{\frac{m-2l(1-\delta)}{2}}\left(1+|y|^2\right)^{\frac{-2l'+\tau}{2}}. \end{aligned} \tag{2.218}$$

Because (2.215) implies that the right-hand side of (2.218) is an element in $L^1\left(\mathbb{R}^n\times\mathbb{R}^n\right)$, Lemma 2.4.13 and the dominated convergence theorem leads to the existence of

$$\begin{aligned} O_s-\int_{\mathbb{R}^n}\int_{\mathbb{R}^n} e^{-iy\cdot\eta}a(\eta,y)\,dyd\eta &= \lim_{\varepsilon\to 0} I_\varepsilon \\ &= \int_{\mathbb{R}^n}\int_{\mathbb{R}^n} e^{-iy\cdot\eta}\left(1+|y|^2\right)^{-l'}(1-\Delta_\eta)^{l'}\{(1+|\eta|^2)^{-l} \\ &\qquad\times(1-\Delta_y)^l a(\eta,y)\}\,dyd\eta. \end{aligned} \tag{2.219}$$

The representation (2.219) does not involve the function χ, thus the value of

$$O_s - \int_{\mathbb{R}^n}\int_{\mathbb{R}^n} e^{-iy\cdot\eta} a(\eta, y)\, dy d\eta$$

is independent of $\chi \in S(\mathbb{R}^n \times \mathbb{R}^n)$ whenever $\chi(0,0) = 1$, and the theorem is proved. □

Remark 2.4.15. Lemma 2.4.13 and Theorem 2.4.14 are taken from Chapter 1.6 of the book [189] of H.Kumano-go.

Note that the proof of Theorem 2.4.14 gives also the following formula of integration by parts

$$O_s - \int_{\mathbb{R}^n}\int_{\mathbb{R}^n} e^{-iy\cdot\eta}\eta^\alpha a(\eta, y)\, dy d\eta = O_s - \int_{\mathbb{R}^n}\int_{\mathbb{R}^n} e^{-iy\cdot\eta} D_y^\alpha a(\eta, y)\, dy d\eta. \tag{2.220}$$

Now we may introduce double symbols.

Definition 2.4.16. Let $\psi \in \Lambda$ and $m, m' \in \mathbb{R}$. The class $S_0^{m,m',\psi}(\mathbb{R}^n)$ of *double symbols* of order m and m' consists of all C^∞-functions $q : \mathbb{R}^n \times \mathbb{R}^n \times \mathbb{R}^n \times \mathbb{R}^n \to \mathbb{C}$ satisfying

$$|\partial_\xi^\alpha \partial_x^\beta \partial_{\xi'}^{\alpha'} \partial_{x'}^{\beta'} q(x, \xi; x', \xi')| \le c_{\alpha\beta\alpha'\beta'} (1 + \psi(\xi))^{m/2} (1 + \psi(\xi'))^{m'/2} \tag{2.221}$$

for all α, β, α', $\beta' \in \mathbb{N}_0^n$.

For $q \in S_0^{m,m',\psi}(\mathbb{R}^n)$ we define on $S(\mathbb{R}^n)$ the operator

$$\begin{aligned} &q(x, D_x; x', D_{x'})u(x) \\ &= (2\pi)^{\frac{-3n}{2}} \int_{\mathbb{R}^n}\int_{\mathbb{R}^n}\int_{\mathbb{R}^n} e^{i(x-x')\cdot\xi + ix'\cdot\xi'} q(x, \xi; x', \xi')\hat{u}(\xi')\, d\xi' dx' d\xi. \end{aligned} \tag{2.222}$$

Theorem 2.4.17. *Let $\psi \in \Lambda$ and $q \in S_0^{m,m',\psi}(\mathbb{R}^n)$. Then for $u \in S(\mathbb{R}^n)$ the iterated integral (2.222) exists and defines a pseudo-differential operator in the class $\Psi_0^{m+m',\psi}(\mathbb{R}^n)$. Moreover*

$$q_L(x, \xi) := O_s - (2\pi)^{-n} \int_{\mathbb{R}^n}\int_{\mathbb{R}^n} e^{-iy\cdot\eta} q(x, \xi + \eta; x + y, \xi)\, dy d\eta \tag{2.223}$$

is a symbol in $S_0^{m+m',\psi}(\mathbb{R}^n)$ and defines the same operator, i.e.

$$q(x, D_x; x', D_{x'})u = q_L(x, D)u \tag{2.224}$$

for all $u \in S(\mathbb{R}^n)$. *In particular, this implies that* $q(x, D_x; x', D_{x'})$ *maps continuously* $S(\mathbb{R}^n)$ *into itself.*

Definition 2.4.18. The symbol $q_L(x,\xi)$ in (2.223) is called the *simplified symbol* of the double symbol $q(x,\xi;x',\xi')$.

Proof of Theorem 2.4.17: First observe that by Peetre's inequality we have

$$\frac{(1+\psi(\xi))^{s/2}}{(1+\psi(\eta))^{s/2}} \leq 2^{|s|/2}(1+\psi(\xi-\eta))^{|s|/2}, \quad s \in \mathbb{R}, \tag{2.225}$$

and since $(1+\psi(\xi)) \leq c_\psi(1+|\xi|^2)$ we get for all $\alpha, \beta \in \mathbb{N}_0^n$

$$\begin{aligned}|\partial_\eta^\alpha \partial_y^\beta q(x,\xi+\eta;x+y,\xi)| &\leq c(1+\psi(\xi+\eta))^{m/2}(1+\psi(\xi))^{m'/2}\\ &\leq \tilde{c}(1+\psi(\xi))^{\frac{m+m'}{2}}(1+\psi(\eta))^{|m|/2} \leq c'_\xi(1+|\eta|^2)^{|m|/2}.\end{aligned}$$

Therefore, for x, ξ fixed, $(\eta, y) \mapsto q(x,\xi+\eta;x+y,\xi)$ belongs to the class $\mathcal{A}$ and it follows that the oscillatory integral (2.223) exists. Next we claim that $q_L(x,\xi)$ belongs to the class $S_0^{m+m',\psi}(\mathbb{R}^n)$. For this we use the representation (2.219) to find

$$\begin{aligned}\partial_\xi^\alpha \partial_x^\beta q_L(x,\xi) = (2\pi)^{-n}\partial_\xi^\alpha \partial_x^\beta \int_{\mathbb{R}^n}\int_{\mathbb{R}^n} e^{-iy\cdot\eta}\left(1+|y|^2\right)^{-l'}\\ \times(1-\Delta_\eta)^{l'}\left\{\left(1+|\eta|^2\right)^{-l}(1-\Delta_y)^l q(x,\xi+\eta;x+y,\xi)\right\}dyd\eta.\end{aligned}$$

Taking l and l' sufficiently large, we may differentiate under the integral sign to get with (2.221) and (2.225)

$$\begin{aligned}|\partial_\xi^\alpha \partial_y^\beta q_L(x,\xi)| \leq c_{\alpha\beta}\int_{\mathbb{R}^n}\int_{\mathbb{R}^n}&(1+|\eta|^2)^{-l}(1+|y|^2)^{-l'}\\ &\times(1+\psi(\xi+\eta))^{m/2}(1+\psi(\xi))^{m'/2}\,dyd\eta\\ &\leq \tilde{c}_{\alpha\beta}(1+\psi(\xi))^{\frac{m+m'}{2}},\end{aligned}$$

i.e. $q_L(x,\xi)$ belongs to $S_0^{m+m',\psi}(\mathbb{R}^n)$. Now we will prove (2.224). For this let $\chi \in S(\mathbb{R}^n\times\mathbb{R}^n)$ such that $\chi(0,0)=1$. For $0<\varepsilon\leq 1$ we put

$$q_\varepsilon(x,\xi;x',\xi') = \chi(\varepsilon(\xi-\xi'),\varepsilon(x-x'))q(x,\xi;x',\xi')$$

and using Leibniz' rule and Lemma 2.4.13 we get for all $\alpha,\beta,\alpha',\beta' \in \mathbb{N}_0^n$ with a constant $c_{\alpha\beta\alpha'\beta'}$ independent of $\varepsilon \in (0,1]$

$$|\partial_\xi^\alpha \partial_x^\beta \partial_{\xi'}^{\alpha'}\partial_{x'}^{\beta'} q_\varepsilon(x,\xi;x',\xi')| \leq c_{\alpha\beta\alpha'\beta'}(1+\psi(\xi))^{m/2}(1+\psi(\xi'))^{m'/2}. \tag{2.226}$$

Define

$$q_{u,\varepsilon}(x,\xi;x',\xi') := q_\varepsilon(x,\xi;x',\xi')\hat{u}(\xi'),$$
$$s_{u,\varepsilon}(x,\xi;x') := (2\pi)^{-n/2}\int_{\mathbb{R}^n} e^{ix'\cdot\xi'} q_{u,\varepsilon}(x,\xi;x',\xi')\,d\xi',$$
$$r_{u,\varepsilon} := (2\pi)^{-n/2}\int_{\mathbb{R}^n} e^{-ix'\cdot\xi} s_{u,\varepsilon}(x,\xi;x')\,dx'$$

and fix $l, n_0 \in \mathbb{N}$ such that $2l > n + m^+$, $m^+ = m \vee 0$, and $2n_0 > n$. Using once again the identity

$$e^{ix'\cdot\xi'} = \left(1+|x'|^2\right)^{-n_0}(1-\Delta_{\xi'})^{n_0}\, e^{ix'\cdot\xi'}$$

we find for all $\beta' \in \mathbb{N}_0^n$, $|\beta'| \le 2l$, by integration by parts and Leibniz' rule

$$\begin{aligned}
&|\partial_{x'}^{\beta'} s_{u,\varepsilon}(x,\xi;x')| \\
&\le (2\pi)^{-n/2}\left|\partial_{x'}^{\beta'}\int_{\mathbb{R}^n}\left(1+|x'|^2\right)^{-n_0} e^{ix'\cdot\xi'}(1-\Delta_{\xi'})^{n_0} q_{u,\varepsilon}(x,\xi;x',\xi')\,d\xi'\right| \\
&\le c_{q,u,l,n_0}\,(1+\psi(\xi))^{m/2}\left(1+|x'|^2\right)^{-n_0}
\end{aligned} \tag{2.227}$$

with c_{q,u,l,n_0} independent of ε. It follows that $r_{u,\varepsilon}(x,\xi)$ is well defined and since $(1+\psi(\xi)) \le c_\psi\left(1+|\xi|^2\right)$ we have

$$\begin{aligned}
&|r_{u,\varepsilon}(x,\xi)| \\
&\le (2\pi)^{-n/2}\left|\left(1+|\xi|^2\right)^{-l}\int_{\mathbb{R}^n} e^{-ix'\cdot\xi}(1-\Delta_{x'})^l\, s_{u,\varepsilon}(x,\xi;x')\,dx'\right| \\
&\le c_{q,u,l,n_0}\,(1+\psi(\xi))^{m/2}\left(1+|\xi|^2\right)^{-l}\tilde{c}_{q,u,l,n_0,\psi}\left(1+|\xi|^2\right)^{\frac{-2l+m}{2}}.
\end{aligned} \tag{2.228}$$

Thus the integral

$$q_\varepsilon(x,D_x;x',D_{x'})u(x) = (2\pi)^{-n/2}\int_{\mathbb{R}^n} e^{ix\cdot\xi} r_{u,\varepsilon}(x,\xi)\,d\xi$$

exists and the integral (2.222) is well-defined. Taking into account that the estimates (2.227) and (2.228) are uniform with respect to $\varepsilon \in (0,1]$, we may successively apply the dominated convergence theorem to find with

$$q_u(x,\xi;x',\xi') = q(x,\xi;x',\xi')\hat{u}(\xi')$$

that

$$
\begin{aligned}
&q(x, D_x; x', D_{x'})u(x) \\
&= (2\pi)^{-3n/2} \int_{\mathbb{R}^n} \int_{\mathbb{R}^n} \int_{\mathbb{R}^n} e^{i(x-x')\cdot\xi + ix'\cdot\xi} q_u(x, \xi; x', \xi')\, d\xi' dx d\xi \\
&= \lim_{\varepsilon \to 0} (2\pi)^{-3n/2} \int_{\mathbb{R}^n} \int_{\mathbb{R}^n} \int_{\mathbb{R}^n} e^{i(x-x')\cdot\xi + ix'\cdot\xi'} q_{u,\varepsilon}(x, \xi; x', \xi')\, d\xi' dx d\xi \\
&= \lim_{\varepsilon \to 0} q_\varepsilon(x, D_x; x', D_{x'})u(x).
\end{aligned} \tag{2.229}
$$

On the other hand, for $\varepsilon > 0$ we define

$$
q_{L,\varepsilon}(x, \xi) = (2\pi)^{-n} \int_{\mathbb{R}^n} \int_{\mathbb{R}^n} e^{-iy\cdot\eta} \chi(\varepsilon\eta, \varepsilon y) q(x, \xi+\eta; x+y, \xi)\, dy d\eta, \tag{2.230}
$$

implying by the definition of an oscillatory integral

$$
\lim_{\varepsilon \to 0} q_{L,\varepsilon}(x, \xi) = q_L(x, \xi), \tag{2.231}
$$

and for $l_1, l' \in \mathbb{N}$ such that $2l_1 > |m| + n$, $2l_1' > n$ integration by parts yields

$$
\begin{aligned}
|q_{L,\varepsilon}(x, \xi)| = c' \Big| &\int_{\mathbb{R}^n} \int_{\mathbb{R}^n} e^{-iy\cdot\eta} \left(1 + |\eta|^2\right)^{-l_1} (1 - \Delta_y)^{l_1} \\
&\times \left\{ \left(1 + |y|^2\right)^{-l_1'} (1 - \Delta_\eta)^{l_1'} \chi(\varepsilon\eta, \varepsilon y) q(x, \xi + \eta; x + y, \xi) \right\} dy d\eta \Big| \\
\leq c'' &\int_{\mathbb{R}^n} \int_{\mathbb{R}^n} \left(1 + |\eta|^2\right)^{-l_1} \left(1 + |y|^2\right)^{-l_1'} \\
&\times (1 + \psi(\xi + \eta))^{m/2} (1 + \psi(\xi))^{m'/2}\, dy d\eta \\
\leq \tilde{c} &\int_{\mathbb{R}^n} \left(1 + |\eta|^2\right)^{-l_1 + |m|/2} (1 + \psi(\xi))^{\frac{m+m'}{2}}\, d\eta \\
= c\, &(1 + \psi(\xi))^{\frac{m+m'}{2}}
\end{aligned}
$$

uniformly in $\varepsilon \in (0, 1]$. Hence by (2.231) we have

$$
\lim_{\varepsilon \to 0} q_{L,\varepsilon}(x, D)u(x) = q_L(x, D)u(x) \tag{2.232}
$$

for all $u \in S(\mathbb{R}^n)$. The substitution $x' = x(y) = x + y$ and $\xi = \xi(\eta) = \xi' + \eta$

shows however

$$
\begin{aligned}
&q_\varepsilon(x, D_x; x', D_{x'})u(x) \\
&= (2\pi)^{-3n/2} \int_{\mathbb{R}^n} \int_{\mathbb{R}^n} \int_{\mathbb{R}^n} e^{i(x-x')\cdot\xi + ix'\cdot\xi'} q_\varepsilon(x, \xi; x', \xi')\hat{u}(\xi')\, d\xi' dx' d\xi \\
&= (2\pi)^{-3n/2} \int_{\mathbb{R}^n} \int_{\mathbb{R}^n} \int_{\mathbb{R}^n} e^{ix\cdot\xi'} e^{-iy\cdot\eta} \chi(\varepsilon\eta, \varepsilon y) \\
&\quad \times q(x, \xi' + \eta; x + y, \xi')\hat{u}(\xi')\, d\xi' dy d\eta \\
&= (2\pi)^{-n/2} \int_{\mathbb{R}^n} e^{ix\cdot\xi'} q_{L,\varepsilon}(x, \xi')\hat{u}(\xi')\, d\xi' \\
&= q_{L,\varepsilon}(x, D)u(x).
\end{aligned}
\tag{2.233}
$$

Combining (2.229), (2.232) and (2.233) finally gives

$$
q(x, D_x; x', D_{x'})u(x) = q_L(x, D)u(x),
$$

and the theorem is proved. □

As a first consequence of Theorem 2.4.17 we have

Corollary 2.4.19. *Let $\psi \in \Lambda$.*

A. If $q_j \in S_0^{m_j,\psi}(\mathbb{R}^n)$, $j = 1, 2$, then $q_1(x, D) \circ q_2(x, D) \in \Psi_0^{m_1+m_2,\psi}(\mathbb{R}^n)$.

B. For any $q \in S_0^{m,\psi}(\mathbb{R}^n)$ there exists $q^* \in S_0^{m,\psi}(\mathbb{R}^n)$, such that

$$
(q(x, D)u, v)_0 = (u, q^*(x, D)v)_0 \tag{2.234}
$$

holds for all $u, v \in S(\mathbb{R}^n)$.

Proof: A. We put $q'(x, \xi; x', \xi') := q_1(x, \xi) \cdot q_2(x', \xi')$ and find $q' \in S_0^{m_1,m_2,\psi}(\mathbb{R}^n)$. Therefore the operator $q'_L(x, D)$ exists and belongs to the class $\Psi_0^{m_1+m_2,\psi}(\mathbb{R}^n)$ and for $u \in S(\mathbb{R}^n)$ it follows that

$$
\begin{aligned}
&q_1(x, D) \circ q_2(x, D)u(x) \\
&= (2\pi)^{-3n/2} \int_{\mathbb{R}^n} e^{ix\cdot\xi} q_1(x, \xi) \int_{\mathbb{R}^n} e^{-ix'\cdot\xi} \int_{\mathbb{R}^n} e^{ix'\cdot\xi'} q_2(x', \xi')\hat{u}(\xi')\, d\xi' dx' d\xi \\
&= q'(x, D_x; x', D_{x'})u(x) = q'_L(x, D)u(x).
\end{aligned}
$$

B. First note that for all $n_0 \in \mathbb{N}$ we have

$$\begin{aligned} &\left| \int_{\mathbb{R}^n} e^{ix'\cdot\xi} q(x',\xi)\overline{v(x')}\,dx' \right| \\ &\qquad = \left(1+|\xi|^2\right)^{-n_0} \left| \int_{\mathbb{R}^n} e^{ix'\cdot\xi} \left(1-\Delta_{x'}\right)^{n_0} q(x',\xi)\overline{v(x')}\,dx' \right| \\ &\qquad \le c\left(1+|\xi|^2\right)^{-n_0} \left(1+\psi(\xi)\right)^{m/2}. \end{aligned}$$

This observation allows us to apply Fubini's theorem in the following calculation provided we take n_0 sufficiently large. We define $\tilde{q}(x,\xi;x',\xi') := \overline{q(x',\xi)}$, i.e. $\tilde{q}$ depends in fact only on two variables. It follows that $\tilde{q} \in S_0^{m,\psi}(\mathbb{R}^n)$ and further we have

$$\begin{aligned} &(q(x,D)u,v)_0 = \int_{\mathbb{R}^n} q(x',D)u(x')\overline{v(x')}\,dx' \\ &= (2\pi)^{-n/2} \int_{\mathbb{R}^n}\int_{\mathbb{R}^n} e^{ix'\cdot\xi} q(x',\xi)\hat{u}(\xi)\,d\xi\,\overline{v(x')}\,dx' \\ &= (2\pi)^{-n} \int_{\mathbb{R}^n}\int_{\mathbb{R}^n} e^{-ix\cdot\xi}u(x)\Big\{\int_{\mathbb{R}^n} e^{ix'\cdot\xi} q(x',\xi)\overline{v(x')}\,dx'\Big\}\,dxd\xi \\ &= (2\pi)^{-n} \int_{\mathbb{R}^n} u(x) \int_{\mathbb{R}^n}\int_{\mathbb{R}^n} e^{-i(x-x')\cdot\xi} q(x',\xi)\overline{v(x')}\,dx'\,d\xi\,dx \\ &= (2\pi)^{-3n/2} \int_{\mathbb{R}^n} u(x) \overline{\int_{\mathbb{R}^n}\int_{\mathbb{R}^n}\int_{\mathbb{R}^n} e^{i(x-x')\cdot\xi+ix'\cdot\xi'}\overline{q(x',\xi)}\hat{v}(\xi')\,d\xi' dx' d\xi dx} \\ &= (u,\tilde{q}(x,D_x;x',D'_x)v)_0, \end{aligned}$$

which proves the assertion with $q^*(x,D) = \tilde{q}_L(x,D)$. □

So fare we have proved

Theorem 2.4.20. *The set* $\Psi_0^{\infty,\psi}(\mathbb{R}^n) := \bigcup_{m\in\mathbb{R}} \Psi_0^{m,\psi}(\mathbb{R}^n)$ *is an algebra of (pseudo-differential) operators with the composition as multiplication and with involution* $*$. *Further, the graded structure of* $\Psi_0^{\infty,\psi}(\mathbb{R}^n)$ *is respected in the sense that we have*

$$\lambda\Psi_0^{m,\psi}(\mathbb{R}^n)+\mu\Psi_0^{m,\psi}(\mathbb{R}^n) \subset \Psi_0^{m,\psi}(\mathbb{R}^n), \quad \lambda,\mu\in\mathbb{C}; \tag{2.235}$$

$$\left(\Psi_0^{m,\psi}(\mathbb{R}^n)\right)^* \subset \Psi_0^{m,\psi}(\mathbb{R}^n), \tag{2.236}$$

and

$$\Psi_0^{m,\psi}(\mathbb{R}^n)\circ\Psi_0^{m',\psi}(\mathbb{R}^n) \subset \Psi_0^{m+m',\psi}(\mathbb{R}^n). \tag{2.237}$$

In every classical symbolic calculus for pseudo-differential operators asymptotic expansions of symbols play a fundamental rôle. In principle they allow us to reduce (up to lower order terms) operations on the level of operators to operations on the level of symbols. Such asymptotic expansions are in general not valid for the symbol class $\bigcup_{m\in\mathbb{R}} S_0^{m,\psi}(\mathbb{R}^n)$, $\psi \in \Lambda$. However for the class $\bigcup_{m\in\mathbb{R}} S_\rho^{m,\psi}(\mathbb{R}^n)$ some asymptotic expansions up to order 2 can be shown. First we prove

Lemma 2.4.21. *Let $\psi \in \Lambda$ and $q \in S_0^{m,m',\psi}(\mathbb{R}^n)$ such that*

$$\partial_\xi^\alpha q(x,\xi;x',\xi') \in S_0^{m-\rho(|\alpha|),m',\psi}(\mathbb{R}^n) \tag{2.238}$$

holds for all $\alpha \in \mathbb{N}_0^n$. For all $N \in \mathbb{N}$ the simplified symbol q_L satisfies

$$q_L(x,\xi) - \sum_{|\alpha|<N} \frac{1}{\alpha!} q_\alpha(x,\xi) \in S_0^{m+m'-\rho(N),\psi}(\mathbb{R}^n), \tag{2.239}$$

where

$$q_\alpha(x,\xi) = D_{x'}^\alpha \partial_\xi^\alpha q(x,\xi;x',\xi')|_{\substack{x'=x\\ \xi'=\xi}} \in S_0^{m+m'-\rho(|\alpha|),\psi}(\mathbb{R}^n). \tag{2.240}$$

Proof: By Taylor's formula we find

$$\begin{aligned} &q(x,\xi+\eta;x+z,\xi) \\ &\quad = \sum_{|\alpha|<N} \frac{\eta^\alpha}{\alpha!} \partial_\xi^\alpha q(x,\xi;x+z,\xi')|_{\xi'=\xi} + N \sum_{|\gamma|=N} \frac{\eta^\gamma}{\gamma!} q_\gamma(x,z;\xi,\eta), \end{aligned}$$

where

$$q_\gamma(x,z;\xi,\eta) = \int_0^1 (1-t)^{N-1} \partial_\xi^\gamma q(x,\xi+t\eta;x+z,\xi')|_{\xi'=\xi}\, dt.$$

Using Theorem 2.4.17, especially (2.223), we find

$$\begin{aligned} &(2\pi)^n q_L(x,\xi) \\ &\quad = \sum_{|\alpha|<N} \frac{1}{\alpha!} O_s - \int_{\mathbb{R}^n}\int_{\mathbb{R}^n} e^{-iz\cdot\eta} \eta^\alpha \partial_\xi^\alpha q(x,\xi;x+z,\xi')|_{\xi'=\xi}\, dz d\eta \\ &\qquad + \sum_{|\gamma|=N} \frac{N}{\gamma!} O_s - \int_{\mathbb{R}^n}\int_{\mathbb{R}^n} e^{-iz\cdot\eta} \eta^\gamma q_\gamma(x,z;\xi,\eta)\, dz d\eta \\ &\quad = \sum_{|\alpha|<N} \frac{1}{\alpha!} I_\alpha(x,\xi) + \sum_{|\gamma|=N} \frac{N}{\gamma!} J_\gamma(x,\xi). \end{aligned}$$

It remains to prove

$$I_\alpha = q_\alpha \in S_0^{m+m'-\rho(|\alpha|),\psi}(\mathbb{R}^n) \quad \text{for } |\alpha| < N, \tag{2.241}$$

and

$$J_\gamma \in S_0^{m+m'-\rho(N),\psi}(\mathbb{R}^n) \quad \text{for } |\gamma| = N. \tag{2.242}$$

Let $|\alpha| < N$ and take $\chi_1, \chi_2 \in C_0^\infty(\mathbb{R}^n)$ such that χ_1 and χ_2 equal 1 in a neighbourhood of $0 \in \mathbb{R}^n$. The definition of I_α and (2.220) yield

$$\begin{aligned}
(2\pi)^n I_\alpha &= O_s - \int_{\mathbb{R}^n}\int_{\mathbb{R}^n} e^{-iz\cdot\eta}\eta^\alpha \partial_\xi^\alpha q(x,\xi;x+z,\xi')|_{\xi'=\xi}\, dz d\eta \\
&= O_s - \int_{\mathbb{R}^n}\int_{\mathbb{R}^n} e^{-iz\cdot\eta} D_z^\alpha \partial_\xi^\alpha q(x,\xi;x+z,\xi')|_{\xi'=\xi}\, dz d\eta \\
&= \lim_{\varepsilon\to 0}\int_{\mathbb{R}^n}\int_{\mathbb{R}^n} e^{-iz\cdot\eta}\chi_1(\varepsilon\eta)\chi_2(\varepsilon z) D_{x'}^\alpha \partial_\xi^\alpha q(x,\xi;x',\xi')\big|_{\substack{x'=x+z\\ \xi'=\xi}}\, dz d\eta \\
&= \lim_{\varepsilon\to 0}(2\pi)^{n/2}\int_{\mathbb{R}^n}\chi_2(\varepsilon z)\varepsilon^{-n}\hat{\chi}_1\left(\frac{z}{\varepsilon}\right) D_{x'}^\alpha \partial_\xi^\alpha q(x,\xi;x',\xi')\big|_{\substack{x'=x+z\\ \xi'=\xi}}\, dz \\
&= (2\pi)^n D_{x'}^\alpha \partial_\xi^\alpha q(x,\xi;x',\xi')|_{\substack{x'=x\\ \xi'=\xi}} = (2\pi)^n q_\alpha(x,\xi),
\end{aligned}$$

because $\chi_2(\varepsilon z)\varepsilon^{-n}\hat{\chi}_1(\frac{z}{\varepsilon})$ converges to $(2\pi)^{-n/2}$-times the unit mass at $0 \in \mathbb{R}^n$ as $\varepsilon \to 0$, and (2.238) gives $q_\alpha \in S_0^{m+m'-\rho(|\alpha|),\psi}(\mathbb{R}^n)$.

Now let $|\gamma| = N$. By (2.238), Peetre's inequality and the estimate $1+\psi(\xi) \leq c_\psi\left(1+|\xi|^2\right)$ we find

$$\begin{aligned}
&|\partial_x^\alpha \partial_z^{\alpha'} \partial_\xi^\beta \partial_\eta^{\beta'} q_\gamma(x,z;\xi,\eta)| \\
&= \left|\int_0^1 (1-t)^{N-1}\partial_x^\alpha \partial_z^{\alpha'} \partial_\xi^\beta \partial_\eta^{\beta'}\left(\partial_\xi^\gamma q(x,\xi+t\eta;x+z,\xi')|_{\xi'=\xi}\right) dt\right| \\
&\leq c_{\alpha,\alpha',\beta,\beta',\gamma}\int_0^1 (1+\psi(\xi+t\eta))^{\frac{m-\rho(N)}{2}}(1+\psi(\xi))^{m'/2}\, dt \\
&\leq c_{\alpha,\alpha',\beta,\beta',\gamma}(1+\psi(\xi))^{\frac{m-\rho(N)}{2}}(1+\psi(\xi))^{m'/2}\int_0^1 (1+\psi(t\eta))^{\frac{|m-\rho(N)|}{2}}\, dt \\
&\leq \tilde{c}_{\alpha,\alpha',\beta,\beta',\gamma}(1+\psi(\xi))^{\frac{m+m'-\rho(N)}{2}}(1+|\eta|^2)^{\frac{|m-\rho(N)|}{2}}.
\end{aligned}$$

Using (2.219) for $l, n_0 \in \mathbb{N}$ such that $2l > N + |m-\rho(N)| + n$ and $2n_0 > n$ it follows that

$$\begin{aligned}
&|\partial_\xi^\alpha \partial_x^\beta J_\gamma(x,\xi)| \\
&= \Big| \int_{\mathbb{R}^n} \int_{\mathbb{R}^n} e^{-iz\cdot\eta} \left(1+|\eta|^2\right)^{-l} (1-\Delta_z)^l \\
&\quad \times \Big\{ \left(1+|z|^2\right)^{-n_0} (1-\Delta_\eta)^{n_0} \{\eta^\gamma \partial_\xi^\alpha \partial_x^\beta q_\gamma(x,z;\xi,\eta)\} \Big\} dz d\eta \Big| \\
&\leq c_{l,\alpha,\beta} \int_{\mathbb{R}^n} \int_{\mathbb{R}^n} \left(1+|\eta|^2\right)^{\frac{-2l+N+|m-\rho(N)|}{2}} \left(1+|z|^2\right)^{-n_0} \\
&\quad \times \left(1+\psi(\xi)\right)^{\frac{m+m'-\rho(N)}{2}} dz d\eta \\
&\leq c_{l,\alpha,\beta} \left(1+\psi(\xi)\right)^{\frac{m+m'-\rho(N)}{2}},
\end{aligned}$$

and (2.242) is proved. □

Remark 2.4.22. The proof of Lemma 2.4.21 shows that q_α and the remainder $q_L - \sum_{|\alpha|<N} \frac{1}{\alpha!} q_\alpha$ are in the class $S_0^{m+m'-\rho(|\alpha|),\psi}(\mathbb{R}^n)$ and $S_0^{m+m'-\rho(N),\psi}(\mathbb{R}^n)$, respectively. Moreover they satisfy (2.199) with constants depending only on $c_{\alpha,\beta,\alpha',\beta'}$ and the double symbol $q(x,\xi;x',\xi')$.

Corollary 2.4.23. *Let $\psi \in \Lambda$.*

A. For $q_1 \in S_\rho^{m_1,\psi}(\mathbb{R}^n)$ and $q_2 \in S_\rho^{m_2,\psi}(\mathbb{R}^n)$ the symbol q of the operator $q(x,D) := q_1(x,D) \circ q_2(x,D)$ is given by

$$q(x,\xi) = q_1(x,\xi)\cdot q_2(x,\xi) + \sum_{j=1}^n \partial_{\xi_j} q_1(x,\xi) D_{x_j} q_2(x,\xi) + q_{r_1}(x,\xi) \tag{2.243}$$

with $q_{r_1}(x,\xi) \in S_0^{m+m'-2,\psi}(\mathbb{R}^n)$.

B. For $q \in S_\rho^{m,\psi}(\mathbb{R}^n)$ the symbol of $q^*(x,D)$ is given by

$$q^*(x,D) = \overline{q(x,\xi)} + \sum_{j=1}^n \partial_{\xi_j} D_{x_j} q(x,\xi) + q_{r_2}(x,\xi) \tag{2.244}$$

with $q_{r_2}(x,\xi) \in S_0^{m-2,\psi}(\mathbb{R}^n)$.

Proof: We only have to apply Lemma 2.4.21 to the double symbols $q'(x,\xi;x',\xi') := q_1(x,\xi) q_2(x',\xi)$ and $\tilde{q}(x,\xi;x',\xi) := \overline{q(x',\xi)}$, respectively, introduced in the proof of Corollary 2.4.19. □

Next we want to introduce the Friedrichs symmetrization.

Definition 2.4.24. Let $r \in C_0^\infty(\mathbb{R}^n)$ be a fixed non-negative function which is even, supported in the unit ball $B_1(0)$ and satisfies $\int_{\mathbb{R}^n} r^2(\xi)\,d\xi = 1$. For $\psi \in \Lambda$ we define

$$F(\xi,\zeta) := (1+\psi(\xi))^{-n/8}\, r\left((\zeta-\xi)\,(1+\psi(\xi))^{-1/4}\right). \tag{2.245}$$

For any $q \in S_0^{m,\psi}(\mathbb{R}^n)$ we define its *Friedrichs symmetrization* to be the double symbol

$$q_F(\xi;x',\xi') := \int_{\mathbb{R}^n} F(\xi,\zeta)q(x',\zeta)F(\xi',\zeta)\,d\zeta. \tag{2.246}$$

(Note that q_F is independent of x.)

Our aim is to prove

Theorem 2.4.25. *For the Friedrichs symmetrization of $q \in S_0^{m,\psi}(\mathbb{R}^n)$, $\psi \in \Lambda$, we have the estimates*

$$|\partial_{x'}^{\alpha'}\partial_\xi^\beta\partial_{\xi'}^{\beta'} q_F(\xi;x',\xi')| \leq c_{\alpha',\beta,\beta'}\,(1+\psi(\xi))^{\frac{m-\frac{1}{2}\rho(\beta)}{2}}\,(1+\psi(\xi'))^{\frac{-\rho(\beta')}{2}}. \tag{2.247}$$

In particular $q_F \in S_0^{m,0,\psi}(\mathbb{R}^n)$ and the simplified symbol $q_{F,L}$ belongs to $S_0^{m,\psi}(\mathbb{R}^n)$. Moreover, if $q \in S_\rho^{m,\psi}(\mathbb{R}^n)$, then

$$q - q_{F,L} \in S_0^{m-1,\psi}(\mathbb{R}^n) \tag{2.248}$$

holds.

We need some auxiliary results.

Lemma 2.4.26. *For all $\beta \in \mathbb{N}_0^n$ we have with $\varphi_{\beta,\gamma,\gamma_1} \in S_0^{-\frac{1}{2}\rho(|\beta|),\psi}(\mathbb{R}^n)$*

$$\begin{aligned}\partial_\xi^\beta F(\xi,\zeta) = (1+\psi(\xi))^{-n/8} &\sum_{\substack{|\gamma|\leq|\beta|\\ \gamma_1\leq\gamma}} \varphi_{\beta,\gamma,\gamma_1}(\xi)\left((\xi-\zeta)\,(1+\psi(\xi))^{-\frac{1}{4}}\right)^{\gamma_1}\\ &\times(\partial^\gamma r)\left((\xi-\zeta)\,(1+\psi(\xi))^{-\frac{1}{4}}\right).\end{aligned} \tag{2.249}$$

Proof: For $\beta = 0$ we have (2.249) with $\varphi_{0,0,0} = 1$. Now we proceed further by induction using the identity

$$\partial_{\xi_j}(1+\psi(\xi))^{m/2} = m\,(1+\psi(\xi))^{m/2}\,(1+\psi(\xi))^{-1/2}\,\partial_{\xi_j}(1+\psi(\xi))^{1/2}$$

and differentiating in (2.249):

$$\partial_{\xi_j}\partial_\xi^\beta F(\xi,\zeta) = (1+\psi(\xi))^{-n/8} \sum_{\substack{|\gamma|\le|\beta| \\ \gamma_1\le\gamma}} \{[\chi^{(1)}_{\beta,\gamma,\gamma_1}(\xi) + \chi^{(2)}_{\beta,\gamma,\gamma_1}(\xi)]$$
$$\times \left((\xi-\zeta)(1+\psi(\xi))^{-\frac{1}{4}}\right)^{\gamma_1} (\partial^\gamma r)\left((\xi-\zeta)(1+\psi(\xi))^{-\frac{1}{4}}\right)$$
$$+\chi^{(3)}_{\beta,\gamma,\gamma_1}(\xi)\left((\xi-\zeta)(1+\psi(\xi))^{-\frac{1}{4}}\right)^{\gamma_1-\varepsilon_j} (\partial^\gamma r)\left((\xi-\zeta)(1+\psi(\xi))^{-\frac{1}{4}}\right)$$
$$+\chi^{(4)}_{\beta,\gamma,\gamma_1}(\xi)\left((\xi-\zeta)(1+\psi(\xi))^{-\frac{1}{4}}\right)^{\gamma_1} (\partial^{\gamma+\varepsilon_j} r)\left((\xi-\zeta)(1+\psi(\xi))^{-\frac{1}{4}}\right)$$
$$+\chi^{(5)}_{\beta,\gamma,\gamma_1}(\xi)\sum_{k=1}^n \left((\xi-\zeta)(1+\psi(\xi))^{-\frac{1}{4}}\right)^{\gamma_1+\varepsilon_k}$$
$$(\partial^{\gamma+\varepsilon_k} r)\left((\xi-\zeta)(1+\psi(\xi))^{-\frac{1}{4}}\right) \tag{2.250}$$

with

$$\chi^{(1)}_{\beta,\gamma,\gamma_1}(\xi) = -\varphi_{\beta,\gamma,\gamma_1}(\xi)\left(\frac{n}{4}+\frac{|\gamma_1|}{2}\right)(1+\psi(\xi))^{-1/2}\,\partial_{\xi_j}(1+\psi(\xi))^{1/2},$$
$$\chi^{(2)}_{\beta,\gamma,\gamma_1}(\xi) = \partial_{\xi_j}\varphi_{\beta,\gamma,\gamma_1}(\xi),$$
$$\chi^{(3)}_{\beta,\gamma,\gamma_1}(\xi) = \gamma_{1,j}\,(1+\psi(\xi))^{-1/4}\,\varphi_{\beta,\gamma,\gamma_1}(\xi),$$
$$\chi^{(4)}_{\beta,\gamma,\gamma_1}(\xi) = (1+\psi(\xi))^{-1/4}\,\varphi_{\beta,\gamma,\gamma_1}(\xi)$$

and

$$\chi^{(5)}_{\beta,\gamma,\gamma_1}(\xi) = -\frac{1}{2}(1+\psi(\xi))^{-1/2}\,\partial_{\xi_j}(1+\psi(\xi))\,\varphi_{\beta,\gamma,\gamma_1}(\xi).$$

Note that (2.250) is an expression for $\partial_{\xi_j}\partial_\xi^\beta F(\xi,\zeta)$ claimed in (2.249) and we have to check that $\chi^{(l)}_{\beta,\gamma,\gamma_1} \in S_0^{-\frac{1}{2}\rho(|\beta|+1),\psi}(\mathbb{R}^n)$, $l=1,\dots,5$. Lemma 2.4.2 implies $(1+\psi(\cdot))^{-1/4} \in S_0^{-1/2,\psi}(\mathbb{R}^n)$ and $(1+\psi(\cdot))^{-1/2}\,\partial_{\xi_j}(1+\psi(\cdot))^{1/2} \in S_0^{-1,\psi}(\mathbb{R}^n)$. Since $\varphi_{0,0,0}=1$, we see for $\beta=0$ that $\chi^{(l)}_{\beta,\gamma,\gamma_1} \in \operatorname{span}\{(1+\psi(\cdot))^{-1/4}, (1+\psi(\cdot))^{-1/2}\partial_{\xi_j}(1+\psi(\cdot))^{1/2}\} \in S_0^{-1/2,\psi}(\mathbb{R}^n)$, which also gives that $\varphi_{\beta,\gamma,\gamma_1} \in S_0^{-1/2,\psi}(\mathbb{R}^n)$ for $|\beta|=1$. Next note that $\partial_{\xi_k}(1+\psi(\cdot))^{-1/4} \in S_0^{-3/2,\psi}(\mathbb{R}^n)$ for $|\beta|=1$. Thus from Lemma 2.4.9.B we conclude for $|\beta|=1$ that $\chi^{(l)}_{\beta,\gamma,\gamma_1} \in S_0^{-1,\psi}(\mathbb{R}^n)$. Since $S_0^{-1,\psi}(\mathbb{R}^n)$ is stable under taking derivatives, we may apply Lemma 2.4.9 once again to get by induction $\chi^{(l)}_{\beta,\gamma,\gamma_1} \in S_0^{-1,\psi}(\mathbb{R}^n)$ for all $|\beta|\ge 2$. □

Lemma 2.4.27. *Let $q \in S_0^{m,\psi}(\mathbb{R}^n)$, $t \in \mathbb{R}$ and $\sigma \in \mathbb{R}^n$. Then we have with $\chi_{\beta,\gamma,\gamma_1} \in S_0^{0,\psi}(\mathbb{R}^n)$*

$$\begin{aligned}
&\partial_x^\alpha \partial_\xi^\beta \left(q(x, \xi + t(1+\psi(\xi))^{1/4}\sigma)\right) \\
&= \sum_{\substack{|\gamma|\le|\beta| \\ \gamma_1\le\gamma}} \chi_{\beta,\gamma,\gamma_1}(\xi)\left(\partial_x^\alpha \partial_\xi^\beta q\right)(x, \xi + t\,(1+\psi(\xi))^{1/4}\,\sigma)(t\sigma)^{\gamma_1}.
\end{aligned} \tag{2.251}$$

Proof: Since for $\alpha \in \mathbb{N}_0^n$ we also have $\partial_x^\alpha q(x,\xi) \in S_0^{m,\psi}(\mathbb{R}^n)$, we may replace $\partial_x^\alpha q(x,\xi)$ by $q(x,\xi)$, or assume $\alpha = 0$. With $\chi_{0,0,0} = 1$ there is nothing to prove for $\beta = 0$. Let $\tilde{\xi} = \xi + t\,(1+\psi(\xi))^{1/4}\,\sigma$. Then by induction it follows that

$$\begin{aligned}
\partial_{\xi_i}\partial_\xi^\beta q(x,\tilde{\xi}) = \sum_{\substack{|\gamma|\le|\beta| \\ \gamma_1\le\gamma}} &\{\partial_{\xi_i}\chi_{\beta,\gamma,\gamma_1}(\xi)(\partial_\xi^\gamma q)(x,\tilde{\xi})(t\sigma)^{\gamma_1} \\
&+ \chi_{\beta,\gamma,\gamma_1}(\xi)(\partial_\xi^{\gamma+\varepsilon_i} q)(x,\tilde{\xi})(t\sigma)^{\gamma_1} \\
&+ \sum_{k=1}^n \chi_{\beta,\gamma,\gamma_1}(\xi)(\partial_\xi^{\gamma+\varepsilon_k} q)(x,\tilde{\xi})\partial_{\xi_i}\,(1+\psi(\xi))^{1/4}\,(t\sigma)^{\gamma_1+\varepsilon_k}\},
\end{aligned}$$

which proves the lemma, since $\partial_{\xi_i}\,(1+\psi(\xi))^{1/4} \in S_0^{0,\psi}(\mathbb{R}^n)$. □

Proof of Theorem 2.4.25: First we prove (2.247). Using Lemma 2.4.26 and the support properties of r we find

$$\begin{aligned}
|\partial_{x'}^{\alpha'}\partial_{\xi'}^{\beta'}\partial_\xi^\beta q_F(\xi;x',\xi')| &= \left|\int_{\mathbb{R}^n} \partial_\xi^\beta F(\xi,\zeta)\partial_{x'}^{\alpha'} q(x',\zeta)\partial_{\xi'}^{\beta'} F(\xi',\zeta)\,d\zeta\right| \\
&\le (1+\psi(\xi))^{-n/8}(1+\psi(\xi'))^{-n/8} \sum_{\substack{|\gamma|\le|\beta| \\ \gamma_1\le\gamma}} \sum_{\substack{|\gamma'|\le|\beta'| \\ \gamma_1'\le\gamma'}} \left|\varphi_{\beta,\gamma,\gamma_1}(\xi)\varphi_{\beta',\gamma',\gamma_1'}(\xi')\right| \\
&\quad\times \left| \int_{\substack{|\xi'-\zeta|\le(1+\psi(\xi'))^{1/4} \\ |\xi-\zeta|\le(1+\psi(\xi))^{1/4}}} \left(\frac{\xi-\zeta}{(1+\psi(\xi))^{1/4}}\right)^{\gamma_1} \left(\frac{\xi'-\zeta}{(1+\psi(\xi'))^{1/4}}\right)^{\gamma_1'} \right. \\
&\quad\times \left. (\partial^\gamma r)\left(\frac{\xi-\zeta}{(1+\psi(\xi))^{1/4}}\right)(\partial^{\gamma'} r)\left(\frac{\xi'-\zeta}{(1+\psi(\xi'))^{1/4}}\right)\partial_{x'}^{\alpha'} q(x',\zeta)\,d\zeta \right| \\
&\le c_{\alpha',\beta',\beta}\,(1+\psi(\xi))^{-n/8}\,(1+\psi(\xi'))^{-n/8} \\
&\quad\times (1+\psi(\xi))^{-\frac{\rho(|\beta|)}{4}}\,(1+\psi(\xi'))^{-\frac{\rho(|\beta|)}{4}}\cdot I,
\end{aligned} \tag{2.252}$$

where

$$I = \int_{\substack{|\xi'-\zeta|\le(1+\psi(\xi'))^{1/4}\\ |\xi-\zeta|\le(1+\psi(\xi))^{1/4}}} \left|\partial_{x'}^{\alpha'} q(x',\zeta)\right| d\zeta.$$

Now, for $\sigma \in \mathbb{R}^n$, $|\sigma| \le 1$, we have using the triangle inequality for $\psi^{1/2}$

$$\begin{aligned} &\left(1+\psi(\xi + (1+\psi(\xi))^{1/4}\sigma)\right)^{1/2} \\ &\quad\le (1+\psi(\xi))^{1/2} + \left(1+\psi((1+\psi(\xi))^{1/4}\sigma)\right)^{1/2} \\ &\quad\le (1+\psi(\xi))^{1/2} + c\left(1+(1+\psi(\xi))^{1/4}|\sigma|\right) \\ &\quad\le c'(1+\psi(\xi))^{1/2}. \end{aligned} \tag{2.253}$$

Thus, using the substitution $\zeta = \xi + (1+\psi(\xi))^{1/4}\sigma$ is follows from the Cauchy–Schwarz inequality that

$$\begin{aligned} |I| &\le \left(\int_{|\xi-\zeta|\le(1+\psi(\xi))^{1/4}} |\partial_{x'}^{\alpha'} q(x',\zeta)|^2\, d\zeta\right)^{1/2} \left(\int_{|\xi'-\zeta|\le(1+\psi(\xi'))^{1/4}} 1\, d\zeta\right)^{1/2} \\ &= (1+\psi(\xi))^{n/8} \left(\int_{|\sigma|\le1} |\partial_{x'}^{\alpha'} q(x',\xi+(1+\psi(\xi))^{1/4}\sigma)|^2\, d\sigma\right)^{1/2} \\ &\quad\times (1+\psi(\xi'))^{n/8} \left(\int_{|\sigma|\le1} 1\, d\sigma\right)^{1/2} \\ &\le c\,(1+\psi(\xi))^{n/8} (1+\psi(\xi'))^{n/8} (1+\psi(\xi))^{m/2}, \end{aligned}$$

which together with (2.252) gives (2.247). Now we prove (2.248). We use expression (2.239) from Lemma 2.4.21, replacing ρ by $\frac{1}{2}\rho$, and (2.247). It follows that

$$q_{F,L} - q_{F,0} - \sum_{|\alpha|=1} q_{F,\alpha} \in S_0^{m-1,\psi}(\mathbb{R}^n).$$

Thus it remains to show

$$q_{F,\alpha} \in S_0^{m-1,\psi}(\mathbb{R}^n) \quad \text{for } |\alpha| = 1 \tag{2.254}$$

and

$$q_{F,0} - q \in S_0^{m-1,\psi}(\mathbb{R}^n). \tag{2.255}$$

Let $|\alpha| = 1$. It follows that

$$\begin{aligned}
\partial_\xi^\alpha F(\xi,\eta) &= \partial_\xi^\alpha \left((1+\psi(\xi))^{-n/8} r\left((\eta-\xi)(1+\psi(\xi))^{-1/4} \right) \right) \\
&= (1+\psi(\xi))^{-n/8} \left(-\frac{n}{4} r \right) \left((\eta-\xi)(1+\psi(\xi))^{-1/4} \right) \\
&\quad \times (1+\psi(\xi))^{-1/2} \partial_\xi^\alpha (1+\psi(\xi))^{1/2} \\
&\quad + \sum_{k=1}^n (\partial_k r)\left((\eta-\xi)(1+\psi(\xi))^{-1/4} \right)(\eta_k-\xi_k)\partial_\xi^\alpha (1+\psi(\xi))^{-1/4} \\
&\quad - (\partial^\alpha r)\left((\eta-\xi)(1+\psi(\xi))^{-1/4} \right)(1+\psi(\xi))^{-1/4},
\end{aligned}$$

and consequently with $\sigma = (\eta-\xi)(1+\psi(\xi))^{-1/4}$ we find

$$\begin{aligned}
q_{F,\alpha}(x,\xi) &= D_{x'}^\alpha \partial_\xi^\alpha q_F(\xi; x', \xi')|_{\substack{x'=x \\ \xi'=\xi}} = \int_{\mathbb{R}^n} \partial_\xi^\alpha F(\xi,\eta) D_x^\alpha q(x,\eta) F(\xi,\eta)\, d\eta \\
&= -\frac{n}{4} (1+\psi(\xi))^{-1/2} \partial_\xi^\alpha (1+\psi(\xi)) \\
&\quad \times \int_{\mathbb{R}^n} r(\sigma) D_x^\alpha q(x, \xi + (1+\psi(\xi))^{1/4}\sigma)\, d\sigma \\
&\quad + \sum_{k=1}^n (1+\psi(\xi))^{1/4} \partial_\xi^\alpha (1+\psi(\xi))^{-1/4} \\
&\quad \times \int_{\mathbb{R}^n} \sigma_k \partial_k r(\sigma) D_x^\alpha q(x, \xi + (1+\psi(\xi))^{1/4}\sigma)\, d\sigma \\
&\quad - (1+\psi(\xi))^{-1/4} \int_{\mathbb{R}^n} (\partial^\alpha r)(\sigma) r(\sigma) D_x^\alpha q(x, \xi + (1+\psi(\xi))^{1/4}\sigma)\, d\sigma \\
&= I_1 + I_2 + I_3.
\end{aligned}$$

We estimate each term separately. Since

$$\begin{aligned}
&\left| \partial_x^\delta \partial_\xi^\beta \int_{\mathbb{R}^n} r^2(\sigma) D_x^\alpha q\left(x, \xi + (1+\psi(\xi))^{1/4}\sigma \right) d\sigma \right| \\
&= \left| \int_{\mathbb{R}^n} r^2(\sigma) \partial_x^\delta \partial_\xi^\beta D_x^\alpha q\left(x, \xi + (1+\psi(\xi))^{1/4}\sigma \right) d\sigma \right| \\
&\le \sum_{\substack{|\gamma| \le |\beta| \\ \gamma_1 \le \gamma}} |\chi_{\beta,\gamma,\gamma_1}| \int_{\mathbb{R}^n} r^2(\sigma) \left| \left(\partial_x^\delta \partial_\xi^\beta D_x^\alpha q \right)\left(x, \xi + (1+\psi(\xi))^{1/4}\sigma \right) \right| |\sigma^{\gamma_1}|\, d\sigma \\
&\le c' \int_{\mathbb{R}^n} r^2(\sigma) \left(1 + \psi\left(\xi + (1+\psi(\xi))^{1/4}\sigma \right) \right)^{m/2} d\sigma
\end{aligned}$$

$$\leq c'' \int_{\mathbb{R}^n} r^2(\sigma)\, d\sigma\, (1+\psi(\xi))^{m/2} = c\,(1+\psi(\xi))^{m/2},$$

where we used Lemma 2.4.27, $\operatorname{supp} r \subset B_1(0)$, and (2.253), and further $(1+\psi(\xi))^{-1/2}\, \partial_\xi^\alpha\, (1+\psi(\xi)) \in S_0^{-1,\psi}(\mathbb{R}^n)$. Thus we get $I_1 \in S_0^{m-1,\psi}(\mathbb{R}^n)$. Analogously we find $(1+\psi(\xi))^{1/4}\, \partial_\xi^\alpha\, (1+\psi(\xi))^{-1/4} \in S_0^{-1,\psi}(\mathbb{R}^n)$ and

$$\int_{\mathbb{R}^n} \sigma_k \partial_k r(\sigma)\cdot r(\sigma) D_x^\alpha q\left((x, \xi + (1+\psi(\xi))^{1/4}\,\sigma\right)\, d\sigma \in S_0^{m,\psi}(\mathbb{R}^n),$$

implying that $I_2 \in S_0^{m-1,\psi}(\mathbb{R}^n)$. In order to estimate I_3 we use Taylor's formula to get

$$\begin{aligned}
&\int_{\mathbb{R}^n} (\partial^\alpha r(\sigma))\, r(\sigma) D_x^\alpha q\left(x, \xi + (1+\psi(\xi))^{1/4}\,\sigma\right)\, d\sigma \\
&\quad = \int_{\mathbb{R}^n} (\partial^\alpha r(\sigma))\, r(\sigma)\, d\sigma\, D_x^\alpha q(x,\xi) + (1+\psi(\xi))^{1/4} \int_{\mathbb{R}^n} (\partial^\alpha r(\sigma))\, r(\sigma) \\
&\qquad \times \sum_{k=1}^n \sigma_k \Big\{ \int_0^1 (\partial_{\xi_k} D_x^\alpha q)\left(x, \xi + (1+\psi(\xi))^{1/4}\, t\sigma\right)\, dt \Big\} d\sigma.
\end{aligned}$$

By the symmetry of r the first term vanishes and for the second term we get, using once again Lemma 2.4.27,

$$\begin{aligned}
&\left| \partial_x^\delta \partial_\xi^\beta \int_{\mathbb{R}^n} (\partial^\alpha r(\sigma))\, r(\sigma)\sigma_k \cdot \int_0^1 (\partial_{\xi_k} D_x^\alpha q)\left(x, \xi + (1+\psi(\xi))^{1/4} t\sigma\right)\, dt d\sigma \right| \\
&\leq \int_{\mathbb{R}^n} |\, (\partial^\alpha r(\sigma))\, r(\sigma)\sigma_k| \int_0^1 \left| \partial_x^\delta \partial_\xi^\beta\, (\partial_{\xi_k} D_x^\alpha q)\, (x, \xi + (1+\psi(\xi))^{1/4} t\sigma) \right| dt d\sigma \\
&\leq c' \int_{\mathbb{R}^n} |(\partial^\alpha r(\sigma))\, r(\sigma)\sigma_k| \int_0^1 \left(1 + \psi\left(\xi + (1+\psi(\xi))^{1/4}\, t\sigma\right)\right)^{\frac{m-1}{2}}\, dt d\sigma \\
&\leq c\,(1+\psi(\xi))^{\frac{m-1}{2}}.
\end{aligned}$$

Thus $\int_{\mathbb{R}^n} (\partial^\alpha r(\sigma)) r(\sigma) D_x^\alpha q(x, \xi + (1+\psi(\xi))^{1/4}\,\sigma) d\sigma$ belongs to the symbol class $S_0^{m-1/2,\psi}(\mathbb{R}^n)$ which implies $I_3 \in S_0^{m-1,\psi}(\mathbb{R}^n)$, and (2.254) is proved. In order to see (2.255), we use Taylor's formula to find

$$\begin{aligned}
q_{F,0}(x,\xi) &= \int_{\mathbb{R}^n} r(\sigma)^2 q\left(x, \xi + (1+\psi(\xi))^{1/4}\,\sigma\right)\, d\sigma \\
&= \int_{\mathbb{R}^n} r(\sigma)^2 \Big(q(x,\xi) + \sum_{k=1}^n (1+\psi(\xi))^{1/4}\, \sigma_k \partial_{\xi_k} q(x,\xi)
\end{aligned}$$

$$+\int_0^1 (1-t)\sum_{|\gamma|=2}\frac{2}{\gamma!}\left(1+\psi(\xi)\right)^{1/2}\sigma^\gamma(\partial_\xi^\gamma q)\left(x,\xi+t(1+\psi(\xi))^{1/4}\sigma\right)dt\Big)d\sigma.$$

The symmetry of r implies that the first order term vanishes and therefore

$$q_{F,0}(x,\xi)-q(x,\xi)=\sum_{|\gamma|=2}\frac{2}{\gamma!}\left(1+\psi(\xi)\right)^{1/2}$$
$$\times\int_{\mathbb{R}^n}\int_0^1 (1-t)r(\sigma)^2\sigma^\gamma(\partial_\xi^\gamma q)\left(x,\xi+t\left(1+\psi(\xi)\right)^{1/4}\sigma\right)dtd\sigma. \tag{2.256}$$

Now, from Lemma 2.4.27, $\operatorname{supp} r\subset B_1(0)$ and (2.253), we may conclude that the right hand side of (2.256) defines a symbol in $S_0^{m-2,\psi}(\mathbb{R}^n)$ and (2.255) is proved. □

Now we can prove

Theorem 2.4.28. *Let $q\in S_0^{m,\psi}(\mathbb{R}^n)$ be real-valued. Then $q_F(D_x;x',D_{x'})$ with domain $S(\mathbb{R}^n)$ is a symmetric operator in $L^2(\mathbb{R}^n)$. If in addition $q(x,\xi)$ is non-negative, then $q_F(D_x;x',D_{x'})$ is a non-negative operator.*

Proof: For $u,v\in S(\mathbb{R}^n)$ it follows that

$$(q_F(D_x;x',D_{x'})u,v)_0$$
$$=c\int_{\mathbb{R}^n}F^{-1}_{\xi\to x}\left(\int_{\mathbb{R}^n}\int_{\mathbb{R}^n}e^{-ix'\cdot\xi+ix'\cdot\xi'}q_F(\xi;x',\xi')\hat{u}(\xi')\,d\xi'dx'\right)(x)\overline{v(x)}\,dx$$
$$=c\int_{\mathbb{R}^n}\int_{\mathbb{R}^n}\int_{\mathbb{R}^n}e^{-ix'\cdot\xi+ix'\cdot\xi'}\int_{\mathbb{R}^n}F(\xi,\eta)q(x',\eta)$$
$$\times F(\xi',\eta)\,d\eta\,\hat{u}(\xi')\,d\xi'dx'\overline{\hat{v}(x)}\,d\xi$$
$$=c\int_{\mathbb{R}^n}\int_{\mathbb{R}^n}q(x',\eta)\Big(\int_{\mathbb{R}^n}e^{ix'\cdot\xi'}F(\xi',\eta)\hat{u}(\xi')\,d\xi'\Big)$$
$$\times\overline{\Big(\int_{\mathbb{R}^n}e^{ix'\cdot\xi}F(\xi,\eta)\hat{v}(\xi)\,d\xi\Big)}\,d\eta dx'$$

and the theorem is proved. □

2.5 Estimates for Pseudo-Differential Operators with Negative Definite Symbols Using the Symbolic Calculus

The aim of this section is to use the symbolic calculus developed in the last section to obtain estimates for the operator $q(x, D)$ in related anisotropic Sobolev spaces. These estimates will enable us later on to construct Feller and sub-Markovian semigroups. We will need the theorem of A.P.Calderón and R.Vaillancourt which we will prove first following their paper [41]. We start with some auxiliary results. The first is a *Cotlar-type lemma.*

Lemma 2.5.1. *Let $(Z, \mathcal{A}, \mu)$ be a measure space and $A_z : L^2(\mathbb{R}^n) \to L^2(\mathbb{R}^n)$, $z \in Z$, be a weakly measurable family of uniformly bounded operators, i.e. the mapping $z \mapsto A_z$ is weakly measurable and for all $z \in Z$ we have*

$$\|A_z\| \leq M_0. \tag{2.257}$$

Here, as usual, $\|\cdot\|$ denotes the operator norm. In addition suppose that $k : Z \times Z \to \mathbb{R}_+ \cup \{0\}$ is a non-negative function that give rise to a bounded integral operator K_{op} on $L^2(Z;\mu)$, i.e. $K_{op} : L^2(Z;\mu) \to L^2(Z;\mu)$,

$$K_{op}u(z) = \int_Z k(z, z')u(z')\,dz',$$

with operator norm

$$\|K_{op}\| = M. \tag{2.258}$$

If the inequalities

$$\|A_z A^*_{z'}\| \leq k^2(z, z') \quad \text{and } \|A^*_z A_{z'}\| \leq k^2(z, z') \tag{2.259}$$

holds, then the operator

$$A := \int_Z A_z\,dz \tag{2.260}$$

is bounded in $L^2(\mathbb{R}^n)$ with norm $\|A\| \leq M$.

Proof: For $z_1, \ldots, z_{2m} \in Z$, $m \in \mathbb{N}$, define

$$T_m := \|A_{z_1} A^*_{z_2} A_{z_3} \ldots A^*_{z_{2m}}\|.$$

Multiplying the two inequalities

$$T_m \le \|A_{z_1} A^*_{z_2}\| \cdot \ldots \cdot \|A_{z_{2m-1}} A^*_{z_{2m}}\|$$

and

$$T_m \le \|A_{z_1}\| \|A^*_{z_2} A_{z_3}\| \cdot \ldots \cdot \|A^*_{z_{2m-2}} A_{z_{2m-1}}\| \|A^*_{z_{2m}}\|$$

we obtain

$$T_m^2 \le \|A_{z_1}\| \|A_{z_1} A^*_{z_2}\| \|A^*_{z_2} A_{z_3}\| \cdot \ldots \cdot \|A_{z_{2m-1}} A^*_{z_{2m}}\| \|A^*_{z_{2m}}\|.$$

Using (2.259) we get further

$$\begin{aligned} T_m &\le M_0^{1/2} k(z_1, z_2) k(z_2, z_3) \cdot \ldots \cdot k(z_{2m-1}, z_{2m}) M_0^{1/2} \\ &= M_0 k(z_1, z_2) k(z_2, z_3) \cdot \ldots \cdot k(z_{2m-1}, z_{2m}). \end{aligned}$$

For a measurable set $N \subset Z$ such that $\mu(N) < \infty$ and for $u \in L^2(\mathbb{R}^n)$ such that $\|u\|_0 \le 1$ we find

$$\begin{aligned} &\Big\| \Big[\Big(\int_N A_z \, dz \Big) \Big(\int_N A_z \, dz \Big)^* \Big]^m u \Big\|_0^{1/m} \\ &\qquad = \Big\| \Big(\int_N A_{z_1} \, dz_1 \Big) \Big(\int_N A_{z_2} \, dz_2 \Big)^* \ldots \\ &\qquad\quad \Big(\int_N A_{z_{2m-1}} \, dz_{2m-1} \Big) \Big(\int_N A_{z_{2m}} \, dz_{2m} \Big)^* u \Big\|_0^{1/m} \\ &\qquad = \Big\| \int_N \cdots \int_N A_{z_1} A^*_{z_2} \ldots A_{z_{2m-1}} A^*_{z_{2m}} \, dz_1 \ldots dz_{2m} u \Big\|_0^{1/m} \\ &\qquad \le \Big(\int_N \cdots \int_N \Big\| A_{z_1} A^*_{z_2} \ldots A_{z_{2m-1}} A^*_{z_{2m}} \Big\| \, dz_1 \ldots dz_{2m} \Big)^{1/m} \\ &\qquad \le \Big| M_0 \int_N \int_N \Big\{ \int_Z \cdots \int_Z k(z_1, z_2) \ldots \\ &\qquad\quad k(z_{2m-1}, z_{2m}) \, dz_2 \ldots dz_{2m-1} \Big\} \, dz_1 dz_{2m} \Big|^{1/m} \\ &\qquad = \Big| M_0 \int_N \int_N k^{(2m-1)}(z_1, z_{2m}) \, dz_1 dz_{2m} \Big|^{1/m}, \end{aligned}$$

where $k^{(l)}$ denotes the l-times iterated kernel function. Thus we find

$$\Big\| \Big[\Big(\int_N A_z \, dz \Big) \Big(\int_N A_z \, dz \Big)^* \Big]^m u \Big\|_0^{1/m}$$

$$\le \Big(M_0 \int_Z \int_Z \chi_N(z_1) k^{(2m-1)}(z_1, z_{2m}) \chi(z_{2m})\, dz_{2m} dz_1\Big)^{1/m}$$
$$\le \Big(M_0 \mu(N) M^{2m-1}\Big)^{1/m},$$

which gives

$$\Big\| \Big[\Big(\int_N A_z\, dz\Big)\Big(\int_N A_z\, dz\Big)^*\Big]^m u \Big\|_0^{1/m} \le \Big(M_0 \mu(N) M^{2m-1}\Big)^{1/m}$$

and as $m \to \infty$ it follows that

$$\Big\| \int_N A_z\, dz \Big\|^2 \le M^2,$$

i.e.

$$\Big\| \int_Z A_z\, dz \Big\| \le M,$$

and the lemma is proved. □

Our next auxiliary result is

Lemma 2.5.2. *The function*

$$E(x) := \frac{1}{2} x^2 e^{-x} \chi_{\{x \ge 0\}}(x) \tag{2.261}$$

defined on $\mathbb{R}$ *is a fundamental solution of the differential operator* $\big(1 + \frac{d}{dx}\big)^3$, *i.e.*

$$\left(1 + \frac{d}{dx}\right)^3 E = \varepsilon_0 \tag{2.262}$$

in the sense of distributions.

Proof: From the definition of E it follows for all $\varphi \in C_0^\infty(\mathbb{R})$ and all $\varepsilon > 0$ that

$$\Big\langle \left(1 + \frac{d}{dx}\right)^3 E, \varphi \Big\rangle = \Big\langle E, \left(1 - \frac{d}{dx}\right)^3 \varphi \Big\rangle$$
$$= \int_0^\infty \frac{1}{2} x^2 e^{-x} \left(1 - \frac{d}{dx}\right)^3 \varphi(x)\, dx$$
$$= \int_\varepsilon^\infty \left(1 + \frac{d}{dx}\right)^3 \left(\frac{1}{2} x^2 e^{-x}\right) \varphi(x)\, dx + \int_0^\varepsilon \frac{1}{2} x^2 e^{-x} \left(1 - \frac{d}{dx}\right)^3 \varphi(x)\, dx$$
$$+ \varphi(\varepsilon) e^{-\varepsilon} - 2\varepsilon e^{-\varepsilon} \{\varphi(\varepsilon) + \varphi'(\varepsilon)\} + \frac{1}{2} \varepsilon^2 e^{-\varepsilon} \{\varphi(\varepsilon) + \varphi''(\varepsilon)\}.$$

Since $\left(1+\frac{d}{dx}\right)^3\left(\frac{1}{2}x^2e^{-x}\right)=0$ for $x>0$, it follows as $\varepsilon \to 0$ that

$$\left\langle \left(1+\frac{d}{dx}\right)^3 E, \varphi\right\rangle = \varphi(0),$$

which is of course equivalent to (2.262). □

Now we can prove the **Calderón–Vaillancourt theorem:**

Theorem 2.5.3. *Let $q:\mathbb{R}^n\times\mathbb{R}^n \to \mathbb{C}$ be a function such that for all $\alpha,\beta\in\mathbb{N}_0^n$, $|\alpha|,|\beta|\le 3$, the partial derivatives $\partial_x^\beta\partial_\xi^\alpha q(x,\xi)$ exist, are continuous and satisfy the estimates*

$$|\partial_x^\beta\partial_\xi^\alpha q(x,\xi)| \le c_{\alpha,\beta}. \tag{2.263}$$

Then the pseudo-differential operator $q(x,D)$ which is defined on $S(\mathbb{R}^n)$ by

$$q(x,D)u(x) = (2\pi)^{-n/2}\int_{\mathbb{R}^n} e^{ix\cdot\xi}q(x,\xi)\hat{u}(\xi)\,d\xi$$

extends to a bounded operator from $L^2(\mathbb{R}^n)$ to itself.

Proof: We prove the result for the case $n=1$. For this define

$$g(x,\xi) := \left(1+\frac{\partial}{\partial x}\right)^3\left(1+\frac{\partial}{\partial\xi}\right)^3 q(x,\xi). \tag{2.264}$$

Using the function E from Lemma 2.5.2 we find

$$q(x,\xi) = \int_{\mathbb{R}}\int_{\mathbb{R}} g(s,t)E(x-s)E(\xi-t)\,dsdt. \tag{2.265}$$

Thus on $S(\mathbb{R})$ the operator $q(x,D)$ has the representation

$$\begin{aligned} &q(x,D)u(x)\\ &\quad= (2\pi)^{-1/2}\int_{\mathbb{R}} e^{ix\cdot\xi}\int_{\mathbb{R}}\int_{\mathbb{R}} g(s,t)E(x-s)E(\xi-t)\,dsdt\,\hat{u}(\xi)\,d\xi\\ &\quad= (2\pi)^{-1/2}\int_{\mathbb{R}}\int_{\mathbb{R}} g(s,t)\int_{\mathbb{R}} e^{ix\cdot\xi}E(x-s)E(\xi-t)\hat{u}(\xi)\,d\xi dtds. \end{aligned} \tag{2.266}$$

Defining the operators

$$\begin{aligned} A_{st}u(x) &:= (2\pi)^{-1/2}\int_{\mathbb{R}} e^{ix\cdot\xi}E(x-s)E(\xi-t)\hat{u}(\xi)\,d\xi\\ &= E(x-s)(2\pi)^{-1/2}\int_{\mathbb{R}} e^{ix\cdot\xi}E(\xi-t)\hat{u}(\xi)\,d\xi, \end{aligned}$$

we find

$$q(x,D)u(x) = \int_{\mathbb{R}}\int_{\mathbb{R}} g(s,t)A_{st}u(x)\,dsdt.$$

Put $z = (s,t) \in \mathbb{R}^2$ and observe that $z = (s,t) \mapsto g(s,t)$ is measurable and bounded. Hence it is sufficient to prove that the operator $A_z = A_{st}$ satisfies the assumptions of Lemma 2.5.1 with $(Z,\mathcal{A},\mu) = \left(\mathbb{R}^2, \mathcal{B}^{(2)}, \lambda^{(2)}\right)$. Since $\|E\|_\infty \leq 2e^{-2}$ it follows that

$$\begin{aligned} \|A_z u\|_0^2 &\leq \|E\|_\infty^2 \|F^{-1}(E(\cdot - t)\hat{u}(\cdot))\|_0^2 \\ &= \|E\|_\infty^2 \|E(\cdot - t)\hat{u}(\cdot)\|_0^2 \\ &\leq \|E\|_\infty^4 \|\hat{u}\|_0^2 = \|E\|_\infty^4 \|u\|_0^2, \end{aligned}$$

or

$$\|A_z\| \leq 4e^{-4} =: M_0. \tag{2.267}$$

Next we define the function

$$k^2(s,t) := c_0^2(1+|s|)^{-3}(1+|t|)^{-3}$$

with a constant c_0 which will be determined later. Note that $k \in L^1\left(\mathbb{R}^2\right)$ and we put

$$M := c_0\Big(\int_{\mathbb{R}} (1+|t|)^{-3/2}\,dt\Big)^2. \tag{2.268}$$

Now, let (s,t), $(s',t') \in \mathbb{R}^2$ and note that

$$\begin{aligned} &H_{st,s't'}(x,y) \\ &\quad := (2\pi)^{-1/2}\int_{\mathbb{R}} e^{i(x-y)\cdot\xi}E(x-s)E(\xi-t)E(y-s')E(\xi-t')\,d\xi \\ &\quad = E(x-s)E(y-s')(2\pi)^{-1/2}\int_{\mathbb{R}} e^{i(x-y)\cdot\xi}E(\xi-t)E(\xi-t')d\xi \end{aligned} \tag{2.269}$$

is the kernel of the operator $A_{st}A^*_{s't'}$. With the transformation

$$\omega := x-y, \quad \eta := \xi - t \quad \text{and } \theta_t := t - t' \geq 0$$

we find for the integral in (2.269)

$$\begin{aligned}(2\pi)^{-1/2}e^{i\omega t}&\int_{\mathbb{R}} e^{i\omega\eta}E(\eta)E(\eta+\theta_t)\,d\eta\\&=(2\pi)^{-1/2}\frac{1}{4}e^{i\omega t-\theta_t}\int_0^\infty e^{i\omega\eta}\eta^2(\eta+\theta_t)^2e^{-2\eta}\,d\eta\\&=(2\pi)^{-1/2}\frac{1}{4}e^{i\omega t-\theta_t}\left(-i\frac{d}{d\omega}\right)^2\left(-i\frac{d}{d\omega}+\theta_t\right)^2\left(\frac{1}{2-i\omega}\right).\end{aligned}$$

Analogously we may handle the case $\theta_t<0$ and we obtain a similar expression with $-\theta_t$. Thus it follows that

$$|H_{st,s't'}(x,y)|\le ce^{-|\theta_t|/2}(1+|x-y|)^{-3}E(x-s)E(y-s'),$$

implying that

$$\begin{aligned}\int_{\mathbb{R}}\int_{\mathbb{R}}&|H_{st,s't'}(x,y)|^2\,dxdy\\&\le c'\,e^{-|\theta_t|}\int_{\mathbb{R}}\int_{\mathbb{R}}(1+|x-y|)^{-6}E^2(x-s)E^2(y-s')\,dxdy.\end{aligned}$$

The substitution $u+v=x-s$ and $u-v=y-s'$ for $\theta_s:=s-s'\ge 0$ yields

$$\begin{aligned}\int_{\mathbb{R}}\int_{\mathbb{R}}&(1+|x-y|)^{-6}E^2(x-s)E^2(y-s')\,dxdy\\&=2\int_{\mathbb{R}}\int_{\mathbb{R}}(1+|2v+\theta_s|)^{-6}\,E^2(u+v)E^2(u-v)\,dudv\\&=\frac{1}{8}\int_{\mathbb{R}}(1+|2v+\theta_s|)^{-6}\int_{u\ge|v|}(u^2-v^2)^2e^{-4u}\,dudv\\&\le c\int_{\mathbb{R}}(1+|2v+\theta_s|)^{-6}\int_{u\ge|v|}e^{-u}\,dudv\\&=c\int_{\mathbb{R}}(1+|2v+\theta_s|)^{-6}\,e^{-|v|}\,dv.\end{aligned}\tag{2.270}$$

Since we have the Peetre-type estimate

$$(1+|2v+\theta_s|)^{-6}\le\tilde{c}\,(1+|\theta_s|)^{-6}\,(1+|2v|)^6\,,$$

it follows further from (2.270) that

$$\begin{aligned}\int_{\mathbb{R}}\int_{\mathbb{R}}&(1+|x-y|)^{-6}E^2(x-s)E^2(y-s')\,dxdy\\&\le c'\,(1+|\theta_s|)^{-6}\int_{\mathbb{R}}(1+|2v|)^6\,e^{-|v|}\,dv\le c''\,(1+|\theta_s|)^{-6}\,.\end{aligned}$$

Now we arrive at

$$\begin{aligned}\|A_{st}A^*_{s't'}\|^2 &\leq \int_{\mathbb{R}}\int_{\mathbb{R}} |H_{st,s't'}(x,y)|^2 \, dxdy \\ &\leq c\,(1+|s-s'|)^{-6}\, e^{-2|t-t'|} \\ &\leq c_0^4\,(1+|s-s'|)^{-6}\,(1+|t+t'|)^{-6} \\ &= k^4(s-s',t-t'),\end{aligned}$$

which gives

$$\|A_{st}A^*_{s't'}\|^2 \leq k^4(s-s',t-t'). \tag{2.271}$$

Since the kernel of $A^*_{st}A_{s't'}$ is given by

$$(2\pi)^{-1/2}E(\xi-t)E(\eta-t')\int_{\mathbb{R}} e^{-ix(\xi-\eta)}E(x-s)E(x-s')\,dx$$

which is nothing but $\overline{H_{ts,t's'}(\xi,\eta)}$, it follows further that

$$\|A^*_{st}A_{s't'}\|^2 \leq k^2(s-s',t-t'). \tag{2.272}$$

Thus the operators $(A_z)_{z=(s,t)\in\mathbb{R}^2}$ satisfy the assumptions of Lemma 2.5.1 with $M_0 = 2e^{-2}$ and M given by (2.268). Hence we have proved the theorem for $n=1$. But the general case follows now by observing that we may handle each pair of variables (x_j,ξ_j), $1\leq j\leq n$, separately. □

Now we may return to the class of pseudo-differential operators considered in Section 2.4, and we will follow once again closely the work [127] and [128] of W.Hoh. First we prove

Theorem 2.5.4. *Let $q \in S_0^{m,\psi}(\mathbb{R}^n)$ and let $q(x,D)$ be the corresponding pseudo-differential operator. For all $s\in\mathbb{R}$ the operator $q(x,D)$ maps the space $H^{\psi,m+s}(\mathbb{R}^n)$ continuously into the space $H^{\psi,s}(\mathbb{R}^n)$ and for all $u\in H^{\psi,m+s}(\mathbb{R}^n)$ we have the estimate*

$$\|q(x,D)u\|_{\psi,s} \leq c\|u\|_{\psi,m+s}. \tag{2.273}$$

Proof: First note that it is sufficient to prove (2.273) for all $u\in S(\mathbb{R}^n)$. For $s=m=0$, i.e. $q\in S_0^{0,\psi}(\mathbb{R}^n)$, the assumptions of the Calderón–Vaillancourt theorem, Theorem 2.5.3, are fulfilled which implies that

$$\|q(x,D)u\|_0 \leq c\|u\|_0$$

holds for all $u \in L^2(\mathbb{R}^n)$ and c depends on the constants $\tilde{c}_{\alpha,\beta}$, $|\alpha|$, $|\beta| \leq 3$, from (2.199). Next let $s = 0$ and m arbitrary. It follows that

$$\begin{aligned} &q(x,D)u(x) \\ &\quad = (2\pi)^{-n/2} \int_{\mathbb{R}^n} e^{ix\xi} q(x,\xi) \left(1+\psi(\xi)\right)^{-m/2} \left(1+\psi(\xi)\right)^{m/2} \hat{u}(\xi)\, d\xi \end{aligned}$$

and $q(x,\xi)\left(1+\psi(\xi)\right)^{-m/2}$ belongs to $S_0^{0,\psi}(\mathbb{R}^n)$ which yields the estimate

$$\|q(x,D)u\|_0 \leq c\| \left(1+\psi(D)\right)^{m/2} u\|_0 = c\|u\|_{\psi,m}.$$

Now, by Corollary 2.4.19 we know that $\left(1+\psi(D)\right)^{s/2} \circ q(x,D) \in \Psi_0^{s+m,\psi}(\mathbb{R}^n)$ which leads to

$$\|q(x,D)u\|_{\psi,s} = \| \left(1+\psi(D)\right)^{s/2} q(x,D)u\|_0 \leq c\|u\|_{\psi,s+m},$$

and the theorem is proved. □

Next we give a *sharp Gårding inequality.*

Theorem 2.5.5. *Let $q \in S_\rho^{m,\psi}(\mathbb{R}^n)$ be non-negative, i.e. $q(x,\xi) \geq 0$ for all $x \in \mathbb{R}^n$, $\xi \in \mathbb{R}^n$. Then there exists a constant $K \geq 0$ such that*

$$\operatorname{Re}\left(q(x,D)u,u\right)_0 \geq -K\|u\|^2_{\psi,\frac{m-1}{2}} \tag{2.274}$$

holds for all $u \in S(\mathbb{R}^n)$.

Proof: Denote as before by $q_F(D_x;x',D_{x'})$ the Friedrichs symmetrization of $q(x,D)$. From Theorem 2.4.25 we know that $q - q_{F,L} \in S_0^{m-1,\psi}(\mathbb{R}^n)$, hence by Theorem 2.5.4 the operator $q(x,D) - q_F(D_x;x',D_{x'})$ maps $H^{\psi,m-1+s}(\mathbb{R}^n)$ continuously into $H^{\psi,s}(\mathbb{R}^n)$, $s \in \mathbb{R}$. Moreover, by Theorem 2.4.28 the operator $q_F(D_x;x',D_{x'})$ is non-negative. This implies now

$$\begin{aligned} \operatorname{Re}\left(q(x,D)u,u\right)_0 &= \operatorname{Re}\left(q_F(D_x;x',D_{x'})u,u\right)_0 \\ &\quad + \operatorname{Re}\Big(\left(q(x,D) - q_F(D_x;x',D_{x'})\right)u,u\Big)_0 \\ &\geq \operatorname{Re}\Big(\left(1+\psi(D)\right)^{-\frac{m-1}{4}} \left(q(x,D)\right. \\ &\quad \left. - q_F(D_x;x',D_{x'})\right)u, \left(1+\psi(D)\right)^{\frac{m-1}{4}} u\Big)_0 \\ &\geq -\|\left(q(x,D) - q_F(D_x;x',D_{x'})\right)u\|_{\psi,-\frac{m+1}{2}} \|u\|_{\psi,\frac{m-1}{2}} \\ &\geq -K\|u\|^2_{\psi,\frac{m-1}{2}}. \quad \Box \end{aligned}$$

We want to examine the sesquilinear form

$$B(u,v) := (q(x,D)u,v)_0 \tag{2.275}$$

more closely.

Theorem 2.5.6. *Let $q \in S_\rho^{m,\psi}(\mathbb{R}^n)$ be real-valued and $m > 0$. It follows that*

$$|B(u,v)| \leq c\|u\|_{\psi,\frac{m}{2}}\|v\|_{\psi,\frac{m}{2}} \tag{2.276}$$

holds for all $u, v \in S(\mathbb{R}^n)$. Hence the sesquilinear form B has a continuous extension onto $H^{\psi,m/2}(\mathbb{R}^n)$. If in addition for all $x \in \mathbb{R}^n$

$$q(x,\xi) \geq \gamma_0 (1+\psi(\xi))^{m/2} \quad \text{for } |\xi| \geq R \tag{2.277}$$

for some $\gamma_0 > 0$ and $R \geq 0$ and

$$\lim_{|\xi|\to\infty} \psi(\xi) = \infty \tag{2.278}$$

hold, then we have for all $u \in H^{\psi,m/2}(\mathbb{R}^n)$ the Gårding inequality

$$\operatorname{Re} B(u,u) \geq \frac{\gamma_0}{2}\|u\|^2_{\psi,\frac{m}{2}} - \lambda_0\|u\|_0^2. \tag{2.279}$$

Proof: Using Theorem 2.5.4 we find

$$\begin{aligned} |B(u,v)| &= |(q(x,D)u,v)_0| \\ &= \left|\left((1+\psi(D))^{-m/4} q(x,D)u, (1+\psi(D))^{m/4} v\right)_0\right| \\ &\leq c\|u\|_{\psi,\frac{m}{2}}\|v\|_{\psi,\frac{m}{2}}, \end{aligned}$$

which proves (2.276). Now if (2.277) holds, for λ sufficiently large we have

$$q(x,\xi)+\lambda \geq \gamma_0 (1+\psi(\xi))^{m/2}$$

for all $x \in \mathbb{R}^n$ and $\xi \in \mathbb{R}^n$. Hence the symbol

$$r(x,\xi) = q(x,\xi)+\lambda-\gamma_0 (1+\psi(\xi))^{m/2}$$

is non-negative and Theorem 2.5.5 implies

$$\operatorname{Re} B(u,u) - \gamma_0\|u\|^2_{\psi,\frac{m}{2}} + \lambda\|u\|_0^2 = \operatorname{Re}(r(x,D)u,u)_0 \geq -K\|u\|^2_{\psi,\frac{m-1}{2}}$$

or

$$\operatorname{Re} B(u,u) \geq \gamma_0 \|u\|^2_{\psi,\frac{m}{2}} - K\|u\|^2_{\psi,\frac{m-1}{2}} - \lambda\|u\|_0^2. \tag{2.280}$$

For $m-1 \leq 0$ we have $\|u\|_{\psi,\frac{m-1}{2}} \leq \|u\|_0$ or $-\|u\|_{\psi,\frac{m-1}{2}} \geq -\|u\|_0$ which yields

$$\operatorname{Re} B(u,u) \geq \gamma_0 \|u\|^2_{\psi,\frac{m}{2}} - (K+\lambda)\|u\|_0^2. \tag{2.281}$$

But for $m-1 > 0$ it follows from (2.278) that for any $\varepsilon > 0$ we have

$$(1+\psi(\xi))^{\frac{m-1}{2}} \leq \varepsilon^2 (1+\psi(\xi))^{m/2} + c^2(\varepsilon),$$

which leads to

$$\|u\|^2_{\psi,\frac{m-1}{2}} \leq \varepsilon \|u\|^2_{\psi,\frac{m}{2}} + c(\varepsilon)\|u\|_0^2. \tag{2.282}$$

Taking $\varepsilon = \frac{\gamma_0}{2K}$ we arrive once again at (2.279). □

Remark 2.5.7. Note that the proof of (2.279) also yields the estimate

$$\operatorname{Re} B(u,u) \geq \frac{\gamma_0}{2}\|u\|^2_{\psi,\frac{m}{2}} - \lambda_1 \|u\|^2_{\psi,\frac{m-1}{2}}. \tag{2.283}$$

In fact, for $m-1 \geq 0$ we have $\|u\|_0 \leq \|u\|_{\psi,\frac{m-1}{2}}$, or $-\|u\|_0 \geq -\|u\|_{\psi,\frac{m-1}{2}}$, and (2.280) gives (2.283) for $m-1 \geq 0$. In case $m-1 < 0$ but $m > 0$ condition (2.278) leads to

$$1 \leq \varepsilon^2 (1+\psi(\xi))^{m/2} + \frac{1}{4\varepsilon^2}(1+\psi(\xi))^{\frac{m-1}{2}},$$

which implies also (2.283).

It is also possible to prove a lower bound for the operator $q(x,D)$.

Theorem 2.5.8. *Let $q \in S^{m,\psi}_\rho(\mathbb{R}^n)$ be real-valued and assume (2.277) as well as (2.278). Then for $s > -m$ we have*

$$\frac{\gamma_0^2}{2}\|u\|_{\psi,m+s} \leq \|q(x,D)u\|^2_{\psi,s} + c\|u\|^2_{\psi,m+s-\frac{1}{2}} \tag{2.284}$$

for all $u \in H^{\psi,s+m}(\mathbb{R}^n)$.

Proof: We set $r_s(x,\xi) := q(x,\xi)^2 \left(1+\psi(\xi)\right)^s$ and observe that

$$r_s(x,\xi) \geq \gamma_0^2 \left(1+\psi(\xi)\right)^{m+s} \quad \text{for } |\xi| \geq R.$$

From Corollary 2.4.23.B we know that the leading term in the expansion of the symbol $q^*(x,\xi)$ is given by $\overline{q(x,\xi)} = q(x,\xi)$. Thus we get

$$\begin{aligned} \|q(x,D)u\|_{\psi,s}^2 &= \left(\left(1+\psi(D)\right)^{s/2} q(x,D)u, \left(1+\psi(D)\right)^{s/2} q(x,D)u \right)_0 \\ &= \left(q^*(x,D)\left(1+\psi(D)\right)^s q(x,D)u, u\right)_0 \\ &= \operatorname{Re}\left(r_s(x,D)u,u\right)_0 + \operatorname{Re}\left(\tilde{r}(x,D)u,u\right)_0 \end{aligned}$$

with $\tilde{r}(x,D) \in \Psi_0^{2(s+m)-1,\psi}(\mathbb{R}^n)$. Now we may apply Theorem 2.5.6 in form of Remark 2.5.7 to obtain

$$\|q(x,D)u\|_{\psi,s}^2 \geq \frac{\gamma_0^2}{4}\|u\|_{\psi,m+s}^2 - c\|u\|_{\psi,m+s-\frac{1}{2}}^2 - c'\|u\|_{\psi,m+s-\frac{1}{2}}^2$$

and the theorem is proved. □

In order to use Theorem 2.5.8 to obtain regularity result for certain solutions of the equation $q_\lambda(x,D)u = f$, we have to examine the Friedrichs mollifier more closely in the context of the symbolic calculus introduced in Section 2.4. For this let j be as in (I.2.77) and $j_\varepsilon(x) := \varepsilon^{-n} j(\frac{x}{\varepsilon})$, $\varepsilon > 0$. The Friedrichs mollifier is given by

$$J_\varepsilon(u)(x) = (j_\varepsilon * u)(x) = (2\pi)^{-n/2} \int_{\mathbb{R}^n} e^{ix\xi} \hat{j}(\varepsilon\xi)\hat{u}(\xi)\,d\xi. \tag{2.285}$$

Hence J_ε is a pseudo-differential operator with symbol $(\hat{j}_\varepsilon)(\xi) = \hat{j}(\varepsilon\xi)$.

Lemma 2.5.9. *For any continuous negative definite function $\psi \in \Lambda$, see Definition 2.4.3, and any $\varepsilon > 0$ the function $\xi \mapsto \hat{j}(\varepsilon\xi)$ belongs to the symbol class $S_\rho^{0,\psi}(\mathbb{R}^n)$.*

Proof: Since $\partial_\xi^\alpha \hat{j}(\varepsilon\xi) = \varepsilon^{|\alpha|}\partial_\eta^\alpha \hat{j}(\eta)|_{\eta=\varepsilon\xi}$ we find for $|\xi| \leq 1$ and $\alpha \in \mathbb{N}_0^n$

$$|\partial_\xi^\alpha \hat{j}(\varepsilon\xi)| \leq \varepsilon^{|\alpha|} \sup_{|\xi|\leq 1} \left(1+\psi(\xi)\right)^{\rho(|\alpha|)/2} \|\partial_\eta^\alpha \hat{j}\|_\infty \left(1+\psi(\xi)\right)^{-\rho(|\alpha|)/2}.$$

But for $|\xi| \geq 1$ it follows since $\hat{j} \in S(\mathbb{R}^n)$ that

$$\begin{aligned}|\partial_\xi^\alpha \hat{j}(\varepsilon\xi)| &\leq \varepsilon^{|\alpha|} \frac{(1+\psi(\xi))^{\rho(|\alpha|)/2}}{1+|\varepsilon\xi|^2} (1+\psi(\xi))^{-\rho(|\alpha|)/2} \\ &\leq \varepsilon^{|\alpha|} c_{\psi,\alpha} \frac{1+|\xi|^2}{1+|\varepsilon\xi|^2} (1+\psi(\xi))^{-\rho(|\alpha|)/2} \\ &\leq c_{\psi,\varepsilon,\alpha} (1+\psi(\xi))^{-\rho(|\alpha|)/2},\end{aligned}$$

which proves the lemma. □

For $|\xi| \geq 1$ and $0 < \varepsilon \leq 1$ we find for $|\alpha| = 1$

$$\begin{aligned}\frac{\varepsilon^{|\alpha|} (1+\psi(\xi))^{\rho(|\alpha|)/2}}{(1+|\varepsilon\xi|^2)^{1/2}} &\leq c_\psi \frac{\varepsilon (1+|\xi|^2)^{1/2}}{(1+|\varepsilon\xi|^2)^{1/2}} \\ &\leq c_\psi \frac{(1+|\varepsilon\xi|^2)^{1/2}}{(1+|\varepsilon\xi|^2)^{1/2}} = c_\psi.\end{aligned}$$

But for $|\alpha| \geq 2$ it follows that

$$\begin{aligned}\frac{\varepsilon^{|\alpha|} (1+\psi(\xi))^{\rho(|\alpha|)/2}}{(1+|\varepsilon\xi|^2)^{|\alpha|/2}} &\leq c_\psi \frac{\varepsilon^{|\alpha|} + \varepsilon^{|\alpha|}|\xi|^2}{(1+|\varepsilon\xi|^2)^{|\alpha|/2}} \\ &\leq c_\psi \frac{\varepsilon^{|\alpha|} (1+|\xi|^2)}{(\varepsilon^2+\varepsilon^2|\xi|^2)^{|\alpha|/2}} \leq c'_\psi \frac{1+|\xi|^2}{(1+|\xi|^2)^{|\alpha|/2}} \leq 2c'_\psi.\end{aligned}$$

Thus we have

Corollary 2.5.10. *For any $\psi \in \Lambda$ the Friedrichs mollifier is a pseudo-differential operator with symbol $\hat{j}(\varepsilon\xi) \in S_\rho^{0,\psi}(\mathbb{R}^n)$ and for $0 < \varepsilon \leq 1$ the constants in the estimates* (2.198) *are independent of ε.*

Now take $q \in S_\rho^{m,\psi}(\mathbb{R}^n)$ and consider the commutator

$$[q(x,D), J_\varepsilon] = q(x,D)J_\varepsilon - J_\varepsilon q(x,D) \tag{2.286}$$

which has the double symbol

$$q(x,\xi)\hat{j}(\varepsilon\xi') - \hat{j}(\varepsilon\xi)q(x',\xi'). \tag{2.287}$$

From the asymptotic expansion (2.239) it follows that $[q(x,D), J_\varepsilon]$ has order $m-1$, i.e. it maps $H^{\psi,m-1+s}(\mathbb{R}^n)$ continuously into $H^{\psi,s}(\mathbb{R}^n)$ and the estimates for the remainder terms are uniform for $0 < \varepsilon \leq 1$. For these reasons we have proved

Theorem 2.5.11. *For $q \in S_\rho^{m,\psi}(\mathbb{R}^n)$ and $s \in \mathbb{R}$ there is a constant c independent of ε, $0 < \varepsilon \leq 1$, such that*

$$\|[q(x,D), J_\varepsilon]u\|_{\psi,s} \leq c\|u\|_{\psi,m+s-1} \tag{2.288}$$

holds for all $u \in H^{\psi,m-1+s}(\mathbb{R}^n)$.

For solving the equation $q_\lambda(x,D)u = f$ we may now proceed as in Section 2.3 and set

$$B_\lambda(u,v) = (q(x,D)u,v)_0 + \lambda(u,v)_0. \tag{2.289}$$

From Theorem 2.5.6 we know that B_λ extends to a continuous sesquilinear form on $H^{\psi,m/2}(\mathbb{R}^n)$ which we denote once again by B_λ. As in Definition 2.3.26 we call $u \in H^{\psi,m/2}(\mathbb{R}^n)$ a *variational solution* to the equation

$$q_\lambda(x,D)u = q(x,D)u + \lambda u = f \tag{2.290}$$

for all $\lambda \in \mathbb{R}$ and $f \in L^2(\mathbb{R}^n)$ if

$$B_\lambda(u,\varphi) = (\varphi, f) = \overline{(f,\varphi)_0} \tag{2.291}$$

holds for all $\varphi \in C_0^\infty(\mathbb{R}^n)$, or equivalently for all $\varphi \in H^{\psi,m/2}(\mathbb{R}^n)$.

Theorem 2.5.12. *Suppose that $\varphi \in \Lambda$ fulfills (2.278) and that $q \in S_\rho^{m,\psi}(\mathbb{R}^n)$, $m > 0$, is real -valued and fulfills (2.277). Then for all $\lambda \geq \lambda_0$, λ_0 taken from (2.279), there exists for all $f \in L^2(\mathbb{R}^n)$ a unique variational solution to (2.290).*

Proof: First we note that for $f \in L^2(\mathbb{R}^n)$ a linear continuous functional on $H^{\psi,m/2}(\mathbb{R}^n)$ is given by $\varphi \mapsto (\varphi, f)_0$. This follows from the obvious estimate

$$|(\varphi,f)_0| \leq \|\varphi\|_0\|f\|_0 \leq \|f\|_0\|\varphi\|_{\psi,m/2}.$$

The rest of the proof is completely analogous to the proof of the Theorem 2.3.27. □

Theorem 2.5.13. *Let $q \in S^{m,\psi}(\mathbb{R}^n)$, $m \geq 1$, be as in Theorem 2.5.8. For $f \in H^{\psi,s}(\mathbb{R}^n)$, $s \geq 0$, any variational solution $u \in H^{\psi,m/2}(\mathbb{R}^n)$ to (2.290) belongs to $H^{\psi,m+s}(\mathbb{R}^n)$.*

Proof: For $0 < \varepsilon \leq 1$ set $u_\varepsilon := J_\varepsilon(u)$. Then we have by Proposition 2.3.15 that $u_\varepsilon \in H^{\psi,t+m}(\mathbb{R}^n)$ for all $t \in \mathbb{R}$ and for $t \leq s$ we find

$$\begin{aligned}(q_\lambda(x,D)u_\varepsilon,\varphi)_0 &= B_\lambda(u,J_\varepsilon(\varphi)) + ([q_\lambda(x,D),J_\varepsilon]u,\varphi)_0 \\ &= (J_\varepsilon(\varphi),f)_0 + ([q_\lambda(x,D),J_\varepsilon]u,\varphi)_0 \\ &= (\varphi,J_\varepsilon(f))_0 + ([q_\lambda(x,D),J_\varepsilon]u,\varphi)_0\end{aligned}$$

or

$$\|q_\lambda(x,D)u_\varepsilon\|_{\psi,t} \leq \|J_\varepsilon(f)\|_{\psi,t} + \|[q_\lambda(x,D),J_\varepsilon]u\|_{\psi,t}\,.$$

From Theorem 2.5.8 applied to $q_\lambda(x,D)$ we deduce that

$$\begin{aligned}\|u_\varepsilon\|_{\psi,t+m} &\leq c\|q_\lambda(x,D)u_\varepsilon\|_{\psi,t} + c_1\|u_\varepsilon\|_{\psi,t+m-\frac{1}{2}} \\ &\leq c_2\left(\|f\|_{\psi,t+m} + \|u_\varepsilon\|_{\psi,t+m-\frac{1}{2}} + \|[q_\lambda(x,D),J_\varepsilon]u\|_{\psi,t}\right).\end{aligned}$$

Now, for $t = t_0 := \frac{1-m}{2}$ we find by Theorem 2.5.11, Corollary 2.5.10 and Theorem 2.5.4 that

$$\|u_\varepsilon\|_{\psi,\frac{m}{2}+\frac{1}{2}} \leq c_2\left(\|f\|_{\psi,s} + \|u\|_{\psi,\frac{m}{2}} + \|u\|_{\psi,\frac{m}{2}-\frac{1}{2}}\right),$$

which yields by Proposition 2.3.15 that $u \in H^{\psi,\frac{m}{2}+\frac{1}{2}}(\mathbb{R}^n)$. But now we may repeat our argument with $t_1 = \frac{2-m}{2} \leq t_0 + \frac{1}{2}$ to obtain $u \in H^{\psi,\frac{m}{2}+1}(\mathbb{R}^n)$, and further recursive applications yield the theorem. □

2.6 Feller Semigroups and Sub-Markovian Semigroups Generated by Pseudo-Differential Operators

In this section we will see how it is possible to use the estimates for the operators considered in Section 2.3–2.5 to extend these operators as generators of a Feller semigroup and certain L^p-sub-Markovian semigroups, respectively. Our aim is to use the Hille–Yosida–Ray Theorem, Theorem I.4.5.3, to get the Feller semigroup. Then we will apply Theorem I.4.6.20 in order to get the L^2-sub-Markovian semigroup. Finally we will use some interpolation techniques to construct the L^p-sub-Markovian semigroups for $2 < p < \infty$. In case of symmetric operators we get also the L^p-sub-Markovian semigroups for $1 < p < 2$.

For this purpose let $q(x,\xi)$ be a negative definite symbol and consider on $C_0^\infty(\mathbb{R}^n;\mathbb{R})$ the operator $-q(x,D)$. Since $C_0^\infty(\mathbb{R}^n;\mathbb{R})$ is dense in $C_\infty(\mathbb{R}^n;\mathbb{R})$ and $-q(x,D)$satisfies on $C_0^\infty(\mathbb{R}^n;\mathbb{R})$ the positive maximum principle, see Theorem I.4.5.6, an application of Theorem I.4.5.3 "only" requires to solve for some $\lambda \geq 0$ the equation

$$q_\lambda(x,D)u = q(x,D)u + \lambda u = f \tag{2.292}$$

for a dense set (in $C_\infty(\mathbb{R}^n;\mathbb{R})$) of right-hand sides f in the space $C_0^\infty(\mathbb{R}^n;\mathbb{R})$. This problem is too hard to attack. Our strategy is to consider $q(x,D)$ on a large domain where (2.292) is easier to handle. For this it is necessary to prove that $-q(x,D)$ also satisfies the positive maximum principle on a larger domain.

Theorem 2.6.1. *Let $D(A) \subset C_\infty(\mathbb{R}^n;\mathbb{R})$ and suppose that $A : D(A) \to C_\infty(\mathbb{R}^n;\mathbb{R})$ is a linear operator. In addition assume that $C_0^\infty(\mathbb{R}^n;\mathbb{R}) \subset D(A)$ is an* operator core *of A in the sense that to every $u \in D(A)$ there exists a sequence $(\varphi_k)_{k\in\mathbb{N}}$, $\varphi_k \in C_0^\infty(\mathbb{R}^n;\mathbb{R})$, such that*

$$\lim_{k\to\infty} \|\varphi_k - u\|_\infty = \lim_{k\to\infty} \|A\varphi_k - Au\|_\infty = 0.$$

If $A|_{C_0^\infty}$ satisfies the positive maximum principle on $C_0^\infty(\mathbb{R}^n;\mathbb{R})$, then it satisfies the positive maximum principle also on $D(A)$.

Proof: We have to prove that for any $u \in D(A)$ such that $u(x_0) = \sup_{x\in\mathbb{R}^n} u(x) \geq 0$ it follows that $Au(x_0) \leq 0$. For u and x_0 as above choose $\varphi \in C_0^\infty(\mathbb{R}^n;\mathbb{R})$ such that $\varphi(x_0) = 1$ but $\varphi(x) < 1$ for all $x \neq x_0$. For $\eta > 0$ we have now

$$\sup_{x\in\mathbb{R}^n} (u + \eta\varphi)(x) = u(x_0) + \eta > 0.$$

Since $C_0^\infty(\mathbb{R}^n;\mathbb{R})$ is a core of the operator A, there exists a sequence $(\varphi_k)_{k\in\mathbb{N}}$, $\varphi_k \in C_0^\infty(\mathbb{R}^n;\mathbb{R})$, such that

$$\lim_{k\to\infty} \|\varphi_k^\eta - (u+\eta\varphi)\|_\infty = 0 \quad \text{and} \lim_{k\to\infty} \|A\varphi_k^\eta - A(u+\eta\varphi)\|_\infty = 0. \tag{2.293}$$

For each $k \in \mathbb{N}$ take a point $x_k \in \mathbb{R}^n$ where the function φ_k^η attains its maximum. By the uniform convergence of $(\varphi_k^\eta)_{k\in\mathbb{N}}$ to $u + \eta\varphi$ we find

$$\lim_{k\to\infty} \varphi_k^\eta(x_k) = u(x_0) + \eta. \tag{2.294}$$

Suppose that there is a neighbourhood $U(x_0)$ of x_0 such that $x_k \in U(x_0)$ for all k. Then there is an $\varepsilon > 0$ such that

$$(u+\eta\varphi)(x) < u(x_0)+\eta-\varepsilon$$

for all $x \in \mathbb{R}^n \backslash U(x_0)$, which contradicts (2.294). Thus there is a subsequence of $(x_k)_{k\in\mathbb{N}}$ converging to x_0. Without loss of generality we may assume that $x_k \to x_0$ and that $\varphi_k^\eta(x_k) \geq 0$ by (2.294). Since A satisfies the positive maximum principle on $C_0^\infty(\mathbb{R}^n;\mathbb{R})$ we get

$$\begin{aligned} Au(x_0) &= A(u+\eta\varphi)(x_0) - \eta A\varphi(x_0) \\ &= \lim_{k\to\infty} A\varphi_k^\eta(x_k) - \eta A\varphi(x_0) \leq -\eta A\varphi(x_0), \end{aligned}$$

but $\eta > 0$ was arbitrary, thus we have $Au(x_0) \leq 0$, i.e. A satisfies the positive maximum principle on $D(A)$. □

Remark 2.6.2. Theorem 2.6.1 was proved in concrete situation in [154] and [162]. The proof presented here is taken from W.Hoh [128] who, however, modeled it after these papers.

Now let us come back to equation (2.292). Note that for solving (2.292) we will not use the positive maximum principle, hence we may consider complex-valued functions, i.e. we may work in spaces of complex-valued functions.

The following strategy for solving (2.292) was suggested in [153]. Let us suppose that we may extend $q_\lambda(x, D)$ to some space $H^{\psi,s}(\mathbb{R}^n)$ where ψ is a fixed real-valued continuous negative definite function. In addition suppose that ψ satisfies

$$\psi(\xi) \geq c_0|\xi|^{r_0} \tag{2.295}$$

for some $c_0 > 0$, $r_0 > 0$ and all ξ, $|\xi| \geq R$. Then it follows from Theorem I.3.10.12 that for $s > \frac{n}{r_0}$

$$H^{\psi,s}(\mathbb{R}^n) \hookrightarrow C_\infty(\mathbb{R}^n) \tag{2.296}$$

and

$$\|u\|_\infty \leq c_{s,r_0,n}\|u\|_{\psi,s} \tag{2.297}$$

holds. Thus assuming (2.295) and having in mind some properties of the operators considered in Section 2.3 to Section 2.5 we may consider operators $q_\lambda(x, D)$ satisfying

$$q_\lambda(x, D) : H^{\psi,t_0+s}(\mathbb{R}^n) \to H^{\psi,t_0}(\mathbb{R}^n) \tag{2.298}$$

and

$$\|q_\lambda(x, D)u\|_{\psi,t_0} \leq c\|u\|_{\psi,t_0+2}, \tag{2.299}$$

where $t_0 > \frac{n}{r_0}$ (and $C_0^\infty(\mathbb{R}^n)$ being an operator core). Since now we have $H^{\psi,t_0+2}(\mathbb{R}^n) \subset H^{\psi,t_0}(\mathbb{R}^n) \subset C_\infty(\mathbb{R}^n)$ it follows that $q_\lambda(x, D)$, i.e. $-q_\lambda(x, D)$ with domain $H^{\psi,t_0+2}(\mathbb{R}^n)$ is a densely defined operator on $C_\infty(\mathbb{R}^n)$. In virtue of Theorem 2.6.1 the operator $-q_\lambda(x, D)$ satisfies on $H^{\psi,t_0+2}(\mathbb{R}^n)$ also the positive maximum principle. So far we have reduced an application of the Hille–Yosida–Ray Theorem, Theorem I.4.5.3, to solving the equation $q_\lambda(x, D)u = f$ in the space $H^{\psi,t_0+2}(\mathbb{R}^n)$ for (all) $f \in H^{\psi,t_0}(\mathbb{R}^n)$. The advantage is that we may now work in the scale of Hilbert spaces $H^{\psi,s}(\mathbb{R}^n)$ and we may consider first variational solutions. Suppose that for some $\lambda \geq 0$ there exists for all $f \in L^2(\mathbb{R}^n)$ a variational solution $u \in H^{\psi,1}(\mathbb{R}^n)$ to (2.292), i.e. we may extend the sesquilinear form

$$B_\lambda(u, \varphi) := (q(x, D)u, v)_0 + \lambda(u, v)_0 \tag{2.300}$$

to $H^{\psi,1}(\mathbb{R}^n)$ and that for any $f \in L^2(\mathbb{R}^n)$ there exists $u \in H^{\psi,1}(\mathbb{R}^n)$ such that

$$B_\lambda(u, \varphi) = (\varphi, f)_0 \tag{2.301}$$

holds for all $\varphi \in H^{\psi,1}(\mathbb{R}^n)$. Then we may try to get some "regularity" theory of elliptic type, but now in the scale $H^{\psi,s}(\mathbb{R}^n)$: Prove that $f \in H^{\psi,s_0}(\mathbb{R}^n)$ always implies for a variational solution $u \in H^{\psi,1}(\mathbb{R}^n)$ that $u \in H^{\psi,s_0+2}(\mathbb{R}^n)$. But a sufficient condition for the existence of variational solutions is beside the continuity of B_λ the validity of a Gårding inequality, and the regularity result might be deduced from certain lower bounds for $q(x, D)$. It turns out that we may apply this strategy described above in several situations. First we consider the operators handled in Section 2.3. Let $\psi : \mathbb{R}^n \to \mathbb{R}$ be a fixed continuous negative definite function satisfying

$$\psi(\xi) \geq c_0|\xi|^{r_0} \tag{2.302}$$

for some $c_0 > 0$, $r_0 > 0$ and all $|\xi| \geq 1$. Further recall the definition of the constants $\tilde{c}_{n,k}$, $\tilde{\gamma}_{m,n}$ and c_ψ given in (2.130)–(2.132). We want to apply Theorem 2.3.30 with t_0 such that $H^{\psi,t_0}(\mathbb{R}^n) \hookrightarrow C_\infty(\mathbb{R}^n)$ holds. Thus we have to take $t_0 > \frac{n}{r_0}$, say

$$t_0 := \left[\frac{n}{r_0}\right] + 1. \tag{2.303}$$

Assumption 2.6.3. We assume that the function $q : \mathbb{R}^n \times \mathbb{R}^n \to \mathbb{C}$ is a continuous negative definite symbol having the decomposition

$$q(x,\xi) = q_1(\xi) + q_2(x,\xi) \tag{2.304}$$

such that with ψ as in (2.302) the following conditions hold:

A.1. The function q_1 is assumed to be continuous and negative definite and to satisfy with $\gamma_0 > 0$ and $\gamma_1, \gamma_2 \geq 0$

$$\gamma_0\psi(\xi) \leq \operatorname{Re} q_1(\xi) \leq \gamma_1\psi(\xi) \quad \text{for all } |\xi| \geq 1, \tag{2.305}$$

and

$$|\operatorname{Im} q_1(\xi)| \leq \gamma_2 \operatorname{Re} q_1(\xi) \quad \text{for all } \xi \in \mathbb{R}^n. \tag{2.306}$$

A.2.m_0. Set $m_0 = t_0 + n + 2 = \left[\frac{n}{r_0}\right] + n + 3$, note $t_0 \geq 1$. We assume that $x \mapsto q_2(x,\xi)$ belongs to $C^{m_0}(\mathbb{R}^n)$ for all $\xi \in \mathbb{R}^n$ and that we have the estimate

$$|\partial_x^\alpha q_2(x,\xi)| \leq \varphi_\alpha(x)\,(1+\psi(\xi)) \tag{2.307}$$

for all $\alpha \in \mathbb{N}_0^n$, $|\alpha| \leq m_0$, with functions $\varphi_\alpha \in L^1(\mathbb{R}^n)$.

A.3.m_0. With κ_2 from (2.181) and $\tilde{c}_{n,m_0,\frac{t_0}{2},\psi}$ from (2.154) we require

$$\sum_{|\alpha|\leq m_0} \|\varphi_\alpha\|_{L^1} \leq \frac{1}{4}\gamma_0\Big(\frac{1}{\kappa_2} \wedge \frac{1}{\tilde{c}_{n,m_0,\frac{t_0}{2},\psi}}\Big). \tag{2.308}$$

Note that A.3.m_0 implies that (2.162) holds with $\delta_0 = \frac{1}{4}\gamma_0$ and (2.184) holds with $\delta_1 = \frac{1}{4}\gamma_0$, too.

Theorem 2.6.4. *Suppose that Assumption* 2.6.3 *holds with* $t_0 = \left[\frac{n}{r_0}\right] + 1$. *Then* $-q(x,D)$ *extends to a generator of a Feller semigroup.*

Proof: Consider the operator $(A, D(A))$ on $C_\infty(\mathbb{R}^n;\mathbb{R})$ with domain $D(A) = H^{\psi,t_0+2}(\mathbb{R}^n;\mathbb{R})$ and $A = -q(x,D)$. It follows by Proposition 2.3.6 and Theorem 2.3.11 that $Au \in H^{\psi,t_0}(\mathbb{R}^n;\mathbb{R})$ if $u \in D(A)$, and therefore $Au \in C_\infty(\mathbb{R}^n;\mathbb{R})$ and $(A, D(A))$ is a well defined operator on $C_\infty(\mathbb{R}^n;\mathbb{R})$ with a dense domain. Moreover, by Theorem 2.6.1 it satisfies on $D(A)$ the positive maximum principle. Further we may apply Theorem 2.3.30, thus for $\lambda \geq \tilde{\lambda}_0$, $\tilde{\lambda}_0$ taken from (2.170) we find for every $f \in H^{\psi,t_0}(\mathbb{R}^n;\mathbb{R})$ a (unique) $u \in H^{\psi,t_0+2}(\mathbb{R}^n;\mathbb{R})$ satisfying

$$(A-\lambda)u = -q_\lambda(x,D)u = f.$$

The theorem follows now by the Hille–Yosida–Ray Theorem, Theorem I.4.5.3.

Remark 2.6.5. For real-valued symbol Theorem 2.6.4. was proved in [155], an extension to complex-valued symbol was given in [156].

The next step is to use our estimates in order to prove that $-q_\lambda(x,D)$, λ sufficiently large, extends to a generator of an L^2-sub-Markovian semigroup. The idea is to prove that $-q_\lambda(x,D)$ extends to a generator of an L^2-contraction semigroup and then to apply Theorem I.4.6.20.

Theorem 2.6.6. *Under the assumption of Theorem* 2.3.27 *and* (2.162) *with* $s = 2$ *the operator* $-q_\lambda(x,D)$ *defined on* $H^{\psi,2}(\mathbb{R}^n;\mathbb{R})$ *is for* $\lambda \geq \tilde{\lambda}_0$, $\tilde{\lambda}_0$ *from* (2.179), *a generator of a strongly continuous contraction semigroup on* $L^2(\mathbb{R}^n;\mathbb{R})$.

Proof: We will apply the Hille–Yosida theorem, Theorem I.4.1.33. From Section 2.3 we know that $\left(-q_\lambda(x,D), H^{\psi,2}(\mathbb{R}^n;\mathbb{R})\right)$ is a densely defined closed operator on $L^2(\mathbb{R}^n;\mathbb{R})$ and for $\lambda \geq \tilde{\lambda}_0$ the equation $q_\lambda(x,D)u = f$ is uniquely solvable for all $f \in L^2(\mathbb{R}^n;\mathbb{R})$ with a solution $u \in H^{\psi,2}(\mathbb{R}^n;\mathbb{R})$ by Theorem 2.3.28. It remains to prove that $-q_\lambda(x,D)$, $\lambda \geq \tilde{\lambda}_0$ is dissipative. But for all $\varphi \in C_0^\infty(\mathbb{R}^n;\mathbb{R})$ we have

$$(q_\lambda(x,D)\varphi,\varphi)_0 \geq 0$$

which extends to all $u \in H^{\psi,2}(\mathbb{R}^n;\mathbb{R})$. Now the dissipativity of $-q_\lambda(x,D)$, $\lambda \geq \tilde{\lambda}_0$, follows from

$$\|\tau u+q_\lambda(x,D)u\|_0^2 = \tau^2\|u\|_0^2+\|q_\lambda(x,D)\|_0^2+2\tau(u,q_\lambda(x,D)u)$$

and the theorem is proved. □

Corollary 2.6.7. *Under the assumptions of Theorem* 2.6.6 *and Assumption* 2.6.3 *with* $t_0 = \left[\frac{n}{r_0}\right] + 3$ *the operator* $\left(-q_\lambda(x,D), H^{\psi,2}(\mathbb{R}^n;\mathbb{R})\right)$ *is for all* $\lambda \geq \tilde{\lambda}_0$ *a Dirichlet operator and generates an* L^2*-sub-Markovian semigroup.*

Proof: We only have to apply Theorem I.4.6.20 for $p = 2$ with $U = H^{\psi,t_0+2}(\mathbb{R}^n;\mathbb{R}) \subset D(A^{(\infty)})$ and to note that $[q_\lambda(x,D)]^{-1}\left(H^{\psi,t_0+2}(\mathbb{R}^n;\mathbb{R})\right) = H^{\psi,t_0+4}(\mathbb{R}^n;\mathbb{R})$. □

Remark 2.6.8. We have seen that $-q_\lambda(x,D)$ extends to a generator of a Feller semigroup $(T_t^{(\infty)})_{t\geq 0}$ and $-q_\lambda(x,D)$, $\lambda \geq \tilde{\lambda}_0$, extends to a generator of an L^2-sub-Markovian semigroup $(T_t^{(2),\lambda})_{t\geq 0}$ as well as to a Feller semigroup $(T_t^{(\infty),\lambda})_{t\geq 0}$. Note that $T_t^{(\infty),\lambda} = e^{-\lambda t}T_t^{(\infty)}$. Clearly $(e^{\lambda t}T_t^{(2),\lambda})_{t\geq 0}$ is a strongly continuous L^2-semigroup and its generator is $\left(-q(x,D), H^{\psi,t_0+2}(\mathbb{R}^n;\mathbb{R})\right)$. However, in general we are not able to prove that this semigroup consists of L^2-contractions. We only know the estimate $\|e^{\lambda t}T_t^{(2),\lambda}\|_0 \leq e^{\lambda t}\|u\|_0$, i.e. following Definition I.4.1.8, it has growth bound λ. In case that we have constructed a symmetric Feller semigroup we may improve this estimate by Theorem I.4.6.25. In fact, this theorem yields that $(T_t^{(\infty)}|_{L^2\cap C_\infty})_{t\geq 0}$ extends to an L^2-sub-Markovian semigroup $(T_t^{(2)})_{t\geq 0}$ with generator being an extension of $-q(x,D)$ and it is clear that the semigroups $(T_t^{(2)})_{t\geq 0}$ and $(e^{\lambda t}T_t^{(2),\lambda})_{t\geq 0}$ must coincide. Thus we find that $e^{\lambda t}T_t^{(2),\lambda} = T_t^{(2),\lambda}$ is an L^2-contraction. Note that it follows further that $(q(x,D)u,v)_0$ extends to a symmetric Dirichlet form with domain $H^{\psi,1}(\mathbb{R}^n;\mathbb{R})$. Therefore we have $B(u,u) \geq 0$ for all $u \in H^{\psi,1}(\mathbb{R}^n;\mathbb{R})$, B being the extension of $(q(x,D)u,v)_0$, which yields the improved Gårding inequality

$$B(u,u) + \varepsilon\|u\|_0^2 \geq c_\varepsilon\|u\|_{\psi,1}^2, \quad \varepsilon > 0. \tag{2.309}$$

Next we turn our attention to the class of pseudo-differential operators considered by W.Hoh in [127]–[130]. The estimates provided in Section 2.5 allows us to follow the same strategy as before. First we get

Theorem 2.6.9. *Let* $\psi : \mathbb{R}^n \to \mathbb{R}$ *be a continuous negative definite function in the class* Λ, *see Definition* 2.4.3, *and suppose in addition* (2.295) *to hold. If* $q(x,\xi)$ *is a negative definite symbol belonging to* $S_\rho^{2,\psi}(\mathbb{R}^n)$ *and satisfies*

$$q(x,\xi) \geq \delta\,(1+\psi(\xi)) \tag{2.310}$$

for some $\delta > 0$ *and all* $\xi \in \mathbb{R}^n$, $|\xi|$ *sufficiently large, then* $-q(x,D)$ *defined on* $C_0^\infty(\mathbb{R}^n;\mathbb{R})$ *is closable in* $C_\infty(\mathbb{R}^n;\mathbb{R})$ *and its closure is a generator of a Feller semigroup.*

Proof: We may argue as in the proof of Theorem 2.6.4. We choose $t_0 > \frac{n}{r_0}$ and find by Theorem 2.5.4 that $-q(x,D) : H^{\psi,t_0+2}(\mathbb{R}^n;\mathbb{R}) \to H^{\psi,t_0}(\mathbb{R}^n;\mathbb{R}) \subset C_\infty(\mathbb{R}^n;\mathbb{R})$ is a densely defined operator on $C_\infty(\mathbb{R}^n;\mathbb{R})$ with domain $H^{\psi,t_0+2}(\mathbb{R}^n;\mathbb{R})$, and further by Theorem 2.6.1 it satisfies on $H^{\psi,t_0+2}(\mathbb{R}^n;\mathbb{R})$ the positive maximum principle. Moreover, by Theorem 2.5.12 for $\lambda \geq \lambda_0$, λ_0 taken from (2.279), there exists for all $f \in L^2(\mathbb{R}^n;\mathbb{R})$ a unique variational solution $u \in H^{\psi,1}(\mathbb{R}^n;\mathbb{R})$ to the equation $q_\lambda(x,D)u = f$ and Theorem 2.5.13 tells us that for $f \in H^{\psi,t_0}(\mathbb{R}^n;\mathbb{R})$ this solution belongs to $H^{\psi,t_0+2}(\mathbb{R}^n;\mathbb{R})$. Hence all conditions of Hille–Yosida–Ray Theorem are satisfied and the theorem is proved. □

As in the case before we obtain also L^2-sub-Markovian semigroups.

Theorem 2.6.10. *Under the assumption of Theorem* 2.6.9 *the operator* $\left(-q_\lambda(x,D), H^{\psi,2}(\mathbb{R}^n;\mathbb{R})\right)$, $\lambda > \lambda_0$, λ_0 *as in* (2.279), *is the generator of an* L^2*-contraction semigroup which is in virtue of Theorem I.4.6.20 an* L^2*-sub-Markovian semigroup. Hence* $\left(-q_\lambda(x,D), H^{\psi,2}(\mathbb{R}^n;\mathbb{R})\right)$, *is a Dirichlet operator and* $\left(B_\lambda, H^{\psi,1}(\mathbb{R}^n;\mathbb{R})\right)$ *is a semi-Dirichlet form.*

Remark 2.6.11. In case of a symmetric operator $-q(x,D)$ all considerations in Remark 2.6.8 apply also to the operators handled in Theorem 2.6.9 and 2.6.10.

So far we construct for two classes of pseudo-differential operators Feller semigroups and L^2-sub-Markovian semigroups. In both cases certain (disturbing) restriction show up. Either the symbol has "coefficients" $x \mapsto q(x,\xi)$ which have only a small and controlled oscillation, or the function $\xi \mapsto q(x,\xi)$ must have certain smoothness properties which in general a continuous negative definite function does not share. However starting with one of these two classes and applying perturbation theory we may enlarge the class of pseudo-differential operators generating Feller and (or) L^2-sub-Markovian semigroups. We will come back to this question in Section 2.8.

In the remaining of this section we want to discuss the question how (if possible) to get L^p-sub-Markovian semigroups for $p \in [1,\infty)$ starting with the Feller and L^2-sub-Markovian semigroups just constructed. The first part of these considerations is of more abstract nature and it will be applied later on also to the semigroups obtained via a perturbation of the semigroups just constructed. Although we may use Theorem I.4.6.20 to get some L^p-intermediate semigroups, we prefer to use some interpolation result developed in [86].

Let $-q_\lambda(x, D)$ be first defined on $C_0^\infty(\mathbb{R}^n;\mathbb{R})$ and suppose that it extends to a generator $A_\lambda^{(\infty)}$ of a Feller semigroup $(T_t^{(\infty),\lambda})_{t\geq 0}$ and to a generator $A_\lambda^{(p_0)}$, $p_0 > 1$, of an L^{p_0}-sub-Markovian semigroup $(T_t^{(p_0),\lambda})_{t\geq 0}$, respectively. In order to interpolate the operators $T_t^{(\infty),\lambda}$ and $T_t^{(p_0),\lambda}$ we have to overcome a difficulty, namely the fact that $T_t^{(\infty),\lambda}u$ is pointwise defined whereas $T_t^{(p_0),\lambda}u$ is only almost everywhere defined. The following considerations are taken from our joint paper [86] with W.Farkas and R.L.Schilling.

A function $u \in B_b(\mathbb{R}^n;\mathbb{R})$ is uniquely determined on $\mathbb{R}^n$ whereas an element $v \in L^p(\mathbb{R}^n;\mathbb{R})$ is an equivalence class of functions which may differ on a set of Lebesgue measure zero. By writing $u \in B_b(\mathbb{R}^n;\mathbb{R}) \cap L^p(\mathbb{R}^n;\mathbb{R})$ we mean the uniquely determined element in $B_b(\mathbb{R}^n;\mathbb{R})$. Let $S^{(p)} : L^p(\mathbb{R}^n;\mathbb{R}) \to L^p(\mathbb{R}^n;\mathbb{R})$ and $\tilde{S}^{(\infty)} : B_b(\mathbb{R}^n;\mathbb{R}) \to B_b(\mathbb{R}^n;\mathbb{R})$ be two linear operators and let $u \in B_b(\mathbb{R}^n;\mathbb{R}) \cap L^p(\mathbb{R}^n;\mathbb{R})$. Then $\tilde{S}^{(\infty)}u \in B_b(\mathbb{R}^n;\mathbb{R})$ and $S^{(p)}u \in L^p(\mathbb{R}^n;\mathbb{R})$. For the latter we have $S^{(p)}u = S^{(p)}v$ almost everywhere whenever $v = u$ a.e., i.e. although $u \in B_b(\mathbb{R}^n;\mathbb{R}) \cap L^p(\mathbb{R}^n;\mathbb{R})$ can be constructed as a uniquely determined function, $S^{(p)}u$ is still an equivalence class of functions and all $v \in L^p(\mathbb{R}^n;\mathbb{R})$ such that $u = v$ almost everywhere are mapped via $S^{(p)}$ into the equivalence class of $S^{(p)}u$.

Definition 2.6.12. We write

$$S^{(p)}|_{L^p\cap B_b} = \tilde{S}^{(\infty)}_{L^p\cap B_b} \tag{2.311}$$

if and only if for $u \in L^p(\mathbb{R}^n;\mathbb{R}) \cap B_b(\mathbb{R}^n;\mathbb{R})$ we have

$$S^{(p)}u = \tilde{S}^{(\infty)}u \quad a.e. \tag{2.312}$$

Since $\tilde{S}^{(\infty)}u$ is uniquely determined, this allows the interpretation $S^{(p)}u \in B_b(\mathbb{R}^n;\mathbb{R}) \cap L^p(\mathbb{R}^n;\mathbb{R})$, i.e. we may choose $\tilde{S}^{(\infty)}u$ as representative for $S^{(p)}u$.

Theorem 2.6.13. *Let $S^{(p_0)}$ be a linear contraction on $L^{p_0}(\mathbb{R}^n)$ and $\tilde{S}^{(\infty)}$ a linear contraction on $B_b(\mathbb{R}^n)$ such that*

$$S^{(p_0)}|_{L^{p_0}\cap B_b} = \tilde{S}^{(\infty)}_{L^{p_0}\cap B_b}. \tag{2.313}$$

Then there exists for every p, $1 < p_0 < p < \infty$, a linear contraction $S^{(p)}$ on $L^{p_0}(\mathbb{R}^n)$ such that

$$S^{(p)}|_{L^{p_0}\cap L^p\cap B_b} = \tilde{S}^{(\infty)}_{L^{p_0}\cap L^p\cap B_b}, \tag{2.314}$$

and therefore

$$S^{(p)}|_{L^{p_0}\cap L^p} = \tilde{S}^{(p_0)}_{L^{p_0}\cap L^p}. \tag{2.315}$$

Remark 2.6.14. Theorem 2.6.13 is a type of Riesz–Thorin interpolation theorem, compare Theorem I.2.8.1, with the space $L^\infty(\mathbb{R}^n)$ of the interpolation couple being substituted by $B_b(\mathbb{R}^n)$.

We follow the proof from [86]; we do not know another precise reference for this situation. The proof, however, follows the standard proof, see for example the monograph [24] of C.Bennett and R.Sharpley, Theorem 2.2 on page 196.

Proof of Theorem 2.6.13: We denote by S the operator

$$S := S^{(p_0)}|_{L^{p_0}\cap B_b} = \tilde{S}^{(\infty)}_{L^{p_0}\cap B_b}.$$

Since for $\frac{1}{p}+\frac{1}{p'}=1$

$$\begin{aligned}\|S\|_{L^p-L^p} &= \sup\{\|Su\|_{L^p};\quad u\in L^p(\mathbb{R}^n) \quad\text{and}\quad \|u\|_{L^p}=1\}\\ &= \sup\left\{\left|\int_{\mathbb{R}^n}(Su)(x)v(x)\,dx\right|\right\},\end{aligned}$$

where the supremum is taken over all $u\in L^p(\mathbb{R}^n)$ with $\|u\|_{L^p}=1$ and all $v\in L^{p'}(\mathbb{R}^n)$ with $\|v\|_{L^{p'}}=1$, it is sufficient to show that

$$\left|\int_{\mathbb{R}^n}(Su)(x)v(x)\,dx\right|\le 1 \tag{2.316}$$

for all simple function u,v satisfying $\|u\|_{L^p}=\|v\|_{L^{p'}}=1$. Such functions u and v have a representation

$$u=\sum_{j=1}^{J}a_j\chi_{A_j} \quad\text{and}\quad v=\sum_{k=1}^{K}b_k\chi_{B_k}, \tag{2.317}$$

where the sets A_j, $j=1,\ldots,J$, and B_k, $k=1,\ldots,K$, are two families of pairwise disjoint Borel sets with finite Lebesgue measure, and the coefficients a_j and b_k satisfy

$$\sum_{j=1}^{J}|a_j|^p\lambda^{(n)}(A_j)=\sum_{k=1}^{K}|b_k|^{p'}\lambda^{(n)}(B_k)=1.$$

For each $z\in\mathbb{C}$ let $\alpha(z):=\frac{1-z}{p_0}$ and let $\theta\in(0,1)$ be such that $\alpha(\theta)=\frac{1}{p}$, i.e. $\frac{1-\theta}{p_0}=\frac{1}{p}$. Further, for $z\in\mathbb{C}$ define

$$u_z:=|u|^{\alpha(z)/\alpha(\theta)}e^{i\arg u},\quad v_z:=|u|^{\frac{1-\alpha(z)}{1-\alpha(\theta)}}e^{i\arg v}$$

and set

$$F(z) := \int_{\mathbb{R}^n} (Su_z)(x)v_z(x)\,dx.$$

Since

$$F(\theta) = \int_{\mathbb{R}^n} (Su_\theta)(x)v_\theta(x)\,dx = \int_{\mathbb{R}^n} (Su)(x)v(x)\,dx$$

the desired estimate (2.316) will follow from Hadamard three lines theorem, Theorem I.2.5.1.B, if we show that F is analytic in $\Omega := \{x+iy;\ 0<x<1$ and $y \in \mathbb{R}\}$ and continuous and bounded in $\overline{\Omega}$ with

$$|F(iy)| \le 1 \quad \text{and } |F(1+iy)| \le 1 \text{ for all } y \in \mathbb{R}.$$

First note that

$$F(z) = \sum_{j=1}^{J}\sum_{k=1}^{K} |a_j|^{\alpha(z)/\alpha(\theta)}|b_k|^{\frac{1-\alpha(z)}{1-\alpha(\theta)}} \int_{\mathbb{R}^n} (S\chi_{A_j}(x)\chi_{B_k}(x)e^{i(\arg a_j+\arg b_k)}\,dx$$

which shows that $z \mapsto F(z)$ is an entire function. It also reveals that F is bounded in $\overline{\Omega}$ since the real part of $\alpha(z)$ is bounded there. Applying Hölder's inequality and using the fact that S is an L^{p_0}-contraction we have

$$\begin{aligned} |F(iy)| &\le \int_{\mathbb{R}^n} |(Su_{iy})(x)v_{iy}(x)|\,dx \\ &\le \|Su_{iy}\|_{L^{p_0}}\|v_{iy}\|_{L^{p_0'}} \le \|u_{iy}\|_{L^{p_0}}\|v_{iy}\|_{L^{p_0'}}. \end{aligned} \tag{2.318}$$

Moreover, since $\alpha(\theta) = \frac{1}{p}$ and $\operatorname{Re}\alpha(iy) = \frac{1}{p_0}$, we have

$$\|u_{iy}\|_{L^{p_0}}^{p_0} = \sum_{j=1}^{J}\left||a_j|^{\alpha(z)/\alpha(\theta)}\right|^{p_0}\lambda^{(n)}(A_j) = \sum_{j=1}^{J}|a_j|^p\lambda^{(n)}(A_j) = 1, \tag{2.319}$$

and similarly, since $1-\alpha(\theta) = \frac{1}{p_0}$,

$$\begin{aligned} \|v_{iy}\|_{L^{p_0'}}^{p_0'} &= \sum_{k=1}^{K}\left||b_k|^{(1-\alpha(z))/(1-\alpha(\theta))}\right|^{p_0}\lambda^{(n)}(B_k) \\ &= \sum_{k=1}^{K}|b_k|^{p'}\lambda^{(n)}(B_k) = 1. \end{aligned} \tag{2.320}$$

From (2.318)–(2.320) we conclude that $|F(iy)| \leq 1$. In order to estimate $|F(1+iy)|$ observe that

$$|u_{1+iy}(x)| = \left| |u(x)|^{\alpha(1+iy)/\alpha(\theta)} \right| = 1 \tag{2.321}$$

since $\alpha(\theta) = \frac{1}{p}$ and $\operatorname{Re} \alpha(1+iy) = 0$, and

$$|v_{1+iy}(x)| = \left| |v(x)^{(1-\alpha(1+iy))/(1-\alpha(\theta))} \right| = |v(x)|^{p'} \tag{2.322}$$

because of $1 - \alpha(\theta) = 1 - \frac{1}{p} = \frac{1}{p'}$ and $\operatorname{Re}(1 - \alpha(1+iy)) = 1$. Using (2.321), (2.322) and the fact that S is also a contraction on $B_b(\mathbb{R}^n)$ we arrive at

$$\begin{aligned} |F(1+iy)| &\leq \int_{\mathbb{R}^n} |(Su_{1+iy})(x) v_{1+iy}(x)| \, dx \\ &\leq \|Su_{1+iy}\|_{L^\infty} \int_{\mathbb{R}^n} |(v_{1+iy}(x)| \, dx \leq \int_{\mathbb{R}^n} |v(x)|^{p'} \, dx = 1, \end{aligned}$$

and the theorem is proved. □

Remark 2.6.15. A. Clearly the consideration of Theorem 2.6.13 holds for $B_b(\mathbb{R}^n)$ substituted by $L^\infty(\mathbb{R}^n)$.

B. Let $(T_t^{(p_0)})_{t\geq 0}$ be an L^p-sub-Markovian semigroup on $L^{p_0}(\mathbb{R}^n;\mathbb{R})$. For $u \in L^{p_0}(\mathbb{R}^n;\mathbb{R}) \cap L^\infty(\mathbb{R}^n;\mathbb{R})$, $p_0 > 1$, we find using the decomposition $u = u^+ - u^-$ for $t > 0$ that

$$\begin{aligned} |T_t^{(p_0)} u| &= T_t^{(p_0)} u^+ - T_t^{(p_0)} u^- \leq T_t^{(p_0)} u^+ + T_t^{(p_0)} u^- \\ &= T_t^{(p_0)} (u^+ + u^-) = T_t^{(p_0)} |u| \end{aligned}$$

implying for $u \in L^{p_0}(\mathbb{R}^n;\mathbb{R}) \cap L^\infty(\mathbb{R}^n;\mathbb{R})$ further

$$\frac{1}{\|u\|_{L^\infty}} |T_t^{(p_0)} u| = \left| T_t^{(p_0)} \left(\frac{u}{\|u\|_{L^\infty}} \right) \right| \leq T_t^{(p_0)} \left(\frac{|u|}{\|u\|_{L^\infty}} \right) \leq 1$$

which yields

$$\|T_t^{(p_0)} u\|_{L^\infty} \leq \|u\|_{L^\infty}. \tag{2.323}$$

Since every $u \in L^\infty(\mathbb{R}^n;\mathbb{R})$, $u \geq 0$ a.e., is a monotone limit of a sequence $(u_\nu)_{\nu\in\mathbb{N}}$, $u_\nu \in L^{p_0}(\mathbb{R}^n;\mathbb{R}) \cap L^\infty(\mathbb{R}^n;\mathbb{R})$ and $u_\nu \geq 0$ a.e., we may extend $T_t^{(p_0)}$ to $L^\infty(\mathbb{R}^n;\mathbb{R})$ and it is easily seen that this extension is independent of the special choice of approximating sequences for u^+ and u^-. This extension $(T_t^{(\infty)})_{t\geq 0}$ is a contraction semigroup of sub-Markovian operator on $L^\infty(\mathbb{R}^n;\mathbb{R})$, but it is in general not strongly continuous. Hence according to part A of this remark all intermediate semigroups $(T_t^{(p)})_{t\geq 0}$, $p_0 < p < \infty$, do exist as L^p-sub-Markovian semigroups.

Now let $-q_\lambda(x,D)$ be a pseudo-differential operator with negative definite symbol which extends to a generator $A_\lambda^{(\infty)}$ of a Feller semigroup $(T_t^{(\infty),\lambda})_{t\geq 0}$ as well as to a generator $A_\lambda^{(2)}$ of L^2-sub-Markovian semigroup $(T_t^{(2),\lambda})_{t\geq 0}$ with $D(A_\lambda^{(2)}) = H^{\psi,2}(\mathbb{R}^n;\mathbb{R})$ for some continuous negative definite function $\psi : \mathbb{R}^n \to \mathbb{R}$. Moreover denote by $(\tilde{T}_t^{(\infty),\lambda})_{t\geq 0}$ the extension of $(T_t^{(\infty),\lambda})_{t\geq 0}$ to the space $B_b(\mathbb{R}^n;\mathbb{R})$ which according to Theorem I.4.8.1 and Remark I.4.8.2 always exists.

Corollary 2.6.16. *Let $-q_\lambda(x,D)$, $\lambda \geq 0$, be as above and suppose that for all $t \geq 0$ the operator $\tilde{T}_t^{(\infty),\lambda}|_{L^2\cap B_b}$ and $T_t^{(2),\lambda}|_{L^2\cap B_b}$ coincide. Then $\tilde{T}_t^{(2),\lambda}|_{L^2\cap B_b}$ extends for all $2 < p < \infty$ to a contraction $T_t^{(p),\lambda}$ on $L^p(\mathbb{R}^n;\mathbb{R})$. The family $(T_t^{(p),\lambda})_{t\geq 0}$ is an L^p-sub-Markovian semigroup and $-q_\lambda(x,D)$ extends to an L^p-Dirichlet operator $A_\lambda^{(p)}$.*

Remark 2.6.17. A. Note that for the operators considered in Theorem 2.6.6. and Theorem 2.6.10 the assumptions of Corollary 2.6.16 are fulfilled which follows for example from the proof of Theorem I.4.6.20.

B. It remains to determine the domain of $A_\lambda^{(p)}$ in terms of function spaces, see however Section 3.2 and 3.3.

2.7 Further Analytic Approaches for Constructing Feller and Sub-Markovian Semigroups

In Section 2.3–Section 2.6 we considered pseudo-differential operators with negative definite symbols. In particular our intention was to work with rather general continuous negative definite functions ψ (or $\xi \mapsto q(x,\xi)$) and to develop tools to handle the associated operators. However, another way is also possible. One may use existing "classical" theorems and calculi for the pseudo-differential operators and restrict oneself to symbols $q(x,\xi)$ having the additional property that $x \mapsto q(x,\xi)$ is a continuous negative definite function. In this section we briefly want to discuss some results in this direction, but we will omit often the proofs and refer to the literature. We will handle three approaches. First we apply the strategy of Section 2.6 to symbols in the class $S^m_{\rho,\delta}(\mathbb{R}^n)$. Then we indicate how to construct a parametrix (by using the method of E.E.Levi [198]) to the "parabolic" problem $\left(\frac{\partial}{\partial t} + q(x,D)\right) u = 0$, $u(0,x) = f(x)$, and how this parametrix can be used to get a transition function and hence a (Feller) semigroup. Finally we shortly mention results related to a Weyl calculus for pseudo-differential operators. Note that generators which are fractional powers of given generators will be examined in Section 2.9 within the general context of subordination.

Our main reference for handling operators with symbols in the class $S^m_{\rho,\delta}(\mathbb{R}^n)$ is the monograph [189] of H.Kumano-go.

Definition 2.7.1. Let $m \in \mathbb{R}$, $0 \le \delta \le \rho \le 1$, $\delta < 1$, and $q : \mathbb{R}^n \times \mathbb{R}^n \to \mathbb{C}$ an arbitrarily often differentiable function. We say that $q(x,\xi)$ belongs to the *symbols class* $S^m_{\rho,\delta}(\mathbb{R}^n)$ if for any pair of multiindices $\alpha, \beta \in \mathbb{N}^n_0$ there exists a constant $c_{\alpha,\beta}$ such that

$$|\partial^\alpha_\xi D^\beta_x q(x,\xi)| \le c_{\alpha\beta}(1+|\xi|^2)^{\frac{m+\delta|\beta|-\rho|\alpha|}{2}} \tag{2.324}$$

holds.

Example 2.7.2. A. For $m \in \mathbb{N}_0$ and $a_\alpha \in C^\infty_b(\mathbb{R}^n;\mathbb{R})$ the function $q(x,\xi) = \sum_{|\alpha|\le m} a_\alpha(x)\xi^\alpha$ belongs to the class $S^m_{1,0}(\mathbb{R}^n)$.

B. For any $r \in \mathbb{R}$ the mapping $\xi \mapsto \left(1+|\xi|^2\right)^{r/2}$ is a symbol in the class $S^r_{1,0}(\mathbb{R}^n)$.

C. The function $q(\xi) = i\xi_1 + \sum_{j=2}^n \xi_j^2$ being given, we find that for $\varphi \in C^\infty(\mathbb{R}^n)$ such that $\varphi|_{B_1(0)} = 0$ and $\varphi|_{B^c_2(0)} = 1$ the function $\varphi\frac{1}{q}$ belongs to $S^{-1}_{1/2,0}(\mathbb{R}^n)$.

Given $q \in S^m_{\rho,\delta}(\mathbb{R}^n)$ we may define on $S(\mathbb{R}^n)$ the operator

$$q(x,D)u(x) = (2\pi)^{-n/2}\int_{\mathbb{R}^n} e^{ix\cdot\xi} q(x,\xi)\hat{u}(\xi)\,d\xi. \tag{2.325}$$

Definition 2.7.3. Let $q \in S^m_{\rho,\delta}(\mathbb{R}^n)$, $\delta < \rho$. The symbol $q(x,\xi)$ satisfies *Condition H.m′* if

$$|q(x,\xi)| \ge \tilde{\gamma}_0\left(1+|\xi|^2\right)^{m'/2} \quad \text{for some } \tilde{\gamma}_0 > 0,\ m' \le m,\ |\xi| \ge R, \tag{2.326}$$

and

$$\left|\frac{\partial_x^\beta D_\xi^\alpha q(x,\xi)}{q(x,\xi)}\right| \leq \gamma_{0,\alpha,\beta}\left(1+|\xi|^2\right)^{\frac{\delta|\beta|-\rho|\alpha|}{2}} \tag{2.327}$$

holds for all $\alpha, \beta \in \mathbb{N}_0^n$ and $|\xi| \geq R$ with some $\gamma_{0,\alpha,\beta} \geq 0$.

Using the symbolic calculus explained in [189] we obtain the following estimates for operators with symbols in the class $S^m_{\rho,\delta}(\mathbb{R}^n)$.

Theorem 2.7.4. *A. Let $q \in S^m_{\rho,\delta}(\mathbb{R}^n)$, $\delta < \rho$, and $s \in \mathbb{R}$. Then $q(x,D)$ is a continuous operator from $H^{s+m}(\mathbb{R}^n)$ into $H^s(\mathbb{R}^n)$ and we have the estimate*

$$\|q(x,D)u\|_s \leq c\|u\|_{s+m}. \tag{2.328}$$

B. Suppose in addition that $q(x,\xi)$ satisfies Condition H.m. Then the (sharp) Gårding inequality

$$\|q(x,D)u\|_s \geq \gamma_0\|u\|_{s+m} - \tilde{\lambda}\|u\|_0 \tag{2.329}$$

holds for $u \in H^{s+m}(\mathbb{R}^n)$.

Theorem 2.7.5. *Let $q \in S^m_{\rho,\delta}(\mathbb{R}^n)$, $\delta < \rho$, be real-valued and suppose that* $\operatorname{Re} q(x,\xi)$ *satisfies Condition H.m. Then we have for all $u \in H^m(\mathbb{R}^n)$*

$$|(q(x,D)u,v)_0| \leq c\|u\|_{m/2}\|v\|_{m/2} \tag{2.330}$$

and

$$\operatorname{Re} B(u,u) \geq \gamma_1\|u\|^2_{m/2} - \lambda_1\|u\|_0^2. \tag{2.331}$$

Remark 2.7.6. Although we give no proofs of these results and refer to H.Kumano-go [189], it should be clear to the reader that in principle we may use the considerations of Section 2.4 and Section 2.5 to get these estimates.

Definition 2.7.7. We say that $q(x,\xi)$ belongs to the class $CNS^m_{\rho,\delta}(\mathbb{R}^n)$ if $q(x,\xi)$ belongs to $S^m_{\rho,\delta}(\mathbb{R}^n)$ and if $\xi \mapsto q(x,\xi)$ is for all $x \in \mathbb{R}^n$ a continuous negative definite function.

Clearly, for $q \in CNS^m_{\rho,\delta}(\mathbb{R}^n)$ we must have $0 \leq m \leq 2$.

Example 2.7.8. Suppose that $\sum_{j,k=1}^n a_{jk}(x)\xi_j\xi_k$, $a_{kj} = a_{jk} \in C_b^\infty(\mathbb{R}^n;\mathbb{R})$ is for all $x \in \mathbb{R}^n$ positive definite, i.e.

$$\sum_{j,k=1}^n a_{jk}(x)\xi_j\xi_k \geq 0 \tag{2.332}$$

for all $x \in \mathbb{R}^n$ and $\xi_j \in \mathbb{R}$. For all $0 < \alpha < 1$ the function $(x,\xi) \mapsto \left(\sum_{j,k=1}^n a_{jk}(x)\xi_j\xi_k + \tau_0^{1/\alpha}\right)^\alpha - \tau_0$ is for any $\tau_0 > 0$ an element in the class $CNS^{2\alpha}_{\rho,\delta}(\mathbb{R}^n)$, compare Corollary I.3.6.9.

Now, with the same arguments as in Section 2.6. we have

Theorem 2.7.9. *Let $q \in CNS^m_{\rho,\delta}(\mathbb{R}^n)$, $\delta < \rho$, be a real-valued symbol satisfying Condition H.m. Then the operator $-q(x,D)$ extends to the generator of a Feller semigroup $(T_t^{(\infty)})_{t\geq 0}$. Further, for $\lambda \geq \lambda_1$, λ_1 taken from (2.331), $-q(x,D)$ extends to the generator of an L^p-sub-Markovian semigroup $(T_t^{(p),\lambda})_{t\geq 0}$, $2 \leq p < \infty$.*

Remark 2.7.10. Note that we may apply Theorem 2.7.9 to the symbols in Example 2.7.8.

Remark 2.7.11. Let $q \in CNS^m_{\rho,\delta}(\mathbb{R}^n)$ be a real-valued symbol satisfying the assumptions of Theorem 2.7.9. According to a result of V.P.Maslov and I.A.Shishmarev [210] the operator $T_t^{(\infty)}|_{C_0^\infty(\mathbb{R}^n)}$ is a pseudo-differential operator with symbol $p(t;x,\xi)$ satisfying for $t \to 0$ the asymptotic relation

$$p(t;x,\xi) = e^{-q(x,\xi)t} + o(1) \tag{2.333}$$

uniformly for $x \in K$, $K \subset \mathbb{R}^n$ compact, and $\xi \in \mathbb{R}^n$.

Another way of constructing Feller or sub-Markovian semigroups starting with a pseudo-differential operator is to construct a fundamental solution to the problem

$$\frac{\partial u(t,x)}{\partial t} + q(x,D)u(t,x) = 0 \quad \text{and } u(0,x) = f(x). \tag{2.334}$$

We will first explain how to use such a fundamental solution to construct the semigroup(s), then we will briefly discuss the construction of fundamental solutions for some classes of pseudo-differential operators generating Feller and/or sub-Markovian semigroups. In the papers we will refer to, mostly the case of time-dependent coefficients is handled. However, due to the scope of this treatise we state results only for stationary coefficients.

Definition 2.7.12. Let $q : \mathbb{R}^n \times \mathbb{R}^n \to \mathbb{C}$ be a continuous negative definite symbol in the sense of Definition 2.3.1. We call a function $\Gamma : (0,\infty) \times \mathbb{R}^n \times \mathbb{R}^n \to \mathbb{R}$ a *fundamental solution* to the problem (2.334) if for fixed $y \in \mathbb{R}^n$ the function $(t,x) \mapsto \Gamma(t,x,y)$ satisfies the equation

$$\frac{\partial \Gamma(t,x,y)}{\partial t} + q(x,D_x)\Gamma(t,x,y) = 0 \tag{2.335}$$

and

$$\lim_{t\searrow 0}\int_{\mathbb{R}^n}\Gamma(t,x,y)f(y)\,dy = f(x), \quad x\in\mathbb{R}^n. \tag{2.336}$$

Remark 2.7.13. Since $q(x,\xi)$ is independent of t it follows for all $s>0$ that

$$\lim_{t\searrow s}\int_{\mathbb{R}^n}\Gamma(t-s,x,y)f(y)\,dy = f(x), \quad x\in\mathbb{R}^n. \tag{2.337}$$

Example 2.7.14. Let $q(x,\xi)=\psi(\xi)$ be x-independent and suppose that for all $t>0$ the function $\xi\mapsto e^{-t\psi(\xi)}$ belongs to the Wiener algebra, see Definition I.3.2.8. Then a fundamental solution to the problem (2.334) with $\psi(D)$ instead of $q(x,D)$ is given by

$$\Gamma(t,x,y)=(2\pi)^{-n/2}\int_{\mathbb{R}^n} e^{i(x-y)\cdot\xi}e^{-t\psi(\xi)}\,d\xi. \tag{2.338}$$

Indeed, let $f\in S(\mathbb{R}^n)$ (for simplicity). It follows that

$$\begin{aligned}\frac{\partial\Gamma(t,x,y)}{\partial t} &= (2\pi)^{-n/2}\int_{\mathbb{R}^n} e^{i(x-y)\cdot\xi}(-\psi(\xi))e^{-t\psi(\xi)}\,d\xi\\ &= -\psi(D_x)\Gamma(t,x,y),\end{aligned}$$

thus $\frac{\partial\Gamma(t,x,y)}{\partial t}+\psi(D_x)\Gamma(t,x,y)=0$. Further we know that

$$\int_{\mathbb{R}^n}\Gamma(t,x,y)f(y)\,dy=(2\pi)^{-n/2}\int_{\mathbb{R}^n} e^{-t\psi(\xi)}\hat{f}(\xi)e^{ix\xi}\,d\xi$$

which leads for all $x\in\mathbb{R}^n$ to

$$\lim_{t\to 0}\int_{\mathbb{R}^n}\Gamma(t,x,y)f(y)\,dy=f(x).$$

Now let $\Gamma(t,x,y)$ be a fundamental solution to (2.334). Suppose in addition $\Gamma(t,x,y)\geq 0$ for all $t>0$ and $x,y\in\mathbb{R}^n$. Moreover we assume that $(x,y)\mapsto\Gamma(t,x,y)$ is measurable, $\int_{\mathbb{R}^n}\Gamma(t,x,y)\,dy\leq 1$, and that the *Chapman–Kolmogorov equations*

$$\int_{\mathbb{R}^n}\Gamma(t,x,y)\Gamma(s,y,z)\,dy=\Gamma(t+s,x,z) \tag{2.339}$$

do hold. In this case $p_t(x,A):=\int_A\Gamma(t,x,y)\,dy$ gives a transition function, in particular $p_t(x,dy)=\Gamma(t,x,y)\lambda^{(n)}(dy)$ is a sub-probability measure. Hence we may define the operator

$$T_tf(x)=\int_{\mathbb{R}^n}f(y)p_t(x,dy)=\int_{\mathbb{R}^n}f(y)\Gamma(t,x,y)\,dy. \tag{2.340}$$

It is clear that $(T_t)_{t\geq 0}$ has on $B_b(\mathbb{R}^n)$ the semigroup property and each of the operators T_t is a positivity preserving contraction on $(B_b(\mathbb{R}^n;\mathbb{R}), \|\cdot\|_\infty)$. Whether $(T_t)_{t\geq 0}$ gives a Feller semigroup or an L^p-sub-Markovian semigroup dependents of course on properties of $\Gamma(t,x,y)$. In case that we may identify $(T_t)_{t\geq 0}$ with a certain strongly continuous contraction semigroup, it is reasonable to expect that its generator is some extension of $-q(x,D)$ with an appropriate domain. Thus we are concerned with the problem of constructing a fundamental solution to problem (2.334). When working with a symbolic calculus in Section 2.4 (or in the beginning of this section), the idea was that $\frac{1}{q(x,\xi)}$ is the symbol of a first approximation of the operator $[q(x,D)]^{-1}$. This idea led us to the concept of a parametrix for $q(x,D)$. The construction of a fundamental solution to (2.334) uses a different idea of a parametrix, namely to consider a fundamental solution of an operator with constant coefficients as first approximation of the fundamental solution of the original operator. This idea is due to E.E.Levi [198] for handling parabolic differential operators and it is explained in detail in A.Friedman's monograph [94] or that of S.Itô [147]. For classical pseudo-differential operators such a construction was done first by C.Tsutsumi [273], see also her paper [148] published under the name C.Iwasaki, and and is presented in the monograph [189] of H.Kumano-go. In connection with Markov processes and related semigroups it seems that A.N.Kochubei [176] was the first using this method of handle certain pseudo-differential operators.

Let $q(x,D)$ be a pseudo-differential operator with a continuous negative definite symbol and for $x_0 \in \mathbb{R}^n$ denote by $q(x_0,D)$ the operator obtained from $-q(x,D)$ by "freezing" the coefficients at x_0. As indicated in Example 2.7.14 a fundamental solution to the problem

$$\frac{\partial u}{\partial t} + q(x_0,D)u = 0, \quad u(0,x) = f(x), \tag{2.341}$$

is given by

$$\Gamma^{x_0}(t,x,y) = (2\pi)^{-n/2} \int_{\mathbb{R}^n} e^{i(x-y)\cdot\xi} e^{-tq(x_0,\xi)}\, d\xi. \tag{2.342}$$

Let us define

$$Z(t,x,x_0) := \Gamma^{x_0}(t,x,y). \tag{2.343}$$

Levi's method consists in the Ansatz to find a fundamental solution $\Gamma(t,x,y)$

to the problem

$$\frac{\partial u}{\partial t}+q(x,D)u=0, \quad u(0,x)=f(x), \tag{2.344}$$

in the representation

$$\Gamma(t,x,y)=Z(t,x,y)+\int_0^t\int_{\mathbb{R}^n} Z(t,x,z)\Phi(z,r,y)\,dzdr \tag{2.345}$$

and $\Phi:\mathbb{R}^n\times(0,\infty)\times\mathbb{R}^n\to\mathbb{C}$ (or $\mathbb{R}$) is a function to be determined. Suppose that

$$\Big(\frac{\partial}{\partial t}+q(x,D_x)\Big)\Gamma(t,x,y)=0.$$

Then (2.345) gives an equation for the function Φ, namely

$$\Big(\frac{\partial}{\partial t}+q(x,D_x)\Big)\Big(Z(t,x,y)+\int_0^t\int_{\mathbb{R}^n} Z(t,x,z)\Phi(z,r,y)\,dzdr\Big)=0,$$

and by a formal calculation we find

$$\begin{aligned}\Phi(x,t,y)&=\Big(\frac{\partial}{\partial t}+q(x,D_x)\Big)Z(t,x,y)\\&\quad+\int_0^t\int_{\mathbb{R}^n}[(q(x,D_x)-q(x_0,D_x))Z(t,x,y)]\Phi(z,r,y)\,dzdr.\end{aligned} \tag{2.346}$$

Thus the task is to make for concrete operators (or symbols) these calculations correct and most of all to solve equation (2.346) in order to determine with the help of Φ a fundamental solution $\Gamma(t,x,y)$ to problem (2.344).

In the following we discuss two classes of operators where this method was successfully applied to, first we have a look to A.N.Kochubei's results obtained in [176], then we introduce the result of V.N.Kolokoltsov [180]. In both cases, as already mentioned in the beginning of this section, we do not present the proofs.

In [176] A.N.Kochubei used the parametrix method to construct a fundamental solution to the problem

$$\frac{\partial u}{\partial t}+q(x,D)u=0, \quad u(0,x)=f(x), \tag{2.347}$$

where the symbol of $q(x,D)$ is given by

$$q(x,\xi)=\sum_{j=0}^m q_j(x,\xi), \tag{2.348}$$

and each $q_j(x,\xi)$ is homogeneous of some degree γ_j, $\gamma_0 > 1$ and $0 < \gamma_j < \gamma_0$ for $j = 1, \ldots, n$. Of course the symbol $q_j(x,\xi)$ have to fulfill additional hypotheses, in particular hypotheses which guarantee that the theory of *hypersingular integral operators* developed by S.G.Samko is applicable, compare [243] and the references given therein. The following theorem shall give the flavour of the type of results obtained by A.N.Kochubei. For more general results and details we refer to his paper [176]. In particular it should be noted that he treated in his paper the case of time-dependent coefficients, i.e. symbols $q_j(x,t,\xi)$.

Assumption 2.7.15. A. The symbol $q_0(x,\xi)$ is assumed to be homogeneous with respect to ξ of degree $\gamma_0 \in (1,2)$. For $N \in \mathbb{N}$, $N \geq 2n+3$, suppose that $q_0(x,\cdot) \in C^N(\mathbb{R}^n \backslash \{0\})$ and assume for $|\alpha| \leq N$

$$|D_\xi^\alpha q_0(x,\xi)| \leq C_N |\xi|^{\gamma_0 - |\alpha|}, \quad \xi \in \mathbb{R}^n \backslash \{0\}, \tag{2.349}$$

as well as

$$|D_\xi^\alpha (q_0(x,\xi) - q_0(y,\xi))| \leq C_N |x-y|^\lambda |\xi|^{\gamma_0 - |\alpha|} \tag{2.350}$$

for all $x, y \in \mathbb{R}^n$, $\xi \in \mathbb{R}^n \backslash \{0\}$ and $|\alpha| \leq N$ with some $\lambda \in (0,1)$. In addition we suppose the ellipticity condition

$$\operatorname{Re} q_0(x,\xi) \geq C_0 > 0 \quad \text{for all } x \in \mathbb{R}^n,\ \xi \in S^{n-1}. \tag{2.351}$$

B. For the symbol $q_j(x,\xi)$ $1 \leq j \leq m$, it is assumed that they are with respect to ξ homogeneous of degree γ_j, $0 < \gamma_j < \gamma_0$, and $\gamma_j \neq 1$. Further for $|\alpha| \leq N$, $\xi \in \mathbb{R}^n \backslash \{0\}$ and $x, y \in \mathbb{R}^n$ the estimates

$$|D_\xi^\alpha q_j(x,\xi)| \leq C |\xi|^{\gamma_j - |\alpha|} \tag{2.352}$$

and

$$|D_\xi^\alpha (q_j(x,\xi) - q_j(y,\xi))| \leq C_N |x-y|^\lambda |\xi|^{\gamma_j - |\alpha|} \tag{2.353}$$

are required.

The homogeneity assumption on $q_j(x,\xi)$, $j = 0, 1, \ldots, m$ allows to represent each of the pseudo-differential operators $q_j(x,D)$ as a *hypersingular integral operator*. Thus we have

$$q_j(x,D)u(x) = c_{n,l,\gamma_j} \int_{\mathbb{R}^n} \Omega_j\Big(x, \frac{h}{|h|}\Big) \frac{(\Delta_h^l u)(x)}{|h|^{n+\gamma_j}}\, dh, \tag{2.354}$$

where $(\Delta_h^l u)(x) = \sum_{k=0}^{l}(-1)^k\binom{l}{k}u(x-kh)$ with $l \in \mathbb{N}$, and it follows that

$$q_j(x,\xi) = c_{n,l,\gamma_j}\int_{\mathbb{R}^n}\frac{(1-e^{-i\xi h})^l}{|h|^{n+\gamma_j}}\Omega_j\left(x,\frac{h}{|h|}\right)dh. \tag{2.355}$$

Following S.G.Samko, we will call $\Omega_j\left(x,\frac{h}{|h|}\right)$ the *characteristic* of the (hyper-singular integral) operator $q_j(x,D)$.

Theorem 2.7.16 (A.N.Kochubei). *Suppose that Assumption* 2.7.15 *is fulfilled and that the operator* $q(x,D) = \sum_{j=0}^{m} q_j(x,D)$ *has a representation as sum of hypersingular integral operators each with a characteristic being even and non-negative. Then a fundamental solution to the problem* (2.347) *exists and gives a transition function of a Feller semigroup.*

Remark 2.7.17. Note that the assumption of Theorem 2.7.16 imply that $\xi \mapsto q(x,\xi)$ is a continuous negative definite function.

In his paper [180] V.N.Kolokoltsov constructed Markov processes, hence semigroups, by starting with pseudo-differential operators $q(x,\xi)$ with symbol given by

$$q(x,\xi) = i(b(x),\xi)+\int_0^\infty\int_{S^{n-1}}\left(e^{-i\xi\cdot y}-1+\frac{i\xi\cdot y}{1+|y|^2}\right)\frac{d|y|}{|y|^{1+\alpha}}\,\tilde{\mu}(x,d\eta) \tag{2.356}$$

where $\alpha \in (0,2)$, $y = |y|\eta$ and $\tilde{\mu}(x,d\eta)$ is a kernel on $\mathbb{R}^n\times\mathcal{B}(S^{n-1})$. Clearly $\xi \mapsto q(x,\xi)$ is a continuous negative definite function and it has the representation

$$q(x,\xi) = |\xi|^\alpha\int_{S^{n-1}}\left|\left(\frac{\xi}{|\xi|},\eta\right)\right|^\alpha\mu(x,d\eta). \tag{2.357}$$

Assumption 2.7.18. We suppose that for the symbol $q(x,\xi)$ we have the representation (2.357)and that for the kernel $\mu(x,d\eta)$

$$C_1 \leq \int_{S^{n-1}}\left|\left(\frac{\xi}{|\xi|},\eta\right)\right|^\alpha\mu(x,d\eta) \leq C_2 \tag{2.358}$$

holds. Further assume that for $\alpha \in \mathbb{N}_0^n$, $|\alpha| \leq m$, $m \geq 2$, a real-valued measure is given for each fixed $x \in \mathbb{R}^n$ by $\partial_x^\alpha\mu(x,d\eta)$ which is of uniform bounded total variation with respect to x.

Theorem 2.7.19 (V.N.Kolokoltsov). *Under Assumption* 2.7.18 *there exists a fundamental solution* $\Gamma(t,x,y)$ *to the problem* $\frac{\partial u}{\partial t}+q(x,D)u = 0$, $u(0,x) = f(x)$, *which determines a Feller semigroup by setting*

$$T_t^{(\infty)}u(x) = \int_{\mathbb{R}^n}\Gamma(t,x,y)y(y)\,dy. \tag{2.359}$$

For details of the proof and estimates for $\Gamma(t,x,y)$ we refer to the paper [180] of V.N.Kolokoltsov as well as his lecture notes volume [181] and the most recent related paper [58] of V.G.Danilov and S.M.Frolovitchev.

Finally in this section we will briefly discuss results due to F.Baldus [13], [14], who used the Weyl calculus of pseudo-differential operators to construct Feller semigroups. Besides an appropriate formulation of the Weyl calculus he used the theory of inverse-closed Fréchet operator algebras as introduced by B.Gramsch in [109]. Unfortunately we need quite a lot of special notions to state the results. In some sense the following considerations are more or less an extended summary, for a detailed discussion we refer to [13] and [14], but also to L.Hörmander's original paper [140] and his monograph [141].

Denote by the *standard symplectic form* on $\mathbb{R}^n \times \mathbb{R}^n$, i.e.

$$\sigma((x,\xi),(y,\eta)) = y\cdot\xi - x\cdot\eta, \tag{2.360}$$

and for a positive definite quadratic form γ on $\mathbb{R}^n \times \mathbb{R}^n$ we set

$$\begin{aligned}&\gamma^\sigma((x,\xi),(x,\xi))\\ &:= \sup\left\{\frac{\sigma((x,\xi),(y,\eta))^2}{\gamma((y,\eta),(y,\eta))};\quad (y,\eta)\in\mathbb{R}^n\times\mathbb{R}^n,\quad (y,\eta)\neq(0,0)\right\}.\end{aligned} \tag{2.361}$$

In the following a *metric on* $\mathbb{R}^m$ does mean a family $\gamma = (\gamma_x)_{x\in\mathbb{R}^m}$, of positive definite quadratic forms on $\mathbb{R}^m$, which we may interpret as Riemannian metric and denote it sometimes by $\gamma(dx,dx)$. Given a metric γ on $\mathbb{R}^n \times \mathbb{R}^n$. We say that it *splits* if we have for each $(y,\eta)\in\mathbb{R}^n\times\mathbb{R}^n$

$$\gamma_{(y,\eta)}((x,\xi),(x,\xi)) = \gamma^{(1)}_{(y,\eta)}(x,x)+\gamma^{(2)}_{(y,\eta)}(\xi,\xi) \tag{2.362}$$

with two metrics $\gamma^{(1)}$ and $\gamma^{(2)}$ on $\mathbb{R}^n$.

Definition 2.7.20. A. A metric γ on $\mathbb{R}^m$ is called a *slowly varying metric* if there exists a constant c_γ such that for $x,y\in\mathbb{R}^m$ satisfying $\gamma_x(x-y,x-y)\leq \frac{1}{c_\gamma}$ it follows that

$$\frac{1}{c_\gamma}\gamma_x(z,z)\leq\gamma_y(z,z)\leq c_\gamma\gamma_x(z,z) \tag{2.363}$$

holds for all $z\in\mathbb{R}^m$.

B. Let γ be a slowly varying metric on $\mathbb{R}^m$. A function $M:\mathbb{R}^m\to\mathbb{R}_+$ is called *γ-slowly varying* if there is a constant c_M such that for all $x,y\in\mathbb{R}^m$

with $\gamma_x(x-y,x-y) \le \frac{1}{c_M}$ we have

$$\frac{1}{c_M} M(x) \le M(y) \le c_M M(x). \tag{2.364}$$

Next we may introduce the notion of a Hörmander metric and that of (sub-) admissible weight functions.

Definition 2.7.21. A. A slowly varying metric γ on $\mathbb{R}^n \times \mathbb{R}^n$ is called a *Hörmander metric* if there exists constants $c_\gamma > 0$ and $N_\gamma \in \mathbb{N}$ such that for all $(x,\xi) \in \mathbb{R}^n \times \mathbb{R}^n$ we have

$$\gamma_{(x,\xi)}(\cdot,\cdot) \le \gamma^\sigma_{(x,\xi)}(\cdot,\cdot) \tag{2.365}$$

and for all $(x,\xi),(y,\eta) \in \mathbb{R}^n \times \mathbb{R}^n$

$$\left(\frac{\gamma_{(x,\xi)}(\cdot,\cdot)}{\gamma_{(y,\eta)}(\cdot,\cdot)}\right)^{\pm 1} \le c_\gamma \Big(1+\gamma^\sigma_{(y,\eta)}((y-x,\eta-\xi),(y-x,\eta-\xi))^{N_\gamma}\Big). \tag{2.366}$$

B. Let γ be a slowly varying metric on $\mathbb{R}^n \times \mathbb{R}^n$. We call $M : \mathbb{R}^n \times \mathbb{R}^n \to \mathbb{R}_+$ a *γ-admissible weight function* if M is γ-slowly varying and satisfies with $c_M > 0$ and $N_M \in \mathbb{N}$

$$\left(\frac{M(x,\xi)}{M(y,\eta)}\right)^{\pm 1} \le c_M \left(1+\gamma^\sigma_{(y,\eta)}((y-x,\eta-\xi),(y-x,\eta-\xi))^{N_M}\right) \tag{2.367}$$

for all $(x,\xi),(y,\eta) \in \mathbb{R}^n \times \mathbb{R}^n$.

Denoting for a metric γ on $\mathbb{R}^n \times \mathbb{R}^n$ the function h_γ by

$$h_\gamma((x,\xi)) := \sup\left\{\frac{\gamma_{(x,\xi)}((y,\eta),(y,\eta))}{\gamma^\sigma_{(x,\xi)}((y,\eta),(y,\eta))};\quad (y,\eta) \in \mathbb{R}^n \times \mathbb{R}^n,\quad (y,\eta) \ne (0,0)\right\}, \tag{2.368}$$

we give further

Definition 2.7.22. Given a Hörmander metric γ on $\mathbb{R}^n \times \mathbb{R}^n$. A function $M : \mathbb{R}^n \times \mathbb{R}^n \to \mathbb{R}_+$ is called *sub-γ-admissible weight function* if there exists a γ-admissible weight function M_0 such that $M \le M_0$ and for some $m \in \mathbb{N}$ and $c > 0$ it follows that

$$h^m_\gamma M_0 \le cM. \tag{2.369}$$

If M and $\frac{1}{M}$ are both sub-γ-admissible we call M an *invertible sub-γ-admissible weight function.*

For $(y,\eta) \in \mathbb{R}^n \times \mathbb{R}^n$ and $u : \mathbb{R}^n \times \mathbb{R}^n \to \mathbb{C}$ we set

$$\partial_{(y,\eta)} u(x,\xi) = \left((y,\eta), \nabla_{2n} u(x,\xi)\right),$$

where ∇_{2n} is the gradient in $\mathbb{R}^n \times \mathbb{R}^n$ and $(\cdot,\cdot)$ the scalar product. Further we set for a metric γ on $\mathbb{R}^n \times \mathbb{R}^n$ and $k \in \mathbb{N}$,

$$\begin{aligned} &|u|_k^{\gamma(y,\eta)}(x,\xi) \\ &:= \sup\left\{ \frac{|\partial_{(t_1,\tau_1)} \dots \partial_{(t_k,\tau_k)} u(x,\xi)|}{\prod_{j=1}^k \gamma^{(y,\eta)}\left((t_j,\tau_j),(t_j,\tau_j)\right)^{1/2}};\ (t_j,\tau_j) \in (\mathbb{R}^n \times \mathbb{R}^n \backslash \{0\}) \right\}. \end{aligned} \tag{2.370}$$

Definition 2.7.23. Given a metric γ on $\mathbb{R}^n \times \mathbb{R}^n$ and a weight function $M : \mathbb{R}^n \times \mathbb{R}^n \to \mathbb{R}_+$. The *symbol class* $S(M,\gamma)$ consists of all functions $q \in C^\infty(\mathbb{R}^n \times \mathbb{R}^n)$ which satisfy for all $k \in \mathbb{N}_0$

$$\sup\left\{ \frac{|q|_l^{\gamma(x,\xi)}}{M(x,\xi)};\quad (x,\xi) \in \mathbb{R}^n \times \mathbb{R}^n \text{ and } 0 \le l \le k \right\} < \infty. \tag{2.371}$$

Now we can introduce the operator associated with $S(M,\gamma)$.

Definition 2.7.24. Given $q \in S(M,\gamma)$. We define the associated *Weyl-pseudo-differential operator* $q^W(x,D) : S(\mathbb{R}^n) \to S'(\mathbb{R}^n)$ by

$$q^W(x,D)u(x) = (2\pi)^{-n} \int_{\mathbb{R}^n} \int_{\mathbb{R}^n} e^{i(x-y)\cdot\xi} q\Big(\frac{x+y}{2}, \xi\Big) u(y)\, dy d\xi. \tag{2.372}$$

The set of all operators $q^W(x,D)$ with symbol $q \in S(M,\gamma)$is denoted by $\Psi(M,\gamma)$.

Example 2.7.25. For $0 \le \delta \le \rho \le 1$, $\delta < 1$, a Hörmander metric is given by

$$\gamma_{(x,\xi)}(dx,d\xi) = \frac{(dx)^2}{(1+|\xi|^2)^{\delta/2}} + \frac{(d\xi)^2}{(1+|\xi|^2)^{\rho/2}}.$$

Taking in addition the weight function $M(x,\xi) = \left(1+|\xi|^2\right)^{m/2}$ we find

$$S(M,\gamma) = S^m_{\rho,\delta}(\mathbb{R}^n). \tag{2.373}$$

Note that the Weyl-pseudo-differential operator $q^W(x,D)$ we are interested in can be transformed into the "usual" form

$$q(x,D)(u) = (2\pi)^{-n/2} \int_{\mathbb{R}^n} e^{ix\cdot\xi} q(x,\xi) \hat{u}(\xi)\, d\xi \tag{2.374}$$

see [141], Theorem 18.5.10, or [13], Theorem 4.7.

Let us denote by $B\left(L^2(\mathbb{R}^n)\right)^{-1}$ the set of all bounded linear operators $A : L^2(\mathbb{R}^n) \to L^2(\mathbb{R}^n)$ which have a bounded inverse, and denote by $\Psi(M,\gamma)^{-1}$ the set of all $q^W(x,D) \in \Psi(M,\gamma)$ with inverse in $\Psi(M,\gamma)$. Now we can state the result of F.Baldus [13]:

Theorem 2.7.26. *Let γ be a Hörmander metric on $\mathbb{R}^n \times \mathbb{R}^n$ which splits and assume*

$$\Psi(1,\gamma) \cap B\left(L^2(\mathbb{R}^n)\right)^{-1} = \Psi(1,\gamma)^{-1}. \tag{2.375}$$

Further let M be an invertible sub-γ-admissible weight function satisfying with some $k \in \mathbb{N}$ and $c_M > 0$

$$(m\, h_\gamma)^k \le \frac{c_M}{M}, \tag{2.376}$$

where h_γ is given by (2.368). If $q \in S(M,\gamma)$ and satisfies in addition

$$\sup\left\{ \frac{|q|_{l+1}^{\gamma(x,\xi)}(x,\xi)}{M(x,\xi)};\quad (x,\xi) \in \mathbb{R}^n \times \mathbb{R}^n \text{ and } 0 \le l \le k \right\} < \infty \tag{2.377}$$

for all $k \in \mathbb{N}_0$,

$$|\lambda + q(x,\xi) + c_q| \ge \tilde{c}_q(\lambda + M(x,\xi)) \tag{2.378}$$

for all $(x,\xi) \in \mathbb{R}^n \times \mathbb{R}^n$, $\lambda \ge \lambda_q \ge 0$ and c_q, $\tilde{c}_q > 0$, and

$$\xi \mapsto q(x,\xi) \tag{2.379}$$

is a negative definite function, then the operator $-q(x,D) : C_0^\infty(\mathbb{R}^n;\mathbb{R}) \to C_\infty(\mathbb{R}^n;\mathbb{R})$ is a densely defined operator on $C_\infty(\mathbb{R}^n;\mathbb{R})$ which extends to a generator of a Feller semigroup.

Remark 2.7.27. A. As we will see by the following examples, Theorem 2.7.26 is the most general result so far for constructing Feller semigroups generated by a pseudo-differential operator with a symbol $q \in C^\infty(\mathbb{R}^n \times \mathbb{R}^n)$ where $\xi \mapsto q(x,\xi)$ is negative definite.

B. Condition (2.375) is often called the *spectral invariance* of $\Psi(1,\gamma)$ in $B\left(L^2(\mathbb{R}^n)\right)$.

Example 2.7.28. A. We know already that the class $S^m_{\rho,\sigma}(\mathbb{R}^n)$ is included in Baldus' calculus, thus Theorem 2.7.26 induces Theorem 2.7.9.

B. Symbols of the mixed homogeneity are partly included.

C. The class $S^{m,\psi}_\rho(\mathbb{R}^n)$ considered by W.Hoh in [127], see Definition 2.4.4, is included when working with the metric

$$\gamma_{(x,\xi)}(dx,d\xi) = \sum_{j=1}^n (dx_j)^2 + \frac{(d\xi_j)^2}{(1+\psi(\xi))^{1/2}}.$$

(To cover this example a slight extension of Theorem 2.7.26 is needed.) But note that Hoh's class $S^{m,\psi}_\rho(\mathbb{R}^n)$ originate from non-smooth negative definite functions and symbols and the whole scope of his approach will become clear after we dealt with perturbations in Section 2.8.

D. Some classes of pseudo-differential operators introduces by R.Beals and C.Fefferman [21], [22] are included. For this we refer also to the paper [43] of O.Caps.

Without going into details we would like to mention two more recent applications of (generalized) pseudo-differential operator theory to generate semigroups: the theory of generalized Mehler semigroups, see the works of V.Bogachev, M.Fuhrman, P.Lescot, M.Röckner and B.Schmuland [34], [96] and [197], and the extension to the infinite dimensional case, again initiated by V.Bogachev, P.Lescot and M.Röckner in [33].

2.8 Some Perturbation Results

In this section we want to extend the class of pseudo-differential operators generating a Feller or an L^p-sub-Markovian semigroup by perturbing a generator $-q(x,D)$. Let us denote the perturbation by $-p(x,D)$. In order to assure that $-q(x,D) - p(x,D)$ is a generator we want to apply (variants of) Theorem I.4.4.3. For this we have to assume that $-p(x,D)$ is dissipative and $(-q(x,D))$-bounded. However, to get a Feller or an L^p-sub-Markovian semigroup we need more, namely that $-q(x,D)-p(x,D)$ satisfies also the positive maximum principle or is an L^p-Dirichlet operator, respectively. Thus we may immediately formulate the following result:

Theorem 2.8.1. *A. Let $-q(x,D)$ be a pseudo-differential operator extending to a generator of a Feller semigroup. Denote the domain of this extension* *by* $D^{(\infty)}(q(x,D)) \subset C_\infty(\mathbb{R}^n;\mathbb{R})$. *Let* $-p(x,D)$ *be a pseudo-differential operator defined on* $D^{(\infty)}(q(x,D))$ *which maps* $D^{(\infty)}(q(x,D))$ *into* $C_\infty(\mathbb{R}^n;\mathbb{R})$. *If* $-p(x,D)$ *is* $C_\infty(\mathbb{R}^n;\mathbb{R})$*-dissipative,* $(-q(x,D))$*-bounded, and if in addition* $-q(x,D)-p(x,D)$ *satisfies on* $D^{(\infty)}(q(x,D))$ *the positive maximum principle, then* $(-q(x,D)-p(x,D), D^{(\infty)}(q(x,D)))$ *is the generator of a Feller semigroup.*

B. Let $-q(x,D)$ *be a pseudo-differential operator extending to a generator of an* L^p*-sub-Markovian semigroup,* $1 < p < \infty$*, and denote the domain of this extension by* $D^{(p)}(q(x,D)) \subset L^p(\mathbb{R}^n;\mathbb{R})$. *If* $-p(x,D)$ *is* L^p*-dissipative,* $(-q(x,D))$*-bounded, and if in addition* $(-q(x,D)-p(x,D), D^{(p)}(q(x,D)))$ *is an* L^p*-Dirichlet operator, then* $(-q(x,D)-p(x,D), D^{(p)}(q(x,D)))$ *is the generator of an* L^p*-sub-Markovian semigroup.*

Using Proposition I.4.3.25 we get the following corollary:

Corollary 2.8.2. *Let* $0 < \alpha < 1$ *and* $f_\alpha : [0,\infty) \to \mathbb{R}$ *the Bernstein function* $f_\alpha = x^\alpha$.

A. If $(-q(x,D), D^{(\infty)}(q(x,D)))$ *is a pseudo-differential operator generating a Feller semigroup, then the operator* $(-q(x,D) - q^{f_\alpha}(x,D), D^{(\infty)}(q(x,D)))$ *is also a generator of Feller semigroup.*

B. Suppose that the pseudo-differential operator $(-q(x,D), D^{(p)}(q(x,D)))$ *is a generator of an* L^p*-sub-Markovian semigroup. It follows that the operator* $-q(x,D) - q^{f_\alpha}(x,D)$ *with domain* $D^{(p)}(q(x,D))$ *is also the generator of an* L^p*-sub-Markovian semigroup.*

Remark 2.8.3. A. In contrary to classical pseudo-differential operators, and in particular to partial differential operators, we do not have the notion of a principal part for the operators under consideration, i.e. there is no obvious way to define the principal symbol of $q(x,\xi)$ if $q(x,\cdot) : \mathbb{R}^n \to \mathbb{C}$ is just a continuous negative definite function. Hence there is no obvious notion of "lower order terms". The content of Corollary 2.8.2 is that using fractional powers we may construct to a given pseudo-differential operator certain operators which behave as lower order perturbations.

B. Note further that in Theorem 2.8.1 it is clearly not necessary to assume that $-p(x,D)$ satisfies the positive maximum principle or is an L^p-Dirichlet operator, respectively.

C. It is possible to prove Corollary 2.8.2 for every Bernstein function f satisfying $\lim\limits_{x\to\infty} \frac{f(x)}{x} < 1$.

In order to apply Theorem 2.8.1 in concrete situations we are faced with several problems

— we need to find appropriate $(-q(x,D))$-bounds for $-p(x,D)$;

— we have to prove that $-p(x,D)$ is dissipative in the space $C_\infty(\mathbb{R}^n;\mathbb{R})$ or $L^p(\mathbb{R}^n;\mathbb{R})$, respectively;

— we have to show that $-q(x,D)-p(x,D)$ satisfies the positive maximum principle or is an L^p-Dirichlet operator, respectively.

The last two problems may be overcome by the assumption that $-p(x,D)$ itself satisfies on $D^{(\infty)}(q(x,D))$ the positive maximum principle or is an L^p-Dirichlet operator on $D^{(p)}(q(x,D))$, respectively. Indeed, it is obvious that if $-q(x,D)$ and $-p(x,D)$ are operators satisfying on $D^{(\infty)}(q(x,D))$ the positive maximum principle or are both Dirichlet operators on $D^{(p)}(q(x,D))$, then their sum $-q(x,D)-p(x,D)$ has the same property. Further, as an operator satisfying the positive maximum principle the operator $-p(x,D)$ is dissipative on $C_\infty(\mathbb{R};\mathbb{R})$, compare Lemma I.4.5.2, and as an L^p-Dirichlet operator $-p(x,D)$ is dissipative on $L^p(\mathbb{R};\mathbb{R})$, see Proposition I.4.6.11 together with Proposition I.4.6.12.

We know already the structure of the operators satisfying the positive maximum principle. In order to construct interesting perturbations we will now use a very recent structure result for L^p-Dirichlet operators which is due to R.Schilling [250] and is interesting in itself.

Theorem 2.8.4. *Let $(T_t^{(p)})_{t\geq 0}$ be for some $p>1$ an L^p-sub-Markovian semigroup on $L^p(\mathbb{R}^n;\mathbb{R})$ with generator $(A^{(p)}, D(A^{(p)}))$. If $C_0^\infty(\mathbb{R}^n;\mathbb{R})\subset D(A^{(p)})$ and if $A: C_0^\infty(\mathbb{R}^n;\mathbb{R})\to C(\mathbb{R}^n;\mathbb{R})$, then $-A|_{C_0^\infty(\mathbb{R}^n;\mathbb{R})}$ is a pseudo-differential operator with symbol $p:\mathbb{R}^n\times\mathbb{R}^n\to\mathbb{C}$ such that $\xi\mapsto p(x,\xi)$ is for all $x\in\mathbb{R}^n$ a continuous negative definite function. Furthermore, if A defined on $C_0^\infty(\mathbb{R}^n)$ is symmetric, satisfies the positive maximum principle and maps $C_0^\infty(\mathbb{R}^n)$ into $L^2(\mathbb{R}^n)$ then A satisfies on $C_0^\infty(\mathbb{R}^n;\mathbb{R})$*

$$\int_{\mathbb{R}^n}(Au)(x)((u(x)-1)^+)^{p-1}\,dx\leq 0, \tag{2.380}$$

i.e. $(A, C_0^\infty(\mathbb{R}^n;\mathbb{R})$ is a Dirichlet operator. In particular every symmetric pseudo-differential operator $-p(x,D)$ with a symbol $p:\mathbb{R}^n\times\mathbb{R}^n\to\mathbb{C}$ such that $\xi\mapsto p(x,\xi)$ is a continuous negative definite function which maps $C_0^\infty(\mathbb{R}^n;\mathbb{R})$ into $L^2(\mathbb{R}^n;\mathbb{R})\cap C(\mathbb{R}^n;\mathbb{R})$, is a Dirichlet operator.

In [250] R.Schilling proved a more general version of this theorem. The importance of this theorem in our context of perturbation results is that it makes the assumption on the perturbation $-p(x,D)$ to be a pseudo-differential operator with symbol $-p(x,\xi)$ where $\xi \mapsto p(x,\xi)$ is a continuous negative definite function also in case of Dirichlet operators very natural.

Now, once we restrict ourselves to perturbations $-p(x,D)$ as described above, it remains to prove their $-q(x,D)$-boundedness. According to Definition I.4.4.1 an operator $(Q, D(Q))$ on a Banach space $(X, \|\cdot\|_X)$ is A-bounded where $(A, D(A))$ is a further linear operator on X if $D(A) \subset D(Q)$ and

$$\|Q\,u\|_X \leq \alpha\|A\,u\|_X + \beta\|u\|_X, \quad u \in D(A), \tag{2.381}$$

holds for some $\alpha \in [0,1)$ and $\beta \geq 0$. For concrete operators we will have to work in some spaces $H_p^{\psi,s}(\mathbb{R}^n;\mathbb{R})$, for $p=2$ see Section I.3.10 and for general $p \in (1,\infty)$ see Section 3.3 for the definition of these spaces. In particular, we usually have a result stating that the graph norm of $-q(x,D)$ considered as an operator on $L^p(\mathbb{R}^n;\mathbb{R})$ is equivalent to the norm $\|\cdot\|_{H_p^{\psi,2}}$. If the interpolation estimate

$$\|u\|_{H_p^{\psi,s}} \leq \varepsilon\|u\|_{H_p^{\psi,2}} + c(\varepsilon)\|u\|_{L^p}, \quad 0 \leq s < 2, \tag{2.382}$$

holds for any $\varepsilon > 0$, it is clear that any pseudo-differential operator $p(x,D)$ mapping $H_p^{\psi,s}(\mathbb{R}^n;\mathbb{R})$, $0 \leq s < 2$, continuously into $L^p(\mathbb{R}^n;\mathbb{R})$ is automatically $-q(x,D)$-bounded provided

$$\|q(x,D)u\|_{L^p} + \|u\|_{L^p} \sim \|u\|_{H_p^{\psi,2}}.$$

But we know that (2.382) holds from Proposition I.4.3.25, especially (I.4.180). Indeed, since $\|u\|_{H_p^{\psi,2}} = \|(\mathrm{id} + \psi(D))^{s/2}u\|_{L^p}$ we have to apply Proposition I.4.3.25 with $\alpha = s/2$. As already mentioned, so far no general results on the L^p-boundedness of the class of pseudo-differential operators we are interested in are known. However the situation is dramatically different in the case $p=2$. In fact the operators considered in Section 2.3-2.5 and 2.7 are in general "L^2-continuous", i.e. map some spaces $H_2^{\psi,2}(\mathbb{R}^n;\mathbb{R})$ into $L^2(\mathbb{R}^n;\mathbb{R})$. Let us state a typical result in this direction.

Theorem 2.8.5. *Let $q_\lambda(x,\xi)$ be as in Theorem 2.6.10. In particular $q \in S_\rho^{2,\psi}$. In addition let $p \in S_\rho^{2-\eta,\psi}$ for some $0 < \eta \leq 2$ be a symbol such that $\xi \mapsto p(x,\xi)$ is continuous negative definite function and suppose that $-p(x,D)$ maps $C_0^\infty(\mathbb{R}^n;\mathbb{R})$ into $L^2(\mathbb{R}^n;\mathbb{R}) \cap C(\mathbb{R}^n;\mathbb{R})$. Then the operator $-q(x,D) - p(x,D)$ extends from $C_0^\infty(\mathbb{R}^n;\mathbb{R})$ to a generator of a L^2-sub-Markovian semigroup.*

So far our discussion was straightforward: just adapting the case we are interested in, i.e. generators of Feller or L^p-sub-Markovian semigroups, to general perturbation results. The following discussion is a bit more sophisticated.

Its aim is to enlarge the class of Hoh's symbolic calculus. We follow closely W.Hoh [128], see also [130]. The underlying idea is that a continuous negative definite function with a Lévy measure having a bounded support is arbitrarily often differentiable and admissible for constructing a symbolic calculus in the sense of Section 2.4. Now, given any continuous negative definite function ψ with representation

$$\psi(\xi) = \int_{\mathbb{R}^n\setminus\{0\}} (1-\cos y\cdot\xi)\nu(d\xi) \tag{2.383}$$

then the decomposition

$$\begin{aligned}\psi(\xi) &= \psi_R(\xi) + \tilde{\psi}_R(\xi)\\ &= \int_{B_R(0)\setminus\{0\}} (1-\cos y\cdot\xi)\nu(d\xi) + \int_{B_R^c(0)} (1-\cos y\cdot\xi)\nu(d\xi)\end{aligned}$$

yields a smooth continuous negative definite function and a bounded one. Thus, if ψ satisfies a certain minimal growth condition so will ψ_R, implying that $\tilde{\psi}_R(D)$ will be a $\psi_R(D)$-bounded perturbation.

Now let $p \in S_\rho^{2,\psi}$ with representation

$$p(x,\xi) = \int_{\mathbb{R}^n\setminus\{0\}} (1-\cos y\cdot\xi)N(x,dy). \tag{2.384}$$

A decomposition as

$$\begin{aligned}p(x,\xi) = &\int_{B_R(0)\setminus\{0\}} (1-\cos y\cdot\xi)N(x,dy)\\ &+ \int_{B_R^c(0)} (1-\cos y\cdot\xi)N(x,dy)\end{aligned}$$

promises a decomposition of $-p(x,D)$ into an operator $p_R(x,D)$ with symbol in some class $S_\rho^{2,\psi}$ and a bounded perturbation $\tilde{p}_R(x,D)$ such that under certain conditions $-p_R(x,D)$ generates a Feller semigroup as well as the original operator $-p(x,D) = -p_R(x,D) - \tilde{p}_R(x,D)$ by a perturbation argument. We are going to make this argument precise.

As in Section 2.4 let $\psi : \mathbb{R}^n \to \mathbb{R}$ be a continuous negative definite function such that

$$\psi(\xi) \geq c|\xi|^{r/2}, \quad |\xi| \text{ large and } 0 < r \leq 2, \tag{2.385}$$

holds. Recall the definition of the symbol class $S_\rho^{m,\psi}$, see Definition 2.4.4. Let $p \in S_\rho^{2,\psi}$ and assume in addition that $\xi \mapsto p(x,\xi)$ is a continuous negative definite function. Moreover we suppose that $p(x,\xi)$ has the Lévy–Khinchin representation

$$p(x,\xi) = \int_{\mathbb{R}^n \setminus \{0\}} (1 - \cos y \cdot \xi) N(x, dy) \tag{2.386}$$

and satisfies the uniform estimate

$$p(x,\xi) \leq c(1 + \psi(\xi)). \tag{2.387}$$

In order to decompose $p(x,\xi)$ let $\theta \in C_0^\infty(\mathbb{R}^n)$ be such that $0 \leq \theta \leq 1$ and $\theta = 1$ in a neighbourhood of 0 and define

$$N_1(x, dy) := \theta(y) N(x, dy) \quad \text{and} \quad N_2(x, dy) := (1 - \theta(y)) N(x, dy). \tag{2.388}$$

Clearly, $N_1(x, dy)$ and $N_2(x, dy)$ are both Lévy kernels each being associated for a fixed x to a real-valued continuous negative definite function. We decompose $p(x,\xi)$ now according

$$p_1(x,\xi) = \int_{\mathbb{R}^n \setminus \{0\}} (1 - \cos y \cdot \xi) N_1(x, dy) \tag{2.389}$$

and

$$p_2(x,\xi) = \int_{\mathbb{R}^n \setminus \{0\}} (1 - \cos y \cdot \xi) N_2(x, dy). \tag{2.390}$$

i.e. $p(x,\xi) = p_1(x,\xi) + p_2(x,\xi)$. For $x \in \mathbb{R}^n$ fixed the support of the measure $N_1(x,\cdot)$ is bounded, hence $\xi \mapsto p(x,\xi)$ is a smooth function. Thus we are close to the situation discussed in Theorem 2.6.9.

Theorem 2.8.6. *Assume* (2.385) *and that* $p_1 \in S_\rho^{2,\psi}$. *In addition suppose that for some* $\delta > 0$ *and* τ *sufficiently large*

$$p_1(x,\xi) + \tau \geq \delta(1 + \psi(\xi)) \tag{2.391}$$

holds for all $x \in \mathbb{R}^n$. *Then* $-p_1(x, D)$ *extends to a generator of a Feller semigroup.*

Our aim is to prove that in the case of Theorem 2.8.6 also $-p(x,D) = -p_1(x,D) - p_2(x,D)$ generates a Feller semigroup. We need some auxiliary results.

Lemma 2.8.7. *There is a bounded measure μ on $\mathbb{R}^n\backslash\{0\}$ such that*

$$\int_{\mathbb{R}^n\backslash\{0\}} (1-\cos y\cdot\xi)\,\mu(d\xi) = \frac{|y|^2}{1+|y|^2} \tag{2.392}$$

and

$$\int_{\mathbb{R}^n\backslash\{0\}} (1+|\xi|^2)\,\mu(d\xi) < \infty \tag{2.393}$$

holds.

Proof: According to Corollary I.3.6.13 the function $y \mapsto m(y) = \frac{1}{1+|y|^2}$ is positive definite, and since $m(0) = 1$ by Bochner's theorem, Theorem I.3.5.7 there is a probability measure $\tilde{\mu}$ on $\mathbb{R}^n$ such that

$$\frac{1}{1+|y|^2} = \int_{\mathbb{R}^n} \cos y\cdot\xi\,\tilde{\mu}(d\xi). \tag{2.394}$$

In particular (2.394) implies that $\tilde{\mu}$ has no atom at 0 and therefore we have

$$\int_{\mathbb{R}^n\backslash\{0\}} (1-\cos y\cdot\xi)\,\tilde{\mu}(d\xi) = 1-\frac{1}{1+|y|^2} = \frac{|y|^2}{1+|y|^2}$$

and (2.392) follows with $\mu = \tilde{\mu}|_{\mathbb{R}^n\backslash\{0\}}$. Since μ is bounded, in order to prove (2.393) we need to show that

$$\int_{|\xi|\geq 1} |\xi|^2\,\mu(d\xi) < \infty.$$

For $0 < \alpha < 2$ we find with

$$c_{n,\alpha} = \alpha\cdot 2^{\alpha-1}\Gamma\Big(\frac{\alpha+n}{2}\Big)\Big/\pi^{n/2}\Gamma\Big(1-\frac{\alpha}{2}\Big) \tag{2.395}$$

the Lévy–Khinchin representation

$$|\xi|^\alpha = c_{n,\alpha}\int_{\mathbb{R}^n\backslash\{0\}} (1-\cos y\cdot\xi)|y|^{-\alpha-n}\,dy \tag{2.396}$$

which yields

$$\begin{aligned}
&\int_{\mathbb{R}^n\setminus\{0\}} |\xi|^\alpha \,\mu(d\xi) \\
&\quad = c_{n,\alpha}\int_{\mathbb{R}^n\setminus\{0\}}\int_{\mathbb{R}^n}(1-\cos y\cdot\xi)|y|^{-\alpha-n}\,dy\mu(d\xi) \\
&\quad = c_{n,\alpha}\int_{\mathbb{R}^n}\frac{|y|^2}{1+|y|^2}|y|^{-\alpha-n}\,dy \\
&\quad \le c_{n,\alpha}\int_{|y|\le 1}|y|^{2-\alpha-n}\,dy + c_{n,\alpha}\int_{|y|>1}|y|^{-\alpha-n}\,dy \\
&\quad = \frac{\alpha\cdot 2^{\alpha-1}\Gamma\left(\frac{\alpha+n}{2}\right)\omega_n}{\pi^{n/2}}\frac{1}{\Gamma\left(\frac{2-\alpha}{2}\right)}\left(\frac{1}{2-\alpha}+\frac{1}{\alpha}\right)
\end{aligned}$$

with ω_n being surface area of S^{n-1}. Now, taking into account that the Γ-function has a pole of order one at 0 we obtain by a monotone convergence arguments that

$$\infty > \lim_{\alpha\uparrow 2}\int_{\mathbb{R}^n\setminus\{0\}}|\xi|^\alpha\,\mu(d\xi) \ge \lim_{\alpha\uparrow 2}\int_{|\xi|\ge 1}|\xi|^\alpha\,\mu(d\xi) = \int_{|\xi|\ge 1}|\xi|^2\,\mu(d\xi),$$

and the lemma is proved. □

This proof is taken from W.Hoh [128]. In fact in [165] together with R.Schilling it was proved that

$$\frac{|y|^2}{1+|y|^2} = \int_{\mathbb{R}^n}(1-e^{-ix\cdot\xi})g(\xi)d\xi$$

with

$$g(\xi) = \frac{1}{2}\int_0^\infty (2\pi\lambda)^{-n/2}e^{-|\xi|^2/2\lambda}e^{-\lambda/2}d\lambda.$$

Proposition 2.8.8. *If* (2.387) *holds then* $N_2(x,dy)$ *has with respect to* $x\in\mathbb{R}^n$ *a uniformly bounded mass.*

Proof: As in Lemma 2.8.7 denote by μ the measure satisfying (2.392) and

(2.393). It follows that

$$\begin{aligned} N_2(x, \mathbb{R}^n \backslash \{0\}) &= \int_{\mathbb{R}^n \backslash \{0\}} (1 - \theta(y)) N(x, dy) \\ &\leq c \int_{\mathbb{R}^n \backslash \{0\}} \Big(1 - \frac{1}{1+|y|^2}\Big) N(x, dy) \\ &= c \int_{\mathbb{R}^n \backslash \{0\}} \int_{\mathbb{R}^n} (1 - \cos y \cdot \xi) \, \mu(d\xi) N(x, dy) \\ &= c \int_{\mathbb{R}^n} p(x, \xi) \, \mu(d\xi) \\ &\leq c' \int_{\mathbb{R}^n} (1 + |\xi|^2) \, \mu(d\xi) < \infty \end{aligned}$$

proving the proposition. □

Proposition 2.8.9. *The operator* $p_2(x, D)$ *extends to a continuous operator from* $C_b(\mathbb{R}^n)$ *into itself as well as from* $B_b(\mathbb{R}^n)$ *into itself. In each case it has the operator norm* $2 \sup_{x \in \mathbb{R}^n} N_2(x, \mathbb{R}^n \backslash \{0\})$.

Proof: We use the finiteness of $N_2(x, \cdot)$ and the representation (2.390) of $p_2(x, \xi)$ to find for $u \in C_0^\infty(\mathbb{R}^n)$

$$-p(x, D)u(x) = \int_{\mathbb{R}^n \backslash \{0\}} (u(x+y) - u(x)) \, N_2(x, dy). \tag{2.397}$$

Further, for $\varphi \in C_0^\infty(\mathbb{R}^n \backslash \{0\})$, $\varphi(0) = 0$, it follows that

$$\begin{aligned} &\int_{\mathbb{R}^n \backslash \{0\}} \varphi(y) N_2(x, dy) \\ &\qquad = -(2\pi)^{-n/2} \int_{\mathbb{R}^n \backslash \{0\}} \int_{\mathbb{R}^n} (1 - \cos y \cdot \xi) \hat{\varphi}(\xi) \, d\xi N_2(x, dy) \\ &\qquad = -(2\pi)^{-n/2} \int_{\mathbb{R}^n} p_2(x, \xi) \hat{\varphi}(\xi) \, d\xi. \end{aligned}$$

Since $x \mapsto p_2(x, \xi)$ is continuous we deduce now that $N_2(x, dy)$ depends continuously on x with respect to the vague topology. This implies already the extension properties claimed, whereas the representation (2.397) yields the bound $2 \sup_{x \in \mathbb{R}^n} N_2(x, \mathbb{R}^n \backslash \{0\})$. □

Since $p_2(x, D)$ is a nonlocal operator we may not deduce that $u \in C_\infty(\mathbb{R}^n)$ implies that $p_2(x, D)u \in C_\infty(\mathbb{R}^n)$. Suppose that $N_2(x, dy)$ has the property that for every $\varepsilon > 0$ there exists $R > 0$ such that

$$N_2(x, B_R^c(0)) < \varepsilon \tag{2.398}$$

uniformly in $x \in \mathbb{R}^n$. For $u \in C_0^\infty(\mathbb{R}^n)$ clearly $p_2(x, D)u$ is continuous and further, for $x \notin \operatorname{supp} u$ it follows that

$$\begin{aligned} p_2(x, D)u(x) &= \int_{\mathbb{R}^n\setminus\{0\}} (u(x+y) - u(x))N_2(x, dy) \\ &= \int_{\mathbb{R}^n\setminus\{0\}} u(x+y)N_2(x, dy) \end{aligned}$$

and further

$$\begin{aligned} |p_2(x, D)u(x)| &= \left| \int_{\mathbb{R}^n\setminus\{0\}} u(x+y)N_2(x, dy) \right| \\ &\leq \|u\|_\infty N_2(x, \operatorname{supp} u(x + \cdot)). \end{aligned}$$

But if (2.398) holds it follows that $N_2(x, \operatorname{supp} u(x + \cdot)) \to 0$ as $|x| \to \infty$.

Proposition 2.8.10. *Let $p_2(x, \xi)$ and $N_2(x, dy)$ be as before. Then*

$$\sup_{x\in\mathbb{R}^n} p_2(x, \xi) \to 0 \quad \text{as } \xi \to 0 \tag{2.399}$$

implies (2.398).

Proof: Assume that (2.399) holds true and take $\varphi \in C_0^\infty(\mathbb{R}^n)$, $0 \leq \varphi \leq 1$, $\varphi(0) = 1$, $\operatorname{supp} \varphi \subset B_1(0)$ and set $\varphi_R(x) = \varphi(\frac{x}{R})$ for $R > 1$. Then we have

$$\begin{aligned} N_2(x, B_R^c(0)) &\leq \int_{\mathbb{R}^n\setminus\{0\}} (\varphi_R(0) - \varphi_R(y))N_2(x, dy) \\ &= (2\pi)^{-n/2} \int_{\mathbb{R}^n\setminus\{0\}} \int_{\mathbb{R}^n} (1 - \cos y \cdot \xi)\hat{\varphi}_R(\xi)N_2(x, dy) \\ &\leq (2\pi)^{-n/2} \int_{\mathbb{R}^n} p_2(x, \xi)|\hat{\varphi}_R(\xi)|\, d\xi \\ &\leq (2\pi)^{-n/2} \int_{|\xi|\leq\frac{1}{\sqrt{R}}} p_2(x, \xi)R^n|\hat{\varphi}(R\xi)|\, d\xi + c\int_{|\xi|>\frac{1}{\sqrt{R}}} p_2(x, \xi)|\hat{\varphi}_R(\xi)|\, d\xi \\ &\leq (2\pi)^{-n/2} \int_{|\xi|\leq\frac{1}{\sqrt{R}}} p_2(x, \xi)R^n|\hat{\varphi}(R\xi)|\, d\xi \end{aligned}$$

$$+ c' \int_{|\xi| > \frac{1}{\sqrt{R}}} (1 + |\xi|^2) R^n |R\xi|^{-(n+3)} \, d\xi$$

$$\leq \tilde{c} \big(\sup_{|\xi| \leq \frac{1}{\sqrt{R}}} p_2(x, \xi) \big) \|\hat{\varphi}\|_{L^1} + c'' R^{-3/2},$$

where we used that $\hat{\varphi} \in S(\mathbb{R}^n)$. Thus we proved that

$$N_2(x, B_R^c(0)) \to 0 \quad \text{as } R \to \infty$$

implying of course (2.398). □

Corollary 2.8.11. *If* (2.399) *holds then* $p_2(x, D)$ *maps* $C_\infty(\mathbb{R}^n)$ *into itself.*

Proof: We know already that $p_2(x, D)$ maps $C_b(\mathbb{R}^n)$ continuously into itself. Since $C_0^\infty(\mathbb{R}^n)$ is dense in $C_\infty(\mathbb{R}^n)$ the corollary follows from Proposition 2.8.10. □

Remark 2.8.12. A. Note that (2.399) is equivalent to

$$\sup_{x \in \mathbb{R}^n} (p(x, \xi) - p(x, 0)) \to 0 \quad \text{as } |\xi| \to 0 \tag{2.400}$$

since from (2.389) it always follows by the uniform boundedness of the Lévy measure $N_1(x, \cdot)$ that

$$\sup_{x \in \mathbb{R}^n} (p_1(x, \xi) - p_1(x, 0)) \to 0 \quad \text{as } |\xi| \to 0.$$

B. As proved in W.Hoh [128], condition (2.399) is in fact equivalent to (2.398). An independent proof is given in R.Schilling [248].

Summing up we arrive at the following result due to W.Hoh [128].

Theorem 2.8.13. *Let* $p \in S_\rho^{2,\psi}$ *satisfy* (2.387) *and* (2.400). *Assume further that* $\xi \mapsto p(x, \xi)$ *is negative definite and has the decomposition* (2.389) *and* (2.390). *If* $-p_1(x, D)$ *extends to a generator of a Feller semigroup, so will* $-p(x, D)$.

For discussing an analogous perturbation result in $L^2(\mathbb{R}^n; \mathbb{R})$ for a symbol

$$p(x, \xi) = \int_{\mathbb{R}^n \setminus \{0\}} (1 - \cos y \cdot \xi) N(x, dy)$$

we make an assumption on the kernel $N(x, dy)$ in (2.384), namely that

$$N(x, dy) = n(x, y)\mu(dy) \tag{2.401}$$

where μ is a fixed Lévy measure on $\mathbb{R}^n \backslash \{0\}$. Now we decompose $N(x, dy)$ according to

$$N_1(x, dy) = \theta(y)n(x, y)\mu(dy) = n_1(x, y)\mu(dy) = n(x, y)\mu_1(dy) \tag{2.402}$$

and

$$N_2(x, dy) = (1-\theta(y))n(x, y)\mu(dy) = n_2(x, y)\mu(dy) = n(x, y)\mu_2(dy) \tag{2.403}$$

where θ is as in (2.388). The corresponding decomposition of $p(x, \xi)$ is given by

$$\begin{aligned} p(x, \xi) = p_1(x, \xi) + p_2(x, \xi) = &\int_{\mathbb{R}^n \backslash \{0\}} (1 - \cos y \cdot \xi) n_1(x, y)\mu(dy) \\ &+ \int_{\mathbb{R}^n \backslash \{0\}} (1 - \cos y \cdot \xi) n_2(x, y)\mu(dy) \end{aligned} \tag{2.404}$$

Proposition 2.8.14. *Assume that $n(x, y)$ is bounded, i.e. $n(x, y) \leq M$ for some $M \geq 0$. Then $p_2(x, D)$ is a bounded operator from $L^p(\mathbb{R}^n; \mathbb{R}^n)$ to $L^p(\mathbb{R}^n; \mathbb{R})$ for $1 < p < \infty$, i.e.*

$$\|p_2(x, D)u\|_{L^p} \leq c\|u\|_{L^p} \tag{2.405}$$

holds for all $u \in L^p(\mathbb{R}^n; \mathbb{R})$.

Proof: It is sufficient to prove (2.405) for all $u \in C_0^\infty(\mathbb{R}^n; \mathbb{R})$. Observe that for $u \in C_0^\infty(\mathbb{R}^n; \mathbb{R})$

$$\begin{aligned} -p_2(x, D)u(x) &= \int_{\mathbb{R}^n \backslash \{0\}} (u(x+y) - u(x)) n_2(x, y)\mu(dy) \\ &= \int_{\mathbb{R}^n \backslash \{0\}} u(x+y) n_2(x, y)\mu(dy) - \int_{\mathbb{R}^n \backslash \{0\}} n_2(x, y)\mu(dy) u(x). \end{aligned}$$

Since $\int_{\mathbb{R}^n \backslash \{0\}} n_2(x, y)\mu(dy) \leq c$, it follows that we only need to discuss the first term. For $v \in L^{p'}(\mathbb{R}^n;\mathbb{R})$, $p' = \frac{p}{p-1}$ we find that

$$\begin{aligned}
&\Big| \int_{\mathbb{R}^n} \int_{\mathbb{R}^n\setminus\{0\}} u(x+y) n_2(x,y) \mu(dy) v(x)\, dx \Big| \\
&\quad \le M \int_{\mathbb{R}^n\setminus\{0\}} \int_{\mathbb{R}^n} |u(x+y)||v(x)|\, dx \mu_2(dy) \\
&\quad \le M \int_{\mathbb{R}^n\setminus\{0\}} \Big(\int_{\mathbb{R}^n} |u(x+y)|^p\, dx \Big)^{1/p} \Big(\int_{\mathbb{R}^n} |v(x)|^{p'}\, dx \Big)^{1/p'} \mu_2(dy) \\
&\quad = M \mu_2(\mathbb{R}^n\setminus\{0\}) \|u\|_{L^p} \|v\|_{L^{p'}}.
\end{aligned}$$

Dividing by $\|v\|_{L^{p'}}$ and taking the supremum over all $v \in L^{p'}(\mathbb{R}^n;\mathbb{R})$ with $\|v\|_{L^{p'}} = 1$, estimate (2.405) follows. □

Since we have good L^2-bounds within Hoh's calculus we can derive the following perturbation result which is again due to W.Hoh [130].

Theorem 2.8.15. *Let $p(x,\xi) = p_1(x,\xi) + p_2(x,\xi)$ be as in (2.404) and assume that $p_1 \in S_\rho^{2,\psi}(\mathbb{R}^n)$ is such that for some $\delta > 0$ and τ large*

$$p_1(x,\xi) + \tau \ge \delta(1+\psi(\xi)) \tag{2.406}$$

holds for all $x \in \mathbb{R}^n$. Then we have on $C_0^\infty(\mathbb{R}^n;\mathbb{R})$ the estimate

$$\|p(x,D)u\|_{L^2} \le c\|u\|_{H^{\psi,2}}$$

and

$$\|u\|_{H^{\psi,2}} \le c(\|p(x,D)u\|_{L^2} + \|u\|_{L^2}),$$

and $-p(x,D)$ extends to $H^{\psi,2}(\mathbb{R}^n;\mathbb{R})$ as a generator of an L^2-sub-Markovian semigroup.

Proof: For $p_1(x,D)$ we know all estimates stated in Section 2.5, especially we know that the graph norm of $p_1(x,D)$ is equivalent to $\|\cdot\|_{H^{\psi,2}}$. Now the boundedness of $p_2(x,D)$ in $L^2(\mathbb{R}^n;\mathbb{R})$ implies the theorem. □

Let us give a sufficient condition on $n(x,y)$ in (2.401) implying that $p_1(x,\xi)$ satisfies the assumptions of Theorem 2.8.15.

Proposition 2.8.16. *Suppose that for $n(x,y)$ in (2.401) we have for all $\beta \in \mathbb{N}_0^n$ the estimate*

$$|\partial_x^\beta n(x,y)| \le M_\beta \tag{2.407}$$

with M_β *independent of* $x, y \in \mathbb{R}^n$, *and assume*

$$n(x,y) \geq \eta > 0 \quad \textit{for all } x, y \in \mathbb{R}^n, \tag{2.408}$$

then $p_1 \in S_\rho^{\psi,2}(\mathbb{R}^n;\mathbb{R})$*and* (2.406) *holds.*

Proof: For $\beta \in \mathbb{N}_0^n$ and $|\alpha| \geq 1$ we have

$$\begin{aligned}|\partial_\xi^\alpha \partial_x^\beta p_1(x,\xi)| &= \Big|\partial_\xi^\alpha \partial_x^\beta \int_{\mathbb{R}^n\setminus\{0\}} (1-\cos y\cdot\xi) n(x,y)\mu_1(dy)\Big| \\ &\leq M_\beta \int_{\mathbb{R}^n\setminus\{0\}} |y^\alpha \cos^{(|\alpha|)}(y\cdot)|\mu_1(dy),\end{aligned}$$

and further by the Cauchy–Schwarz inequality for $|\alpha| = 1$

$$\begin{aligned}|\partial_\xi^\alpha \partial_x^\beta p_1(x,\xi)| &\leq M_\beta \|y^\alpha\|_{L^2(\mu_1)} \Big(\int_{\mathbb{R}^n\setminus\{0\}} \sin^2(y\cdot\xi)\mu_1(dy)\Big)^{1/2} \\ &\leq c_{\alpha\beta}\Big(\int_{\mathbb{R}^n\setminus\{0\}} (1-\cos(y\cdot\xi))\mu_1(dy)\Big)^{1/2} \leq c_{\alpha\beta}(1+\psi(\xi))^{1/2}\end{aligned}$$

and for $|\alpha| \geq 2$

$$|\partial_\xi^\alpha \partial_x^\beta p_1(x,\xi)| \leq M_\beta \int_{\mathbb{R}^n\setminus\{0\}} |y|^{|\alpha|}\mu_1(dy) = c_{\alpha\beta}.$$

For $\alpha = 0$ we have always

$$|\partial_x^\beta p_1(x,\xi)| \leq M_0(1+\psi(\xi)),$$

thus $p_1 \in S_\rho^{\psi,2}(\mathbb{R}^n;\mathbb{R})$. Now, in order to prove (2.406) we use (2.408) to find

$$\begin{aligned}p_1(x,\xi) &= \int_{\mathbb{R}^n\setminus\{0\}} (1-\cos y\cdot\xi) n(x,y)\mu_1(dy) \\ &\geq \eta \int_{\mathbb{R}^n\setminus\{0\}} (1-\cos y\cdot\xi)\mu_1(dy) \geq \eta\psi(\xi) - 2\eta\mu_2(\mathbb{R}^n\setminus\{0\})\end{aligned}$$

which gives for τ sufficiently large the the derived estimate (2.406). □

Remark 2.8.17. It is clear that we can combine the conditions of Theorem 2.8.13 and 2.8.15 in such a way that we will obtain L^p-sub-Markovian semigroup for $2 \leq p < \infty$ with generator $-p(x,D)$.

The following example is taken from W.Hoh [128].

Example 2.8.18. Consider

$$p(x,\xi) = \sum_{j=1}^{N} b_j(x)\psi_j(\xi) \tag{2.409}$$

where $N \in \mathbb{N}$, $b_j : \mathbb{R}^n \to \mathbb{R}_+$ are bounded functions, and $\psi_j : \mathbb{R}^n \to \mathbb{R}$ are continuous negative defined function having the Lévy–Khinchin representation

$$\psi_j(\xi) = \int_{\mathbb{R}^n\setminus\{0\}} (1-\cos y\cdot\xi)\tilde{\mu}_j(dy).$$

Then $\mu := \sum_{j=1}^{N} \tilde{\mu}_j$ is a Lévy measure with corresponding continuous negative definite function $\psi(\xi) = \sum_{j=1}^{N} \psi_j(\xi)$. We may use the Radon–Nikodym theorem to find functions $h_j : \mathbb{R}^n \to \mathbb{R}_+$, $|h_j| \le 1$, such that $\tilde{\mu}_j = h_j\mu$. This yields a representation of $p(x,\xi)$ as

$$p(x,\xi) = \int_{\mathbb{R}^n\setminus\{0\}} (1-\cos y\cdot\xi) \sum_{j=1}^{N} b_j(x)h_j(y)\mu(dy),$$

i.e. we have (2.401) with $n(x,y) = \sum_{j=1}^{N} b_j(x)h_j(y)$. Therefore, if $b_j \in C_b^\infty(\mathbb{R}^n;\mathbb{R})$, $1 \le j \le N$, it follows that $p_1 \in S_\rho^{\psi,2}(\mathbb{R})$ and (2.407) holds. In addition $b_j(x) \ge \eta > 0$ then

$$\begin{aligned} p_1(x,\xi) &= \sum_{j=1}^{N} \int_{\mathbb{R}^n\setminus\{0\}} (1-\cos y\cdot\xi)b_j(x)\theta(y)h_j(y)\mu(dy) \\ &\ge \eta \int_{\mathbb{R}^n\setminus\{0\}} (1-\cos y\cdot\xi)\theta(y)\mu(dy) \\ &\eta\psi(\xi) - 2\eta((1-\theta)\mu)(\mathbb{R}^n\setminus\{0\}), \end{aligned}$$

hence we have also (2.406).

Remark 2.8.19. Some of the pseudo-differential operators $q(x,D)$ generating Markov processes play a fundamental rôle in quantum physics, for example the *relativistic Hamiltonian* $(-\Delta + m^2)^{1/2} - m$. Therefore it is natural to look at perturbations through potentials, i.e. at operators of the form $q(x,D) + V(x)$ with a suitable potential $V(x)$. A central question is first of all whether for a symmetric $q(x,D)$ the operator $q(x,D) + V(x)$ is selfadjoint. This gives a restriction for the potential class, for example $V(x)$ should belong to some Kato-type potential class defined with the help of $q(x,D)$ (or the corresponding Markov process), compare for example M.Demuth's and J.van Casteren's monograph [68] or K.-Th.Sturm [260] and [261], and the references therein. Using a related *Feynman–Kac formula* a stochastic spectral analysis for the operator $q(x,D)+V(x)$ can be developed. We will come back to this question in volume III, the major reference is of course the monograph [68] of M.Demuth and J.van Casteren.

Finally let us discuss non-local perturbations of second order elliptic differential operators in $L^p(\mathbb{R}^n;\mathbb{R})$. Assume that

$$L(x,D)=\sum_{k,l=1}^{n} a_{kl}(x)\frac{\partial^2}{\partial x_k\partial x_l}+\sum_{j=1}^{n} b_j(x)\frac{\partial}{\partial x_j}+c(x) \tag{2.410}$$

is a second order differential operator satisfying $a_{kl}=a_{lk}\in C_b^2(\mathbb{R}^n;\mathbb{R})$, $b_j\in C_b^1(\mathbb{R}^n;\mathbb{R})$ and $c\in C_b(\mathbb{R}^n;\mathbb{R})$ such that $c(x)\leq 0$ for all $x\in\mathbb{R}^n$, and for all $x\in\mathbb{R}^n$, $\xi\in\mathbb{R}^n$ we require with some $\lambda_0>0$ that $\sum_{k,l=1}^n a_{kl}(x)\xi_k\xi_l\geq\lambda_0|\xi|^2$. In addition let us suppose $\sum_{j=1}^n\frac{\partial}{\partial x_j}(b_j(x)-\sum_{k=1}^n\frac{\partial a_{kj}(x)}{\partial x_k})\geq 0$. In this case it follows that that $(-L(x,D),W^{2,p}(\mathbb{R}^n;\mathbb{R}))$ is a closed operator on $L^p(\mathbb{R}^n;\mathbb{R})$ hence it is also an L^p-Dirichlet operator. Moreover we have the estimates

$$\|L(x,D)u\|_{L^p}\leq c\|u\|_{W^{2,p}} \tag{2.411}$$

and

$$\|u\|_{W^{2,p}}\leq c(\|L(x,D)u\|_{L^p}+\|u\|_{L^p}) \tag{2.412}$$

for all $u\in W^{2,p}(\mathbb{R}^n;\mathbb{R})$, compare Section 2.1

Now let $p(x,D)$ be a pseudo-differential operator with negative definite symbol. In order to handle $-p(x,D)$ as an L^p-perturbation of $L(x,D)$ it is more convenient to use its Lévy-type-operator representation. We are interested in operators

$$p(x,D)u(x)=\int_{\mathbb{R}^n\setminus\{0\}}(u(x)-u(x-y))\nu(x,dy) \tag{2.413}$$

where for some $0<\alpha<1$

$$\int_{\mathbb{R}^n\setminus\{0\}}(1\wedge|y|^\alpha)\nu(x,dy)\leq c_{\nu,\alpha}<\infty \tag{2.414}$$

holds with $c_{\nu,\alpha}$ independent of $x \in \mathbb{R}^n$, or in operators

$$p(x,D)u(x) = \int_{\mathbb{R}^n} \Big(u(x)-u(x-y)-\sum_{j=1}^{n} \frac{y_j}{1+|y|^2} \frac{\partial u(x)}{\partial x_j}\Big)\nu(x,dy) \quad (2.415)$$

with

$$\int_{\mathbb{R}^n\setminus\{0\}} (1\wedge|y|^{\alpha+1})\nu(x,dy) \leq c_{\nu,1+\alpha} < \infty \quad (2.416)$$

for some $0 < \alpha < 1$ and $c_{\nu,1+\alpha}$ independent of x. It turns out that L^p-estimates for the operators (2.413) and (2.415) can be obtained in *Sobolev–Slobodeckij spaces* which are special Besov spaces. For $s \in \mathbb{R}_+\setminus\mathbb{N}_0$ and $1 < p < \infty$ we set

$$W^{s,p}(\mathbb{R}^n;\mathbb{R}) := B^s_{p,q}(\mathbb{R}^n;\mathbb{R}) \quad (2.417)$$

where the *Besov spaces* $B^s_{p,q}(\mathbb{R}^n)$ had been defined in Section 4.11. For $s \in \mathbb{N}$ the spaces are the spaces defined in Section 2.1. In $W^{s,p}(\mathbb{R}^n;\mathbb{R})$, $s \in \mathbb{R}_+\setminus\mathbb{N}_0$, we have the equivalent norm

$$\begin{aligned}\|u\|_{W^{s,p}} &= \|u\|_{W^{[s],p}} \\ &+ \sum_{|\alpha|=[s]} \Big(\int_{\mathbb{R}^n}\int_{\mathbb{R}^n} \frac{|\partial^\alpha u(x) - \partial^\alpha u(y)|^p}{|x-y|^{(s-[s])p}} \frac{1}{|x-y|^n}\,dxdy\Big)^{1/p}.\end{aligned} \quad (2.418)$$

From the interpolation properties of Besov spaces we can deduce that for $s > s_0 \geq 0$ and $\varepsilon > 0$ the estimate

$$\|u\|_{W^{s_0,p}} \leq \varepsilon\|u\|_{W^{s,p}} + c(\varepsilon)\|u\|_{L^p} \quad (2.419)$$

holds, compare H.Triebel [267], p.186, Remark 2.4.2.6.

Proposition 2.8.20. *Suppose that $p(x,D)$ is given by (2.413) with a kernel $\nu(x,dy) = g(x,y)\lambda^{(n)}(dy)$ where $g : \mathbb{R}^n \times \mathbb{R}^n \to \overline{\mathbb{R}}_+$ is a measurable function satisfying with some $0 < \kappa < \mu < 1$ the estimate $|g(x,y)| \leq \frac{M}{|y|^{n+\kappa}}$. Then we have for all $u \in W^{\mu,p}(\mathbb{R}^n;\mathbb{R})$ the estimate*

$$\|p(x,D)u\|_{L^p} \leq c\|u\|_{W^{\mu,p}}. \quad (2.420)$$

Proof: For $u \in W^{\mu,p}(\mathbb{R}^n;\mathbb{R})$, $0<\kappa<\mu<1$, we get

$$\begin{aligned}|p(x,D)u(x)|^p &= \Big|\int_{\mathbb{R}^n\setminus\{0\}} (u(x-y)-u(x))g(x,y)\,dy\Big|^p\\ &\le 2^p\Big|\int_{0<|y|\le 1}(u(x-y)-u(x))g(x,y)\,dy\Big|^p\\ &\quad + 2^p\Big|\int_{|y|\ge 1}(u(x-y)-u(x))g(x,y)\,dy\Big|^p\\ &= 2^p(I_1(u)(x)+I_2(u)(x)).\end{aligned}$$

For $I_1u(x)$ it follows by Hölder's inequality, $\frac{1}{p}+\frac{1}{p'}=1$,

$$\begin{aligned}&\Big|\int_{0<|y|\le 1}(u(x-y)-u(x))g(x,y)\,dy\Big|^p\\ &\le \Big|\int_{0<|y|\le 1}\frac{|(u(x-y)-u(x))|}{|y|^{\mu}}\frac{M}{|y|^{n+\kappa-\mu}}\,dy\Big|^p\\ &\le \Big(\int_{0<|y|\le 1}\frac{|(u(x-y)-u(x))|^p}{|y|^{\mu p}}\frac{M}{|y|^{n+\kappa-\mu}}\,dy\Big)\\ &\quad\times\Big(\int_{0<|y|\le 1}\frac{1}{|y|^{n+\kappa-\mu}}\,dy\Big)^{p/p'}\\ &= c_{n,\mu,\kappa,p}\int_{0<|y|\le 1}\frac{|(u(x-y)-u(x))|^p}{|y|^{\mu p}}\frac{|y|^{\mu-\kappa}}{|y|^n}\,dy\\ &\le c_{n,\mu,\kappa,p}\int_{\mathbb{R}^n}\frac{|(u(x-y)-u(x))|^p}{|y|^{\mu p}}\frac{dy}{|y|^n},\end{aligned}$$

which leads to

$$\begin{aligned}\int_{\mathbb{R}^n}|I_1(u)(x)|\,dx &\le c_{n,\mu,\kappa,p}\int_{\mathbb{R}^n}\int_{\mathbb{R}^n}\frac{|(u(x-y)-u(x))|^p}{|y|^{\mu p}}\frac{dy}{|y|^n}dx\\ &\le c_{n,\mu,\kappa,p}\|u\|_{W^{\mu,p}}.\end{aligned}\tag{2.421}$$

In order to estimate $I_2(u)(x)$ consider for $v\in L^{p'}(\mathbb{R}^n;\mathbb{R})$ the form

$$\begin{aligned}&\Big|\int_{\mathbb{R}^n}\int_{|y|\ge 1}(u(x-y)-u(x))g(x,y)\,dy\,v(x)\,dx\Big|\\ &= \Big|\int_{\mathbb{R}^n}\int_{\mathbb{R}^n}(u(x-y)-u(x))g(x,y)\chi_{|y|\ge 1}(y)v(x)\,dydx\Big|\\ &\le M\int_{\mathbb{R}^n}\int_{\mathbb{R}^n}|u(x-y)-u(x)||v(x)|\frac{\chi_{|y|\ge 1}(y)}{|y|^{n+\kappa}}\,dy\end{aligned}$$

$$\leq M \int_{\mathbb{R}^n} \left(\int_{\mathbb{R}^n} |u(x-y) - u(x)|^p \, dx \right)^{1/p}$$
$$\times \left(\int_{\mathbb{R}^n} |v(x)|^{p'} \, dx \right)^{1/p'} \frac{\chi_{|y|\geq 1}(y)}{|y|^{n+\kappa}} \, dy$$
$$\leq 2^{\frac{p+1}{p}} M \|u\|_{L^p} \|v\|_{L^{p'}} \int_{|y|\geq 1} \frac{1}{|y|^{n+\kappa}} \, dy$$
$$= c_{n,\kappa} \|u\|_{L^p} \|v\|_{L^{p'}},$$

which yields

$$\left\| \int_{|y|\geq 1} (u(\cdot - y) - u(\cdot)) g(\cdot, y) \, dy \right\|_{L^p} \leq c_{n,\kappa} \|u\|_{L^p}, \tag{2.422}$$

and combining (2.421) with (2.422) we arrive at (2.420). □

Proposition 2.8.21. *Suppose that $p(x, D)$ is given by (2.415) with a kernel $\nu(x, dy) = g(x, y)\lambda^{(n)}(dy)$ where $g : \mathbb{R}^n \times \mathbb{R}^n \to \overline{\mathbb{R}}_+$ satisfies for some $\kappa \in (0, 1)$ the estimate $g(x, y) \leq \frac{M}{|y|^{n+1+\kappa}}$. For $0 < \kappa < \mu < 1$ we have for all $u \in W^{1+\mu,p}(\mathbb{R}^n; \mathbb{R})$ the estimate*

$$\|p(x, D)u\|_{L^p} \leq c\|u\|_{W^{1+\mu,p}}. \tag{2.423}$$

Proof: For $u \in W^{1+\mu,p}(\mathbb{R}^n; \mathbb{R})$ (or $C_0^\infty(\mathbb{R}^n; \mathbb{R})$) it follows that

$$p(x, D)u(x) = \int_{\mathbb{R}\setminus\{0\}} \left(u(x) - u(x-y) - \sum_{j=1}^n \frac{y_j}{1+|y|^2} \frac{\partial u(x)}{\partial x_j} \right) \nu(x, dy)$$
$$= \int_{|y|>1} \left(u(x) - u(x-y) - \sum_{j=1}^n \frac{y_j}{1+|y|^2} \frac{\partial u(x)}{\partial x_j} \right) g(x, y) \, dy$$
$$+ \int_{0<|y|\leq 1} \left(u(x) - u(x-y) - \sum_{j=1}^n y_j \frac{\partial u(x)}{\partial x_j} \right) g(x, y) \, dy$$
$$+ \int_{0<|y|\leq 1} \sum_{j=1}^n \left(y_j - \frac{y_j}{1+|y|^2} \right) \frac{\partial u(x)}{\partial x_j} g(x, y) \, dy$$
$$= (J_1 u)(x) + (J_2 u)(x) + (J_3 u)(x).$$

Now, for $J_1 u$ we find with $v \in L^{p'}(\mathbb{R}^n; \mathbb{R})$, $\frac{1}{p} + \frac{1}{p'} = 1$,

$$\left| \int_{\mathbb{R}^n} (J_1 u)(x) v(x) \, dx \right|$$

$$= \left| \int_{\mathbb{R}^n} \int_{|y|>1} \Big(u(x) - u(x-y) - \sum_{j=1}^n \frac{y_j}{1+|y|^2} \frac{\partial u(x)}{\partial x_j} \Big) g(x,y)\, dy\, v(x)\, dx \right|$$
$$= \Big| \int_{\mathbb{R}^n} \int_{\mathbb{R}^n} \Big(u(x) - u(x-y) - \sum_{j=1}^n \frac{y_j}{1+|y|^2} \frac{\partial u(x)}{\partial x_j} \Big)$$
$$v(x) g(x,y) \chi_{\{|y|>1\}}\, dx\, dy \Big|$$
$$\leq M \int_{\mathbb{R}^n} \int_{\mathbb{R}^n} \Big(|u(x) - u(x-y)||v(x)|$$
$$+ \sum_{j=1}^n \left| \frac{\partial u(x)}{\partial x_j} \right| |v(x)| \Big)\, dx \frac{\chi_{\{|y|>1\}}(y)}{|y|^{n+1+\kappa}}\, dy$$
$$\leq M \int_{\mathbb{R}^n} \Big\{ \Big(\int_{\mathbb{R}^n} |u(x) - u(x-y)|^p\, dx \Big)^{1/p} \|v\|_{L^{p'}}$$
$$+ \sum_{j=1}^n \left\| \frac{\partial u}{\partial x_j} \right\|_{L^p} \|v\|_{L^{p'}} \Big) \Big\} \frac{\chi_{\{|y|>1\}}(y)}{|y|^{n+1+\kappa}}\, dy$$
$$\leq 2M\omega_n \Big(\|u\|_{L^p} + \left\| \frac{\partial u}{\partial x_j} \right\|_{L^p} \Big) \|v\|_{L^{p'}} \int_1^\infty r^{-2-\kappa}\, dr$$
$$= \frac{2\omega_n M}{1+\kappa} \|u\|_{W^{1,p}} \|v\|_{L^{p'}},$$

implying

$$\|J_1 u\|_{L^p} \leq c_1 \|u\|_{W^{1,p}}. \tag{2.424}$$

For $J_2 u$ it follows by the mean-value theorem that

$$|J_2 u(x)|^p = \left| \int_{0<|y|\leq 1} \Big(u(x) - u(x-y) - \sum_{j=1}^n y_j \frac{\partial u(x)}{\partial x_j} \Big) g(x,y)\, dy \right|^p$$
$$= \left| \sum_{j=1}^n \int_0^1 \int_{0<|y|\leq 1} y_j \Big(\frac{\partial u(x)}{\partial x_j} - \frac{\partial u(x-ty)}{\partial x_j} \Big) g(x,y)\, dy\, dt \right|^p$$
$$= \left| \sum_{j=1}^n \int_0^1 \int_{0<|z|\leq t} z_j \frac{\frac{\partial u(x)}{\partial x_j} - \frac{\partial u(x-z)}{\partial x_j}}{|z|^\mu} g\Big(x, \frac{z}{t}\Big) \frac{1}{|z|^{-\mu}} t^{-1-n}\, dz\, dt \right|^p$$
$$\leq \Big(\sum_{j=1}^n \int_0^1 \int_{0<|z|\leq t} |z_j| \frac{\left| \frac{\partial u(x)}{\partial x_j} - \frac{\partial u(x-z)}{\partial x_j} \right|}{|z|^\mu} \frac{M}{|z|^{n+1+\kappa-\mu}}\, t^{n+1+\kappa-1-n}\, dz\, dt \Big)^p$$

$$\leq \Big(\sum_{j=1}^{n} M \int_{0<|z|\leq 1} \frac{\left|\frac{\partial u(x)}{\partial x_j} - \frac{\partial u(x-z)}{\partial x_j}\right|}{|z|^{\mu}} \frac{1}{|z|^{n+\kappa-\mu}}\, dz\Big)^p$$

$$\leq \sum_{j=1}^{n} M^p \Big(\int_{0<|z|\leq 1} \frac{\left|\frac{\partial u(x)}{\partial x_j} - \frac{\partial u(x-z)}{\partial x_j}\right|^p}{|z|^{p\mu}} \frac{1}{|z|^{n+\kappa-\mu}}\, dz\Big)$$

$$\Big(\int_{0<|z|\leq 1} \frac{1}{|z|^{n+\kappa-\mu}}\, dz\Big)^{p/p'}$$

$$\leq \sum_{j=1}^{n} M^p \Big(\frac{\omega_n}{\mu-\kappa}\Big)^{p/p'} \int_{0<|z|\leq 1} \frac{\left|\frac{\partial u(x)}{\partial x_j} - \frac{\partial u(x-z)}{\partial x_j}\right|^p}{|z|^{p\mu}} \frac{1}{|z|^n}\, dz,$$

which yields that

$$\int_{\mathbb{R}^n} |J_2 u(x)|^p\, dx \leq \tilde{c} \sum_{j=1}^{n} \int_{\mathbb{R}^n} \int_{0<|z|\leq 1} \frac{\left|\frac{\partial u(x)}{\partial x_j} - \frac{\partial u(x-z)}{\partial x_j}\right|^p}{|z|^{p\mu}} \frac{1}{|z|^n}\, dz\, dx,$$

or

$$\|J_2 u\|_{L^p} \leq c_2 \|u\|_{W^{1+\mu,p}}. \tag{2.425}$$

Finally, to estimate J_3 observe that

$$|J_3 u(x)| = \Big| \int_{0<|y|\leq 1} \sum_{j=1}^{n} \frac{y_j |y|^2}{1+|y|^2} \frac{\partial u(x)}{\partial x_j} g(x,y)\, dy \Big|$$

$$\leq \sum_{j=1}^{n} \Big|\frac{\partial u(x)}{\partial x_j}\Big| \int_{0<|y|\leq 1} \frac{|y_j|\,|y|^2}{1+|y|^2} \frac{M}{|y|^{n+1+\kappa}}\, dy$$

$$\leq M \sum_{j=1}^{n} \Big|\frac{\partial u(x)}{\partial x_j}\Big| \int_{0<|y|\leq 1} |y|^{2-n-\kappa}\, dy$$

$$= \frac{M\omega_n}{2-\kappa} \sum_{j=1}^{n} \Big|\frac{\partial u(x)}{\partial x_j}\Big|,$$

implying

$$\|J_3 u\|_{L^p} \leq c_3 \|u\|_{W^{1,p}}. \tag{2.426}$$

Combining (2.424), (2.425) and (2.426) yields (2.423), i.e. the proposition is proved. □

Thus in each of the cases treated in Proposition 2.8.20 and 2.8.21 the operator $-p(x, D)$ is according to (2.419) an $L(x, D)$-bounded operator on $W^{2,p}(\mathbb{R}^n; \mathbb{R})$ and as before it is clear that $-p(x, D)$ is also an L^p-Dirichlet operator. In particular it is dissipative and $-L(x, D) - p(x, D)$ is an L^p-Dirichlet operator, too. Therefore we arrive at

Theorem 2.8.22. *Let $L(x, D)$ be a second order elliptic differential operator as in* (2.410) *and suppose that $p(x, D)$ is either as in Proposition* 2.8.20 *or Proposition* 2.8.21 *then the operator $(-L(x, D) - p(x, D), W^{2,p}(\mathbb{R}^n; \mathbb{R}))$ is an L^p-Dirichlet operator generating an L^p-sub-Markovian semigroup.*

2.9 On Semigroups Obtained by Subordination

Let $(T_t^{(p)})_{t \geq 0}$ be an L^p-sub-Markovian semigroup, $1 < p < \infty$, or a Feller semigroup for $p = \infty$. In Section I.4.3 we studied the general procedure of subordinating in the sense of Bochner. It was shown that for any Bernstein function f the subordinate semigroup $(T_t^{(p),f})_{t \geq 0}$ is again L^p-sub-Markovian or Feller, respectively, and on the level of generators an operational or functional calculus was developed. The essential result was that if $A^{(p)}$ is the generator of $(T_t^{(p)})_{t \geq 0}$ then the generator $A^{(p),f}$ of $(T_t^{(p),f})_{t \geq 0}$ is given by $A^{(p),f} = -f(-A^{(p)})$.

Clearly we may apply subordination to the semigroups constructed in the last sections and we will obtain automatically new examples. The purpose of this section is to combine the general theory of subordination and its underlying functional calculus with a symbolic calculus for pseudo-differential operators being a generator of an L^p-sub-Markovian or Feller semigroup.

Let us outline the situation in the case of translation invariant semigroups, i.e. semigroups characterized by a fixed continuous negative definite function $\psi : \mathbb{R}^n \to \mathbb{C}$ according to

$$(T_t u)^\wedge(\xi) = e^{-t\psi(\xi)} \hat{u}(\xi) \tag{2.427}$$

for $u \in S(\mathbb{R}^n; \mathbb{R})$. (Since in this case all the operators $T_t^{(p)}$, $t \geq 0$, coincides on $S(\mathbb{R}^n; \mathbb{R})$ for $1 < p \leq \infty$, we omit p and write only T_t, and A for the generator etc.) If f is a Bernstein function with corresponding convolution semigroup $(\eta_t)_{t \geq 0}$, $\operatorname{supp} \eta_t \subset [0, \infty)$, we find for the subordinate semigroup

$$(T_t^f u)^\wedge(\xi) = e^{-t(f \circ \psi)(\xi)} \hat{u}(\xi). \tag{2.428}$$

Further, the generator of $(T_t)_{t\geq 0}$ is given on $S(\mathbb{R}^n;\mathbb{R})$ by

$$-\psi(D)u(x) = -(2\pi)^{-n/2}\int_{\mathbb{R}^n} e^{ix\cdot\xi}\psi(\xi)\hat{u}(\xi)\,d\xi \tag{2.429}$$

and that of $(T_t^f)_{t\geq 0}$ by

$$-\psi^f(D)u(x) = -(f\circ\psi)(D)u(x) = -(2\pi)^{-n/2}\int_{\mathbb{R}^n} e^{ix\cdot\xi} f(\psi(\xi))\hat{u}(\xi)\,d\xi. \tag{2.430}$$

Thus the symbol of A is $\sigma(A)(\xi) = -\psi(\xi)$ and that of A^f is $\sigma(A^f)(\xi) = -f(\psi(\xi))$. Analogous results do hold for the resolvents $(R_\lambda)_{\lambda>0}$ and $(R_\lambda^f)_{\lambda>0}$, respectively,

$$\sigma(R_\lambda)(\xi) = \frac{1}{\lambda+\psi(\xi)} \quad \text{and} \quad \sigma(R_\lambda^f)(\xi) = \frac{1}{\lambda+f(\psi(\xi))},$$

as well as for the semigroups (as stated before)

$$\sigma(T_t)(\xi) = e^{-t\psi(\xi)} \quad \text{and} \quad \sigma(T_t^f)(\xi) = e^{-tf(\psi(\xi))}.$$

Now, suppose that the pseudo-differential operator $-q(x,D)$ with symbol $q : \mathbb{R}^n\times\mathbb{R}^n \to \mathbb{C}$ such that $\xi\mapsto q(x,\xi)$ is continuous negative definite function, extends from $S(\mathbb{R}^n;\mathbb{R})$ to a generator of an L^p-sub-Markovian semigroup, $1 < p < \infty$, and a Feller semigroup for $p=\infty$. We denote all these semigroups again simply by $(T_t)_{t\geq 0}$ and the generator(s) by A. Then it is clear that for the subordinate semigroup $(T_t^f)_{t\geq 0}$, its generator A^f and its resolvent $(R_\lambda^f)_{\lambda>0}$ we will have in general

$$\sigma(T_t^f)(x,\xi) \neq e^{-tf(q(x,\xi))}, \tag{2.431}$$

$$\sigma(A^f)(x,\xi) \neq -f(q(x,\xi)) \tag{2.432}$$

and

$$\sigma(R_\lambda^f)(x,\xi) \neq \frac{1}{\lambda+f(q(x,\xi))}. \tag{2.433}$$

However the question arises whether the operators with symbols as on the right-hand side of (2.431)–(2.433) will give some good approximations of the operators we are interested in. There are two cases of particular interest. First suppose we know for a given semigroup $(T_t)_{t\geq 0}$ that T_t has a kernel representation

$$T_t u(x) = \int_{\mathbb{R}^n} \Gamma_t(x,y)u(y)\,dy$$

with upper and lower estimates for the kernel $\Gamma_t(x,y)$. Then the subordinate semigroup $(T_t^f)_{t\geq 0}$ will also have a kernel given by

$$\Gamma_t^f(x,y) = \int_0^\infty \Gamma_s(x,y)\eta_t(ds),$$

and we will have immediately bounds for $\Gamma_t^f(x,y)$. The second interesting case follows from the observation that by Lemma I.3.9.9 the function

$$\xi \mapsto f(q(x,\xi)) \tag{2.434}$$

is a continuous negative definite function too. Hence the pseudo-differential operator $-(f\circ q)(x,D)$ with the symbol $-f(q(x,\xi))$ is a candidate for an operator having an extension generating an L^p-sub-Markovian semigroup and a Feller semigroup, respectively. Denote this semigroup (if it exists) for a moment by $(S_t)_{t\geq 0}$. Then we should expect that S_t has "approximately" the same properties as T_t. But we will have complete knowledge of the symbol of its generator which (as we will see later in volume III of this treatise) give us a lot of information on S_t as well as on the corresponding stochastic process.

As an example for the first case we will discuss the subordination of certain diffusion semigroups, i.e. semigroups generated by certain second order elliptic differential operators. Then we will turn to the general case where so far only a few first results are known. Finally we will pay more attention to fractional powers where much more results are available. We need some additional background material on second order elliptic differential operators, some results we will just summarise, often results will be provided with some (technical) proofs. The selection is determined by preparing some parts for the discussion in Chapter 3 and volume III.

Our first aim is to understand in principle subordinate diffusion semigroups. More precisely we will restrict our attention to semigroups generated by second order elliptic differential operators in divergence form

$$L(x,D)u(x) = \sum_{k,l=1}^n \frac{\partial}{\partial x_k}\Big(a_{kl}(x)\frac{\partial u(x)}{\partial x_l}\Big) \tag{2.435}$$

where $a_{kl} = a_{lk} \in C_b^3(\mathbb{R}^n;\mathbb{R})$ such that

$$\lambda_0|\xi|^2 \leq \sum_{k,l=1}^n a_{kl}(x)\xi_k\xi_l \leq \lambda_0^{-1}|\xi|^2 \tag{2.436}$$

holds for all $x \in \mathbb{R}^n$ and $\xi \in \mathbb{R}^n$ with some $\lambda_0 > 0$. Since

$$\sum_{k,l=1}^{n} \frac{\partial}{\partial x_k}\Big(a_{kl}(x)\frac{\partial}{\partial x_l}u\Big) = \sum_{k,l=1}^{n}\Big(a_{kl}(x)\frac{\partial^2 u}{\partial x_k \partial x_l} + \frac{\partial a_{kl}(x)}{\partial x_k}\frac{\partial u}{\partial x_l}\Big)$$

it follows with $b_l := \sum_{k=1}^{n} \frac{a_{kl}(x)}{\partial x_k}$ that

$$\sum_{l=1}^{n}\Big(b_l - \sum_{k=1}^{n}\frac{\partial}{\partial x_k}a_{kl}\Big) = \sum_{l=1}^{n}\frac{\partial}{\partial x_l}\Big(\sum_{k=1}^{n}\frac{\partial a_{kl}}{\partial x_k} - \sum_{k=1}^{n}\frac{\partial a_{kl}(x)}{\partial x_k}\Big) = 0,$$

and according to Theorem 2.1.44 the operator $(L(x,D), W^{2,p}(\mathbb{R}^n;\mathbb{R}))$ is a generator of an L^p-sub-Markovian semigroup. In fact it also generates a Feller semigroup. Moreover, the associated parabolic operator $\frac{\partial}{\partial t} - L(x,D)$ has a fundamental solution, (compare Definition 2.7.12).

Definition 2.9.1. A function $\Gamma : (0,\infty)\times\mathbb{R}^n\times\mathbb{R}^n \to \mathbb{R}$ is called *fundamental solution* of the operator $\frac{\partial}{\partial t} - L(x,D)$ if $(t,x,y)\mapsto \Gamma(t,x,y)$ $(=\Gamma_t(x,y))$ is continuous, $\Gamma(t,x,y) > 0$, $x \mapsto \Gamma(t,x,y)$ is a C^2-function and $t\mapsto \Gamma(t,x,y)$ is a C^1-function such that

$$\frac{\partial\Gamma(t,x,y)}{\partial t} - L(x,D_x)\Gamma(t,x,y) = 0, \tag{2.437}$$

and for all $f \in C_b(\mathbb{R}^n;\mathbb{R})$

$$\lim_{t\to 0}\int_{\mathbb{R}^n}\Gamma(t,x,y)f(y)\,dy = f(x) \tag{2.438}$$

uniformly on compact subsets of $\mathbb{R}^n$.

Theorem 2.9.2. *For $a_{kl} = a_{lk} \in C_b^3(\mathbb{R}^n;\mathbb{R})$ satisfying (2.436) the operator $\frac{\partial}{\partial t} - L(x,D)$ has a unique fundamental solution $\Gamma(t,x,y) = \Gamma_t(x,y)$ which satisfies the* Chapman–Kolmogorov equations

$$\Gamma(t+s,x,y) = \int_{\mathbb{R}^n}\Gamma(t,x,z)\Gamma(s,z,y)\,dz,$$

and in addition for $f \in C_b(\mathbb{R}^n;\mathbb{R})\cap L^p(\mathbb{R}^n;\mathbb{R})$ we have

$$T_t^{(p)}f(x) = T_t f(x) = \int_{\mathbb{R}^n}\Gamma(t,x,y)f(y)\,dy, \quad t>0 \tag{2.439}$$

which implies that $T_t^{(p)}$ has a kernel representation for $t>0$.

Remark 2.9.3. The general version of Theorem 2.9.2, i.e. for operators with t-depending coefficients and lower regularity, was obtained by A.M.Il'in, A.S.Kalashnikov and O.A.Oleinik [144]. We used a bit E.B.Dynkin [77], [78] for our formulation. A standard reference is the monograph [94] of A.Friedman, see also that of S.Itô [147].

Of course the heat operator $\frac{\partial}{\partial t} - \Delta_n$ is included in Theorem 2.9.2 and the corresponding fundamental solution is given by

$$\Gamma_0(t,x,y) := (4\pi t)^{-n/2} e^{-|x-y|^2/4t}. \tag{2.440}$$

Moreover, if $a_{kl} = a_{lk} \in \mathbb{R}$ satisfy (2.436) and $Q = (a_{kl})$ denotes the corresponding matrix a fundamental solution of the constant coefficient operator

$$\frac{\partial}{\partial t} - \sum_{k,l=1}^{n} a_{kl} \frac{\partial^2}{\partial x_n \partial x_l}$$

is given by

$$\Gamma^Q(t,x,y) = \frac{1}{(4\pi t)^{n/2}\sqrt{\det Q}} e^{-\frac{\langle Q^{-1}(x-y), x-y\rangle}{4t}}. \tag{2.441}$$

compare (I.3.130). If $Q = \gamma \text{ id}$ we will write Γ^γ instead of $\Gamma^{\gamma \text{ id}}$.

Now, accordingly to our philosophy stated above we should try to compare $\Gamma(t,x,y)$ from Theorem 2.9.2 with $\Gamma^{Q(x_0)}(t,x,y)$, i.e. $\Gamma^{Q(x_0)}(t,x,y)$ should be a first order approximation of $\Gamma(t,x,y)$. However, the *Aronson estimates* we are going to prove, i.e. the estimates

$$\kappa_1 \Gamma^{\gamma_1}(t,x,y) \le \Gamma(t,x,y) \le \kappa_2 \Gamma^{\gamma_2}(t,x,y) \tag{2.442}$$

are even better since they allow us to compare $\Gamma(t,x,y)$ and therefore $(T_t^{(p)})_{t\ge 0}$ with the densities of two convolution semigroups and the corresponding operator semigroups, respectively. Note that (2.442) yields immediately results for subordinate semigroups (which we will discuss later in this section), just recall the formula

$$T_t^f u(x) = \int_0^\infty T_s u(x) \eta_t(ds) = \int_{\mathbb{R}^n} \int_0^\infty \Gamma(s,x,y) \eta_t(ds) u(y)\, dy.$$

We are now going to prove (2.442) following the work [84] of the E.Fabes and D.Stroock, see also D.Stroock [258]. In doing so we will encounter techniques which will become useful later on.

Proposition 2.9.4. *For $p < n$, $p \geq 1$, the* Sobolev inequality

$$\|u\|_{L^{\frac{np}{n-p}}} \leq \frac{(n-1)p}{\sqrt{n}(n-p)}\|\nabla u\|_{L^p} \tag{2.443}$$

holds for all $u \in W^{1,p}(\mathbb{R}^n) = H^1_p(\mathbb{R}^n)$.

Proof: For $p = 1$, $u \in C_0^\infty(\mathbb{R}^n)$ and $x = (x_1, \ldots, x_n) \in \mathbb{R}^n$ it follows that

$$|u(x)| \leq \int_{-\infty}^{x_j} |\frac{\partial}{\partial y_j}u(x_1, \ldots, y_j, \ldots, x_n)|\, dy_j \leq \int_{\mathbb{R}} \left|\frac{\partial u(x)}{\partial x_j}\right| dx_j,$$

with y_j in the jth position. Thus we have

$$|u(x)|^{n/n-1} \leq \Big(\prod_{j=1}^{n} \int_{\mathbb{R}} \frac{\partial}{\partial x_j}u(x)|\, dx_j\Big)^{1/n-1}.$$

Integrating with respect to x_1 and applying Hölder's inequality yield

$$\begin{aligned}
&\int_{\mathbb{R}} |u(x)|^{n/n-1}\, dx_1 \\
&\leq \int_{\mathbb{R}} \Big(\int_{\mathbb{R}} \left|\frac{\partial}{\partial y_1}u(y_1, x_2, \ldots, x_n)\right| dy_1 \prod_{j=2}^{n} \int_{\mathbb{R}} \left|\frac{\partial u(x)}{\partial x_j}\right| dx_j\Big)^{1/n-1} dx_1 \\
&= \Big(\int_{\mathbb{R}} \left|\frac{\partial}{\partial y_1}u(y_1, x_2, \ldots, x_n)\right| dy_1\Big)^{1/n-1} \int_{\mathbb{R}} \Big(\prod_{j=2}^{n} \int_{\mathbb{R}} \left|\frac{\partial u(x)}{\partial x_j}\right| dx_j\Big)^{1/n-1} dx_1 \\
&\leq \Big(\int_{\mathbb{R}} \left|\frac{\partial}{\partial y_1}u(y_1, x_2, \ldots, x_n)\right| dy_1\Big)^{1/n-1} \\
&\quad \prod_{j=2}^{n} \Big(\int_{\mathbb{R}} \int_{\mathbb{R}} \left|\frac{\partial u(x)}{\partial x_j}\right| dx_j\, dx_1\Big)^{1/n-1}.
\end{aligned}$$

Now, continuing with integration with respect to $x_2, \ldots, x_n$ and using the same arguments we arrive at

$$\int_{\mathbb{R}^n} |u(x)|^{n/n-1}\, dx \leq \prod_{j=1}^{n} \Big(\int_{\mathbb{R}} |\frac{\partial u(x)}{\partial x_j}|\, dx\Big)^{1/n-1},$$

and using the geometric-arithmetic mean inequality

$$\|u\|_{L^{n/n-1}} \leq \Big(\int_{\mathbb{R}} |\frac{\partial u(x)}{\partial x_j}|\, dx\Big)^{1/n} \leq \frac{1}{n} \int_{\mathbb{R}} \sum_{j=1}^{n} |\frac{\partial u(x)}{\partial x_j}|\, dx.$$

Finally, since $\sum_{j=1}^n |\frac{\partial u(x)}{\partial x_j}| \leq \sqrt{n}|\nabla u(x)|$ we obtain

$$\|u\|_{L^{n/n-1}} \leq \frac{1}{\sqrt{n}} \|\nabla u\|_{L^1}, \tag{2.444}$$

i.e. (2.443) for $p = 1$. But from (2.444) we can easily derive (2.443). For this let $q > 1$ and observe that

$$\begin{aligned} \||u|^q\|_{L^{n/(n-1)}} &\leq \frac{1}{\sqrt{n}} \int_{\mathbb{R}^n} |\nabla(|u|^q)| \, dx \\ &= \frac{q}{\sqrt{n}} \int_{\mathbb{R}^n} |u|^{q-1} |\nabla u| \, dx \\ &\leq \frac{q}{\sqrt{n}} \||u|^{q-1}\|_{L^{p'}} \|\nabla u\|_{L^p}. \end{aligned} \tag{2.445}$$

Now, take $q := \frac{(n-1)p}{n-p}$ and estimate (2.443) follows first for $u \in C_0^\infty(\mathbb{R}^n)$ but then by density for all $u \in W^{1,p}(\mathbb{R}^n)$. $\square$

Remark 2.9.5. We have taken the proof of Proposition 2.9.4 from W.Ziemer [290], Theorem 2.4.1, the original proof is due to L.Nirenberg [224]. Note that the constant in (2.443) is not optimal. For $p = 1$ the optimal constant is $\frac{1}{n(\omega_n/n)^{1/n}}$, compare W.Ziemer [290], Theorem 2.7.4.

Using the *generalized Hölder inequality*

$$\|u\|_{L^r} \leq \|u\|_{L^s}^\lambda \|u\|_{L^t}^{1-\lambda}, \quad u \in L^t(\mathbb{R}^n), \tag{2.446}$$

which holds for $s \leq r \leq t$ and $\frac{1}{r} = \frac{\lambda}{s} + \frac{1-\lambda}{t}$ we obtain

Corollary 2.9.6. *For $u \in W^{1,p}(\mathbb{R}^n)$, $p < n$, the estimate*

$$\|u\|_{L^r} \leq c_{n,p} \|u\|_{L^1}^\lambda \|\nabla u\|_{L^p}^{1-\lambda} \tag{2.447}$$

holds for $\frac{1}{r} = \lambda + (1-\lambda)(\frac{n-p}{np})$. In particular for $r = p = 2$ we find $\lambda = \frac{2}{n+2}$ which yields

$$\|u\|_{L^2} \leq c_{n,2} \|\nabla u\|_{L^2}^{n/(n+2)} \|u\|_{L^1}^{2/(n+2)},$$

or

$$\|u\|_{L^2}^{2+4/n} \leq \tilde{c}_{n,2} \|\nabla u\|_{L^2}^2 \|u\|_{L^1}^{4/n}. \tag{2.448}$$

Let us return to the operator $L(x, D)$ given by (2.435), (2.436). For $\xi \in \mathbb{R}^n$ denote by $L_\xi(x, D)$ the operator

$$L_\xi(x, D)u := e^{-\xi\cdot x} L(x, D)(e^{x\cdot\xi} u(x)). \tag{2.449}$$

Thus we have for $u \in S(\mathbb{R}^n)$

$$L_\xi(x, D)u(x) = \sum_{k,l=1}^{n} \Big\{ \frac{\partial a_{kl}(x)}{\partial x_k}(\xi_l u(x) + \frac{\partial}{\partial x_l} u(x)) + a_{kl}(x)\Big(\xi_k\xi_l u(x) + \xi_l \frac{\partial u(x)}{\partial x_k} + \xi_k \frac{\partial u(x)}{\partial x_l} + \frac{\partial^2 u(x)}{\partial x_k \partial x_l}\Big)\Big\}. \tag{2.450}$$

With $Q(x) = (a_{kl}(x))_{k,l=1,\ldots,n}$ we find further for $u \in S(\mathbb{R}^n)$, $u(x) \geq 0$, that

$$\begin{aligned}
&\int_{\mathbb{R}^n} (L_\xi(x, D))u^{2p-1}\, dx \\
&= \int_{\mathbb{R}^n} \langle \xi, Q(x)\xi\rangle u^{2p}\, dx + \sum_{k,l=1}^{n} \int_{\mathbb{R}^n} \Big(\frac{\partial a_{kl}}{\partial x_k} \xi_l u^{2p} + \frac{\partial a_{kl}}{\partial x_k}\frac{\partial u}{\partial x_l} u^{2p-1}\Big)\, dx \\
&\quad + \sum_{k,l=1}^{n} \int_{\mathbb{R}^n} \Big(a_{kl}(\xi_l \frac{\partial u}{\partial x_k} + \xi_k \frac{\partial u}{\partial x_l}) u^{2p-1} + a_{kl} \frac{\partial^2 u}{\partial x_k \partial x_l} u^{2p-1}\Big)\, dx \\
&= \int_{\mathbb{R}^n} \langle \xi, Q(x)\rangle u^{2p}\, dx - 2(p-1) \int_{\mathbb{R}^n} \langle \xi, Q(x)\nabla u\rangle u^{2p-1}\, dx \\
&\quad + \sum_{k,l=1}^{n} \int_{\mathbb{R}^n} \Big(a_{kl} \frac{\partial^2 u}{\partial x_k \partial x_l} u^{2p-1} - \frac{1}{2p}\frac{\partial^2}{\partial x_k \partial x_l}(u^{2p})\Big)\, dx \\
&= \int_{\mathbb{R}^n} \langle \xi, Q(x)\xi\rangle u^{2p}\, dx - 2(p-1) \int_{\mathbb{R}^n} \langle \xi, Q(x)\nabla u\rangle u^{2p-1}\, dx \\
&\quad - (2p-1) \int_{\mathbb{R}^n} \langle \nabla u, Q(x)\nabla u\rangle u^{2p-1}\, dx.
\end{aligned}$$

Using the relation $u^{p-1}\nabla u = \frac{1}{p}\nabla(u^p)$ we arrive at

$$\begin{aligned}
\int_{\mathbb{R}^n} (L_\xi(x, D)u)u^{2p-1}\, dx = &\int_{\mathbb{R}^n} \langle \xi, Q(x)\xi\rangle u^{2p}\, dx \\
&- \frac{2(p-1)}{p} \int_{\mathbb{R}^n} \langle \xi, Q(x)\nabla(u^p)\rangle u^p\, dx \\
&- \frac{2(p-1)}{p^2} \int_{\mathbb{R}^n} \langle \nabla(u^p), Q(x)\nabla(u^p)\rangle u^p\, dx.
\end{aligned} \tag{2.451}$$

Now we estimate each term on the right-hand side of (2.451). By (2.436) we find first

$$\int_{\mathbb{R}^n} \langle \xi, Q(x)\xi \rangle u^{2p}\, dx \leq \frac{1}{\lambda_0}|\xi|^2 \int_{\mathbb{R}} u^{2p}\, dx,$$

and next for $\eta > 0$ using the fact that $\langle \xi, Q(x)\zeta \rangle$ is a scalar product and that for $a, b > 0$ we have $a \cdot b \leq \eta a^2 + \frac{1}{4\eta} b^2$

$$\begin{aligned}
&\left|\frac{2(p-1)}{p}\int_{\mathbb{R}^n} \langle \xi, Q(x)\nabla(u^p)\rangle u^p\, dx\right| \\
&\leq \frac{2(p-1)}{p}\Big(\int_{\mathbb{R}^n} \langle \xi, Q(x)\nabla(u^p)\rangle^2\, dx\Big)^{1/2}\Big(\int_{\mathbb{R}^n} u^{2p}\, dx\Big)^{1/2} \\
&\leq \frac{2(p-1)}{p}\Big(\int_{\mathbb{R}^n} \langle \xi, Q(x)\xi\rangle \langle \nabla(u^p), Q(x)\nabla(u^p)\rangle\, dx\Big)^{1/2}\Big(\int_{\mathbb{R}^n} u^{2p}\, dx\Big)^{1/2} \\
&\leq \frac{2(p-1)}{p}\Big(\eta\frac{|\xi|^2}{\lambda_0}\int_{\mathbb{R}^n} \langle \nabla(u^p), Q(x)\nabla(u^p)\rangle\, dx + \frac{1}{4\eta}\int_{\mathbb{R}^n} u^{2p}\, dx\Big) \\
&\leq \frac{2(p-1)}{p}\Big(\eta\frac{|\xi|^2}{\lambda_0^2}\int_{\mathbb{R}^n} |\nabla(u^p)|^2\, dx + \frac{1}{4\eta}\int_{\mathbb{R}^n} u^{2p}\, dx\Big),
\end{aligned}$$

and finally, by (2.434),

$$\begin{aligned}
&-\frac{2(p-1)}{p^2}\int_{\mathbb{R}^n} \langle \nabla(u^p), Q(x)\nabla(u^p)\rangle\, dx \\
&\qquad\qquad \leq -\lambda_0\Big(\frac{2(p-1)}{p}\Big)\int_{\mathbb{R}^n} |\nabla(u^p)|^2\, dx.
\end{aligned}$$

Adding up we find

$$\begin{aligned}
\int_{\mathbb{R}^n} (L_\xi(x,D)u)u^{2p-1}\, dx &\leq \Big(\frac{1}{\lambda_0}|\xi|^2 + \frac{p-1}{2p\eta}\Big)\int_{\mathbb{R}^n} u^{2p}\, dx \\
&\quad + \Big(\frac{2(p-1)}{p}\frac{\eta}{\lambda_0^2}|\xi|^2 - \lambda_0\frac{2(p-1)}{p^2}\Big)\int_{\mathbb{R}^n} |\nabla(u^p)|^2\, dx,
\end{aligned}$$

and with $\eta := \frac{\lambda_0}{2p|\xi|^2}$ we obtain

$$\begin{aligned}
\int_{\mathbb{R}^n} (L_\xi(x,D)u)u^{2p-1}\, dx &\leq -\frac{\lambda_0}{p}\int_{\mathbb{R}^n} |\nabla(u^p)|^2\, dx + \frac{p}{\lambda_0}|\xi|^2\int_{\mathbb{R}^n} u^{2p}\, dx \\
&= -\frac{\lambda_0}{p}\int_{\mathbb{R}^n} |\nabla(u^p)|^2\, dx + \frac{p}{\lambda_0}|\xi|^2\|u\|_{2p}^{2p}.
\end{aligned} \tag{2.452}$$

Now, by (2.448) we have

$$\frac{\|u^p\|_{L^2}^{2+4/n}}{\|u^p\|_{L^1}^{4/n}} \le \tilde{c}_{n,2}\|\nabla(u^p)\|_{L^2}^2,$$

or

$$-\int_{\mathbb{R}^n} |\nabla(u^p)|^2\,dx \le -\frac{\left(\int_{\mathbb{R}^n} u^{2p}\,dx\right)^{1+2/n}}{\tilde{c}_{n,2}\left(\int_{\mathbb{R}^n} u^p\,dx\right)^{4/n}} = -\frac{1}{\tilde{c}_{n,2}}\frac{\|u\|_{L^{2p}}^{2p(1+2/n)}}{\|u^p\|_{L^p}^{4p/n}}.$$

Thus we have proved

Lemma 2.9.7. *For all $u \in S(\mathbb{R}^n)$ and all $p \ge 1$ the estimate*

$$\int_{\mathbb{R}^n} (L_\xi(x,D)u)u^{2p-1}\,dx \le -\frac{c_0}{2p}\frac{\|u\|_{L^{2p}}^{2p(1+2/n)}}{\|u^p\|_{L^p}^{4p/n}} + \frac{|\xi|^2 p}{\lambda_0}\|u\|_{L^{2p}}^{2p} \tag{2.453}$$

holds with $c_0 = c_0(n, \lambda_0)$.

Our next aim is to relate (2.453) to $(T_t)_{t\ge 0}$, the semigroup generated by $L(x,D)$. We use the representation (2.439) for $(T_t)_{t\ge0}$ and we want to estimate

$$\begin{aligned} T_t^\xi u(x) &:= e^{-x\cdot\xi}T_t(e^{(\cdot,\xi)}u(\cdot))(x) \\ &= e^{-x\cdot\xi}\int_{\mathbb{R}^n}\Gamma(t,x,y)e^{y\cdot\xi}u(y)\,dy. \end{aligned} \tag{2.454}$$

Since we have no a priori knowledge on the growth of $y \mapsto \Gamma(t,x,y)$ we have to require that $u \in C_0^\infty(\mathbb{R}^n)$. For these $u \ge 0$ it follows by the fact that T_t is positivity preserving that

$$\begin{aligned} \frac{d}{dt}\|T_t^\xi u\|_{L^{2p}}^{2p} &= \frac{d}{dt}\int_{\mathbb{R}^n} |e^{-x\cdot\xi}T_t(e^{(\cdot,\xi)}u(\cdot))|^{2p}\,dx \\ &= 2p\int_{\mathbb{R}^n}(e^{-x\cdot\xi}T_te^{(\cdot,\xi)}u(\cdot))^{2p-1}e^{-x\cdot\xi}L(x,D)T_t(e^{(\cdot,\xi)}u(\cdot))\,dx \\ &= 2p\int_{\mathbb{R}^n} L_\xi(x,D)(T_t^\xi u)(T_t^\xi u)^{2p-1}\,dx. \end{aligned}$$

Since

$$\frac{d}{dt}\|T_t^\xi u\|_{L^{2p}}^{2p} = 2p\|T_t^\xi u\|_{L^{2p}}^{2p-1}\frac{d}{dt}\|T_t^\xi u\|_{L^{2p}}$$

we obtain from Lemma 2.9.7 that

$$\frac{d}{dt}\|T_t^\xi u\|_{L^{2p}} \le -\frac{c_0}{2p}\|T_t^\xi u\|_{L^{2p}}^{1+4p/n}\|T_t^\xi u\|_{L^p}^{-4p/n} + \frac{|\xi|^2 p}{\lambda_0}\|T_t^\xi u\|_{L^p} \tag{2.455}$$

holds for $t \geq 0$ and $p \geq 1$. Taking $p = 1$ and looking to the less sharp inequality

$$\frac{d}{dt}\|T_t^\xi u\|_{L^2} \leq \frac{|\xi|^2 p}{\lambda_0}\|T_t^\xi u\|_{L^2},$$

an application of Gronwall's lemma, compare [120], p. 169, yields the estimate

$$\|T_t^\xi u\|_{L^2} \leq e^{|\xi|^2 t/\lambda_0}\|u\|_{L^2}. \tag{2.456}$$

Lemma 2.9.8. *Let $w \in C([0,\infty))$ be a non-negative, non-decreasing function and suppose that $u \in C^1([0,\infty))$ is a non-negative function satisfying for $p \in [2,\infty)$*

$$u'(t) \leq -\frac{c_0}{2p}\Big(\frac{t^{(p-2)n/4p}}{\omega(t)}\Big)^{4p/n} u^{1+4p/n}(t) + \frac{|\xi|^2 p}{\lambda_0}u(t) \tag{2.457}$$

for all $t \geq 0$. Then for each $\delta > 0$ there is a constant $K = K(c_0, \delta) < \infty$ such that

$$u(t) \leq (Kp^2)^{n/4p}\omega(t)e^{\delta|\xi|^2 t/\lambda_0 p}\ t^{(1-p)n/4p} \tag{2.458}$$

holds for all $t \geq 0$.

Proof: A straightforward calculation for the function

$$v(t) := e^{-|\xi|^2 pt/\lambda_0}u(t)$$

yields first

$$v'(t) = -\frac{|\xi|^2 p}{\lambda_0}e^{-|\xi|^2 pt/\lambda_0}u(t) + e^{-|\xi|^2 pt/\lambda_0}u'(t)$$

and therefore using (2.458)

$$\begin{aligned}(v^{-4p/n})'(t) &= \frac{4p}{n}e^{\frac{4p^2|\xi|^2 t}{\lambda_0 n}}u(t)^{\frac{-4p-n}{n}}\Big(\frac{|\xi|^2 p}{\lambda_0}u(t) - u'(t)\Big) \\ &\geq \frac{4p}{n}e^{\frac{4p^2|\xi|^2 t}{\lambda_0 n}}u(t)^{\frac{-4p-n}{n}}\Big(\frac{|\xi|^2 p}{\lambda_0}u(t) + \frac{c_0}{2p}\Big(\frac{t^{(p-2)n/4p}}{\omega(t)}\Big)^{4p/n}u^{\frac{n+4p}{n}}(t) \\ &\quad - \frac{|\xi|^2 p}{\lambda_0}u(t)\Big) = \frac{2c_0}{n}e^{\frac{4p^2|\xi|^2 t}{\lambda_0 n}}\frac{t^{(p-2)}}{\omega(t)^{4p/n}}.\end{aligned}$$

Thus we find

$$\frac{d}{dt}\Big(\frac{e^{-\frac{|\xi|^2 pt}{\lambda_0}\cdot\frac{4p}{n}}}{u(t)^{4p/n}}\Big) \geq \frac{2c_0}{n}e^{\frac{4p^2|\xi|^2 t}{\lambda_0 n}}\frac{t^{(p-2)}}{\omega(t)^{4p/n}}$$

or

$$\frac{e^{4|\xi|^2p^2t/n\lambda_0}}{u(t)^{4p/n}} \geq \frac{2c_0}{n}\int_0^t e^{\frac{4p^2|\xi|^2s}{\lambda_0 n}} \frac{s^{p-2}}{\omega(s)^{4p/n}}\,ds$$
$$\geq \frac{2c_0}{n}\frac{1}{\omega(t)^{4p/n}}\int_0^t e^{\frac{4p^2|\xi|^2s}{\lambda_0}} s^{p-2}\,ds,$$

where we used that $u(0) \geq 0$ and that ω is non-decreasing. Since

$$\int_0^t e^{\frac{4p^2|\xi|^2s}{\lambda_0}} s^{p-2}\,ds = \Big(\frac{\lambda_0 tn}{4|\xi|^2p^2}\Big)^{p-1}\int_0^{4|\xi|^2p^2/\lambda_0 n} e^{ts}s^{p-2}\,ds$$
$$\geq \Big(\frac{\lambda_0 tn}{4|\xi|^2p^2}\Big)^{p-1} e^{(4|\xi|^2p^2/n-4\delta|\xi|^2/n)t/\lambda_0}\int_{4|\xi|^2p^2(1-\frac{\delta}{p^2})/\lambda_0 n}^{4|\xi|^2p^2/\lambda_0 n} s^{p-2}\,ds$$
$$= \frac{t^{p-1}}{p-1}(1-(1-\delta/p^2)^{p-1})e^{(4|\xi|^2p^2/n-4\delta|\xi|^2/n)t/\lambda_0},$$

we arrive with

$$K := \frac{c_0}{2n}\inf_{p\geq 2}\{p(1-(1-\delta/p^2)^{p-1})\} > 0$$

at

$$\frac{e^{4|\xi|^2p^2t/n\lambda_0}}{u(t)^{4p/n}} \geq \frac{t^{p-1}}{Kp^2\omega(t)^{4p/n}}e^{(4|\xi|^2p^2/n-4\delta|\xi|^2/n)t/\lambda_0}$$

which implies (2.458). □

Now we set $p_k := 2^k$, $u_k(t) := \|T_t^\xi u\|_{L^{p_k}}$ and

$$\omega_k(t) := \max_{0\leq s\leq t} s^{(p_k-2)\frac{n}{4p_k}}u_k(s). \tag{2.459}$$

Supposing $\|u\|_{L^2} = 1$ it follows from (2.456) that

$$u_1(t) \leq e^{|\xi|^2t/\lambda_0}$$

and (2.455) together with Lemma 2.9.8 yield

$$\frac{\omega_{k+1}(t)}{\omega_k(t)} \leq (4^kK)^{n/2^{k+2}}e^{\delta|\xi|^2t/2^k\lambda_0}. \tag{2.460}$$

But (2.460) implies the existence of $c_1 = c_1(n,\lambda_0,\delta) < \infty$ such that

$$\sup_k \omega_k(t) \leq c_1 e^{(1+\delta)|\xi|^2t/\lambda_0}. \tag{2.461}$$

Combining all these auxiliary results we are in a position to prove an upper estimate for $\Gamma(t,x,y)$.

Theorem 2.9.9. *There exists a constant $A = A(n,\lambda_0) > 0$ and $\alpha = \alpha(\lambda_0) > 0$ such that*

$$\Gamma(t,x,y) \le \frac{A}{t^{n/2}} e^{-|y-x|^2/\alpha t} \tag{2.462}$$

holds for all $t \in (0,\infty)$ and $x, y \in \mathbb{R}^n$.

Proof: Taking into account the definition of $\omega_k(t)$, (2.459) and the fact that $\varliminf_{k\to\infty} u_k(t) = \|T_t^\xi u\|_\infty$, we arrive first at

$$\|T_t^\xi u\|_\infty \le \frac{c}{t^{n/4}} e^{2|\xi|^2 t/\lambda_0} \|u\|_{L^2} \tag{2.463}$$

for $u \in C_0^\infty(\mathbb{R}^n)$. By our assumption $(T_t)_{t\ge 0}$ is a symmetric semigroup implying that $(T_t^\xi)^* u = T_t^\xi u$ and by duality (2.463) yields

$$\|T_t^\xi u\|_{L^2} \le \frac{c}{t^{n/4}} e^{2|\xi|^2 t/\lambda_0} \|u\|_{L^1}. \tag{2.464}$$

Using the semigroup property $T_{2t}^\xi u = (T_t^\xi \circ T_t^\xi)u$ we get

$$\begin{aligned}\|T_{2t}^\xi u\|_{L^\infty} &= \|T_t^\xi (T_t^\xi u)\|_{L^\infty} \\ &\le \frac{c}{t^{n/4}} e^{2|\xi|^2 t/\lambda_0} \|T_t^\xi u\|_{L^2} \\ &\le \frac{c^2}{t^{n/2}} e^{4|\xi|^2 t/\lambda_0} \|u\|_{L^1},\end{aligned}$$

i.e.

$$\sup_{x\in\mathbb{R}^n} \int_{\mathbb{R}^n} e^{-x\cdot\xi}\Gamma(2t,x,y)u(y)e^{y\cdot\xi}\,dy \le \frac{c^2}{t^{n/2}} e^{4|\xi|^2 t/\lambda_0} \int_{\mathbb{R}^n} |u|\,dx$$

for all $u \in C_0^\infty(\mathbb{R}^n)$, hence

$$\left(e^{-x\cdot\xi}\Gamma(2t,x,\cdot)e^{(\cdot,\xi)}, u(\cdot)\right)_0 \le \frac{c^2}{t^{n/2}} e^{4|\xi|^2 t/\lambda_0} \|u\|_{L^1}$$

implying

$$e^{-x\cdot\xi}\Gamma(2t,x,y)e^{y\xi} \le \frac{c^2}{t^{n/2}} e^{4|\xi|^2 t/\lambda_0},$$

or

$$\Gamma(2t,x,y) \leq \frac{c^2}{t^{n/2}} e^{4|\xi|^2 t/\lambda_0} e^{\xi\cdot(x-y)}. \tag{2.465}$$

Now we may take $\xi := \frac{\lambda_0}{8t}(y-x)$ to arrive finally at

$$\Gamma(2t,x,y) \leq \frac{c^2}{t^{n/2}} e^{\frac{4\lambda_0^2 t|x-y|^2}{64t^2\lambda_0} - \frac{\lambda_0|y-x|^2}{8t}} = \frac{c^2}{t^{n/2}} e^{-\lambda_0|y-x|^2/16t}$$

which implies (2.462). □

We will now derive a lower bounds for $\Gamma(t,x,y)$. For this we prove first some lemmata.

Lemma 2.9.10. *For all $\varphi \in C_b^1(\mathbb{R}^n;\mathbb{R})$ the estimate*

$$\int_{\mathbb{R}^n} e^{-\pi|y|^2}\left(\varphi(y) - \int_{\mathbb{R}^n} e^{-\pi|y|^2}\varphi(x)\,dx\right)^2 dy \leq 2\pi \int_{\mathbb{R}^n} e^{-\pi|y|^2}|\nabla\varphi(y)|^2\,dy \tag{2.466}$$

holds.

Proof: With $\nu_t(dy) = \mu_t(y)\lambda^{(n)}(dy) = (4\pi t)^{-n/2}e^{-|y|^2/4t}\lambda^{(n)}(dy)$ where $\lambda^{(n)}(dy)$ is the Lebesgue measure it is clear that (2.466) is equivalent to

$$\int_{\mathbb{R}^n} \varphi^2(y)\nu_{1/2}(dy) \leq \int_{\mathbb{R}^n} |\nabla\varphi(y)|^2\nu_{1/2}(dy) \tag{2.467}$$

for all $\varphi \in C_b^1(\mathbb{R}^n;\mathbb{R})$ such that $\int_{\mathbb{R}^n}\varphi(y)\nu_{1/2}(dy) = 0$. The density of $C_0^\infty(\mathbb{R}^n;\mathbb{R})$ in $L^2(\mathbb{R}^n, \nu_{1/2}(dy))$ implies that it is sufficient to prove (2.467) for all $\varphi \in C_0^\infty(\mathbb{R}^n;\mathbb{R})$ with $\int_{\mathbb{R}^n}\varphi(y)\nu_{1/2}(dy) = 0$. Consider

$$S_t\varphi(x) := \int_{\mathbb{R}^n} \varphi(y)\mu_{\frac{t-e^{-2t}}{2}}(y - e^{-t}x)\,dy, \tag{2.468}$$

i.e. $(S_t)_{t\geq 0}$ is the *Ornstein–Uhlenbeck semigroup*. From properties of the heat semigroup (Brownian semigroup) it follows immediately for $\varphi \in C_0^\infty(\mathbb{R}^n;\mathbb{R})$ that $S_t\varphi \to \varphi$ as $t \to 0$ and $S_t\varphi \to 0$ as $t \to \infty$, both limits being boundedly and pointwise, i.e. for all x we have $\sup_{t>0}|S_t\varphi(x) - \varphi(x)| < \infty$, $\lim_{t\to\infty} S_t\varphi(x) = 0$ and $\lim_{t\to 0} S_t\varphi(x) = \varphi(x)$. In addition we have

$$\frac{\partial}{\partial t}S_t\varphi(x) = (\Delta_x - x\cdot\nabla_x)S_t\varphi(x),$$

which means that $\Delta_x - x\cdot\nabla_x$ is the (pre-)generator of $(S_t)_{t\geq 0}$. It follows that

$$\frac{d}{dt}\int_{\mathbb{R}^n}|S_t\varphi(y)|^2\nu_{1/2}(dy) = 2\int_{\mathbb{R}^n} S_t\varphi(y)(\Delta_y - y\cdot\nabla_y)S_t\varphi(y)\nu_{1/2}(dy),$$

and an integration by parts yields

$$\frac{d}{dt}\int_{\mathbb{R}^n}|S_t\varphi(y)|^2\nu_{1/2}(dy) = -2\int_{\mathbb{R}^n}|\nabla_y S_t\varphi(y)|^2\nu_{1/2}(dy), \tag{2.469}$$

which says that $t\mapsto \|S_t\varphi\|_{L^2(\mathbb{R}^n,\nu_{1/2}(dy))}$ is a non-increasing function for $\varphi\in C_0^\infty(\mathbb{R}^n;\mathbb{R})$ or even $\varphi\in C_b(\mathbb{R}^n;\mathbb{R})$. The latter fact follows from an approximation arguments. Next observe that

$$\begin{aligned}\nabla_x S_t\varphi(x) &= \int_{\mathbb{R}^n}\varphi(y)\nabla_x\mu_{\frac{t-e^{-2t}}{2}}(y-e^{-t}x)\,dy\\ &= e^{-t}\int_{\mathbb{R}^n}(\nabla_y\varphi(y))\mu_{\frac{t-e^{-2t}}{2}}(y-e^{-t}x)\,dy \\ &= e^{-t}S_t(\nabla\varphi)(x)\end{aligned} \tag{2.470}$$

with the obvious interpretation of $S_t(\nabla\varphi)$. Thus using (2.468) and (2.470) we arrive at

$$\begin{aligned}\int_{\mathbb{R}^n}|\varphi(y)|^2\nu_{1/2}(dy) &\leq 2\int_0^\infty e^{-2t}\Big(\int_{\mathbb{R}^n}|S_t|\nabla\varphi|(y)|^2\nu_{1/2}(dy)\Big)\,dt\\ &\leq \int_{\mathbb{R}^n}|\nabla\varphi(y)|^2\nu_{1/2}(dy),\end{aligned}$$

where we used in the last step the monotonicity of $t\mapsto\|S_t\varphi\|_{L^2(\mathbb{R}^n,\nu_{1/2}(dy))}$ and the lemma is proved. □

Lemma 2.9.11. *There is a constant $\gamma_1 = \gamma_1(\lambda_0)$ such that for all $|x|\leq 1$*

$$\int_{\mathbb{R}^n} e^{-\pi|y|^2}\log\Gamma(1,y,x)\,dy \geq -\gamma_1 \tag{2.471}$$

holds.

Proof: For simplicity let us set $v(s,y) := \Gamma(s,y,x)$ and

$$G(s) := \int_{\mathbb{R}^n} e^{-\pi|y|^2}\log v(s,y)\,dy.$$

The fact that $(T_t)_{t\geq 0}$ generated by $L(x,D)$ is a sub-Markovian (and Feller) semigroup implies by (2.439) that

$$\int_{\mathbb{R}^n} v(s,y)\,dy = \int_{\mathbb{R}^n} \Gamma(s,y,x)1\,dy = (T_s1)(x) \leq 1.$$

Indeed, in our case it is also clear that $T_s1 = 1$, i.e. $\int_{\mathbb{R}^n} v(s,y)\,dy = 1$. Thus $G(s) < \infty$ and we seek for a lower bound for $G(1)$. For this we observe that

$$\begin{aligned}
G'(s) &= \frac{d}{ds}\int_{\mathbb{R}^n} e^{-\pi|y|^2}\log v(s,y)\,dy = \int_{\mathbb{R}^n} e^{-\pi|y|^2}\frac{1}{v(s,y)}\frac{\partial}{\partial s}v(s,y)\,dy \\
&= \int_{\mathbb{R}^n} e^{-\pi|y|^2}\frac{1}{v(s,y)}L(y,D_y)v(s,y)\,dy \\
&= -\int_{\mathbb{R}^n} \langle \nabla_y\Big(\frac{e^{-\pi|y|^2}}{v(s,y)}, Q(y)\nabla_y v(s,y)\rangle\,dy \\
&= \int_{\mathbb{R}^n} e^{-\pi|y|^2}\langle 2\pi\frac{y}{v(s,y)} + \frac{\nabla_y v(s,y)}{v^2(s,y)}, Q(y)\nabla_y v(s,y)\rangle\,dy \\
&= 2\pi\int_{\mathbb{R}^n} e^{-\pi|y|^2}\langle y, Q(y)\frac{\nabla_y v(s,y)}{v(s,y)}\rangle\,dy \\
&\quad + \int_{\mathbb{R}^n} e^{-\pi|y|^2}\langle \frac{\nabla_y v(s,y)}{v(s,y)}, Q(y)\frac{\nabla_y v(s,y)}{v(s,y)}\rangle\,dy \\
&= 2\pi\int_{\mathbb{R}^n} e^{-\pi|y|^2}\langle y, Q(y)\nabla_y\log v(s,y)\rangle\,dy \\
&\quad + \int_{\mathbb{R}^n} e^{-\pi|y|^2}\langle \nabla_y\log v(s,y), Q(y)\nabla_y\log v(s,y)\rangle\,dy.
\end{aligned}$$

Since

$$\begin{aligned}
&|2\pi\langle y,Q(y)\nabla_y\log v(s,y)\rangle| \\
&\qquad \leq 2\pi^2\langle y,Q(y)y\rangle + \frac{1}{2}\langle\nabla_y\log v(s,y), Q(y)\nabla_y\log v(s,y)\rangle
\end{aligned}$$

we find

$$\begin{aligned}
2\pi\int_{\mathbb{R}^n} e^{-\pi|y|^2}\langle y,Q(y)\nabla_y\log v(s,y)\rangle\,dy &\geq -2\pi^2\int_{\mathbb{R}^n} e^{-\pi|y|^2}\langle y,Q(y)y\rangle\,dy \\
&\quad -\frac{1}{2}\int_{\mathbb{R}^n} e^{-\pi|y|^2}\langle\nabla_y\log v(s,y), Q(y)\nabla_y\log v(s,y)\rangle\,dy
\end{aligned}$$

implying

$$G'(s) \geq -\gamma + \frac{\lambda_0}{2}\int_{\mathbb{R}^n} e^{-\pi|y|^2}|\nabla_y\log v(s,y)|^2\,dy. \tag{2.472}$$

In particular, (2.472) implies that $G(s)+\gamma s$ is non-decreasing on $[\frac{1}{2},1]$. But Lemma 2.9.10 we know that

$$\int_{\mathbb{R}^n} e^{-\pi|y|^2}(\log v(s,y)-G(s))^2\,dy \leq c\int_{\mathbb{R}^n} e^{-\pi|y|^2}|\nabla_y \log v(s,y)|^2\,dy$$

which gives with (2.472)

$$G'(s) \geq -\gamma + c'\int_{\mathbb{R}^n} e^{-\pi|y|^2}(\log v(s,y)-G(s))^2\,dy \tag{2.473}$$

with γ and c' depending only on λ_0. The function $r \mapsto \frac{(\log r - G(s))^2}{r}$ is a non-increasing function on $[e^{2+G(s)},\infty)$ and further, by Theorem 2.9.9, we know that $\sup_{\frac{1}{2}\leq s\leq 1} v(s,y) \leq K$ with K being an absolute constant. This implies by (2.473) for $s\in[\frac{1}{2},1]$ that

$$G'(s) \geq -\gamma + c'\frac{(\log K - G(\gamma))^2}{K}\int_{v(s,y)\geq e^{2+G(s)}} e^{-\pi|y|^2}v(s,y)\,dy. \tag{2.474}$$

The integral in the right-hand side of (2.474) we can estimate according to

$$\begin{aligned}\int_{v(s,y)\geq e^{2+G(s)}} e^{-\pi|y|^2}v(s,y)\,dy &\geq \int_{\mathbb{R}^n} e^{-\pi|y|^2}v(s,y)\,dy - e^{2+G(s)}\\ &\geq e^{-\pi R^2}\int_{|y|<R} v(s,y)\,dy - e^{2+G(s)}\\ &= e^{-\pi R^2}\Big(1-\int_{|y|\geq R} v(s,y)\,dy\Big) - e^{2+G(s)},\end{aligned}$$

where we used that $\int_{\mathbb{R}^n} v(s,y)\,dy = 1$. From Theorem 2.9.9 we may also derive that there exists $R_{\lambda_0} > 0$ depending only on λ_0 such that

$$\sup_{\frac{1}{2}\geq s\geq 1}\int_{|y|>R_{\lambda_0}} v(s,y)\,dy \leq \frac{1}{2}.$$

Now, since $s\mapsto G(s)+\gamma s$ is non-increasing on $[\frac{1}{2},1]$ we may deduce from (2.473) that there are δ_{λ_0} and M_{λ_0} depending only on λ_0 such that

$$G'(s) \geq \delta_{\lambda_0}G(s)^2 \quad \text{for} \quad s\in[\frac{1}{2},1] \tag{2.475}$$

provided $G(1)\leq -M_{\lambda_0}$. Now, (2.475) implies that $G(1)\geq -\frac{2}{\delta_{\lambda_0}}$ and therefore we have proved that

$$G(1) \geq -\max\Big(\frac{2}{\delta_{\lambda_0}}, M_{\lambda_0}\Big)$$

or by the definition of $G(1)$

$$\int_{\mathbb{R}^n} e^{-\pi|y|^2} \log v(s,y)\,dy \geq -\gamma_1$$

and the lemma is proved. □

Lemma 2.9.12. *For all $x, y \in B_1(0)$ we have*

$$\Gamma(1,x,y) \geq e^{-2\gamma_1}. \tag{2.476}$$

Proof: Observe that the Chapman–Kolmogorov equations yield

$$\begin{aligned}
\log\Gamma(1,x,y) &= \log\Big(\int_{\mathbb{R}^n} \Gamma(\frac{1}{2},x,z)\Gamma(\frac{1}{2},z,y)\,dz\Big) \\
&\geq \log\int_{\mathbb{R}^n} e^{-\pi|y|^2}\Gamma(\frac{1}{2},x,z)\Gamma(\frac{1}{2},z,y)\,dz \\
&= \log\int_{\mathbb{R}^n} \Gamma(\frac{1}{2},x,z)\Gamma(\frac{1}{2},z,y)\nu_{1/2}(dz) \\
&\geq \int_{\mathbb{R}^n} \log(\Gamma(\frac{1}{2},x,z)\Gamma(\frac{1}{2},z,y))\nu_{1/2}(dz) \\
&= \int_{\mathbb{R}^n} e^{-\pi|z|^2} \log\Gamma(\frac{1}{2},x,z)\,dz \\
&+ \int_{\mathbb{R}^n} e^{-\pi|z|^2} \log\Gamma(\frac{1}{2},y,z)\nu_{1/2}(dz) \geq -2\gamma_1,
\end{aligned}$$

where we used Jensen's inequality, the symmetry of Γ, i.e. $\Gamma(t,x,y) = \Gamma(t,y,x)$, and Lemma 2.9.11. But the estimate

$$\log\Gamma(1,x,y) \geq -2\gamma_1$$

is equivalent to (2.476). □

Before we will finally prove a lower estimate for $\Gamma(t,x,y)$ we need a *scaling* result.

Lemma 2.9.13. *Denote by $Q^{r,\xi}(x)$, $r > 0$, $\xi \in \mathbb{R}^n$, the matrix $Q(rx+\xi) = (a_{kl}(rx+\xi))_{k,l=1,\ldots,n}$ and by Γ^Q and $\Gamma^{Q^{r,\xi}}$ the fundamental solutions corresponding to $\frac{\partial}{\partial t} - \nabla(Q\nabla\cdot)$ and to $\frac{\partial}{\partial t} - \nabla(Q^{r,\xi}\nabla\cdot)$, respectively. Then we have*

$$\Gamma^Q(r^2t, rx+\xi, ry+\xi) = r^{-n}\Gamma^{Q^{r,\xi}}(t,x,y). \tag{2.477}$$

Proof: For $\varphi \in C_0^\infty(\mathbb{R}^n)$ set

$$u(t,x) := \int_{\mathbb{R}^n} \varphi(\frac{y-\xi}{r})\Gamma^Q(t,x,y)\,dy$$

and

$$\omega(t,x) := u(r^2t, rx+\xi).$$

It follows that

$$\frac{\partial}{\partial t}\omega(t,x) = (\nabla, Q^{r,\xi}\nabla\omega(t,\cdot))(x) \quad \text{and} \quad \lim_{t\to 0}\omega(t,\cdot) = \varphi(\cdot).$$

But now we may apply the uniqueness result for solutions of the initial value problem for second order parabolic differential equations to find

$$r^n \int_{\mathbb{R}^n} \varphi(y)\Gamma^Q(r^2t, rx+\xi, ry+\xi)\,dy = \omega(t,x) = \int_{\mathbb{R}^n} \varphi(y)\Gamma^{Q^{r,\xi}}(t,x,y)\,dy$$

for all $\varphi \in C_0^\infty(\mathbb{R}^n)$ which yields (2.477). □

We may restate (2.477) as

$$\Gamma^Q(t,x,y) = t^{-n/2}\Gamma^{Q^{t^{1/2},\xi}}\left(1, \frac{x-\xi}{t^{1/2}}, \frac{y-\xi}{t^{1/2}}\right)$$

for all $\frac{x-\xi}{t^{1/2}}, \frac{y-\xi}{t^{1/2}} \in B_1(0)$. In particular, since (2.476) holds also for $\Gamma^{Q^{t^{1/2},\xi}}$ we obtain for $|x-y| \le 2t^{1/2}$ the estimate

$$\Gamma^Q(t,x,y) \ge \frac{1}{t^{n/2}}e^{-2\gamma_1} \tag{2.478}$$

which is of course a lower bound for $\Gamma^Q(t,x,y)$ close to the diagonal $x = y$. Now let $x, y \in \mathbb{R}^n$ be arbitrary points with $|x-y| \ge 1$. Determine $k \in \mathbb{N}$ such that $k \le |x-y|^2 < k+1$ and set

$$x_m := x + \frac{m}{k+1}(y-x) \quad B_m := B_{\frac{1}{4\sqrt{k}}}(x_m), \quad m = 0, \dots, k+1.$$

For $\xi_m \in B_m$ and $\xi_{m-1} \in B_{m-1}$ it follows that

$$\begin{aligned}|\xi_m - \xi_{m-1}| &\le |\xi_m - x_m| + |x_m - x_{m-1}| + |x_{m-1} - \xi_{m-1}| \\ &\le \frac{1}{4\sqrt{k}} + \frac{1}{\sqrt{k+1}} + \frac{1}{4\sqrt{k}} = \frac{1}{2\sqrt{k}} + \frac{1}{\sqrt{k+1}} \le \frac{2}{\sqrt{k+1}}\end{aligned}$$

implying by (2.478) that

$$\Gamma^Q\Big(\frac{1}{k+1},\xi_{m-1},\xi_m\Big) \geq (k+1)^{n/2}e^{-2\gamma_1} \geq k^{n/2}e^{-2\gamma_1}.$$

But now the Chapman–Kolmogorov equations yield

$$\begin{aligned}\Gamma^Q(1,x,y) &\geq \int_{B_1}\cdots\int_{B_k}\Gamma^Q\Big(\frac{1}{k+1},x,\xi_1\Big)\Gamma^Q\Big(\frac{1}{k+1},\xi_1,\xi_2\Big)\cdots\\ &\qquad\cdots\Gamma^Q\Big(\frac{1}{k+1}\xi_k,y\Big)\,d\xi_1\ldots d\xi_k\\ &\geq (k^{n/2}e^{-2\gamma_1})^{k+1}\left(\omega_n\Big(\frac{1}{4\sqrt{k}}\Big)^{n/2}\right)^k\\ &= k^{n/2}e^{-2\gamma_1}\left(\frac{\omega_n e^{-2\gamma_1}}{2^n}\right)^k.\end{aligned}$$

Thus taking $\beta > 0$ such that $\frac{\omega_n e^{-2\gamma_1}}{2^n} \geq e^{-\beta}$ we find, remember that $k \leq |x-y|^2 < k+1$,

$$\Gamma^Q(1,x,y) \geq e^{-2\gamma_1}e^{-\beta k} \geq e^{-2\gamma_1}e^{-\beta|x-y|^2}. \tag{2.479}$$

As before we may restate (2.479) first for $\Gamma^{Q^{t^{1/2},\xi}}$ and then rescale as before to arrive finally at

Theorem 2.9.14. *There exist constants $B = B(n,\lambda_0) > 0$ and $\beta = \beta(\lambda_0,n) > 0$ such that*

$$\Gamma^Q(t,x,y) \geq \frac{B}{t^{n/2}}e^{-|x-y|^2/\beta t} \tag{2.480}$$

holds for $t > 0$ and all $x,y \in \mathbb{R}^n$.

Clearly, Theorem 2.9.9 and Theorem 2.9.14 yields *Aronson's theorem*, i.e. the estimates (2.442) with κ_1,κ_2 and γ_1,γ_2 depending only on λ_0 and n. It is important to note that both Γ^{γ_1} and Γ^{γ_2} are kernels representing convolution semigroups, more precisely we have that

$$T_{t,\gamma_j}u(x) = \int_{\mathbb{R}^n}\Gamma^{\gamma_j}(t,x,y)u(y)\,dy$$

is the semigroup generated by (an extension of) $\gamma_j\Delta_n$.

Remark 2.9.15. As stated before when deriving D.Aronson's estimates we followed the paper [84] and [258] and not Aronson's original paper [7]. This choice is motivated by the fact that we later on in Chapter 3, we rely much more on *Nash-type inequalities* and [84] as well as [258] have taken up the ideas of J.Nash [221] in their proofs.

In order to calculate the symbol of the generator of $(T_t^f)_{t\geq 0}$ we need to have a closer look at $(T_t)_{t\geq 0}$. Define

$$\sigma(T_t)(x,\xi) := e^{-ix\xi}T_t(e^{i(\cdot,\xi)})(x) = \int_{\mathbb{R}^n} \Gamma(t,x,y)e^{-i(x-y)\cdot\xi}\,dy.$$

Since for $u \in C_0^\infty(\mathbb{R}^n)$

$$\begin{aligned}(2\pi)^{-n/2} &\int_{\mathbb{R}^n} e^{ix\cdot\xi}\sigma(T_t)(x,\xi)\hat{u}(\xi)\,d\xi \\ &= (2\pi)^{-n/2}\int_{\mathbb{R}^n} e^{ix\cdot\xi}e^{-ix\cdot\xi}T_t(e^{i(\cdot,\xi)})(x)\hat{u}(\xi)\,d\xi \\ &= (2\pi)^{-n/2}\int_{\mathbb{R}^n}\int_{\mathbb{R}^n} \Gamma(t,x,y)e^{iy\cdot\xi}\hat{u}(\xi)\,dy\,d\xi \\ &= \int_{\mathbb{R}^n}\Gamma(t,x,y)u(y)\,dy = (T_tu)(x)\end{aligned}$$

holds, we find that T_t is a pseudo-differential operator with symbol

$$\sigma(T_t)(x,\xi) = \int_{\mathbb{R}^n}\Gamma(t,x,y)e^{-i(x-y)\cdot\xi}\,dy \tag{2.481}$$

and by the Aronson estimates, (2.481) is well defined. We want to use (2.481) to re-calculate the symbol of $L(x,D)$.

We find using the symmetry of $L(x,D)$ and the Aronson estimates

$$\begin{aligned}\frac{d}{dt}\sigma(T_t)(x,\xi) &= \frac{d}{dt}\int_{\mathbb{R}^n}\Gamma(t,x,y)e^{-i(x-y)\cdot\xi}\,dy \\ &= \int_{\mathbb{R}^n}\frac{\partial}{\partial t}\Gamma(t,x,y)e^{-i(x-y)\cdot\xi}\,dy \\ &= \int_{\mathbb{R}^n}(L(x,D)\Gamma(t,x,y))e^{-i(x-y)\cdot\xi}\,dy \\ &= \int_{\mathbb{R}^n}\Gamma(t,x,y)e^{iy\cdot\xi}(L(x,D)e^{-ix\cdot\xi})\,dy \\ &= \int_{\mathbb{R}^n}\Gamma(t,x,y)e^{iy\cdot\xi}e^{-ix\cdot\xi}\sigma(L(x,D))(x,\xi)\,dy\end{aligned}$$

$$= \sigma(L(x,D))(x,\xi) \int_{\mathbb{R}^n} \Gamma(t,x,y) e^{-i(x-y)\cdot\xi}\, dy.$$

Since

$$\lim_{t\to 0} \int_{\mathbb{R}^n} \Gamma(t,x,y) e^{-i(x-y)\cdot\xi}\, dy = 1$$

we arrive at

$$\sigma(L(x,D))(x,\xi) = \frac{d}{dt}\sigma(T_t)(x,\xi)|_{t=0}. \tag{2.482}$$

Now we may turn to subordinate diffusion semigroups. Let f be a Bernstein function with associated convolution semigroup $(\eta_t)_{t\geq 0}$, $\operatorname{supp}\eta_t \subset [0,\infty)$, and suppose for simplicity that f has the representation

$$f(s) = \int_{0+}^{\infty} (1 - e^{-sr})\, d\mu(dr),$$

compare (I.3.229) in Theorem I.3.9.4. From the Aronson estimates it follows that the symmetric semigroup $(T_t)_{t\geq 0}$ is an L^p-sub-Markovian semigroup for $1 < p < \infty$ as well as a Feller semigroup on $C_\infty(\mathbb{R}^n;\mathbb{R})$ which extends to C_b-Feller semigroup in the sense of Definition I.4.8.6. These properties will also hold for the subordinate semigroup $(T_t^f)_{t\geq 0}$, f being a Bernstein function. In addition we find for $u \in C_b(\mathbb{R}^n;\mathbb{R})$

$$T_t^f u(x) = \int_0^\infty T_s u(x) \eta_t(ds) = \int_0^\infty \int_{\mathbb{R}^n} \Gamma(t,x,y) u(y)\, dy \eta_t(ds).$$

Now, in order to handle

$$\int_0^\infty \int_{\mathbb{R}^n} \Gamma(t,x,y) u(y)\, dy \eta_t(ds)$$

note that $\Gamma^{\gamma_j}(t,x,y) = \frac{1}{(4\pi\gamma_j t)^{n/2}} e^{-|x-y|/4\gamma_j t}$, i.e. it is the density of a Gaussian convolution semigroup $(\nu_t^j)_{t\geq 0}$. Therefore

$$\int_0^\infty \nu_s^j \eta_t(ds) = \nu_t^{j,f}$$

is a further convolution semigroup as was proved in Proposition I.3.9.10. In particular

$$\int_{\mathbb{R}^n} u(y) \nu_t^{j,f}(dy) < \infty$$

for all $u \in C_b(\mathbb{R}^n;\mathbb{R})$ and $(\nu_t^{j,f})_{t\geq 0}$ is just the convolution semigroup associated with the continuous negative definite function $\xi \mapsto f(\gamma_j|\xi|^2)$. Therefore we find

$$T_t^f u(x) = \int_0^\infty T_s u(x)\eta_t(ds) = \int_{\mathbb{R}^n}\int_0^\infty \Gamma(s,x,y)\eta_t(ds)u(y)\,dy$$

and we have proved

Proposition 2.9.16. *Suppose the $L(x,D)$ and $\Gamma(t,x,y)$ are as in the Theorem 2.9.2 and denote by $(T_t)_{t\geq 0}$ the semigroup(s) generated by (extensions of) $L(x,D)$. For any Bernstein function f the subordinate semigroup $(T_t^f)_{t\geq 0}$ has a kernel representation*

$$T_t^f u(x) = \int_{\mathbb{R}^n} \Gamma^f(t,x,y)u(y)\,dy \tag{2.483}$$

with

$$\Gamma^f(t,x,y) = \int_0^\infty \Gamma(s,x,y)\eta_t(ds) \tag{2.484}$$

and for Γ^f we have the estimates

$$\kappa_1 \int_0^\infty \Gamma^{\gamma_1}(s,x,y)\eta_t(ds) \leq \Gamma^f(t,x,y) \leq \kappa_2 \int_0^\infty \Gamma^{\gamma_2}(s,x,y)\eta_t(ds) \tag{2.485}$$

where $\Gamma^{\gamma_j}(t,x,y) = \frac{1}{(4\pi\gamma_j t)^{n/2}} e^{-|x-y|/4\gamma_j t}$.

Further, since the convolution semigroup $(\nu_t^{j,f})_{t\geq 0}$ give rise to Feller semigroups we may extend $(T_t^f)_{t\geq 0}$ by taking monotone limits to $C_b(\mathbb{R}^n;\mathbb{R})$. Thus it makes sense to consider

$$\sigma(T_t^f)(x,\xi) = e^{-ix\xi}T_t^f(e^{i(\cdot,\xi)})(x) = \int_{\mathbb{R}^n} \Gamma^f(t,x,y)e^{-i(x-y)\cdot\xi}\,dy.$$

Assuming for convenience that $(T_t^f)_{t\geq 0}$ is even a C_b-Feller semigroup then we may easily argue as before to get

Proposition 2.9.17. *If $(T_t^f)_{t\geq 0}$ is a C_b-Feller semigroup then the operator T_t^f is a pseudo-differential operator with symbol*

$$\sigma(T_t^f)(x,\xi) = e^{-ix\xi}T_t^f(e^{i(\cdot,\xi)})(x). \tag{2.486}$$

Next let us have a look at the generator A^f of $(T_t^f)_{t\geq 0}$, in particular $A^f|_{C_0^\infty(\mathbb{R}^n;\mathbb{R})}$. Using the representation (I.4.132) from Theorem I.4.3.5 we find

$$A^f u(x) = \int_0^\infty (T_s u - u)\mu(ds).$$

Since we may apply T_s to function in $C_b(\mathbb{R}^n;\mathbb{R})$ we find further

$$\begin{aligned} e^{-ix\cdot\xi} A^f (e^{i(\cdot,\xi)})(x) &= \int_0^\infty e^{-ix\cdot\xi}((T_s e^{i(\cdot,\xi)})(x) - e^{ix\cdot\xi})\,\mu(ds) \\ &= \int_0^\infty \Big\{ \int_{\mathbb{R}^n} e^{-ix\cdot\xi}\Gamma(s,x,y)e^{iy\cdot\xi}\,dy - 1 \Big\}\,\mu(ds), \end{aligned}$$

thus we find

$$\sigma(A^f)(x,\xi) = \int_0^\infty (\sigma(T_s)(x,\xi) - 1)\mu(ds). \tag{2.487}$$

Note that if $\log(\sigma(T_s)(x,\xi))$ is well defined then (2.487) reads as

$$\sigma(A^f)(x,\xi) = -\int_0^\infty (1 - e^{-s(\frac{1}{s}\log(\sigma(T_s)(x,\xi)))})\,\mu(ds). \tag{2.488}$$

In particular in the translation invariant case, i.e.

$$\sigma(T_s)(x,\xi) = e^{-t\langle\xi,Q\xi\rangle}$$

we find

$$\sigma(A^f)(x,\xi) = -f(\langle\xi,Q\xi\rangle)$$

as we should expect from the general theory. In general different "approximations" of $\sigma(T_s)(x,\xi)$, i.e. of $\Gamma(s,x,y)$, will lead now to different "approximation" of $\sigma(A^f)(x,\xi)$. For example one may try to approximate $\Gamma(s,x,y)$ as done in E.E.Levi's [198] parametrix method for constructing $\Gamma(s,x,y)$, see also A.Friedman [94] for details.

Let us turn to the general situation, i.e. let $A = -p(x,D)$ be a pseudo differential operator generating a Feller or an L^p-sub-Markovian semigroup and assume for simplicity that $C_0^\infty(\mathbb{R}^n;\mathbb{R}) \subset D(A)$. However, for f let us suppose that it is a *complete Bernstein function*, compare Definition I.3.9.27 with representation

$$f(s) = \int_0^\infty \frac{s}{s+r}\rho(dr) \tag{2.489}$$

where ρ is a measure such that $\int_0^\infty \frac{1}{1+r}\rho(dr) < \infty$, compare Theorem I.3.9.29. According to Theorem I.4.3.10 and its Corollary I.4.3.11 in this case we have

$$A^f u = \int_0^\infty A\, R_\lambda u \rho(d\lambda). \tag{2.490}$$

Introducing the operator

$$r_\lambda(x,D)u(x) := (2\pi)^{-n/2} \int_{\mathbb{R}^n} e^{ix\cdot\xi} \frac{1}{\lambda + p(x,\xi)} \hat{u}(\xi)\, d\xi \tag{2.491}$$

it follows that

$$\begin{aligned} A^f u &= \int_0^\infty ((A-\lambda)+\lambda) R_\lambda u \rho(d\lambda) \\ &= \int_0^\infty (-\operatorname{id} + \lambda R_\lambda) u \rho(d\lambda) \\ &= \int_0^\infty (\lambda r_\lambda(x,D) - \operatorname{id}) u \rho(d\lambda) + \int_0^\infty \lambda (R_\lambda - r_\lambda(x,D)) u \rho(d\lambda). \end{aligned}$$

Since for $u \in C_0^\infty(\mathbb{R}^n;\mathbb{R})$

$$\begin{aligned} &\int_0^\infty (\lambda r_\lambda(x,D) - \operatorname{id}) u \rho(d\lambda) \\ &\qquad = (2\pi)^{-n/2} \int_0^\infty \int_{\mathbb{R}^n} e^{ix\cdot\xi} \Big(\frac{\lambda}{\lambda + p(x,\xi)} - 1 \Big) \hat{u}(\xi)\, d\xi \rho(d\lambda) \\ &\qquad = (2\pi)^{-n/2} \int_{\mathbb{R}^n} e^{ix\cdot\xi} \int_0^\infty \frac{p(x,\xi)}{\lambda + p(x,\xi)} \rho(d\lambda) \hat{u}(\xi)\, d\xi \\ &\qquad = -[(f\circ p)(x,D)]u, \end{aligned}$$

we find with $p_\lambda(x,D) = p(x,D) + \lambda = \lambda - A$ and

$$K_\lambda(x,D) = \operatorname{id} - p_\lambda(x,D) \circ r_\lambda(x,D) \tag{2.492}$$

that

$$A^f u = -[(f\circ p)(x,D)]u + \int_0^\infty \lambda R_\lambda K_\lambda(x,D) u \rho(d\lambda). \tag{2.493}$$

Thus in order to claim that the symbol of A^f is given by

$$\sigma(A^f)(x,\xi) = -f(p(x,\xi)) + \text{“lower order terms”} \tag{2.494}$$

we need to estimate

$$\int_0^\infty \lambda R_\lambda K_\lambda(x,D)u\rho(d\lambda)$$

more carefully. Note that for $p(x,\xi) = p(\xi)$ we have $K_\lambda(x,D) = 0$! First of all let us remark that the problem lies already in defining "lower order terms". Note that, in general, $p(x,\xi)$ has no expansion into homogeneous symbols, i.e. in general we cannot expect that $p(x,\xi) \sim \sum_{j=0}^\infty p_j(x,\xi)$ with p_j being with respect to ξ homogeneous of some degree m_j, $(m_j)_{j\geq 0}$ being strictly decreasing (to $-\infty$ if $(m_j)_{j\geq 0}$ is an infinite sequence). But even if such an expansion would hold, $p_0(x,\xi)$ would in general not be our candidate for a "principal symbol" as seen by symbols of mixed homogeneity such as $(\xi_1,\xi_2) \mapsto |\xi_1|^{\alpha_1} + |\xi_2|^{\alpha_2}$, $0 < \alpha_1 < \alpha_2 \leq 2$.

The definition of $K_\lambda(x,D)$ makes it obvious where our problem lies. We have to control the composition $p_\lambda(x,D) \circ r_\lambda(x,D)$, $r_\lambda(x,D)$ given by (2.491). It is clear that in case where we are "closest" to a classical symbolic calculus we should expect the best result. The calculus we will look at first is of course F.Baldus' Weyl calculus, see [13], which we discussed briefly at the end of Section 2.7. Taking into account some observations from [163] he could prove the following result, where we use the notation of Section 2.7.

Let γ be a Hörmander metric on $\mathbb{R}^n \times \mathbb{R}^n$ which splits and assume that $\Psi(1,\gamma) \cup B(L^2(\mathbb{R}^n))^{-1} = \Psi(1,\gamma)^{-1}$. Further let M be an invertible sub-γ-admissible weight function such that there exist constant $\alpha > 0$ and $c_M \geq 1$ with $\frac{1}{c_M} < M \leq c_M h_\gamma^{-\alpha}$, h_γ given by (2.368). In addition let $p \in S(M,\gamma)$ be such that $\xi \mapsto p(x,\xi)$ is a continuous negative definite function satisfying $|p(x,\xi)| \geq c\,M(x,\xi)$, $c > 0$. For $p(x,D) : C_0^\infty(\mathbb{R}^n;\mathbb{R}) \to C_\infty(\mathbb{R}^n;\mathbb{R})$ we assume

$$(p(x,D)u,v)_0 = (u,p(x,D)v)_0 \tag{2.495}$$

and

$$(p(x,D)u,u)_0 \geq \lambda_0 \|u\|_0^2 \tag{2.496}$$

for some $\lambda_0 > 0$. For the complete Bernstein function f, $f(s) = \int_0^\infty \frac{s}{s+r}\rho(dr)$, we assume in addition

$$\int_1^\infty \frac{1}{\lambda^{2\sigma\eta}}\rho(d\lambda) < \infty \tag{2.497}$$

where $\eta := 1 \wedge \frac{1}{\alpha}$, $0 \leq \sigma \leq \frac{1}{2}$, compare [163], Theorem 6.1.

Theorem 2.9.18 (F.Baldus). *Under all the assumptions stated above on* $p(x,\xi)$ *and* f, $-p(x,D)$ *extends to a generator* A *of a Feller semigroup and if we denote the generator of the subordinate semigroup by* A^f *we have with* $M_f := 1+f\circ p$ *that* $f\circ p \in S(M_f,\gamma)$, $-(f\circ p)(x,D) : C_0^\infty(\mathbb{R}^n;\mathbb{R}) \to C_\infty(\mathbb{R}^n;\mathbb{R})$ *and*

$$A^f = -(f\circ p)(x,D)+p_1(x,D) \tag{2.498}$$

with $p_1(x,D) \in \Psi(h_\gamma^{1-2\sigma},\gamma) \subset \Psi(1,\gamma)$.

Corollary 2.9.19. *If the operator* $L(x,D)$ *in* (2.435) *has coefficients* $a_{kl} \in C_b^\infty(\mathbb{R}^n;\mathbb{R})$, *then Theorem* 2.9.18 *is applicable and in particular we have that*

$$\sigma(A^f)(x,\xi) = -f(\sigma(L(x,D))(x,\xi)+q_1(x,\xi) \tag{2.499}$$

where

$$\lim_{|\xi|\to\infty} \frac{q_1(x,\xi)}{f(\sigma(L(x,D))(x,\xi)} = 0 \tag{2.500}$$

provided that $\lim_{s\to\infty} f(s) = \infty$.

For the proof of Theorem 2.9.18 and its corollary we refer to F.Baldus [13], Section 6.5.

The symbolic calculus of W.Hoh, compare Section 2.4, is closer to our philosophy since combined with his perturbation results, see Section 2.8, it allows to start with some quite general continuous negative definite symbol and then to consider an analysis in function spaces related to their symbol (or a modification of it). We are now going to prove a result analogous to Theorem 2.9.18 within Hoh's calculus, and now we will use the notation of Section 2.4. As a first result recall that if $p \in S_\rho^{m,\psi}(\mathbb{R}^n)$ and f is any Bernstein function, then $f\circ p$ is also a symbol within Hoh's calculus. More precisely, we have by Example 2.4.7.

Example 2.9.20 (W.Hoh). Let f be a Bernstein function satisfying $f(s) \leq c_1 s^r$ for all $s \geq 1$ and some $0 < r \leq 1$, note that $r = 1$ is always possible. Let $p \in S_\rho^{m,\psi}(\mathbb{R}^n)$, $m \geq 2$, where ψ is in the class Λ, i.e. satisfying (2.196) for all $\alpha \in \mathbb{N}_0^n$, and suppose that $p(x,\xi)$ is real-valued and elliptic, i.e.

$$p(x,\xi) \geq c_0(1+\psi(\xi))^{m/2} \tag{2.501}$$

for some $c_0 > 0$. Then $f\circ p \in S_\rho^{m,\psi^r}(\mathbb{R}^n) = S_\rho^{mr,\psi}(\mathbb{R}^n)$. In particular, if in addition $\xi \mapsto p(x,\xi)$ is a continuous negative definite function then $f\circ p$ is an elliptic negative definite symbol in $S_\rho^{m,\psi^r}(\mathbb{R}^n)$.

Remark 2.9.21. According to the results in Sections 2.4–2.5 we know in case that $p(x,\xi)$ is an elliptic continuous negative definite symbol in $S_\rho^{m,\psi}(\mathbb{R}^n)$ and if $f(s) = s^r$, $0 < r < 1$, then $-p^f(x,\xi) = -(p(x,\xi))^r$ is the symbol of an operator extending from $C_0^\infty(\mathbb{R}^n;\mathbb{R})$ to a generator of a Feller semigroup which we denote for a moment by $(S_t)_{t\geq 0}$. Thus it is of interest to compare $(S_t)_{t\geq 0}$ and $(T_t^f)_{t\geq 0}$, $(T_t)_{t\geq 0}$ being the semigroup generated by $-p(x,D)$. We will come back to this problem later.

In order to estimate $K_\lambda(x,D)$ for $p \in S_\rho^{m,\psi}(\mathbb{R}^n)$ it is convenient to extend Hoh's calculus to a parameter dependent calculus. It turns out that for our special purpose a very simple version of such a calculus is sufficient and we state the results needed without proofs, a rigorous parameter dependent symbolic calculus including all proofs can be found in a forthcoming paper [134] with W.Hoh and A.G.Tokarev. Note that once the symbolic calculus of Section 2.4 is established a parameter dependent calculus can easily be developed along the lines of the considerations of G.Grubb [111], see also her highly readable survey [112].

Using Corollary 2.4.23.A we find that for an elliptic continuous negative definite symbol $p \in S_\rho^{2,\psi}(\mathbb{R}^n)$ the symbol of $K_\lambda(x,D)$ is given by

$$K_\lambda(x,\xi) = -\sum_{j=1}^{n} \frac{\partial_{\xi_j} p(x,\xi) D_{x_j} p(x,\xi)}{(p(x,\xi)+\lambda)^2} + p_R(x,\xi,\lambda), \tag{2.502}$$

where $p \in S_\rho^{2,\psi}(\mathbb{R}^n)$ implies that $\dfrac{\partial_{\xi_j} p(x,\xi) D_{x_j} p(x,\xi)}{(p(x,\xi)+\lambda)^2}$ is a symbol in $S_\rho^{1,\psi}(\mathbb{R}^n)$ and $p_R(x,\xi,\lambda)$ belongs to the class $S_\rho^{0,\psi}(\mathbb{R}^n)$, but it is important to note that $\lambda^2 p_R(x,\xi,\lambda)$ is bounded. Now, in order to estimate the L^2-norm of $K_\lambda(x,D)$ we handle in detail the term

$$-\sum_{j=1}^{n} \Big(\frac{\partial_{\xi_j} p(x,\xi) D_{x_j} p(x,\xi)}{(p(x,\xi)+\lambda)^2}\Big)(x,D)u.$$

For $j = 1,\ldots,n$ it follows that

$$\begin{aligned}&\Big\|\Big(\frac{(\partial_{\xi_j}p)(D_{x_j}p)}{(p+\lambda)^2}\Big)(x,D)u\Big\|_0 = \\ &= \Big\|\Big(\frac{(\partial_{\xi_j}p)(D_{x_j}p)}{(p+\lambda)^2}\Big)(x,D)(1+\lambda+\psi(D))^2(1+\psi(D))^{-3/2} \\ &\qquad \circ (1+\lambda+\psi(D))^{-2}(1+\psi(D))^{3/2}u\Big\|_0\end{aligned}$$

$$\leq c\left\|\left(\frac{(1+\psi(\cdot))^{3/2}}{(1+\lambda+\psi(\cdot))^2}\right)(D)u\right\|_0,$$

where we used the fact that

$$(x,\xi)\mapsto\frac{\partial_{\xi_j}p(x,\xi)D_{x_j}p(x,\xi)(1+\lambda+\psi(\xi))^2}{(p(x,\xi)+\lambda)^2(1+\psi(\xi))^{3/2}}$$

is a symbol of order zero uniformly bounded in λ, and thus we may apply Theorem 2.5.4. Now

$$\frac{(1+\psi(\xi))^{3/2}}{(1+\lambda+\psi(\xi))^2}\leq\frac{1}{(1+\lambda)^{1/2}} \tag{2.503}$$

implying the estimate

$$\left\|\left(\frac{(\partial_{\xi_j}p)(D_{x_j}p)}{(p+\lambda)^2}\right)(x,D)u\right\|_0\leq\frac{1}{(1+\lambda)^{1/2}}\|u\|_0. \tag{2.504}$$

Note that (2.503) can be changed to

$$\frac{(1+\psi(\xi))^{3/2}}{(1+\lambda+\psi(\xi))^2}\leq\frac{1}{(1+\lambda)^{\eta}}\frac{1}{(1+\psi(\xi))^{1/2-\eta}} \tag{2.505}$$

for any $0\leq\eta\leq\frac{1}{2}$ which yields

$$\left\|\left(\frac{(\partial_{\xi_j}p)(D_{x_j}p)}{(p+\lambda)^2}\right)(x,D)u\right\|_0\leq\frac{1}{(1+\lambda)^{\eta}}\|u\|_{\psi,\eta-1/2}. \tag{2.506}$$

The structure of the symbol $p_r(x,\xi,\lambda)$ implies in particular that (2.504) and (2.506), respectively, do hold for the corresponding pseudo-differential operator too. Thus we have established

Lemma 2.9.22. *Let $p(x,\xi)\in S^{2,\psi}_{\rho}(\mathbb{R}^n)$ be an elliptic continuous negative definite symbol and define*

$$K_\lambda(x,D)=\mathrm{id}-p_\lambda(x,D)\circ r_\lambda(x,D)$$

with $p_\lambda(x,D)=p(x,D)+\lambda\,\mathrm{id}$ and

$$r_\lambda(x,D)u(x)=(2\pi)^{-n/2}\int_{\mathbb{R}^n}e^{ix\cdot\xi}\frac{1}{p(x,\xi)+\lambda}\hat{u}(\xi)\,d\xi.$$

Then for all $0\leq\eta\leq\frac{1}{2}$ we have the estimate

$$\|K_\lambda(x,D)u\|_0\leq\frac{1}{(1+\lambda)^{\eta}}\|u\|_{\psi,\eta-1/2}. \tag{2.507}$$

Note that $K_\lambda(x,D)$ is an operator of order $-\frac{1}{2}$, i.e. we have $\|K_\lambda(x,D)u\|_0 \leq c'\|u\|_{\psi,-1/2}$, but we need some of its "regularity improving order" in order to get an estimate with some decay in λ. Using Lemma 2.9.22 we may prove an estimate for $A^f - (-f\circ p)(x,D)$ which is given by (2.493).

Theorem 2.9.23. *Let $p \in S_\rho^{2,\psi}(\mathbb{R}^n)$ be an elliptic continuous negative definite symbol generating an L^2-sub-Markovian semigroup. Suppose in addition that $f(s) = \int_0^\infty \frac{s}{s+r}\rho(dr)$ is complete Bernstein function with representing measure ρ satisfying in addition $\int_0^\infty \frac{1}{r^{1/2}}\rho(dr) < \infty$. Then the estimate*

$$\|A^f u - (-f\circ p)(x,D)u\|_0 \leq c\|u\|_0 \tag{2.508}$$

holds.

Proof: From (2.493) it follows by Lemma 2.9.22 that

$$\begin{aligned}\|A^f u + (f\circ p)(x,D)u\|_0 &\leq \int_0^\infty \|\lambda R_\lambda K_\lambda(x,D)u\|_0 \rho(d\lambda)\\ &\leq c'\int_0^\infty \frac{1}{(1+\lambda)^{1/2}}\rho(d\lambda)\|u\|_0 \leq c\|u\|_0,\end{aligned}$$

where we used also the fact that $\|\lambda R_\lambda\| \leq 1$. □

Remark 2.9.24. A. Recall that $s^\alpha = \frac{\sin(\alpha\pi)}{\pi}\int_0^\infty \frac{s}{s+r}r^\alpha\,dr$, $0<\alpha<1$, is a complete Bernstein function and for $0<\alpha<\frac{1}{2}$, $s\mapsto s^\alpha$ satisfies the additional condition of Theorem 2.9.23. Intuitively, and trivially justified for $p(x,\xi) = \psi(\xi)$, for $f(s) = s^\alpha$, $0<\alpha<\frac{1}{2}$, we should expect A^f to have order 2α, i.e. $\|A^f u\|_0 \leq c\|u\|_{\psi,2\alpha}$, note that $0 < 2\alpha < 1$ under our assumptions. The same holds clearly true for $(-f\circ p)(x,D)$. Thus Theorem 2.9.23 says that $A^f - (-(f\circ p))(x,D)u$ is of order lower than A^f and $-(f\circ p)(x,D)$, i.e. it is a lower order perturbation and $-(f\circ p)(x,D)$ should be considered as a first approximation of A^f.

B. Assuming instead of $\int_1^\infty \frac{1}{r^{1/2}}\rho(dr) < \infty$ the condition $\int_1^\infty \frac{1}{r^\eta}\rho(dr) < \infty$ for $0<\eta\leq\frac{1}{2}$, we find by (2.507) the estimate

$$\|A^f u - (-f\circ p)(x,D)u\|_0 \leq \|u\|_{\psi,\eta-\frac{1}{2}}. \tag{2.509}$$

But now we may only allow $f(s) = s^\alpha$ for $0<\alpha<\eta\leq\frac{1}{2}$.

C. Our considerations leading to Theorem 2.9.23 are very close to those given in [163] and [164] with R.Schilling. The only difference is that we used now Hoh's symbolic calculus in order to derive estimates in the scale $H^{\psi,s}(\mathbb{R}^n)$.

As we will discuss in Chapter 3 and in particular in volume III of this treatise we can deduce many properties of the semigroup and also of the corresponding stochastic process generated by an extension of $-p(x, D)$ when knowing the symbol $p(x, \xi)$.

Therefore approximating a Feller or sub-Markovian semigroup and a corresponding stochastic process by a semigroup or stochastic process, respectively, of which we know explicitly the symbol of the generator will give a chance to get approximate properties of the original semigroup or stochastic process, respectively.

This programme should in particular apply to subordinate semigroups and processes. Suppose that $A = -p(x, D)$ generates $(T_t)_{t\geq 0}$ and that f is a Bernstein function. From a functional calculus we may deduce some "abstract" properties of A^f and $(T_t^f)_{t\geq 0}$ but in general we can not determine explicitly $\sigma(A^f)(x, \xi)$ the symbol of A^f. On the other hand, a symbolic calculus may lead to a first approximation to $\sigma(A^f)(x, \xi)$ for example by considering $-(f \circ p)(x, \xi)$. Note that if $\xi \mapsto p(x, \xi)$ is a continuous negative definite function so is $\xi \mapsto (f \circ p)(x, \xi)$, thus we may raise the question whether $-(f \circ p)(x, D)$ extends to a generator of a Feller or a sub-Markovian semigroup too. This is for example true if $p \in S_\rho^{2,\psi}(\mathbb{R}^n)$ is an elliptic continuous negative definite symbol which is due to the monotonicity of the Bernstein function f

By the observations made above we may sometimes consider the operator $A^f - (-f \circ p)(x, D)$ as a "lower order" operator compared to A^f and $(f \circ p)(x, D)$. In the following we will use such a control on $A^f - (-f \circ p)(x, D)$ to indicate how it is possible to estimate the difference of the corresponding semigroups. We restrict ourselves to one example of L^2-sub-Markovian semigroups using Hoh's symbol classes and combine it with the considerations made in [163] with R.Schilling.

Thus let $p(x, \xi) \in S_\rho^{2,\psi}(\mathbb{R}^n)$ be an elliptic continuous negative definite symbol such that $A := -p(x, D)$ generates an L^2-sub-Markovian semigroup $(T_t)_{t\geq 0}$. Further let f be a complete Bernstein function such as in Theorem 2.9.23 and suppose that $-(f \circ p)(x, D)$ extends also to a generator of an L^2-sub-Markovian semigroup which we denote by $(S_t)_{t\geq 0}$. The aim is to estimate $T_t^f - S_t$. By Theorem 2.9.23 we know that

$$Qu := A^f u - (-(f \circ p)(x, D))u = \int_0^\infty \lambda R_\lambda K_\lambda(x, D) u \rho(d\lambda)$$

is a bounded operator on $L^2(\mathbb{R}^n)$. Therefore, $(T_t^f)_{t\geq 0}$ can be considered as a perturbation of $(S_t)_{t\geq 0}$. Denoting the contraction semigroup generated by Q

by $R_t := e^{-Qt}$, $t \geq 0$, note that the power series converges in $L^2(\mathbb{R}^n)$ since Q is bounded, we may apply Trotter's product formula, Corollary I.4.4.15, to obtain

$$T_t^f u := \lim_{n\to\infty} (S_{t/n} R_{t/n})^n u.$$

Furthermore, since

$$T_t^f u - S_t u = \int_0^t T_{t-s}^f (A^f - (-(f \circ p)(x, D)) S_s u \, ds \tag{2.510}$$

we get

Theorem 2.9.25. *In the situation described above the estimate*

$$\|T_t^f - S_t\| \leq c\,t \tag{2.511}$$

holds.

Proof: From (2.510) it follows using (2.508) that

$$\begin{aligned} \|T_t^f u - S_t u\|_0 &\leq \int_0^t \|T_{t-s}^f\| \|A^f - (-f \circ p)(x, D)\| \|S_s\| \|u\|_0 \, ds \\ &\leq c\,t \|u\|_0 \end{aligned}$$

implying (2.511) □

Remark 2.9.26. Suppose that $p_t^f(x, B) := T_t^f \chi_B(x)$ and $p_t(x, B) := S_t \chi_B(x)$, $B \in \mathcal{B}^{(n)}$ with $\lambda^{(n)}(B) < \infty$, have the interpretation of a Markov transition function. Then (2.511) gives the estimate

$$\int_{\mathbb{R}^n} |p_t^f(x, B) - p_t(x, B)|^2 \, dx \leq c\,t \lambda^{(n)}(B) \tag{2.512}$$

i.e. we have a comparison for integrated transition probabilities.

Remark 2.9.27. Using embedding theorems for the spaces $H^{\psi,s}(\mathbb{R}^n)$ it is also possible to obtain pointwise estimate for $p_t^f(x, B) - p_t(x, B)$. For this and many other related estimates for subordinate elliptic diffusion semigroups we refer to the paper [163] and [164] with R.Schilling.

2.10 Pseudo-Differential Operators with Variable Order of Differentiation as Generators of Feller Semigroups

In the last section we explored some one-parameter semigroups and their generators obtained by subordination. We will extend these considerations to the case where the Bernstein function $s \mapsto f(s)$ is substituted by $(x, s) \mapsto s^{r(x)}$ with $r : \mathbb{R}^n \to \mathbb{R}$ being a continuous function such that $0 \leq r(x) \leq 1$ holds.

Let $p : \mathbb{R}^n \times \mathbb{R}^n \to \mathbb{C}$ (for simplicity we will later on take $\mathbb{R}$ instead of $\mathbb{C}$) be a continuous function such that $\xi \mapsto p(x, \xi)$ is a continuous negative definite function. Then it follows by Lemma I.3.9.9 that

$$\xi \mapsto p(x,\xi)^{r(x)} \tag{2.513}$$

is once again a continuous negative definite function implying that the pseudo-differential operator

$$A\,u(x) = -(2\pi)^{-n/2} \int_{\mathbb{R}^n} e^{ix\cdot\xi} p(x,\xi)^{r(x)} \hat{u}(\xi)\, d\xi \tag{2.514}$$

is a candidate for a generator of a Feller semigroup. For $p(x, \xi) = |\xi|^2$ we obtain $(-\Delta)^{r(x)}$, but a bit more convenient is to look at $(1 - \Delta)^{r(x)}$. For $n = 1$ and $p(x, \xi) = \pm i\xi$ we arrive at the operator $\left(\mp \frac{d}{dx}\right)^{r(x)}$, thus the name "operator of variable order of differentiation" is justified. For some "classical" symbols $p(x, \xi)$ operators of variable order of differentiation had been studied by A.Unterberger and J.Bokobza [274], and in particular by H.-G.Leopold [195], [196]. Studies with the symbol $(1 + |\xi|^2)^{r(x)}$ leading to Feller semigroups had been done jointly with H.-G.Leopold [162], and further investigations are due to A.Negoro [222], and in particular to K.Kikuchi and A.Negoro [173], [174]. Note that F.Baldus [13] can also handle some operator of variable order of differentiation within his Weyl–Hörmander calculus.

In this section we follow once again W.Hoh and use his symbolic calculus for pseudo-differential operators with negative definite symbols — but now we consider the case of operators of variable order of differentiation, see [128] and in particular [129]. Using the notation of Section 2.4 we may state the main result of this section.

Theorem 2.10.1 (W.Hoh). *Let $\psi : \mathbb{R}^n \to \mathbb{R}$ be a fixed continuous negative definite function such that its Lévy measure has a compact support and that*

$$\psi(\xi) \geq c_0 |\xi|^r, \quad |\xi| \textit{ large and } r > 0, \tag{2.515}$$

holds. Let $q \in S_\rho^{2,\psi}(\mathbb{R}^n)$ *be a real-valued negative definite symbol which is elliptic, i.e. we have*

$$q(x,\xi) \geq \delta_0(1+\psi(\xi)). \tag{2.516}$$

Further let $m : \mathbb{R}^n \to (0,1]$ *be an element in* $C_b^\infty(\mathbb{R}^n)$ *satisfying*

$$M - \mu < \frac{1}{2} \tag{2.517}$$

where $M := \sup m(x)$ *and* $0 < \mu := \inf m(x)$. *Consider the symbol*

$$(x,\xi) \mapsto p(x,\xi) := q(x,\xi)^{m(x)} \tag{2.518}$$

which has the property that $\xi \mapsto p(x,\xi)$ *is a continuous negative definite function. The operator*

$$-p(x,D)u(x) := -(2\pi)^{-n/2} \int_{\mathbb{R}^n} e^{ix\cdot\xi} p(x,\xi)\hat{u}(\xi)\, d\xi \tag{2.519}$$

maps $C_0^\infty(\mathbb{R}^n;\mathbb{R})$ *into* $C_\infty(\mathbb{R}^n;\mathbb{R})$, *is closable in* $C_\infty(\mathbb{R}^n;\mathbb{R})$ *and its closure is a generator of a Feller semigroup.*

Remark 2.10.2. In volume III of this treatise we will use probabilistic methods to relax some condition in Theorem 2.10.1

The proof of Theorem 2.10.1 requires some preparations.

Proposition 2.10.3. *Let* $p(x,\xi)$ *be as in Theorem* 2.10.1. *Then we have for all* $\varepsilon > 0$ *the estimates*

$$|\partial_\xi^\alpha \partial_x^\beta p(x,\xi)| \leq c_{\alpha,\beta,\varepsilon} p(x,\xi)(1+\psi(\xi))^{-\frac{\rho(|\alpha|)+\varepsilon}{2}} \tag{2.520}$$

implying that $p(x,\xi)$ *belongs to* $S_\rho^{2M+\varepsilon,\psi}(\mathbb{R}^n)$.

Proof: We have to estimate

$$\partial_\xi^\alpha \partial_x^\beta p(x,\xi) = \partial_\xi^\alpha \partial_x^\beta (q(x,\xi)^{m(x)}) = \partial_\xi^\alpha \partial_x^\beta (\exp m(x) \log q(x,\xi)). \tag{2.521}$$

From (I.2.28) it follows that with $l = |\alpha| + |\beta|$

$$|\partial_\xi^\alpha \partial_x^\beta p(x,\xi)| \leq \exp(m(x)\log q(x,\xi)) \times$$

$$\times \Big| \sum_{\substack{\alpha^1+\cdots+\alpha^{l'}=\alpha \\ \beta^1+\cdots+\beta^{l'}=\beta \\ l'=0,1,\ldots,l}} c_{\{\alpha^j,\beta^j\}} \prod_{j=1}^{l'} q_{\alpha^j\beta^j}(x,\xi) \Big|, \tag{2.522}$$

where

$$q_{\alpha^j\beta^j}(x,\xi) = \partial_\xi^{\alpha^j}\partial_x^{\beta^j}(m(x)\log q(x,\xi))$$
$$= \sum_{\overline{\beta}^j \leq \beta^j} \binom{\beta^j}{\overline{\beta}^j} \partial_x^{\beta^j-\overline{\beta}^j} m(x) \partial_\xi^{\alpha^j}\partial_x^{\overline{\beta}^j} \log q(x,\xi). \tag{2.523}$$

Now, from (I.2.26) with $k = |\alpha^j| + |\overline{\beta}^j| > 0$ we may derive

$$\partial_\xi^{\alpha^j}\partial_x^{\overline{\beta}^j} \log q(x,\xi) = \sum_{\substack{\tilde{\alpha}^1+\cdots+\tilde{\alpha}^{l'}=\alpha^j \\ \tilde{\beta}^1+\cdots+\tilde{\beta}^k=\overline{\beta}^j}} c_{\{\tilde{\alpha}^i,\tilde{\beta}^i\}} \prod_{i=1}^k \frac{\partial_\xi^{\tilde{\alpha}^i}\partial_x^{\tilde{\beta}^i} q(x,\xi)}{q(x,\xi)}.$$

By our assumptions $q(x,\xi)$ is an elliptic symbol in $S_\rho^{2,\psi}(\mathbb{R}^n)$ which yields further that

$$|\partial_\xi^{\alpha^j}\partial_x^{\overline{\beta}^j} \log q(x,\xi)| = c_{\alpha^j,\overline{\beta}^j} \sum_{\substack{\tilde{\alpha}^1+\cdots+\tilde{\alpha}^k=\alpha^j \\ \tilde{\beta}^1+\cdots+\tilde{\beta}^k=\overline{\beta}^j}} \prod_{i=1}^k (1+\psi(\xi))^{-\frac{\rho(|\tilde{\alpha}^i|)}{2}}$$
$$\leq c_{\alpha^j,\overline{\beta}^j}(1+\psi(\xi))^{-\frac{\rho(|\alpha^j|)}{2}},$$

where we used the subadditivity of ρ. In addition we have always $|\log q(x,\xi)| \leq c\log(1+\psi(\xi)) \leq c_\varepsilon(1+\psi(\xi))^{\varepsilon/2l}$ with l as in (2.522). For $m \in C_b^\infty(\mathbb{R}^n)$ it follows therefore from (2.523) that

$$|q_{\alpha^j,\beta^j}(x,\xi)| \leq c_{\alpha^j,\beta^j,\varepsilon} \begin{cases} (1+\psi(\xi))^{-\frac{\rho(|\alpha^j|)}{2}}, & \alpha^j \neq 0 \\ (1+\psi(\xi))^{\varepsilon/2l}, & \alpha^j = 0. \end{cases} \tag{2.524}$$

Combining finally (2.524) with (2.522) we arrive at

$$|\partial_\xi^\alpha\partial_x^\beta p(x,\xi)|$$
$$\leq p(x,\xi)c_{\alpha,\beta,\varepsilon} \sum_{\substack{\alpha^1+\cdots+\alpha^{l'}=\alpha \\ \beta^1+\cdots+\beta^{l'}=\beta \\ l'=0,\ldots,l}} \Big(\prod_{\substack{j=1,\ldots,l' \\ \alpha^j\neq 0}} (1+\psi(\xi))^{-\frac{\rho(|\alpha^j|)}{2}} \prod_{\substack{i=1,\ldots,l' \\ \alpha^i=0}} (1+\psi(\xi))^{\varepsilon/2l}\Big)$$
$$\leq p(x,\xi)c_{\alpha,\beta,\varepsilon}(1+\psi(\xi))^{\frac{-\rho(|\alpha|)+\varepsilon}{2}},$$

and (2.520) is proved. The second statement follows now from the estimate $p(x,\xi) \leq c(1+\psi(\xi))^M$ proving the proposition. □

We want to prove Theorem 2.10.1 with the same strategy as we proved for example Theorem 2.6.4, i.e. we will first introduce the notion of weak solutions for $p(x,D)u = f$ and prove that for λ sufficiently large the existence of a unique weak solution to $p_\lambda(x,D)u = p(x,D)u + \lambda u = f$ in some Hilbert space. Then we will prove the regularity of weak solutions, and an application of Theorem 2.6.1 will also give the positive maximum principle for $-p(x,D)$ on a suitable domain.

First we need to construct a suitable Hilbert space for dealing with $p(x,D)$. The idea is taken from the theory of t-coercive operators as considered by Chr.Simader and I.Louhivaara. For $\lambda \geq 0$ we set $p_\lambda(x,\xi) = p(x,\xi) + \lambda$ and further

$$B_\lambda(u,v) := (p_\lambda(x,D)u, v)_0 \quad \text{for } u, v \in C_0^\infty(\mathbb{R}^n). \tag{2.525}$$

Since the proof of Theorem 2.5.6 yields the estimate

$$|(q(x,D)u,v)_0| \leq c\|u\|_{\psi,t}\|v\|_{\psi,t}$$

for all $u, v \in C_0^\infty(\mathbb{R}^n)$ and $q \in S_0^{2t,\psi}(\mathbb{R}^n)$ we find immediately

Lemma 2.10.4. *The bilinear form B_λ has for all $\varepsilon > 0$ a continuous extension to $H^{\psi,M+\frac{\varepsilon}{2}}(\mathbb{R}^n)$ provided $M - \mu + \varepsilon < \frac{1}{2}$. In particular*

$$|B_\lambda(u,v)| \leq c\|u\|_{\psi,M+\frac{\varepsilon}{2}}\|v\|_{\psi,M+\frac{\varepsilon}{2}} \tag{2.526}$$

holds for all $u, v \in H^{\psi,M+\frac{\varepsilon}{2}}(\mathbb{R}^n)$.

In the following we will denote the continuous extension of the bilinear form B_λ to $H^{\psi,M+\frac{\varepsilon}{2}}(\mathbb{R}^n)$ again by B_λ.

We get also a lower bound for B_λ for λ large, but not in the space $H^{\psi,M+\frac{\varepsilon}{2}}(\mathbb{R}^n)$. Since by our assumptions $q(x,\xi)$ is elliptic, there exists $\delta > 0$ such that

$$p(x,\xi) \geq \delta(1+\psi(\xi))^\mu \tag{2.527}$$

for all $\xi \in \mathbb{R}^n$ where μ is the largest lower bound for $m(x)$.

Proposition 2.10.5. *For $\lambda \geq \lambda_0$, λ_0 sufficiently large,*

$$B_\lambda(u,u) \geq \frac{\delta}{2}\|u\|^2_{\psi,\mu} \tag{2.528}$$

holds for all $u \in H^{\psi,\mu}(\mathbb{R}^n)$.

Proof: From (2.527) we know that $Q(x,\xi) := p(x,\xi) - \delta(1+\psi(\xi))^{\mu} \in S^{2M+\varepsilon,\psi}(\mathbb{R}^n)$ and $Q(x,\xi) \geq 0$. We may now argue as in the proofs of Theorem 2.5.5 and Theorem 2.5.6 to find

$$(p(x,D)u,u)_0 - \delta\|u\|^2_{\psi,\mu} \geq -c\|u\|^2_{\psi,M+\frac{\varepsilon}{2}-\frac{1}{2}} \geq \frac{\delta}{2}\|u\|^2_{\psi,\mu} - c(\delta)\|u\|_0^2,$$

where for the last step we used the fact that $M + \frac{\varepsilon}{2} - \frac{1}{2} < \mu$. □

To overcome the difficulty that the upper and lower bounds for B_λ do hold in different norms we work now as Chr.Simader and I.Louhivaara, [200], and [201], with the symmetric part

$$\tilde{B}_\lambda(u,v) = \frac{1}{2}(B_\lambda(u,v) + B_\lambda(v,u)) \tag{2.529}$$

of B_λ on $H^{\psi,M+\frac{\varepsilon}{2}}(\mathbb{R}^n)$, i.e. the smaller space. Obviously we have

$$|\tilde{B}_\lambda(u,v)| \leq c\|u\|_{\psi,M+\frac{\varepsilon}{2}}\|v\|_{\psi,M+\frac{\varepsilon}{2}}, \tag{2.530}$$

and

$$\tilde{B}_\lambda(u,u) \geq \frac{\delta}{2}\|u\|^2_{\psi,\mu} \tag{2.531}$$

for $\lambda \geq \lambda_0$. In particular $\tilde{B}_\lambda$ is a scalar product on $H^{\psi,M+\frac{\varepsilon}{2}}(\mathbb{R}^n)$ and therefore we may consider the closure of $H^{\psi,M+\frac{\varepsilon}{2}}(\mathbb{R}^n) \subset H^{\psi,\mu}(\mathbb{R}^n)$ with respect to this scalar product. Denoting this Hilbert space by $H^{p_\lambda}(\mathbb{R}^n)$ with corresponding norm $\|\cdot\|_{p_\lambda} = \tilde{B}_\lambda^{1/2}(\cdot,\cdot)$ we have the continuous embeddings

$$H^{\psi,M+\frac{\varepsilon}{2}}(\mathbb{R}^n) \hookrightarrow H^{p_\lambda}(\mathbb{R}^n) \hookrightarrow H^{\psi,\mu}(\mathbb{R}^n), \tag{2.532}$$

and $C_0^\infty(\mathbb{R}^n)$ is dense in $H^{p_\lambda}(\mathbb{R}^n)$.

Lemma 2.10.6. *The bilinear form B_λ is continuous on $H^{p_\lambda}(\mathbb{R}^n)$.*

Proof: We may apply Corollary 2.4.23 to find that

$$\begin{aligned}\frac{1}{2}(p_\lambda(x,D) + p_\lambda^*(x,D)) &= \frac{1}{2}(p_\lambda(x,\xi) + \overline{p}_\lambda(x,D)) + r_1(x,D)\\ &= p_\lambda(x,D) + r_1(x,D),\end{aligned}$$

where $r_1 \in S_0^{2M-1+\varepsilon,\psi}(\mathbb{R}^n)$ and we used that $p(x,\xi)$ is real-valued. Now, since

$$\begin{aligned}|B_\lambda(u,v)| &= |(p_\lambda(x,D)u,v)_0|\\ &\leq \frac{1}{2}|((p_\lambda(x,D) + p_\lambda^*(x,D)u,v)_0| + |(r_1(x,D)u,v)_0|\\ &= |\tilde{B}_\lambda(u,v)| + |(r_1(x,D)u,v)_0|,\end{aligned}$$

the continuity of $\tilde{B}_\lambda$ on $H^{p_\lambda}(\mathbb{R}^n)$ and the estimate

$$|(r_1(x,D)u,v)_0| \leq c\|u\|_{\psi,M+\frac{\varepsilon}{2}-\frac{1}{2}}\|v\|_{\psi,M+\frac{\varepsilon}{2}-\frac{1}{2}}$$

which follows from Theorem 2.5.6, imply the result if we note that $M+\frac{\varepsilon}{2}-\frac{1}{2}<\mu$ which yields the continuous embeddings

$$H^{p_\lambda}(\mathbb{R}^n) \hookrightarrow H^{\psi,\mu}(\mathbb{R}^n) \hookrightarrow H^{\psi,M+\frac{\varepsilon}{2}-\frac{1}{2}}(\mathbb{R}^n),$$

i.e. the estimate

$$\|u\|_{\psi,M+\frac{\varepsilon}{2}-\frac{1}{2}} \leq c\|u\|_{p_\lambda}. \quad \square$$

Now we are in a position to prove an existence result.

Theorem 2.10.7. *Let $p(x,\xi)$ be as in Theorem 2.10.1 and $\lambda \geq \lambda_0$ sufficiently large. Then for every $f \in H^{\psi,-\mu}(\mathbb{R}^n) = (H^{\psi,\mu}(\mathbb{R}^n))^*$ there is a unique $H^{\psi,\mu}(\mathbb{R}^n)$ such that*

$$p_\lambda(x,D)u = p(x,D)u + \lambda u = f \tag{2.533}$$

holds.

Proof: Since by Lemma 2.10.6 and Proposition 2.10.5 B_λ is a continuous, coercive bilinear form on H^{p_λ}, by the Lax–Milgram theorem, Theorem I.2.7.41, there exists to any $f \in H^{\psi,-\mu}(\mathbb{R}^n)$ a unique $u \in H^{p_\lambda}(\mathbb{R}^n)$ such that

$$B_\lambda(u,v) = \langle f,v\rangle \tag{2.534}$$

holds for all $v \in H^{p_\lambda}(\mathbb{R}^n)$. Take a sequence $(u_\nu)_{\nu\in\mathbb{N}}$ in $C_0^\infty(\mathbb{R}^n)$ converging in $H^{p_\lambda}(\mathbb{R}^n)$ to u. Since for $v \in C_0^\infty(\mathbb{R}^n)$ the mapping $u \mapsto (u,v)_0$ has a continuous extension to $H^{\psi,\mu-2M-\varepsilon}(\mathbb{R}^n)$ and since $p_\lambda(x,D) : H^{\psi,\mu}(\mathbb{R}^n) \to H^{\psi,\mu-2M-2}(\mathbb{R}^n)$ is continuous, it follows from

$$(p_\lambda(x,D)u_k,v)_0 = B_\lambda(u_k,v), \quad v \in C_0^\infty(\mathbb{R}^n),$$

that for $k \to \infty$

$$\langle p_\lambda(x,D)u,v\rangle = B_\lambda(u,v) = \langle f,v\rangle \tag{2.535}$$

holds for all $v \in C_0^\infty(\mathbb{R}^n)$. Thus $p_\lambda(x,D)u = f$. For two solutions $u_1,u_2 \in H^{p_\lambda}(\mathbb{R}^n)$ we find

$$B_\lambda(u_1-u_2,v) = 0$$

for all $v \in C_0^\infty(\mathbb{R}^n)$ and by the continuity of B_λ on $H^{p_\lambda}(\mathbb{R}^n)$ we have $B_\lambda(u_1 - u_2, v) = 0$ for all $v \in H^{p_\lambda}(\mathbb{R}^n)$, hence

$$\frac{\delta}{2}\|u_1 - u_2\|^2_{\psi,\mu} \leq B_\lambda(u_1 - u_2, u_1 - u_2) = 0$$

implying the uniqueness of u. □

Our next step is to prove that if u is a solution to (2.533) and f is more regular, i.e. belongs to some space $H^{\psi,k}(\mathbb{R}^n)$ with $k > -\mu$, then u is also more regular, i.e. belongs to some space $H^{\psi,l}(\mathbb{R}^n)$ such that $H^{\psi,l}(\mathbb{R}^n) \subsetneqq H^{p_\lambda}(\mathbb{R}^n)$. We need first

Lemma 2.10.8. *The function* $p_\lambda^{-1}(x,\xi) = \dfrac{1}{p(x,\xi) + \lambda}$ *belongs to the class* $S_\rho^{-2\mu+\varepsilon,\psi}(\mathbb{R}^n)$.

Proof: We use (I.2.27) to find with $l := |\alpha| + |\beta|$ that

$$|\partial_\xi^\alpha \partial_x^\beta p_\lambda^{-1}(x,\xi)| \leq \frac{1}{p_\lambda(x,\xi)} \sum_{\substack{\alpha^1+\cdots+\alpha^l=\alpha \\ \beta^1+\cdots+\beta^l=\beta}} c_{\{\alpha^j,\beta^j\}} \prod_{j=1}^{l} \left| \frac{\partial_\xi^{\alpha^j} \partial_x^{\beta^j} p_\lambda(x,\xi)}{p_\lambda(x,\xi)} \right|.$$

Using (2.520) we get for any $\varepsilon > 0$

$$|\partial_\xi^{\alpha^j} \partial_x^{\beta^j} p_\lambda(x,\xi)| \leq \tilde{c}_{\alpha^j,\beta^j}(1+\psi(\xi))^{\frac{-\rho(|\alpha^j|)+\varepsilon}{2}}$$

and the ellipticity assumption (2.516) finally yields by the subadditivity of ρ

$$\left|\partial_\xi^\alpha \partial_x^\beta \frac{1}{p_\lambda(x,\xi)}\right| \leq \tilde{c}_{\alpha,\beta,\varepsilon}(1+\psi(\xi))^{-\mu}(1+\psi(\xi))^{\frac{-\rho(|\alpha|)+\varepsilon}{2}}$$

and the lemma is proved. □

Theorem 2.10.9. *Let* $p(x,\xi)$ *be as in Theorem* 2.10.7 *and let* $u \in H^{p_\lambda}(\mathbb{R}^n) \subset H^{\psi,\mu}(\mathbb{R}^n)$ *be the solution to* (2.533) *for* $f \in H^{\psi,k}(\mathbb{R}^n)$, $k \geq 0$. *Then for all* $\varepsilon > 0$ *it follows that* $u \in H^{\psi,k+2\mu-\varepsilon}(\mathbb{R}^n)$.

Proof: We may apply Corollary 2.4.23 and find

$$p_\lambda^{-1}(x,D) \circ p_\lambda(x,D) = \mathrm{id} + r(x,D) \tag{2.536}$$

with $r \in S_0^{-t,\psi}(\mathbb{R}^n)$ where

$$-t = (-2\mu+\varepsilon)+(2M+\varepsilon)-1 = 2\Big(M-\mu+\varepsilon-\frac{1}{2}\Big) < 0. \tag{2.537}$$

Since $p_\lambda(x,D)u = f$ we deduce from (2.536)

$$u = p_\lambda^{-1}(x,D) \circ p_\lambda(x,D) - r(x,D)u = p_\lambda^{-1}(x,D)f - r(x,D)u.$$

Since further $p_\lambda^{-1}(x,D)f \in H^{\psi,k+2\mu-\varepsilon}(\mathbb{R}^n)$ and $r(x,D)u \subset H^{\psi,\mu+t}(\mathbb{R}^n)$ we find that $u \in H^{\psi,(\mu+t)\wedge(k+2\mu-\varepsilon)}(\mathbb{R}^n)$. A finite number of applications of this argument finally gives $u \in H^{\psi,k+2\mu-\varepsilon}(\mathbb{R}^n)$. □

Now we come to

Proof of Theorem 2.10.1: From (2.515) it follows that for $s > n/2r$ the space $H^{\psi,s}(\mathbb{R}^n)$ is continuously embedded into $C_\infty(\mathbb{R}^n)$. For $\varepsilon > 0$ satisfying $M - \mu + \varepsilon < 1/2$ we have that $p \in S_\rho^{2M+\varepsilon,\psi}(\mathbb{R}^n)$. If we choose $k > n/2r$ it follows that $H^{\psi,k+2M+\varepsilon}(\mathbb{R}^n)$ and $H^{\psi,k}(\mathbb{R}^n)$ can both be considered as dense subspaces of $C_\infty(\mathbb{R}^n)$. Hence

$$-p(x,D) : H^{\psi,k+2M+\varepsilon}(\mathbb{R}^n) \to H^{\psi,k}(\mathbb{R}^n)$$

is a densely definite operator on $C_\infty(\mathbb{R}^n)$ with a continuous negative definite symbol. Theorem 2.6.1 yields that $-p(x,D)$ satisfies also the positive maximum principle on $H^{\psi,k+2M+\varepsilon}(\mathbb{R}^n)$. Finally, combining Theorem 2.10.7 and Lemma 2.10.8, for $\lambda \geq \lambda_0$ sufficiently large, for each $f \in H^{\psi,k+2(M-\mu-\varepsilon)}(\mathbb{R}^n)$ there exists an element $u \in H^{\psi,k+2M+\varepsilon}(\mathbb{R}^n)$ such that $p_\lambda(x,D)u = f$, i.e. $(p_\lambda(x,D), H^{\psi,k+2M+\varepsilon}(\mathbb{R}^n))$ considered as an operator on $C_\infty(\mathbb{R}^n)$ has a dense range. Now the Hille–Yosida–Ray theorem, Theorem I.4.5.3, gives Theorem 2.10.1.

As in Section 2.6 we may argue further (recalling that $B_\lambda(u,u) \geq 0$) to derive

Corollary 2.10.10. *The operator $-p_\lambda(x,D)$ extends to a generator of an L^2-sub-Markovian semigroup and $(B_\lambda, H^{p_\lambda}(\mathbb{R}^n))$ is a semi-Dirichlet form.*

Our interpolation results now yield

Corollary 2.10.11. *For $\lambda \geq \lambda_0$ and $2 \leq p < \infty$ the operator $-p_\lambda(x,D)$ extends to a generator of an L^p-sub-Markovian semigroup.*

Chapter 3

Potential Theory of Semigroups and Generators

Chapter 3 discusses aspects of potential theory. We start with a treatment of capacities and continue with abstract Bessel potential spaces $\mathcal{F}_{r,p}$ associated with an L^p-sub-Markovian semigroup, see Section 3.2. In order to get applications to concrete operators we have to characterize the spaces $\mathcal{F}_{r,p}$ in terms of function spaces which is done in Section 3.3. In Section 3.4 Stein's Littlewood–Paley theory for sub-Markovian semigroups is discussed. Section 3.5 is devoted to global properties of sub-Markovian semigroups. Topics such as invariant sets, transience, recurrence or abstract potential operators are on the agenda. The final section indicates how Nash-type and Sobolev-type inequalities can be used also for non-local operators.

3.1 Capacities and Abstract Bessel Potential Spaces

In this section we first discuss the general theory of Choquet capacities and then we will discuss a lot of concrete capacities associated with sub-Markovian semigroups, Dirichlet forms and kernels. Since we will apply the concept of capacities in several distinct situations we prefer now to work on a general *Polish space* G, i.e. a complete metrizable Hausdorff space with countable base. Of course, as long as it is not stated otherwise, measurability will refer to the Borel-σ field $\mathcal{B}(G)$. We start with introducing analytic subsets of G.

Definition 3.1.1. Let G be a Polish space.

A. A set $E \subset G$ is called a *K_σ-set* if it is a countable union of compact sets.

B. A set $E \subset G$ is called a *$K_{\sigma\delta}$-set* if it is the countable intersection of K_σ-sets.

C. We call a subset $E \subset G$ an *analytic set* if there exists a $K_{\sigma\delta}$-set F in a compact Hausdorff space H and a continuous mapping $f : F \to G$ such that E is the image of F under f, i.e. $E = f(F)$. We write $\mathrm{Ana}(G)$ for the family of all analytic subsets of G.

Example 3.1.2. A. If K is a $K_{\sigma\delta}$-set and $C \subset K$ is a closed subset, then C is also $K_{\sigma\delta}$-set since

$$C = C \cap K = C \cap \Big(\bigcap_{k\in\mathbb{N}} \bigcup_{n\in\mathbb{N}} K_{kn}\Big) = \bigcap_{k\in\mathbb{N}} \bigcup_{n\in\mathbb{N}} (C \cap K_{kn})$$

and each set $C \cap K_{kn}$ is compact.

B. Every compact subset K of G is an analytic set. Indeed, every compact set K is a $K_{\sigma\delta}$-set and therefore, taking as compact Hausdorff space the set K itself and as mapping the identity we see that K is the image of $K_{\sigma\delta}$-set of a compact Hausdorff space.

Remark 3.1.3. A. Our definition of analytic sets is taken from G.Choquet's paper [48]. In more recent treatises such as R.F.Bass [17], J.Bliedtner and W.Hansen [30], or C.Dellacherie and P.-A.Meyer [64] the following definition of analytic sets is given:

A set $E \subset G$ in a Polish space is called *analytic* if there exists a compact Hausdorff space H and a $K_{\sigma\delta}$-set $C \subset H \times G$ such that E is the projection of C onto G.

Since we may embed G into a compact space, it is obvious that analytic sets in the sense of this definition belong to the class $\mathrm{Ana}(G)$. It requires some topological efforts to show that in the cases we are interested in the converse holds too. The essential step is to show that for any Polish space G the projections of closed sets $F \subset \mathbb{N}^{\mathbb{N}} \times G$ onto G coincides with $\mathrm{Ana}(G)$. We refer to the monograph [49] of G.Choquet or the lecture notes of R.Burckel [39].

B. For further definitions and concepts of analytic sets we refer to the work [49] of G.Choquet, Definition 8.12, L.Carleson [45], or the lecture notes [39] of R.Burckel.

C. For a discussion of the relation of *Souslin sets* and analytic sets we refer to R.Burckel, G.Choquet, and especially to K.Jacobs [167], Chapter XIII, and

C.A.Rogers [236].

Lemma 3.1.4. *The family* Ana(G) *is closed under countable unions.*

Proof: Let $(E_n)_{n\in\mathbb{N}}$, $E_n \subset G$, be a sequence of analytic sets. For each of these sets exists a compact Hausdorff space H_n, a $K_{\sigma\delta}$-set $F_n \subset H_n$ and a continuous mapping $f_n : F_n \to G$ such that $f_n(F_n) = E_n$. Without loss of generality we may assume $H_n \cap H_m = \emptyset$ for all $n, m \in \mathbb{N}$. Now, on $\bigcup_{n\in\mathbb{N}} H_n$ we introduce a topology by taking as it base the system of all sets which are open in one of the spaces H_n. Denote by H the one-point-compactification of the topological space $\bigcup_{n\in\mathbb{N}} H_n$. Thus H is a compact Hausdorff space. We define the mapping $f : \bigcup_{n\in\mathbb{N}} F_n \to G$ by $f(x) = f_n(x)$ for $x \in F_n$ which is obviously continuous and maps $\bigcup_{n\in\mathbb{N}} F_n$ onto $\bigcup_{n\in\mathbb{N}} E_n$. It remains to prove that F:=$\bigcup_{n\in\mathbb{N}} F_n$ is a $K_{\sigma\delta}$-set in H. For any $n \in \mathbb{N}$ there exist K_σ-sets F_{nk} such that

$$F = \bigcup_{n\in\mathbb{N}} F_n = \bigcup_{n\in\mathbb{N}} (\bigcap_{k\in\mathbb{N}} F_{kn}).$$

Since by our assumption the spaces H_n are mutually disjoint, we find that $F = \bigcap_{k\in\mathbb{N}}(\bigcup_{n\in\mathbb{N}} F_{nk})$. Since $\bigcup_{n\in\mathbb{N}} F_{nk}$ is a K_σ-set in H it follows that F is a $K_{\sigma\delta}$-set in H implying the lemma. □

Remark 3.1.5. Since every open set is the union of countable many compact sets, Lemma 3.1.4 and Example 3.1.2.B imply that all open sets are analytic.

Lemma 3.1.6. *The family* Ana(G) *is closed under countable intersections.*

Proof: Let $(E_n)_{n\in\mathbb{N}}$, $E_n \subset G$, be a sequence of analytic sets and suppose that this sequence is decreasing (otherwise consider the new sequence $(E'_n)_{n\in\mathbb{N}}$ defined by $E'_1 := E_1$ and $E'_n := E'_{n-1} \cap E_n$). We have to show that

$$E := \bigcap_{n\in\mathbb{N}} E_n \in \text{Ana}(G). \tag{3.1}$$

By assumption each set E_n is the continuous image of a $K_{\sigma\delta}$-set F_n which belongs to a compact Hausdorff space H_n, i.e. $E_n = f_n(F_n)$ with $f_n : H_n \to G$ and $f_n : F_n \to G$ is continuous.

We define the space $H := \prod_{n\in\mathbb{N}} H_n$ and set $F := \prod_{n\in\mathbb{N}} F_n \subset H$. By Tychonov's theorem the space H is compact. Let us introduce the sets $F^*_n := \prod_{n\in\mathbb{N}} Y_j$ where

$Y_j = H_j$ for $j \neq n$ and $Y_n = F_n$. It follows that F_n^* is a $K_{\sigma\delta}$-set in H since H_j is compact and $F_n \subset H_n$ is a $K_{\sigma\delta}$-set. But $F = \bigcap_{n\in\mathbb{N}} F_n^*$, thus F is a $K_{\sigma\delta}$-set in H. Let us denote by $\pi : H \to H_n$ the family of (continuous) projections and define the continuous mappings $\tilde{f}_n$ by $\tilde{f}_n \circ \pi_n : H \to G$. Note that $\tilde{f}_n(F) = E_n$. By continuity of $\tilde{f}_n$ the sets

$$D_{kl} := \{x \in F; \tilde{f}_k(x) = \tilde{f}_l(x)\} \tag{3.2}$$

are for all $k, l \in \mathbb{N}$ closed subsets of F, hence again $K_{\sigma\delta}$-sets in H implying that

$$D := \bigcap_{k,l\in\mathbb{N}} D_{kl} \tag{3.3}$$

is a $K_{\sigma\delta}$-set in H. Further we find for all $k, l \in \mathbb{N}$

$$\tilde{f}_k|_D = \tilde{f}_l|_D. \tag{3.4}$$

Thus the function $\tilde{f} : D \to G$, $\tilde{f} = \tilde{f}_k|_D$ for some $k \in \mathbb{N}$, is a continuous function from the $K_{\sigma\delta}$-set $D \subset H$ into G and the image of D under $\tilde{f}$ is for all $k \in \mathbb{N}$ given by

$$\tilde{f}(D) = \tilde{f}_k(D) \subset \tilde{f}_k(F) = f_k(F_k) = E_k,$$

which yields

$$\tilde{f}(D) \subset \bigcap_{n\in\mathbb{N}} E_n = E. \tag{3.5}$$

On the other hand for $y \in E$ we find for all $k \in \mathbb{N}$ some $x_k \in F_k$ such that $f_k(x_k) = y$ which leads to $x := (x_1, \ldots, x_k, \ldots) \in D$ and $\tilde{f}(x) = y$, i.e. $E = \tilde{f}(D)$ implying the lemma. □

Theorem 3.1.7. *The Borel sets in a Polish space G are analytic sets.*

Proof: By definition the Borel sets form the smallest family of subsets of G containing all compact sets and which is closed with respect to complements and countable unions. Denote by C the system of all analytic subsets of G having an analytic complement. Since compact and open sets are analytic, it follows that C contains all compact sets. Clearly, C is closed with respect to countable union. Hence C contains all Borel sets and the theorem is proved. □

Definition 3.1.8. Let G be a Polish space. A function $\text{cap}^* : \mathcal{P}(G) \to \mathbb{R}_+ \cup \{\infty\}$ is called an *outer Choquet capacity* on G if the following properties are fulfilled

$$\text{if } A_1, A_2 \in \mathcal{P}(G) \text{ and } A_1 \subset A_2, \text{ then } \text{cap}^*(A_1) \leq \text{cap}^*(A_2), \tag{3.6}$$

$$\begin{aligned}&\text{if } (A_\nu)_{\nu\in\mathbb{N}},\ A_\nu \in \mathcal{P}(G), \text{ is an increasing sequence}\\ &\text{then } \text{cap}^*\Big(\bigcup_{\nu\in\mathbb{N}} A_\nu\Big) = \sup_{\nu\in\mathbb{N}} \text{cap}^*(A_\nu);\end{aligned} \tag{3.7}$$

$$\begin{aligned}&\text{for any decreasing sequence } (A_\nu)_{\nu\in\mathbb{N}} \text{ of compact sets}\\ &A_\nu \in \mathcal{P}(G) \text{ it follows that } \text{cap}^*\Big(\bigcap_{\nu\in\mathbb{N}} A_\nu\Big) = \inf_{\nu\in\mathbb{N}} \text{cap}^*(A_\nu).\end{aligned} \tag{3.8}$$

For a given outer Choquet capacity we define the *inner Choquet capacity* on G, $\text{cap}_* : \mathcal{P}(G) \to \mathbb{R}_+ \cup \{\infty\}$, by

$$\text{cap}_*(A) := \sup\{\text{cap}^*(B); B \subset A \quad \text{and } B \text{ compact}\}. \tag{3.9}$$

Note that (3.7) implies

$$\text{cap}_*(A) \leq \text{cap}^*(A) \quad \text{for all } A \subset G. \tag{3.10}$$

Definition 3.1.9. Let G be a Polish space and cap^* an outer Choquet capacity on G. A set $A \subset G$ is called *capacitable* with respect to cap^* if

$$\text{cap}^*(A) = \text{cap}_*(A) \tag{3.11}$$

holds. In this case we write $\text{cap}(A) := \text{cap}^*(A)$ and call $\text{cap}(A)$ the *Choquet capacity* of the set $A \subset G$ (with respect to a given outer capacity).

Remark 3.1.10. A. Instead of outer Choquet capacity, Choquet capacity, etc, we will often write just outer capacity, capacity etc.

B. Clearly, all definitions make sense for a general Hausdorff space and we will use this fact sometimes, i.e. we will study sometimes outer capacities on Hausdorff spaces which are not Polish.

C. Note that there is some redundancy in Definition 3.1.8. Using (3.7) we may deduce (3.6): Consider the increasing sequence $A_1, A_2, A_2, \ldots, A_1 \subset A_2$. It follows that

$$\begin{aligned}\text{cap}^*(A_2) &= \text{cap}^*\Big(\bigcup_{\nu\in\mathbb{N}} A_\nu\Big) = \sup_{\nu\in\mathbb{N}} \text{cap}^*(A_\nu)\\ &= \max\{\text{cap}^*(A_1), \text{cap}^*(A_2)\} \geq \text{cap}^*(A_1).\end{aligned}$$

We decided to put (3.6) into the definition since this is most common in the literature.

Our next aim is to prove that all analytic subsets of G are capacitable, thus in particular all Borel sets are capacitable. For this we need some preparations which we take from R.F.Bass [17].

Lemma 3.1.11. *In a Hausdorff space G all $K_{\sigma\delta}$-sets are for all outer capacities capacitable.*

Proof: Let A be a $K_{\sigma\delta}$-set, i.e. for compact sets A_{kl} we have

$$A = \bigcap_{k\in\mathbb{N}} A_k = \bigcap_{k\in\mathbb{N}} \bigcup_{l\in\mathbb{N}} A_{kl},$$

and without loss of generality we may assume that for each k the family $(A_{kl})_{l\in\mathbb{N}}$ is increasing. Now let cap^* be an outer capacity on G and suppose that $\alpha < \mathrm{cap}^*(A)$. We will construct a compact set $B \subset A$ such that $\mathrm{cap}^*(B) \geq \alpha$ which will prove the lemma. Indeed, suppose that in this situation $\mathrm{cap}_*(A) < \mathrm{cap}^*(A)$ holds. Then there exists α such that

$$\mathrm{cap}_*(A) < \alpha < \mathrm{cap}^*(A),$$

but in addition we have according to (3.9)

$$\alpha \leq \mathrm{cap}^*(B) \leq \sup\{\mathrm{cap}^*(B'); B' \subset A \quad \text{compact}\} = \mathrm{cap}_*(A) < \alpha,$$

which gives a contradiction. By induction we construct first compact sets $B_n \subset A_n$, $n \in \mathbb{N}$, such that for $D_n := A \cap B_1 \cap \ldots \cap B_n$ we have $\mathrm{cap}^*(D_n) > \alpha$. From (3.7) we deduce

$$\mathrm{cap}^*(A) = \mathrm{cap}^*(A \cap A_1) = \sup_{l\in\mathbb{N}} \mathrm{cap}^*(A \cap A_{1l}).$$

We choose now l_1 large enough in order to have that $\mathrm{cap}^*(A \cap A_{1l_1}) > \alpha$ and set $B_1 := A_{1l_1}$ as well as $D_1 := A \cap B_1$. Assume that $B_1, \ldots, B_{n-1}$ and D_{n-1} are already constructed. Since $A \subset A_n$ it follows that

$$\alpha < \mathrm{cap}^*(D_{n-1}) = \mathrm{cap}^*(D_{n-1} \cap A_n)$$

and therefore, once again by (3.7)

$$\mathrm{cap}^*(D_{n-1}) = \sup_{l\in\mathbb{N}} \mathrm{cap}^*(D_{n-1} \cap A_{nl}).$$

Now we take l_n large enough to get $\text{cap}^*(D_{n-1} \cap A_{nl_n}) > \alpha$, and define $B_n := A_{nl_n}$ and $D_n := D_{n-1} \cap B_n$. Thus we find $D_n \subset A_n$ and $\text{cap}^*(D_n) > \alpha$. Now we put $E_n := B_1 \cap \ldots \cap B_n$ and $B := \bigcap_{n\in\mathbb{N}} B_n = \bigcap_{n\in\mathbb{N}} E_n$. Since each set E_n is compact and the family $(E_n)_{n\in\mathbb{N}}$ decreases, it follows that B is compact. Further we have $D_n \subset E_n$ implying $\text{cap}^*(E_n) \geq \text{cap}^*(D_n) > \alpha$. Now we use (3.8) to get $\text{cap}^*(B) \geq \alpha$. But in addition we have $B_n \subset A_n$ which yields $B \subset A$ and the lemma is proved. □

In proving the next result we use the definition of analyticity given in Remark 3.1.3A.

Lemma 3.1.12. *Let $A \subset G$ be an analytic set and let C be a $K_{\sigma\delta}$-set in the space $H \times G$ where H is a compact Hausdorff space having the property that $\pi(C) = A$, here $\pi : H \times G \to G$ is the natural projection. On $\mathcal{P}(H \times G)$ we define the function*

$$c^*(B) := \text{cap}^*(\pi(B)), \quad B \in \mathcal{P}(H\times G). \tag{3.12}$$

Then c^* is an outer capacity on $H \times G$.

Proof: It is clear that c^* satisfies (3.6) and (3.7). We are going to prove (3.8). First note that for any decreasing sequence $(B_n)_{n\in\mathbb{N}}$ of compact sets $B_n \subset H \times G$ it follows that $\pi(\bigcap_{n\in\mathbb{N}} B_n) = \bigcap_{n\in\mathbb{N}} \pi(B_n)$. Since we always have $\pi(\bigcap_{n\in\mathbb{N}} B_n) \subset \bigcap_{n\in\mathbb{N}} \pi(B_n)$ it remains to prove the opposite inclusion. Let $x \in \bigcap_{n\in\mathbb{N}} \pi(B_n)$. We are looking for $y \in \bigcap_{n\in\mathbb{N}} B_n$ such that $\pi(y) = x$. Since for any $x \in \bigcap_{n\in\mathbb{N}} \pi(B_n)$ the family $((\pi^{-1}(x) \cap B_n)_{n\in\mathbb{N}}$ has the finite intersection property, i.e. for any finite sequence we have that $\bigcap_{m=1}^{l}((\pi^{-1}(x)) \cap B_{n_m})$ is non-empty, it follows that $\bigcap_{n=1}^{\infty}(\pi^{-1}(x) \cap B_n)$ is non-empty, and any $y \in \bigcap_{n=1}^{\infty}(\pi^{-1}(x) \cap B_n)$ leads to $\pi(y) = x$. Thus it follows

$$c^*\big(\bigcap_{n\in\mathbb{N}} B_n\big) = \text{cap}^*\,\big(\pi(\bigcap_{n\in\mathbb{N}} B_n)\big) = \text{cap}^*\,\big(\bigcap_{n\in\mathbb{N}} \pi(B_n)\big).$$

The continuity of π implies that for compact sets B_n the sets $\pi(B_n)$ are compact too. Thus we have

$$\text{cap}^*\,\big(\bigcap_{n\in\mathbb{N}} \pi(B_n)\big) = \inf_{n\in\mathbb{N}} \text{cap}^*(\pi(B_n)) = \inf c^*(B_n),$$

proving the lemma. □

Now we are in a position to prove

Theorem 3.1.13. *Let G be a Polish space and* cap^* *any outer Choquet capacity, then every analytic set A is capacitable with respect to* cap^*.

Proof: Since $A \subset G$ is analytic there exists a $K_{\sigma\delta}$-set C in the space $H \times G$ where H is a compact Hausdorff space having the property that $\pi(C) = A$ and $\pi : H \times G \to G$ is the natural projection. From Lemma 3.1.11 and Lemma 3.1.12 it follows that C is capacitable with respect to the outer capacity c^* defined by (3.12). Assume that $\text{cap}^*(A) > \alpha$. If we can show that there exists a compact set $E \subset A$ such that $\text{cap}^*(E) \geq \alpha$ we may conclude as in the proof of Lemma 3.1.11 that $\text{cap}_*(A) \geq \text{cap}^*(A)$, implying that $\text{cap}_*(A) = \text{cap}^*(A)$, i.e. the capacitability of A. Now, since $\text{cap}^*(A) = c^*(C) > \alpha$ and C is capacitable, there exists some compact set $J \subset C$ such that $c^*(J) \geq \alpha$, but then $E := \pi(J)$ satisfy $\text{cap}^*(E) = c^*(J) \geq \alpha$ and $E \subset A$ is compact. □

Corollary 3.1.14. *The Borel sets in a Polish space are for all outer capacities capacitable.*

In our definition of an outer capacity we started with a set function defined on $\mathcal{P}(G)$. Often it happen that we can define easily a function c^* on a subfamily of $\mathcal{P}(G)$. In this case the question arises whether we may extend c^* to the whole family $\mathcal{P}(G)$ as an outer capacity. The following result is taken from the monograph [102] of M.Fukushima et. al., see also C.Dellacherie and P.-A.Meyer [64].

Theorem 3.1.15. *Let G be a Hausdorff space and denote by $\mathcal{O}_G$ the family of its open sets. Assume for a function $c^* : \mathcal{O}_G \to \mathbb{R}_+ \cup \{\infty\}$*

$$\text{if } A, B \in \mathcal{O}_G \text{ and } A \subset B \text{ then } c^*(A) \leq c^*(B); \tag{3.13}$$

$$\text{for all } A, B \in \mathcal{O}_G \text{ we have } c^*(A \cup B) + c^*(A \cap B) \leq c^*(A) + c^*(B); \tag{3.14}$$

$$\text{if } (A_n)_{n\in\mathbb{N}} \text{ is an increasing sequence in } \mathcal{O}_G \text{ then } c^*\Big(\bigcup_{n\in\mathbb{N}} A_n\Big) = \sup_{n\in\mathbb{N}} c^*(A_n). \tag{3.15}$$

Then the function

$$\text{cap}^*(B) := \inf\{c^*(A); A \in \mathcal{O}_G \quad \text{and} \quad B \subset A\}, \quad B \in \mathcal{P}(G), \tag{3.16}$$

is an outer Choquet capacity extending c^* to $\mathcal{P}(G)$. Furthermore cap^* *is countable subadditive.*

Note that (3.15) implies (3.13), but we consider it to be convenient to state it separately.

We prove first an auxiliary result.

Lemma 3.1.16. *Let* c^* *be as in Theorem* 3.1.15 *and denote the family* $\{A \in \mathcal{O}_G; c^*(A) < \infty\}$ *by* $\mathcal{O}_{G,finite}$. *For any pair of sequences* $(A_n)_{n\in\mathbb{N}}$, $(B_n)_{n\in\mathbb{N}}$ *such that* $A_n \subset B_n$ *and* $A_n, B_n \in \mathcal{O}_{G,finite}$ *we have*

$$c^*\Big(\bigcup_{k=1}^{n} B_k\Big) - c^*\Big(\bigcup_{k=1}^{n} A_k\Big) \le \sum_{k=1}^{n}(c^*(B_k) - c^*(A_k)). \tag{3.17}$$

Proof: We prove (3.17) by induction. For $n = 1$ this is just a triviality. Now suppose that (3.17) holds for $n \in \mathbb{N}$. We set

$$\tilde{A}_n := \bigcup_{k=1}^{n} A_n \quad \text{and} \quad \tilde{B}_n := \bigcup_{k=1}^{n} B_n.$$

Since $\tilde{B}_n \cup B_{n+1} = \tilde{B}_n \cup (B_{n+1} \cup \tilde{A}_n)$ and $\tilde{A}_n \subset \tilde{B}_n \cap (B_{n+1} \cup \tilde{A}_n)$ it follows from (3.13) and (3.14) that

$$\begin{aligned} c^*(\tilde{B}_n \cup B_{n+1}) &+ c^*(\tilde{A}_n) \\ &\le c^*(\tilde{B}_n \cup (\tilde{B}_{n+1} \cup \tilde{A}_n)) + c^*(\tilde{B}_n \cap (B_{n+1} \cup \tilde{A}_n)) \\ &\le c^*(\tilde{B}_n) + c^*(B_{n+1} \cup \tilde{A}_n), \end{aligned} \tag{3.18}$$

and further we find by the same reasoning

$$c^*(B_{n+1} \cup \tilde{A}_n) + c^*(A_{n+1}) \le c^*(B_{n+1}) + c^*(\tilde{A}_n \cup A_{n+1}). \tag{3.19}$$

Adding (3.18) and (3.19) yields

$$\begin{aligned} c^*(\tilde{B}_n \cup B_{n+1}) &+ c^*(\tilde{A}_n) + c^*(B_{n+1} \cup \tilde{A}) + c^*(A_{n+1}) \\ &\le c^*(\tilde{B}_n) + c^*(B_{n+1} \cup \tilde{A}_n) + c^*(B_{n+1}) + c^*(\tilde{A}_n \cup A_{n+1}), \end{aligned}$$

or, since $\tilde{B}_n \cup B_{n+1} = \tilde{B}_{n+1}$ and $\tilde{A}_n \cup A_{n+1} = \tilde{A}_{n+1}$, the induction assumption implies

$$\begin{aligned} c^*(\tilde{B}_{n+1}) - c^*(\tilde{A}_{n+1}) &\le c^*(\tilde{B}_n) - c^*(\tilde{A}_n) + c^*(B_{n+1}) - c^*(A_{n+1}) \\ &\le \sum_{k=1}^{n}(c^*(B_k) - c^*(A_k)) + c^*(B_{n+1}) - c^*(A_{n+1}) \\ &= \sum_{k=1}^{n+1}(c^*(B_k) - c^*(A_k)), \end{aligned}$$

and the lemma is proved. □

Proof of Theorem 3.1.15: Clearly cap* is an extension of c^* to $\mathcal{P}(G)$ which is monotone in the sense that

$$\operatorname{cap}^*(B_1) \leq \operatorname{cap}^*(B_2) \quad \text{for } B_1 \subset B_2,$$

and which is subadditive, i.e.

$$\operatorname{cap}^*(B_1 \cup B_2) \leq \operatorname{cap}^*(B_1) + \operatorname{cap}^*(B_2).$$

Next we prove that for any increasing sequence $(A_n)_{n\in\mathbb{N}}$, $A_n \subset G$, it follows that

$$\operatorname{cap}^*\left(\bigcup_{n\in\mathbb{N}} A_n\right) = \sup_{n\in\mathbb{N}} \operatorname{cap}^*(A_n) \tag{3.20}$$

holds. Since cap* is monotone and subadditive it is sufficient to prove under the assumption that $\sup\limits_{n\in\mathbb{N}} \operatorname{cap}^*(A_n) < \infty$ it follows that

$$\operatorname{cap}^*\left(\bigcup_{n\in\mathbb{N}} A_n\right) \leq \sup_{n\in\mathbb{N}} \operatorname{cap}^*(A_n).$$

By the definition of cap* for any $\varepsilon > 0$ there exists $U_n \in \mathcal{O}_{G,finite}$ such that $A_n \subset U_n$ and

$$\operatorname{cap}^*(A_n) \leq c^*(U_n) \leq \operatorname{cap}^*(A_n) + \frac{\varepsilon}{2^n}, \tag{3.21}$$

which yields

$$\lim_{n\to\infty} c^*(U_n) = \lim_{n\to\infty} \operatorname{cap}^*(A_n). \tag{3.22}$$

Furthermore, for $m < n$ it follows that

$$\operatorname{cap}^*(A_m) \leq c^*(U_m \cap U_n) \leq c^*(U_m) \leq \operatorname{cap}^*(A_m) + \frac{\varepsilon}{2^m},$$

and consequently we have

$$\sum_{k=1}^{n} (c^*(U_k) - c^*(U_k \cap U_n)) < \varepsilon.$$

Now we apply Lemma 3.1.16, i.e. (3.17), to get

$$c^*\Big(\bigcup_{k=1}^{n} U_k\Big) - c^*(U_n) = c^*\Big(\bigcup_{k=1}^{n} U_k\Big) - c^*\Big(\bigcup_{k=1}^{n}(U_k \cap U_n)\Big)$$
$$\leq \sum_{k=1}^{n}(c^*(U_k) - c^*(U_k \cap U_n)) < \varepsilon,$$

which leads to

$$\lim_{n\to\infty} c^*\Big(\bigcup_{k=1}^{n} U_k\Big) \leq \lim_{k\to\infty} c^*(U_k)+\varepsilon. \tag{3.23}$$

For the set $U := \bigcup_{k=1}^{\infty} U_k$ we have $U \in \mathcal{O}_G$ and $A \subset U$. Thus (3.15) implies

$$\text{cap}^*(A) \leq c^*(U) = \lim_{n\to\infty} c^*\Big(\bigcup_{k=1}^{n} U_k\Big). \tag{3.24}$$

Combining (3.22) and (3.23) we obtain

$$\text{cap}^*(A) \leq \lim_{n\to\infty} \text{cap}^*(A_n)+\varepsilon,$$

which gives (3.20). Finally we prove for a decreasing sequence of a compact sets $A_n \subset G$, $n \in \mathbb{N}$, that

$$\text{cap}^*\Big(\bigcap_{n=1}^{\infty} A_n\Big) = \inf_{n\in\mathbb{N}} \text{cap}^*(A_n) \tag{3.25}$$

holds. Again it is sufficient to assume that $\text{cap}^*(\bigcap_{n=1}^{\infty} A_n) < \infty$ and to prove an inequality, namely,

$$\inf_{n\in\mathbb{N}} \text{cap}^*(A_n) \leq \text{cap}^*\Big(\bigcap_{n=1}^{\infty} A_n\Big). \tag{3.26}$$

Now, for $\varepsilon > 0$ there exists $U \in \mathcal{O}_{G,finite}$ such that $\bigcap_{n=1}^{\infty} A_n \subset U$ and $c^*(U) \leq \text{cap}^*(\bigcap_{n=1}^{\infty} A_n) + \varepsilon$. But $A_n \subset U$ from some n and (3.26) follows. Finally, let us remark that the subadditivity of cap* and (3.20) imply the countable subadditivity of cap* and the theorem is proved. □

The first class of capacities we are going to consider are capacities which are associated to an L^p-sub-Markovian semigroup $(T_t^{(p)})_{t\geq 0}$. In particular we will use these capacities to control exceptional sets. We start with

Definition 3.1.17. Let $(T_t^{(p)})_{t\geq 0}$ be an L^p-sub-Markovian semigroup on $L^p(\mathbb{R}^n;\mathbb{R})$. We define its *gamma-transform* $(V_r^{(p)})_{r\geq 0}$ by

$$V_r^{(p)}u := \frac{1}{\Gamma(\frac{r}{2})}\int_0^\infty t^{r/2-1}e^{-t}T_t^{(p)}u\,dt, \quad u\in L^p(\mathbb{R}^n;\mathbb{R}). \tag{3.27}$$

Remark 3.1.18. Clearly, by

$$\eta_r(ds) = \chi_{[0,\infty)}(s)\frac{1}{\Gamma(\frac{r}{2})}s^{r/2-1}e^{-s}\lambda^{(1)}(ds) \tag{3.28}$$

a convolution semigroup $(\eta_r)_{r\geq 0}$ with support in $[0,\infty)$ is given, namely the (modified) Γ-semigroup, see Example I.3.9.15. It is associated with the complete Bernstein function

$$f(s) = \frac{1}{2}\log(1+s). \tag{3.29}$$

Thus the gamma-transform $(V_r^{(p)})_{r\geq 0}$ of any L^p-sub-Markovian semigroup $(T_t^{(p)})_{t\geq 0}$ is once again an L^p-sub-Markovian semigroup which is obtained from the original one by subordination. Therefore we have

$$V_{r_1}^{(p)}\circ V_{r_2}^{(p)} = V_{r_1+r_2}^{(p)}; \tag{3.30}$$

$$\|V_r^{(p)}u\|_{L^p} \leq \|u\|_{L^p}; \tag{3.31}$$

$$\lim_{r\to 0} V_r^{(p)}u = u \quad \text{in } L^p(\mathbb{R}^n;\mathbb{R}); \tag{3.32}$$

and

$$0\leq u\leq 1 \quad \text{a.e. implies } 0\leq V_r^{(p)}u\leq 1 \text{ a.e.} \tag{3.33}$$

The following result is taken from our joint paper [86] with W.Farkas and R.Schilling.

Theorem 3.1.19. *Let $(T_t^{(p)})_{t\geq 0}$ be an L^p-sub-Markovian semigroup with generator $(A^{(p)}, D(A^{(p)}))$. For all $r>0$ and all $u\in L^p(\mathbb{R}^n;\mathbb{R})$ we have*

$$V_r^{(p)}u = (\mathrm{id} - A^{(p)})^{-r/2}u. \tag{3.34}$$

In particular, each of the operators $V_r^{(p)}$ is injective.

Proof: Let f be as in Remark 3.1.18 the Bernstein function $f(s) = \frac{1}{2}\log(1+s)$. We know already that $V_r^{(p)} = T_r^{(p),f}$ where $(T_r^{(p),f})_{r\geq 0}$ denotes the semigroup obtained by subordinating $(T_r^{(p)})_{r\geq 0}$ with respect to f (or $(\eta_r)_{r\geq 0}$), see Section I.4.3. On $D(A^{(p)})$ the resolvent of $(A^{(p)}, D(A^{(p)}))$ satisfies for $\lambda > 0$

$$((\lambda+1)\,\mathrm{id} - A^{(p)})^{-1}u = \int_0^\infty e^{-(\lambda+1)t} T_t^{(p)} u\, dt.$$

Using the Dunford calculus for unbounded operators, see Section I.2.7, applied to the operator $\mathrm{id} - A^{(p)}$ we find

$$(\mathrm{id} - A^{(p)})^{-s} = \frac{1}{2\pi i}\int_\gamma \xi^{-s}((\xi+1)\,\mathrm{id} - A^{(p)})^{-1} u\, d\xi. \tag{3.35}$$

Here the path $\gamma \subset \rho(\mathrm{id} - A^{(p)})$ surrounds the spectrum $\sigma(\mathrm{id} - A^{(p)})$, and is negative oriented, i.e. clockwise. Since the integrand is analytic, we can deform it to get

$$(\mathrm{id} - A^{(p)})^{-s} = \frac{\sin s\pi}{\pi}\int_0^\infty \lambda^{-s}((\lambda+1)\,\mathrm{id} - A^{(p)})^{-1} u\, d\lambda. \tag{3.36}$$

By the resolvent formula we have

$$\begin{aligned}(\mathrm{id} - A^{(p)})^{-s} &= \frac{\sin s\pi}{\pi}\int_0^\infty \lambda^{-s}\Big(\int_0^\infty e^{-(\lambda+1)t} T_t^{(p)} u\, dt\Big)\, d\lambda \\ &= \frac{\sin s\pi}{\pi}\int_0^\infty\int_0^\infty \lambda^{-s} e^{-(\lambda+1)t} T_t^{(p)} u\, dt\, d\lambda.\end{aligned}$$

Observe that the term $\lambda^{-s}e^{-(\lambda+1)t}$ is positive. Thus Tonnelli's theorem yields for $s \in (0,1)$

$$\begin{aligned}\int_0^\infty\int_0^\infty \lambda^{-s}e^{-(\lambda+1)t}\, dt\, d\lambda &= \int_0^\infty\int_0^\infty \lambda^{-s}e^{-(\lambda+1)t}\, d\lambda\, dt \\ &- \int_0^\infty\int_0^\infty \Big(\frac{\mu}{t}\Big)^{-s} e^{-\mu} e^{-t}\frac{d\mu}{t}\, dt \\ &= \int_0^\infty \mu^{-s}e^{-\mu}\, d\mu \int_0^\infty t^{s-1}e^{-t}\, dt < \infty.\end{aligned}$$

Since $\|\lambda^{-s}e^{-(\lambda+1)t}T_t^{(p)}u\|_{L^p} \leq \lambda^{-s}e^{-(\lambda+1)t}\|u\|_{L^p}$, we find for $0 < s < 1$ that

$$\begin{aligned}(\mathrm{id} - A^{(p)})^{-s}u &= \frac{\sin s\pi}{\pi}\int_0^\infty \int_0^\infty \lambda^{-s}e^{-(\lambda+1)t}\, d\lambda T_t^{(p)}u\, dt \\ &= \frac{\sin s\pi}{\pi}\int_0^\infty \int_0^\infty \lambda^{-s}e^{-\lambda t}\, d\lambda e^{-t}T_t^{(p)}u\, dt \\ &= \frac{\sin s\pi}{\pi}\int_0^\infty \int_0^\infty \mu^{-s}e^{-\mu}\, d\mu\;\, t^{s-1}e^{-t}T_t^{(p)}u\, dt \\ &= \frac{1}{\Gamma(s)}\int_0^\infty t^{s-1}e^{-t}T_t^{(p)}u\, dt = V_{2s}^{(p)}u,\end{aligned}$$

where we used the formula $\Gamma(s)\Gamma(1-s) = \frac{\pi}{\sin s\pi}$, $0 < s < 1$.

So far we have proved for $0 < r < 2$ that

$$(\mathrm{id} - A^{(p)})^{-r/2}u = V_r^{(p)}u, \quad u \in D(A^{(p)}). \tag{3.37}$$

Using the functional calculus, see Section I.2.7, on the left hand side, and the semigroup property on the right-hand side, we see that (3.37) extends to all $r > 0$. Finally, the strong continuity of the semigroup $(V_r^{(p)})_{r\geq 0}$ yields that (3.37) holds also for $r = 0$ and the theorem is proved. □

Remark 3.1.20. From Example I.4.3.27 it follows that for every strongly continuous contraction semigroup $(T_t)_{t\geq 0}$ its gamma-transform is an analytic semigroup.

Using the injectivity of $V_r^{(p)}$ we may define the following *Bessel potential type spaces*:

$$\mathcal{F}_{r,p}(\mathbb{R}^n;\mathbb{R}) := V_r^{(p)}(L^p(\mathbb{R}^n;\mathbb{R})) \tag{3.38}$$

with norm

$$\|u\|_{\mathcal{F}_{r,p}} := \|v\|_{L^p} \quad \text{for } u = V_r^{(p)}v. \tag{3.39}$$

Clearly, the spaces $(\mathcal{F}_{r,p}(\mathbb{R}^n;\mathbb{R}); \|\cdot\|_{\mathcal{F}_{r,p}})$ are separable Banach spaces. Moreover we have

Corollary 3.1.21. *In the situation of Theorem* 3.1.19 *it holds*

$$\mathcal{F}_{r,p}(\mathbb{R}^n;\mathbb{R}) = D((\mathrm{id} - A^{(p)})^{r/2}). \tag{3.40}$$

Proof: Obviously we have $u \in \mathcal{F}_{r,p}(\mathbb{R}^n;\mathbb{R})$ if and only if $u = V_r^{(p)} v$ for some $v \in L^p(\mathbb{R}^n;\mathbb{R})$. Since $V_r^{(p)} = (\mathrm{id} - A^{(p)})^{-r/2}$ we have $u \in \mathcal{F}_{r,p}(\mathbb{R}^n;\mathbb{R})$ if and only if $u = (\mathrm{id} - A^{(p)})^{-r/2} v$ which implies $D((\mathrm{id} - A^{(p)})^{r/2}) = \mathcal{F}_{r,p}(\mathbb{R}^n;\mathbb{R})$. □

Our next aim is to prove the equivalence of the norms $\|\cdot\|_{\mathcal{F}_{r,p}}$ and $\|(-A^{(p)})^{r/2}\cdot\|_{L^p} + \|\cdot\|_{L^p}$. For this we show first

Lemma 3.1.22. *Let $(A^{(p)}, D(A^{(p)}))$ be the generator of an L^p-sub-Markovian semigroup $(T_t^{(p)})_{t\geq 0}$. For all $r > 0$ and all $u \in D((\mathrm{id} - A^{(p)})^{r/2}) = \mathcal{F}_{r,p}(\mathbb{R}^n;\mathbb{R})$ we have*

$$\|u\|_{L^p} \leq \|(\mathrm{id} - A^{(p)})^{r/2} u\|_{L^p}. \tag{3.41}$$

Moreover, if $s, r \geq 0$ then there is a continuous embedding

$$\mathcal{F}_{r+s,p}(\mathbb{R}^n;\mathbb{R}) \hookrightarrow \mathcal{F}_{r,p}(\mathbb{R}^n;\mathbb{R}). \tag{3.42}$$

Proof: Since $\mathrm{id} = (\mathrm{id} - A^{(p)})^{-r/2}(\mathrm{id} - A^{(p)})^{r/2} = V_r^{(p)}(\mathrm{id} - A^{(p)})^{r/2}$ it follows by the contraction property of $V_r^{(p)}$ that

$$\|u\|_{L^p} = \|V_r^{(p)}(\mathrm{id} - A^{(p)})^{r/2} u\|_{L^p} \leq \|(\mathrm{id} - A^{(p)})^{r/2} u\|_{L^p},$$

whereas (3.42) follows from

$$\begin{aligned}\|(\mathrm{id} - A^{(p)})^{r/2} u\|_{L^p} &= \|(\mathrm{id} - A^{(p)})^{-s/2}(\mathrm{id} - A^{(p)})^{\frac{s+r}{2}} u\|_{L^p} \\ &= \|V_s^{(p)}(\mathrm{id} - A^{(p)})^{\frac{s+r}{2}} u\|_{L^p} \leq \|(\mathrm{id} - A^{(p)})^{\frac{s+r}{2}} u\|_{L^p}.\end{aligned}$$ □

Lemma 3.1.23. *Let A, B be two generators of L^p-sub-Markovian semigroups such that $A - B$ is a bounded operator which extends to $L^p(\mathbb{R}^n;\mathbb{R})$. Then we have*

$$\|(-A)^\alpha - (-B)^\alpha\| \leq \|A - B\|. \tag{3.43}$$

Proof: For $\lambda > 0$ and $u \in L^p(\mathbb{R}^n;\mathbb{R})$ we find

$$\begin{aligned}(-A)R_\lambda^A u - (-B)R_\lambda^B u &= u - \lambda R_\lambda^A u - u + \lambda R_\lambda^B u \\ &= \lambda R_\lambda^B u - \lambda R_\lambda^A u = \lambda R_\lambda^A (B - A) R_\lambda^B u,\end{aligned}$$

which implies

$$\|(-A)R_\lambda^A - (-B)R_\lambda^B\| \leq \|\lambda R_\lambda^A\| + \|\lambda R_\lambda^B\| \leq 2$$

as well as

$$\|(-A)R_\lambda^A-(-B)R_\lambda^B\| \le \|\lambda R_\lambda^A\|\|B-A\|\|R_\lambda^B\| \le \frac{\|A-B\|}{\lambda}.$$

Now, for any $\varepsilon > 0$ we have

$$\lambda \le \frac{\varepsilon}{2}\|A-B\| \quad \text{implies} \quad 2 \le (2+\varepsilon)\frac{\|A-B\|}{\lambda+\|A-B\|}$$

and

$$\lambda \ge \frac{\varepsilon}{2}\|A-B\| \quad \text{implies} \quad \frac{\|A-B\|}{\lambda} \le \frac{2+\varepsilon}{2}\frac{\|A-B\|}{\lambda+\|A-B\|}$$

which yields

$$\begin{aligned}
\|(-A)^\alpha-(-B)^\alpha\| &\le \frac{\sin\alpha\pi}{\pi}\int_0^\infty \lambda^{\alpha-1}\|(-A)R_\lambda^A-(-B)R_\lambda^B\|\,d\lambda \\
&= \frac{\sin\alpha\pi}{\pi}\int_0^{\frac{\varepsilon}{2}\|A-B\|} \lambda^{\alpha-1}\|(-A)R_\lambda^A-(-B)R_\lambda^B\|\,d\lambda \\
&\quad + \frac{\sin\alpha\pi}{\pi}\int_{\frac{\varepsilon}{2}\|A-B\|}^\infty \lambda^{\alpha-1}\|(-A)R_\lambda^A-(-B)R_\lambda^B\|\,d\lambda \\
&\le \frac{\sin\alpha\pi}{\pi}\int_0^{\frac{\varepsilon}{2}\|A-B\|} (2+\varepsilon)\frac{\|A-B\|}{\lambda+\|A-B\|}\lambda^{\alpha-1}\,d\lambda \\
&\quad + \frac{\sin\alpha\pi}{\pi}\int_{\frac{\varepsilon}{2}\|A-B\|}^\infty \left(\frac{2+\varepsilon}{2}\right)\frac{\|A-B\|}{\lambda+\|A-B\|}\lambda^{\alpha-1}d\lambda \\
&\le \frac{\sin\alpha\pi}{\pi}(2+\varepsilon)\frac{1}{\alpha}\left(\frac{\varepsilon}{2}\right)^\alpha\|A-B\|^\alpha \\
&\quad + \left(\frac{2+\varepsilon}{2}\right)\frac{\sin\alpha\pi}{\pi}\int_0^\infty \frac{\|A-B\|}{\lambda+\|A-B\|}\lambda^{\alpha-1}d\lambda,
\end{aligned}$$

and for $\varepsilon \to 0$ we find

$$\|(-A)^\alpha-(-B)^\alpha\| \le \|A-B\|^\alpha$$

and the lemma is proved. □

Corollary 3.1.24. *Let $(A^{(p)}, D(A^{(p)}))$ be the generator of an L^p-sub-Markovian semigroup. Then $(A^{(p)} - \mathrm{id}, D(A^{(p)}))$ is also a generator of an L^p-sub-Markovian semigroup and for $0 < \alpha \le 1$ we have*

$$\|(-A)^\alpha-(\mathrm{id}-A)^\alpha\| \le 1. \tag{3.44}$$

Theorem 3.1.25. *Let $(A^{(p)}, D(A^{(p)}))$ be the generator of an L^p-sub-Markovian semigroup $(T_t^{(p)})_{t\geq 0}$. Further let $0 < r \leq 1$. Then for all $u \in D(A^{(p)})$ we have*

$$\begin{aligned}\frac{1}{3}(\|(-A^{(p)})^r u\|_{L^p} + \|u\|_{L^p}) &\leq \|(\mathrm{id} - A^{(p)})^r u\|_{L^p} \\ &\leq \|(-A^{(p)})^r u\|_{L^p} + \|u\|_{L^p}.\end{aligned} \tag{3.45}$$

In particular we find $D((-A^{(p)})^r) = D((\mathrm{id} - A^{(p)})^r)$.

Proof: Using (3.44) we find

$$\begin{aligned}\|(-A^{(p)})^r u\|_{L^p} - \|(\mathrm{id} - A^{(p)})^r u\|_{L^p} &\leq \|(-A^{(p)})^r u - (\mathrm{id} - A^{(p)})^r u\|_{L^p} \\ &\leq \|u\|_{L^p}\end{aligned}$$

which gives using (3.41)

$$\begin{aligned}\|(-A^{(p)})^r u\|_{L^p} + \|u\|_{L^p} &\leq \|(\mathrm{id} - A^{(p)})^r u\|_{L^p} + 2\|u\|_{L^p} \\ &\leq 3\|(\mathrm{id} - A^{(p)})^r u\|_{L^p}.\end{aligned}$$

On the other hand we have

$$\begin{aligned}\|(\mathrm{id} - A^{(p)})^r u\|_{L^p} - \|(-A^{(p)})^r u\|_{L^p} &\leq \|(\mathrm{id} - A^{(p)})^r u - (-A^{(p)})^r u\|_{L^p} \\ &\leq \|u\|_{L^p},\end{aligned}$$

or

$$\|(\mathrm{id} - A^{(p)})^r u\|_{L^p} \leq \|(-A^{(p)})^r u\|_{L^p} + \|u\|_{L^p}$$

and (3.45) follows. Since $D(A^{(p)}) = D(\mathrm{id} - A^{(p)})$ and since $D((-A^{(p)})^r)$ and $D((\mathrm{id} - A^{(p)})^r)$ are obtained by completion of $D(A^{(p)})$ with respect to the equivalent norms

$$\|(-A^{(p)})^r u\|_{L^p} + \|u\|_{L^p} \quad \text{and} \quad \|(\mathrm{id} - A^{(p)})^r u\|_{L^p},$$

respectively, we have proved the theorem noting that $D(A^{(p)}) = D(\mathrm{id} - A^{(p)})$ is an operator core for both $(-A^{(p)})^r$ and $(\mathrm{id} - A^{(p)})^r$. $\square$

We want to extend Theorem 3.1.25 to $k + r$, $k \in \mathbb{N}_0$, and $0 \leq r < 1$.

Theorem 3.1.26. *Let $(A^{(p)}, D(A^{(p)}))$ be as in Theorem 3.1.25 and let $k \in \mathbb{N}_0$ and $0 \leq r < 1$. Then there exists constants γ_1, γ_2 such that for any $u \in D((-A^{(p)})^{k+1})$*

$$\begin{aligned}\gamma_1(\|u\|_{L^p} + \|(-A^{(p)})^{k+r} u\|_{L^p}) &\leq \|(\mathrm{id} - A^{(p)})^{k+r} u\|_{L^p} \\ &\leq \gamma_2(\|u\|_{L^p} + \|(-A^{(p)})^{k+r} u\|_{L^p}).\end{aligned} \tag{3.46}$$

Furthermore, we have $D((-A^{(p)})^{k+r}) = D((\mathrm{id} - A^{(p)})^{k+r})$.

Proof: We will use a result due to L.Hörmander namely Theorem I.2.7.10. For this assume for a moment that

$$D((-A^{(p)})^s) = D((\mathrm{id} - A^{(p)})^s) \tag{3.47}$$

holds for all $s \geq 0$. It follows that the operators

$$((-A^{(p)})^s, D((-A^{(p)})^s) \quad \text{and} \quad ((\mathrm{id} - A^{(p)})^s, D((\mathrm{id} - A^{(p)})^s)), \quad s > 0,$$

are closed and we obtain from (I.2.154)

$$\|(-A^{(p)})^s u\|_{L^p} \leq c_1(\|(\mathrm{id} - A^{(p)})^s u\|_{L^p} + \|u\|_{L^p}), \quad s > 0,$$

as well as

$$\|(\mathrm{id} - A^{(p)})^s u\|_{L^p} \leq c_2(\|(\mathrm{id} - A^{(p)})^s u\|_{L^p} + \|u\|_{L^p}), \quad s > 0,$$

implying the theorem. Thus it remains to prove (3.47). For any generator $(A, D(A))$ of a strongly continuous contraction semigroup we have for $s, t \geq 0$

$$\begin{aligned} D((-A)^{s+t}) &= \{u \in D((-A)^t); (-A)^t u \in D((-A)^s)\} \\ &= \{u \in D((-A)^s); (-A)^s u \in D((-A)^t)\}. \end{aligned}$$

Thus, assuming that (3.47) holds for $s = k + r$ we find

$$\begin{aligned} D((-A^{(p)})^{k+1+r}) &= \{u \in D(-A^{(p)}); -A^{(p)} u \in D((-A^{(p)})^{k+r})\} \\ &= \{u \in D(\mathrm{id} - A^{(p)}); -A^{(p)} u \in D((\mathrm{id} - A^{(p)})^{k+r})\} \\ &= \{u \in D((-A^{(p)})^{k+r}); -A^{(p)} u \in D(-A^{(p)})\} \\ &= \{u \in D((\mathrm{id} - A^{(p)})^{k+r}); -A^{(p)} u \in D(\mathrm{id} - A^{(p)})\} \\ &= \{u \in D((\mathrm{id} - A^{(p)})^{k+r}); u - A^{(p)} u \in D(\mathrm{id} - A^{(p)})\} \\ &= D((\mathrm{id} - A^{(p)})^{k+r+1}). \end{aligned}$$

This is the induction step k to $k+1$. The beginning of the induction $k = 0$, $0 < r \leq 1$, was given in Theorem 3.1.25, hence the assertion follows. □

Remark 3.1.27. As it will become clear in the following section, (3.45) (and (3.46)) has the interpretation that the norms induced by "Bessel potentials" are equivalent to the norms induced by "Riesz potentials".

Now let us return to the spaces $\mathcal{F}_{r,p}(\mathbb{R}^n;\mathbb{R})$ and discuss a first example.

Example 3.1.28. Let $\psi : \mathbb{R}^n \to \mathbb{R}$ be a fixed continuous negative definite function with associated convolution semigroup $(\mu_t)_{t\geq 0}$ and sub-Markovian semigroup $(T_t^{(p)})_{t\geq 0}$, $1 \leq p < \infty$. The generator of $(T_t^{(p)})_{t\geq 0}$ is the operator $(A^{(p)}, D(A^{(p)}))$ and we know that $S(\mathbb{R}^n) \subset D(A^{(p)})$. On $S(\mathbb{R}^n)$ the generator has the representation

$$A^{(p)}u(x) = -\psi(D)u(x) = -(2\pi)^{(-n/2)} \int_{\mathbb{R}^n} e^{ix\cdot\xi}\psi(\xi)\hat{u}(\xi)\, d\xi. \tag{3.48}$$

This follows from Example I.4.1.13 and the fact that on $S(\mathbb{R}^n)$ the operator $A^{(p)}$ coincide with the operator $A^{(2)}$. Further, for $u \in S(\mathbb{R}^n)$ we find

$$V_r^{(p)}u = (\mathrm{id} - A^{(p)})^{-r/2}u = (\mathrm{id} - A^{(2)})^{-r/2}u = V_r^{(2)}u,$$

and

$$V_r^{(2)}u(x) = (2\pi)^{-n/2} \int_{\mathbb{R}^n} e^{ix\cdot\xi}(1+\psi(\xi))^{-r/2}\hat{u}(\xi)\, d\xi,$$

which implies on $S(\mathbb{R}^n)$ that

$$\|u\|_{\mathcal{F}_{r,p}} = \|(1+\psi(D))^{r/2}u\|_{L^p} = \|F^{-1}((1+\psi(\cdot))^{r/2})\hat{u})\|_{L^p}. \tag{3.49}$$

In particular, for $p = 2$ we have by Plancherel's theorem

$$\|u\|_{\mathcal{F}_{r,2}} = \|(1+\psi(D))^{r/2}u\|_0 = \|(1+\psi(\cdot))^{r/2})\hat{u}\|_0 = \|u\|_{\psi,r}$$

showing that

$$\mathcal{F}_{r,2}(\mathbb{R}^n;\mathbb{R}) = H^{\psi,r}(\mathbb{R}^n;\mathbb{R}). \tag{3.50}$$

Moreover, for $\psi(\xi) = |\xi|^2$ we find

$$\|u\|_{\mathcal{F}_{r,p}} = \|v\|_{L^p} \quad \text{for } u = V_r^{(p)}v = (1-\Delta)^{-r/2}v, \tag{3.51}$$

i.e. in the case $\mathcal{F}_{r,p}(\mathbb{R}^n;\mathbb{R})$ is just the Bessel potential space $H_p^r(\mathbb{R}^n;\mathbb{R})$, see Definition I.3.11.9.

Remark 3.1.29. Let $\psi : \mathbb{R}^n \to \mathbb{R}$ be a continuous negative definite function. In Section 3.3 we will study in detail Bessel potential spaces associated with ψ.

Now we are going to introduce a one-parameter family of capacities associated with a sub-Markovian semigroup $(T_t^{(p)})_{t\geq 0}$ on $L^p(\mathbb{R}^n;\mathbb{R})$, $1 < p < \infty$. For an open set $G \subset \mathbb{R}^n$ we define

$$\mathrm{cap}_{r,p}(G) := \inf\{\|u\|^p_{\mathcal{F}_{r,p}}; u \in \mathcal{F}_{r,p}(\mathbb{R}^n;\mathbb{R}) \quad \text{and } u \geq 1 \text{ a.e. on } G\}, \tag{3.52}$$

where $\mathcal{F}_{r,p}(\mathbb{R}^n;\mathbb{R})$ is as in (3.38). Furthermore, for an arbitrary set $A \subset \mathbb{R}^n$ we set

$$\mathrm{cap}_{r,p}(A) := \inf\{\mathrm{cap}_{r,p}(G); \quad A \subset G \text{ and } G \subset \mathbb{R}^n \text{ open}\}. \tag{3.53}$$

Definition 3.1.30. Let $(T_t^{(p)})_{t\geq 0}$ be an L^p-sub-Markovian semigroup, $1 < p < \infty$, and $r > 0$. We call $\mathrm{cap}_{r,p}$ the (r,p)-*capacity* associated with the semigroup $(T_t^{(p)})_{t\geq 0}$.

We will need some preparations in order to prove that $\mathrm{cap}_{r,p}$ is an outer Choquet capacity.

Theorem 3.1.31. *Let $1 < p < \infty$. For any open set $G \subset \mathbb{R}^n$ with $\mathrm{cap}_{r,p}(G) < \infty$ there exists a unique element $u_G \in \mathcal{F}_{r,p}(\mathbb{R}^n;\mathbb{R})$ such that $u_G \geq 1$ a.e. on G and*

$$\mathrm{cap}_{r,p}(G) = \|u_G\|^p_{\mathcal{F}_{r,p}} \tag{3.54}$$

holds. Moreover, there exists $f \in L^p(\mathbb{R}^n;\mathbb{R})$, $f \geq 0$ a.e. with the property that $u_G = V_r^{(p)} f$.

Proof: Recall that a Banach space $(X, \|\cdot\|_X)$ is called *uniformly convex* if for any $\varepsilon > 0$ there exists a $\delta > 0$ such that $\|x\|_X \leq M$, $\|y\|_X \leq M$ and $\|x-y\|_X \geq \varepsilon$ implies $\|x+y\|_X \leq 2M - \delta$, see K.Yosida [288], p.126-127. By a result of J.A.Clarkson [53] the spaces $L^p(\mathbb{R}^n;\mathbb{R})$, $1 < p < \infty$, are uniformly convex. Since $\mathcal{F}_{r,p}(\mathbb{R}^n;\mathbb{R})$ is the isometric image of $L^p(\mathbb{R}^n;\mathbb{R})$, the spaces $\mathcal{F}_{r,p}(\mathbb{R}^n;\mathbb{R})$, $1 < p < \infty$, are uniformly convex. Now let $u_G^{(1)} \neq u_G^{(2)}$ be two elements in $\mathcal{F}_{r,p}(\mathbb{R}^n;\mathbb{R})$ satisfying the assumptions of the theorem. Clearly, $\frac{1}{2}(u_G^{(1)} + u_G^{(2)}) \in \mathcal{F}_{r,p}$ and $\frac{1}{2}(u_G^{(1)} + u_G^{(2)}) \geq 1$ on G. Now we use (3.52) and Clarkson's Theorem with $M = (\mathrm{cap}_{r,p}(G))^{1/p}$ to find

$$\mathrm{cap}_{r,p}(G)^{1/p} \leq \left\| \frac{u_G^{(1)} + u_G^{(2)}}{2} \right\|_{\mathcal{F}_{r,p}} \leq \frac{1}{2}(2\ \mathrm{cap}_{r,p}(G)^{1/p} - \delta)$$

which is a contradiction thus u_G must be unique. To prove the existence of u_G we use the L^p-version of the *Banach–Saks theorem*, see W.Rudin [240],

Theorem 3.13, for a proof of this theorem, the L^2-version is our Theorem I.2.7.2. Thus we use that for any sequence $(u_\nu)_{\nu\in\mathbb{N}}$, $u_\nu \in L^p(\mathbb{R}^n;\mathbb{R})$ converging weakly to some $u \in L^p(\mathbb{R}^n;\mathbb{R})$ there exists a sub-sequence $(u_{\nu_l})_{l\in\mathbb{N}}$ such that $\left(\frac{1}{N}\sum_{l=1}^N u_{\nu_l}\right)_{N\in\mathbb{N}}$ converges strongly in $L^p(\mathbb{R}^n;\mathbb{R})$ to $u \in L^p(\mathbb{R}^n;\mathbb{R})$. Since we assume that $\mathrm{cap}_{r,p}(G)$ is finite, there exists a minimizing sequence $(u_G^{(\nu)})_{\nu\in\mathbb{N}}$, $u_G^{(\nu)} \in \mathcal{F}_{r,p}(\mathbb{R}^n;\mathbb{R})$ such that

$$\lim_{\nu\to\infty} \|u_G^{(\nu)}\|^p_{\mathcal{F}_{r,p}} = \mathrm{cap}_{r,p}(G).$$

The sequence $(f_G^{(\nu)})_{\nu\in\mathbb{N}}$ defined by $u_G^{(\nu)} = V_r^{(p)} f_G^{(\nu)}$ is bounded in $L^p(\mathbb{R}^n;\mathbb{R})$, hence it has a weakly convergent subsequence and this subsequence has a further subsequence with the property that its arithmetic means converge strongly to the weak limit $f_G \in L^p(\mathbb{R}^n;\mathbb{R})$. We denote this subsequence by $(f_G^{(\nu_l)})_{l\in\mathbb{N}}$. Thus we have

$$\lim_{N\to\infty} \left\| \frac{1}{N}\sum_{l=1}^N f_G^{(\nu_l)} - f_G \right\|^p_{L^p} = 0.$$

The linearity of $V_r^{(p)}$ yields

$$\lim_{N\to\infty} \left\| \frac{1}{N}\sum_{l=1}^N u_G^{(\nu)} - V_r^{(p)} f_G \right\|^p_{\mathcal{F}_{r,p}} = 0.$$

Note that for each $N \in \mathbb{N}$ we have $\frac{1}{N}\sum_{l=1}^N u_G^{(\nu)} \geq 1$ a.e. on G and further $u_G := V_r^{(p)} f_G \in \mathcal{F}_{r,p}(\mathbb{R}^n;\mathbb{R})$, which proves the existence of u_G. It remains to prove that $f_G \geq 0$ a.e. Since $V_r^{(p)}$ is a positivity preserving operator in the sense that $f \geq 0$ a.e. implies $V_r^{(p)} f \geq 0$ a.e., it follows that $V_r^{(p)} f_G \leq V_r^{(p)} f_G^+$ but $\|f_G^+\|_{L^p} \leq \|f_G\|_{L^p}$. Therefore $V_r^{(p)} f_G^+ \geq 1$ a.e. on G and $\|f_G^+\|_{L^p} \leq \mathrm{cap}_{r,p}(G)$, implying that $f_G = f_G^+$ by the uniqueness result, and the theorem is proved. □

Definition 3.1.32. Let $u_G = V_r^{(p)} f_G$ be as in Theorem 3.1.31. We call u_G the (r,p)-*equilibrium potential* of the open set G with respect to the semigroup $(T_t^{(p)})_{t\geq 0}$. The function f_G is called the (r,p)-*capacity function* of the open set G with respect to the semigroup $(T_t^{(p)})_{t\geq 0}$.

Remark 3.1.33. The proof of Theorem 3.1.31 yields that for an open set $G \subset \mathbb{R}^n$ we have

$$\begin{aligned}\operatorname{cap}_{r,p}(G) &= \|f_G\|_{L^p}^p \\ &= \inf\{\|f\|_{L^p}^p;\ f \in L^p(\mathbb{R}^n;\mathbb{R}) \text{ s.th. } f \geq 0 \text{ a.e. and } V_r^{(p)} f \geq 1 \text{ a.e. on } G\}.\end{aligned}$$

Theorem 3.1.34. *Let $(T_t^{(p)})_{t\geq 0}$ be an L^p-sub-Markovian semigroup, $1 < p < \infty$, on $L^p(\mathbb{R}^n;\mathbb{R})$. Its (r,p)-capacities, $r > 0$, have the following properties*

$$\lambda^{(n)}(A) \leq \operatorname{cap}_{r,p}(A) \quad \textit{for all } A \in \mathcal{B}(\mathbb{R}^n); \tag{3.55}$$

$$r_1 < r_2 \quad \textit{implies } \operatorname{cap}_{r_1,p}(A) \leq \operatorname{cap}_{r_2,p}(A) \textit{ for all } A \subset \mathbb{R}^n; \tag{3.56}$$

$$p_1 < p_2 \quad \textit{implies } \operatorname{cap}_{r,p_1}(A) \leq \operatorname{cap}_{r,p_2}(A) \textit{ for all } A \subset \mathbb{R}; \tag{3.57}$$

$$A \subset B \quad \textit{implies } \operatorname{cap}_{r,p}(A) \leq \operatorname{cap}_{r,p}(B); \tag{3.58}$$

$$\operatorname{cap}_{r,p}\Big(\bigcup_{\nu\in\mathbb{N}} A_\nu\Big) \leq \sum_{\nu\in\mathbb{N}} \operatorname{cap}_{r,p}(A_\nu), \quad A_\nu \subset \mathbb{R}^n \tag{3.59}$$

and

$$\operatorname{cap}_{r,p}\Big(\bigcap_{\nu\in\mathbb{N}} K_\nu\Big) = \lim_{\nu\to\infty} \operatorname{cap}_{r,p}(K_\nu) = \inf_{\nu\in\mathbb{N}} \operatorname{cap}_{r,p}(K_\nu) \tag{3.60}$$

for any decreasing sequence of compact sets $K_\nu \subset \mathbb{R}^n$.

Proof: For $u = V_r^{(p)} f$ we find using the contraction property of $(V_r^{(p)})_{r\geq 0}$ and the definition of $\|\cdot\|_{\mathcal{F}_{r,p}}$ that

$$\|u\|_{\mathcal{F}_{r,p}} = \|V_r^{(p)} f\|_{\mathcal{F}_{r,p}} = \|f\|_{L^p} \geq \|V_r^{(p)} f\|_{L^p} = \|u\|_{L^p}.$$

i.e.

$$\|u\|_{L^p} \leq \|u\|_{\mathcal{F}_{r,p}},$$

which implies (3.55). The semigroup property of $(V_r^{(p)})_{r\geq 0}$ and its contraction property yields for $r_1 < r_2$ that $\mathcal{F}_{r_2,p}(\mathbb{R}^n;\mathbb{R}) \subset \mathcal{F}_{r_1,p}(\mathbb{R}^n;\mathbb{R})$ and $\|u\|_{\mathcal{F}_{r_1,p}} \leq \|u\|_{\mathcal{F}_{r_2,p}}$ which lead to (3.56). To prove (3.57) suppose that $p_1 < p_2$ and consider the (r,p_2)-capacity function $f_G \geq 0$ a.e., $u_G = V_r^{(p_2)} f_G$, for an open set G. Using Hölder's inequality we find $1 \leq (V_r^{(p_2)} f_G)^{p_2/p_1} \leq V_r^{(p_2)}(f_G)^{p_2/p_1}$ a.e. on G. Since $f_G^{p_2/p_1} \in L^{p_1}(\mathbb{R}^n;\mathbb{R})$ we arrive at

$$\operatorname{cap}_{r,p_1}(G) \leq \|V_r^{(p_2)}(f_G^{p_2/p_1})\|_{\mathcal{F}_{r,p_1}}^{p_1} = \|f_G\|_{L^{p_2}}^{p_2} = \operatorname{cap}_{r,p_2}(G)$$

implying (3.57).

Obviously (3.58) is trivial.

For proving (3.59) we note first that for open sets $G_\nu \subset \mathbb{R}^n$, $\nu \in \mathbb{N}$, the inequality $\mathrm{cap}_{r,p}\left(\bigcup_{\nu\in\mathbb{N}} G_\nu\right) \le \sum_{\nu\in\mathbb{N}} \mathrm{cap}_{r,p}(G_\nu)$ holds. Now let $A_\nu \subset \mathbb{R}^n$ be arbitrary. For $\varepsilon > 0$ and $\nu \in \mathbb{N}$ there exists $G_\nu \subset \mathbb{R}^n$ open such that

$$\mathrm{cap}_{r,p}(A_\nu) \le \mathrm{cap}_{r,p}(G_\nu) \le \mathrm{cap}_{r,p}(A_\nu) + \frac{\varepsilon}{2^\nu}.$$

Thus we find

$$\begin{aligned}\mathrm{cap}_{r,p}\Big(\bigcup_{\nu\in\mathbb{N}} A_\nu\Big) &\le \mathrm{cap}_{r,p}\Big(\bigcup_{\nu\in\mathbb{N}} G_\nu\Big)\\ &\le \sum_{\nu\in\mathbb{N}} \mathrm{cap}_{r,p}(G_\nu) \le \sum_{\nu\in\mathbb{N}}\Big(\mathrm{cap}_{r,p}(A_\nu) + \frac{\varepsilon}{2^\nu}\Big)\\ &\le \varepsilon + \sum_{\nu\in\mathbb{N}} \mathrm{cap}_{r,p}(A_\nu),\end{aligned}$$

which yields (3.59).

Denote by u_{A_ν} the (r,p)-equilibrium potential for A_ν and set $u_{A_\nu} = V_r^{(p)} f_{A_\nu}$, $f_{A_\nu} \in L^p(\mathbb{R}^n;\mathbb{R})$, $f_{A_\nu} \ge 0$ a.e. and $f := \sup\limits_{\nu\in\mathbb{N}} f_{A_\nu}$. Clearly we have $f^p \le \sum_{\nu\in\mathbb{N}} f_{A_\nu}^p$, thus $f \in L^p(\mathbb{R}^n;\mathbb{R})$ and in addition

$$\|V_r^{(p)} f\|_{\mathcal{F}_{r,p}}^p \le \sum_{\nu\in\mathbb{N}} \mathrm{cap}_{r,p}(A_\nu).$$

Since $V_r^{(p)} f \ge 1$ a.e. on $\bigcup_{\nu\in\mathbb{N}} A_\nu$, it follows further that

$$\mathrm{cap}_{r,p} \bigcup_{\nu\in\mathbb{N}} A_\nu \le \|V_r^{(p)} f\|_{\mathcal{F}_{r,p}}^p$$

which proves (3.59). Finally we prove (3.60). For this let $G \subset \mathbb{R}^n$ be open and $K := \bigcap_{\nu\in\mathbb{N}} K_\nu \subset G$. Thus $K_{\nu_0} \subset G$ for some $\nu_0 = \nu_0(G)$ and it follows

$$\begin{aligned}\mathrm{cap}_{r,p}(K) &\le \lim_{\nu\to\infty} \mathrm{cap}_{r,p}(K_\nu)\\ &\le \inf\{\mathrm{cap}_{r,p}(G); K \subset G \quad \text{and } G \subset \mathbb{R}^n \text{ open}\}\\ &= \mathrm{cap}_{r,p}(K),\end{aligned}$$

and (3.60) is proved. □

Note that it remains only to prove (3.7) in order to identify $\mathrm{cap}_{r,p}$ as an outer Choquet capacity; this will be done in Theorem 3.1.53. For this we need an additional assumption on $\mathcal{F}_{r,p}(\mathbb{R}^n;\mathbb{R})$.

Definition 3.1.35. Let $(T_t^{(p)})_{t\geq 0}$ be an L^p-sub-Markovian semigroup, $1 < p < \infty$, and for $r > 0$ let $\mathcal{F}_{r,p}(\mathbb{R}^n;\mathbb{R})$ be defined as in (3.38).

A. We call $\mathcal{F}_{r,p}(\mathbb{R}^n;\mathbb{R})$ *weakly regular* if $\mathcal{F}_{r,p}(\mathbb{R}^n;\mathbb{R}) \cap C_0(\mathbb{R}^n;\mathbb{R})$ is dense in $\mathcal{F}_{r,p}(\mathbb{R}^n;\mathbb{R})$.

B. The space $\mathcal{F}_{r,p}(\mathbb{R}^n;\mathbb{R})$ is called *regular* if it is weakly regular and if $\mathcal{F}_{r,p}(\mathbb{R}^n;\mathbb{R}) \cap C_0(\mathbb{R}^n;\mathbb{R})$ is dense in $(C_0(\mathbb{R}^n;\mathbb{R}); \|\cdot\|_\infty)$.

C. The space $\mathcal{F}_{r,p}(\mathbb{R}^n;\mathbb{R})$ is *contraction regular* if there exists a subspace $W \subset \mathcal{F}_{r,p}(\mathbb{R}^n;\mathbb{R}) \cap C_0(\mathbb{R}^n;\mathbb{R})$ such that W is dense in $(\mathcal{F}_{r,p}(\mathbb{R}^n;\mathbb{R}), \|\cdot\|_{\mathcal{F}_{r,p}})$ and $(C_0(\mathbb{R}^n;\mathbb{R}), \|\cdot\|_\infty)$, and in addition W has the property that $u \in W$ implies $|u| \in W$.

We give some first results concerning the question whether $\mathcal{F}_{r,p}(\mathbb{R}^n;\mathbb{R})$ is (weakly) regular.

Lemma 3.1.36. *For $k \in \mathbb{N}$ the space $\mathcal{F}_{k+2,p}(\mathbb{R}^n;\mathbb{R})$ is dense in $\mathcal{F}_{k,p}(\mathbb{R}^n;\mathbb{R})$.*

Proof: We know from Corollary 3.1.21 that

$$\mathcal{F}_{k,p}(\mathbb{R}^n;\mathbb{R}) = (\mathrm{id} - A^{(p)})^{-k/2} L^p(\mathbb{R}^n;\mathbb{R}),$$

where $(A^{(p)}, D(A^{(p)}))$ is the generator of the semigroup $(T_t^{(p)})_{t\geq 0}$ defining the scale $\mathcal{F}_{k,p}(\mathbb{R}^n;\mathbb{R})$. Let $u = (\mathrm{id} - A^{(p)})^{-k/2} f \in \mathcal{F}_{k,p}(\mathbb{R}^n;\mathbb{R})$ with $f \in L^p(\mathbb{R}^n;\mathbb{R})$. Since $D(A^{(p)})$ is dense in $L^p(\mathbb{R}^n;\mathbb{R})$ we find for every $\varepsilon > 0$ some $\varphi_\varepsilon \in D(A^{(p)})$ such that $\|f - \varphi_\varepsilon\|_{L^p} < \varepsilon$. Set $h_\varepsilon := (\mathrm{id} - A^{(p)})\varphi_\varepsilon$. Then we have $h_\varepsilon \in L^p(\mathbb{R}^n;\mathbb{R})$ and $\varphi_\varepsilon = (\mathrm{id} - A^{(p)})^{-1} h_\varepsilon$. It follows that

$$(\mathrm{id} - A^{(p)})^{-k/2} \varphi_\varepsilon = (\mathrm{id} - A^{(p)})^{\frac{-k-2}{2}} h_\varepsilon \in \mathcal{F}_{k+2,p}(\mathbb{R}^n;\mathbb{R})$$

and furthermore

$$\begin{aligned}
&\|u - (\mathrm{id} - A^{(p)})^{\frac{-k-2}{2}} h_\varepsilon\|_{\mathcal{F}_{k,p}} \\
&\quad = \|(\mathrm{id} - A^{(p)})^{-k/2}(f - (\mathrm{id} - A^{(p)})^{-1}) h_\varepsilon\|_{\mathcal{F}_{r,p}} \\
&\quad = \|f - (\mathrm{id} - A^{(p)})^{-1} h_\varepsilon\|_{L^p} = \|f - \varphi_\varepsilon\|_{L^p} < \varepsilon,
\end{aligned}$$

which proves the lemma. □

Proposition 3.1.37. *Let $k \in \mathbb{N}$ and suppose that the set $C_0(\mathbb{R}^n;\mathbb{R}) \cap D((-A^{(p)})^k)$ is an operator core for $(-A^{(p)})^k$. Then $\mathcal{F}_{2k,p}(\mathbb{R}^n;\mathbb{R})$ is weakly regular.*

Proof: First note that $D((\mathrm{id} - A^{(p)})^k) = D((-A^{(p)})^k)$, hence we have

$$C_0(\mathbb{R}^n;\mathbb{R}) \cap \mathcal{F}_{2k,p}(\mathbb{R}^n;\mathbb{R}) = C_0(\mathbb{R}^n;\mathbb{R}) \cap D((-A^{(p)})^k).$$

Noting that $C_0(\mathbb{R}^n;\mathbb{R}) \cap \mathcal{F}_{2k,p}(\mathbb{R}^n;\mathbb{R})$ is an operator core for $(-A^{(p)})^k$ we choose for every $u \in \mathcal{F}_{2k,p}(\mathbb{R}^n;\mathbb{R})$ a sequence $(u_\nu)_{\nu\in\mathbb{N}}$, $u_\nu \in C_0(\mathbb{R}^n;\mathbb{R}) \cap \mathcal{F}_{2k,p}(\mathbb{R}^n;\mathbb{R})$, which converges in the graph norm of $(-A^{(p)})^k$ to u. This implies

$$\begin{aligned}\|u_\nu - u\|_{\mathcal{F}_{2k,p}} &= \|(V_{2k}^{(p)})^{-1}(u_\nu - u)\|_{L^p} \\ &= \|(\mathrm{id} - A^{(p)})^k(u_\nu - u)\|_{L^p} \to 0\end{aligned}$$

as $\nu \to \infty$. □

Thus it is possible to reduce the regularity problem for $\mathcal{F}_{r,p}(\mathbb{R}^n;\mathbb{R})$ to find a good operator core for $A^{(p)}$ or $(-A^{(p)})^k$, and we will come back to this fact later on.

Definition 3.1.38. Let $(T_t^{(p)})_{t\geq 0}$ be an L^p-sub-Markovian semigroup, $1 < p < \infty$, and for $r > 0$ let $\mathrm{cap}_{r,p}$ be its (r,p)-capacity.

A. A set $N \subset \mathbb{R}^n$ is called an *(r,p)-exceptional set* with respect to $(T_t^{(p)})_{t\geq 0}$ if $\mathrm{cap}_{r,p}(N) = 0$.

B. A statement is said to hold *(r,p)-quasi-everywhere* with respect to $(T_t^{(p)})_{t\geq 0}$ if there exists an (r,p)-exceptional set N such that the statement holds on $\mathbb{R}^n \backslash N$. We will use the abbreviation (r,p)-q.e. for (r,p)-quasi-everywhere.

C. A real-valued function u defined (r,p)-quasi-everywhere on $\mathbb{R}^n$ is called *(r,p)-quasi-continuous* with respect to $(T_t^{(p)})_{t\geq 0}$ if for any $\varepsilon > 0$ there exists an open set $G \subset \mathbb{R}^n$ such that $\mathrm{cap}_{r,p}(G) < \varepsilon$ and $u|_{G^c}$ is continuous.

D. Let u be as in Part C. We call u *(r,p)-quasi-continuous in the restricted sense* with respect to $(T_t^{(p)})_{t\geq 0}$ if for every $\varepsilon > 0$ there exists an open set $G \subset \mathbb{R}^n \subset \mathbb{R}^n_\infty$ such that $u|_{\mathbb{R}^n_\infty \backslash G}$ is continuous and $\mathrm{cap}_{r,p}(G) < \varepsilon$.

E. A function $\tilde{u} \in \mathcal{F}_{r,p}(\mathbb{R}^n;\mathbb{R})$ is called an *(r,p)-quasi-continuous modification* of $u \in \mathcal{F}_{r,p}(\mathbb{R}^n;\mathbb{R})$ if $\tilde{u}$ is quasi-continuous and $u = \tilde{u}$ almost everywhere.

Proposition 3.1.39. *If $u \in \mathcal{F}_{r,p}(\mathbb{R}^n;\mathbb{R})$ is an (r,p)-quasi-continuous function and $u \geq 0$ a.e. on an open set $G \subset \mathbb{R}^n$, then $u \geq 0$ (r,p)-q.e. on G.*

Proof: Let $\varepsilon > 0$ and take B_ε open such that $\mathrm{cap}_{r,p}(B_\varepsilon) < \varepsilon$ and that $u|_{B_\varepsilon^c}$ is continuous. For $N = \{u < 0\}$ we find by the continuity of $u\big|_{B_\varepsilon^c}$ that $N \cap B_\varepsilon^c = \varnothing$, recall that $u \geq 0$ a.e. Thus it follows that $N \subset B_\varepsilon$ implying

$0 \leq \mathrm{cap}_{r,p}(N) \leq \mathrm{cap}_{r,p}(B_\varepsilon) < \varepsilon$. Taking the limit $\varepsilon \to 0$ we find that N is an (r,p)-exceptional set, i.e. $u \geq 0$ (r,p)-q.e. on G.

To proceed further we first derive a *Chebyshev-type inequality* for capacities.

Lemma 3.1.40. *Let $u \in \mathcal{F}_{r,p}(\mathbb{R}^n;\mathbb{R}) \cap C(\mathbb{R}^n;\mathbb{R})$. For every $\eta > 0$ the estimate*

$$\mathrm{cap}_{r,p}(\{|u| > \eta\}) \leq \frac{2}{\eta^p}\|u\|^p_{\mathcal{F}_{r,p}} \tag{3.61}$$

holds.

Proof: Since $\{|u| > \eta\} \subset \{u > \eta\} \cup \{-u > \eta\}$, i.e.

$$\mathrm{cap}_{r,p}(\{|u| > \eta\}) \leq \mathrm{cap}_{r,p}(\{u > \eta\}) + \mathrm{cap}_{r,p}(\{-u > \eta\})$$

it is sufficient to prove (3.61) with right-hand side $\frac{1}{\eta^p}\|u\|^p_{\mathcal{F}_{r,p}}$ for all $u \geq 0$. Thus let $u \in \mathcal{F}_{r,p}(\mathbb{R}^n;\mathbb{R}) \cap C(\mathbb{R}^n;\mathbb{R})$, $u \geq 0$, be given. For $\eta > 0$ we put $G_{\eta,u} = \{u > \eta\}$. Since u is assumed to be continuous, $G_{\eta,u}$ is an open set. Further we have $G_{\eta,u} = \{u > \eta\} = \{\frac{u}{\eta} > 1\}$ and we find with $v_\eta = \frac{u}{\eta}$ that $v_\eta \in \{v \in \mathcal{F}_{r,p}; v_\eta \geq 1 \text{ a.e. on } G_{\eta,u}\}$, hence

$$\mathrm{cap}_{r,p}(G_{\eta,u}) = \mathrm{cap}_{r,p}\{u > \eta\} \leq \left\|\frac{u}{\eta}\right\|^p_{\mathcal{F}_{r,p}} = \frac{1}{\eta^p}\|u\|^p_{\mathcal{F}_{r,p}},$$

which implies (3.61). □

A reader with a good background in the theory of Dirichlet forms may already have noticed before that some of our proofs for results related to (r,p)-capacities and the spaces $\mathcal{F}_{r,p}(\mathbb{R}^n;\mathbb{R})$ differ a little from the usual proofs as they are given for Dirichlet forms, compare for example the monograph [102] of M.Fukushima, Y.Oshima and M.Takeda. The reason is that in general in $\mathcal{F}_{r,p}(\mathbb{R}^n;\mathbb{R})$ the truncation property does not hold, in particular $u \in \mathcal{F}_{r,p}(\mathbb{R}^n;\mathbb{R})$ does in general not imply that $|u| \in \mathcal{F}_{r,p}(\mathbb{R}^n;\mathbb{R})$. In Lemma 3.1.40 this fact also caused another constant in estimate (3.61), namely 2 instead of 1, which is the correct constant whenever $u \in \mathcal{F}_{r,p}(\mathbb{R}^n;\mathbb{R})$ implies $|u| \in \mathcal{F}_{r,p}(\mathbb{R}^n;\mathbb{R})$. We refer also to P.Malliavin [207], Lemma IV.1.2.5 and Theorem IV.2.2. In Section 3.3 we will discuss more detailed the truncation property.

Theorem 3.1.41. *Let $\mathcal{F}_{r,p}(\mathbb{R}^n;\mathbb{R})$ be weakly regular, then each $u \in \mathcal{F}_{r,p}(\mathbb{R}^n;\mathbb{R})$ admits an (r,p)-quasi continuous modification $\tilde{u}$ which is unique up to (r,p)-quasi everywhere equality. Moreover we have for every $\eta > 0$*

$$\mathrm{cap}_{r,p}(|\tilde{u}| > \eta) \leq \frac{2}{\eta^p}\|u\|^p_{\mathcal{F}_{r,p}}. \tag{3.62}$$

Proof: Given $u \in \mathcal{F}_{r,p}(\mathbb{R}^n;\mathbb{R})$ we take a sequence $(\varphi_\nu)_{\nu\in\mathbb{N}}$ such that $\varphi_\nu \in \mathcal{F}_{r,p}(\mathbb{R}^n;\mathbb{R}) \cap C_0(\mathbb{R}^n;\mathbb{R})$ converges to u with respect to $\|\cdot\|_{\mathcal{F}_{r,p}}$. In addition we may suppose that $\|\varphi_{\nu+1}-\varphi_\nu\|_{\mathcal{F}_{r,p}} \leq 2^{-2\nu/p}$ holds for all $\nu \in \mathbb{N}$ (passing to a subsequence if necessary). Applying Lemma 3.1.40 we arrive at

$$\mathrm{cap}_{r,p}(\{|\varphi_{\nu+1}-\varphi_\nu| > 2^{-\nu/p}\}) \leq 2\frac{\|\varphi_{\nu+1}-\varphi_\nu\|^p_{\mathcal{F}_{r,p}}}{2^{-\nu}} = 2^{-\nu+1}.$$

For $A_m := \bigcup_{\nu\geq m}\{|\varphi_{\nu+1}-\varphi_\nu| > 2^{-\nu/p}\}$ we find by (3.59) in Theorem 3.1.34 that $\mathrm{cap}_{r,p}(A_m) < 2\cdot 2^{-m+1}$. Thus on A_m^c the sequence $(\varphi_\nu)_{\nu\in\mathbb{N}}$ converges uniformly.

Now we define

$$\tilde{u}(x) := \begin{cases} \lim\limits_{\nu\to\infty}\varphi_\nu(x) & \text{for } x \in \left(\bigcap_{m\geq 1} A_m\right)^c \\ 0 & \text{for } x \in \bigcap_{m\geq 1} A_m. \end{cases}$$

Finally, let $u \in \mathcal{F}_{r,p}(\mathbb{R}^n;\mathbb{R})$ with (r,p)-quasi-continuous modification $\tilde{u}$ as constructed before. In particular let $(\varphi_\nu)_{\nu\in\mathbb{N}}$, $\varphi_\nu \in \mathcal{F}_{r,p}(\mathbb{R}^n;\mathbb{R}) \cap C_0(\mathbb{R}^n;\mathbb{R})$, the approximating sequence as before. For $\varepsilon > 0$ there exists an open set G_ε such that $\mathrm{cap}_{r,p}(G_\varepsilon) < \varepsilon$ and $\varphi_\nu \to \tilde{u}$ uniformly on G_ε^c. Now take $\eta > \varepsilon_1 > 0$ and observe that

$$\{|\tilde{u}(x)| > \eta\} \subset \{|\varphi_\nu| > \eta - \varepsilon_1\} \cup G_\varepsilon$$

for ν sufficiently large. Using (3.61) we find

$$\mathrm{cap}_{r,p}\{|\tilde{u}| > \eta\} \leq 2\frac{\|\varphi_\nu\|^p_{\mathcal{F}_{r,p}}}{(\eta-\varepsilon_1)^p} + \varepsilon.$$

For $\nu \to \infty$, $\varepsilon_1 \to 0$ and $\varepsilon \to 0$ we finally arrive at (3.62). □

It is convenient to introduce the notion of a nest.

Definition 3.1.42. Let $(T_t^{(p)})_{t\geq 0}$ be an L^p-sub-Markovian semigroup.

A. A sequence $(F_k)_{k\in\mathbb{N}}$ of closed sets $F_k \subset \mathbb{R}^n$ is called an (r,p)-*nest* if $F_k \subset F_{k+1}$ for all $k \in \mathbb{N}$ and $\lim\limits_{k\to\infty} \mathrm{cap}_{r,p}(F_k^c) = 0$.

B. An (r,p)-nest is called *regular* if for each $k \in \mathbb{N}$ it holds $\mathrm{supp}(\chi_{F_k}\lambda^{(n)}) = F_k$.

Remark 3.1.43. Let $(F_k)_{k\in\mathbb{N}}$ be an (r,p)-nest and set $G := \left(\bigcup_{k=1}^{\infty} F_k\right)^c = \bigcap_{k=1}^{\infty} F_k^c$. Since $F_k \subset F_{k+1}$, i.e. $F_k^c \supset F_{k+1}^c$ it follows that

$$\operatorname{cap}_{r,p}(G) = \operatorname{cap}_{r,p}\Big(\bigcap_{k=1}^{\infty} F_k^c\Big) \le \operatorname{cap}_{r,p}(F_l^c)$$

for any $l \in \mathbb{N}$, hence $\operatorname{cap}_{r,p}(G) = 0$.

Lemma 3.1.44. *Let $(F_k)_{k\in\mathbb{N}}$ be an (r,p)-nest and set $F_k' := \operatorname{supp}\chi_{F_k}$. Then $(F_k')_{k\in\mathbb{N}}$ is a regular (r,p)-nest and $F_k' \subset F_k$ for all $k \in \mathbb{N}$.*

Proof: The last statement is trivial. Further we find

$$\lambda^{(n)}(F_k'^{\,c}\backslash F_k^c) = \int_{\mathbb{R}^n} \chi_{F_k'^{\,c}}(x)\chi_{F_k}(x)\,dx = \int_{F_k'^{\,c}} \chi_{F_k}(x)\,dx = 0$$

since F_k' is the smallest closed set whose complement has with respect to the measure $\chi_{F_k}\lambda^{(n)}$ measure zero. From the definition of $\operatorname{cap}_{r,p}$, see (3.52), we deduce now that

$$\operatorname{cap}_{r,p}(F_k'^{\,c}) = \operatorname{cap}_{r,p}(F_k^c),$$

implying the lemma. □

For a given (r,p)-nest we define

$$C(\{F_k\}) := \{u : \mathbb{R}^n \to \mathbb{R};\quad u|_{F_k} \text{ is for each } k \in \mathbb{N} \text{ continuous}\}, \tag{3.63}$$

hence $u \in C(\{F_k\})$ is quasi everywhere continuous.

Further, extending each function $u : \mathbb{R}^n \to \mathbb{R}$ to $\mathbb{R}^n_\infty$ by setting $u(\infty) = 0$, we put

$$C_\infty(\{F_k\}) := \{u : \mathbb{R}^n_\infty \to \mathbb{R}; u|_{F_k\cup\{\infty\}} \quad \text{is for each } k \in \mathbb{N} \text{ continuous}\}. \tag{3.64}$$

It follows that

$$C_\infty(\{F_k\}) \subset C(\{F_k\})$$

and

$$C_\infty(\mathbb{R}^n;\mathbb{R}) \subset C_\infty(\{F_k\}), \quad C(\mathbb{R}^n;\mathbb{R}) \subset C(\{F_k\}).$$

The next result is essentially adapted from [102], Theorem 2.1.2.

Theorem 3.1.45. *Let $(T_t^{(p)})_{t\geq 0}$ be an L^p-sub-Markovian semigroup.*

A. For a countable family S of (r,p)-quasi continuous functions ((r,p)-quasi continuous functions in the restricted sense) there exists a regular (r,p)-nest $(F_k)_{k\in\mathbb{N}}$ such that $S \subset C(\{F_k\})$ ($S \subset C_\infty(\{F_k\})$).

B. Let $(F_k)_{k\in\mathbb{N}}$ be a regular (r,p)-nest and let $u \in C(\{F_k\})$. If $u \geq 0$ a.e. then $u(x) \geq 0$ for all $x \in \bigcup_{k=1}^\infty F_k$.

Proof: A. For an (r,p)-quasi-continuous u there exists a sequence $(\tilde{F}_k)_{k\in\mathbb{N}}$ of closed sets $\tilde{F}_k \subset \mathbb{R}^n$ such that $\mathrm{cap}_{r,p}(\tilde{F}_k^c) < \frac{1}{k}$ and $u|_{\tilde{F}_k}$ is continuous. It follows that $(F_k)_{k\in\mathbb{N}}$, $F_k := \bigcup_{l=1}^k \tilde{F}_l$, is an (r,p)-nest and $u \in C(\{F_k\})$. Now, for $S = \{u_l; l \in \mathbb{N}\}$, a countable family of (r,p)-quasi-continuous functions u_l, we choose for each $l \in \mathbb{N}$ an (r,p)-nest $(F_k^{(l)})_{k\in\mathbb{N}}$ such that $u_l \in C(\{F_k^{(l)}\})$, $\mathrm{cap}_{r,p}(F_k^{(l),c}) < \frac{1}{2^l}\frac{1}{k}$. We set $F_k := \bigcap_{l=1}^\infty F_k^{(l)}$. It follows that

$$\mathrm{cap}_{r,p}(F_k^c) \leq \sum_{l=1}^\infty \mathrm{cap}_{r,p}(F_k^{(l),c}) \leq \frac{1}{k},$$

thus $(F_k)_{k\in\mathbb{N}}$ is again an (r,p)-nest and $S \subset C(\{F_k\})$. Finally, we may regularize $(F_k)_{k\in\mathbb{N}}$ as in Lemma 3.1.44.

B. Suppose that $u(x) < 0$ for some $x \in F_k$. Then there is an open set $U(x)$ such that $x \in U(x)$ and $u(y) < 0$ for all $y \in U(x) \cap F_k$, recall that $u|_{F_k}$ is continuous. Since $(F_k)_{k\in\mathbb{N}}$ is regular, it follows that $\lambda^{(n)}(U(x) \cap F_k) > 0$, which is an contradiction to $u \geq 0$ a.e. □

Remark 3.1.46. As it is explained in [102], Lemma 2.1.4, the result of Theorem 3.1.45.B can be localized: Let $G \subset \mathbb{R}^n$ be an open set and u (r,p)-quasi continuous on G. If $u \geq 0$ almost everywhere on G, then $u \geq 0$ (r,p)-quasi-everywhere on G.

Extending Lemma 3.1.40 we prove

Theorem 3.1.47. *Let $(T_t^{(p)})_{t\geq 0}$ be an L^p-sub-Markovian semigroup and, assume that $\mathcal{F}_{r,p}(\mathbb{R}^n;\mathbb{R})$ is weakly regular. Then each $u \in \mathcal{F}_{r,p}(\mathbb{R}^n;\mathbb{R})$ admits an (r,p)-quasi continuous modification in the restricted sense.*

Proof: We may follow the proof of Theorem 3.1.41 to find that $(F_k)_{k\in\mathbb{N}}$, $F_k := A_k^c$ is an (r,p)-nest and that for any k the sequence $(\varphi_\nu|_{F_k\cup\{\infty\}})_{\nu\in\mathbb{N}}$ converges uniformly. If we define $\tilde{u}$ as in the proof of Theorem 3.1.41, we deduce that $\tilde{u} \in C_\infty(\{F_k\})$ and $u = \tilde{u}$ a.e. □

Definition 3.1.48. A sequence $(u_\nu)_{\nu\in\mathbb{N}}$, $u_\nu \in \mathcal{F}_{r,p}(\mathbb{R}^n;\mathbb{R})$, is said to *converge (r,p)-quasi-uniformly* to $u \in \mathcal{F}_{r,p}(\mathbb{R}^n;\mathbb{R})$ if for any $\varepsilon > 0$ there exists an open set $G_\varepsilon \subset \mathbb{R}^n$ with $\text{cap}_{r,p}(G_\varepsilon) < \varepsilon$ and $(u_\nu)_{\nu\in\mathbb{N}}$ converges uniformly on G_ε^c to u.

Proposition 3.1.49. *Let $\mathcal{F}_{r,p}(\mathbb{R}^n;\mathbb{R})$ be weakly regular and let $(u_\nu)_{\nu\in\mathbb{N}}$, $u_\nu \in \mathcal{F}_{r,p}(\mathbb{R}^n;\mathbb{R})$, be a sequence converging with respect to $\|\cdot\|_{\mathcal{F}_{r,p}}$ to $u \in \mathcal{F}_{r,p}(\mathbb{R}^n;\mathbb{R})$. Then there exists a subsequence $(u_{\nu_l})_{l\in\mathbb{N}}$ such that $(\tilde{u}_{\nu_l})_{l\in\mathbb{N}}$ converges to $\tilde{u}$ (r,p)-quasi-uniformly.*

Proof: There exists a subsequence $(u_{\nu_l})_{l\in\mathbb{N}}$ such that $\|u_{\nu_l} - u\|^p_{\mathcal{F}_{r,p}} \leq \frac{1}{2^p 2^l}$ which implies by (3.61) that

$$\text{cap}_{r,p}\left(\left\{|\tilde{u}_{\nu_l} - \tilde{u}| > \frac{1}{l}\right\}\right) \leq \frac{2}{l^p}\|u_{\nu_l} - u\|^p_{\mathcal{F}_{r,p}} \leq 2\cdot 2^{-l}.$$

Let $A_m := \bigcup_{l\geq m}\{|\tilde{u}_{\nu_l} - \tilde{u}| > \frac{1}{l}\}$, thus $\text{cap}_{r,p}(A_m) \leq 2\sum_{l\geq m} 2^{-l} =: \varepsilon_m$ and $\lim_{m\to\infty} \varepsilon_m = 0$. Now take an open set such that $A_m \subset G_m$ and $\text{cap}_{r,p}(G_m) < 2\varepsilon_m$. On G_m^c we have for $l \geq m > 1$ that $|\tilde{u}_{\nu_l} - \tilde{u}| < \frac{1}{l}$ which proves the proposition. □

Corollary 3.1.50. *A. If $(u_\nu)_{\nu\in\mathbb{N}}$, $u_\nu \in \mathbb{C}F_{r,p}(\mathbb{R}^n;\mathbb{R})$, is a sequence converging to $u \in \mathcal{F}_{r,p}(\mathbb{R}^n;\mathbb{R})$ in the norm $\|\cdot\|_{\mathcal{F}_{r,p}}$, then a subsequence $(\tilde{u}_{\nu_l})_{l\in\mathbb{N}}$ converges (r,p)-quasi-everywhere to $\tilde{u}$.*

B. If $(u_\nu)_{\nu\in\mathbb{N}}$ is a sequence in $\mathcal{F}_{r,p}(\mathbb{R}^n;\mathbb{R})$ converging in $\mathcal{F}_{r,p}(\mathbb{R}^n;\mathbb{R})$ to $u \in \mathcal{F}_{r,p}(\mathbb{R}^n;\mathbb{R})$, and if $(\tilde{u}_\nu)_{\nu\in\mathbb{N}}$ converges (r,p)-quasi-everywhere to some function v, then $v = \tilde{u}$ (r,p)-q.e.

The next result extends Theorem 3.1.31 to arbitrary sets.

Theorem 3.1.51. *For any set $A \subset \mathbb{R}^n$ with finite (r,p)-capacity there exists a unique element u_A in the set $\{u \in \mathcal{F}_{r,p}(\mathbb{R}^n;\mathbb{R}); \tilde{u} \geq 1$ (r,p)-q.e. on $A\}$ minimizing the norm $\|\cdot\|_{\mathcal{F}_{r,p}}$. This element u_A is non-negative and satisfies*

$$\text{cap}_{r,p}(A) = \|u_A\|^p_{\mathcal{F}_{r,p}}. \tag{3.65}$$

Definition 3.1.52. The function u_A is called the *(r,p)-equilibrium potential* of the set A.

Proof of Theorem 3.1.51: The uniqueness, existence and positivity of u_A can be proved as the corresponding assertions of Theorem 3.1.31. Thus the theorem follows once (3.65) is proved. For $\varepsilon > 0$ there exists an open set $G \subset \mathbb{R}^n$ such that $A \subset G$ and $\operatorname{cap}_{r,p}(A) > \operatorname{cap}_{r,p}(G) - \varepsilon$. Since by Proposition 3.1.39 the (r,p)-equilibrium potential u_G of the open set G belongs to the set $\{u \in \mathcal{F}_{r,p}(\mathbb{R}^n;\mathbb{R}); \tilde{u} \geq 1 \ (r,p)\text{-q.e. on } A\}$ it follows that

$$\operatorname{cap}_{r,p}(G) = \|u\|^p_{\mathcal{F}_{r,p}} \geq \|u_A\|^p_{\mathcal{F}_{r,p}}.$$

Taking the infimum in this inequality with respect to $G \supset A$ we get $\operatorname{cap}_{r,p}(A) \geq \|u_A\|^p_{\mathcal{F}_{r,p}}$. Now we are going to prove the converse inequality. For this let $\tilde{u}_A$ be an (r,p)-quasi-continuous version of u_A and for $\varepsilon > 0$ choose an open set G_ε such that $\operatorname{cap}_{r,p}(G_\varepsilon) < \varepsilon$, $\tilde{u}_A|_{G^c_\varepsilon}$ is continuous and $\tilde{u}_A \geq 1$ on $A \cap G^c_\varepsilon$. We denote by u_{G_ε} the (r,p)-equilibrium potential of the open set G_ε. Further we set

$$B_\varepsilon := \{x \in G^c_\varepsilon; \quad \tilde{u}_A(x) > 1-\varepsilon\} \cup G_\varepsilon.$$

The set B_ε is open and $A \subset B_\varepsilon$. In addition we have $u_A + u_{G_\varepsilon} > 1 - \varepsilon$ a.e. on B_ε. Therefore it follows that

$$\begin{aligned}\operatorname{cap}_{r,p}(A) \leq \operatorname{cap}_{r,p}(B_\varepsilon) &\leq (1-\varepsilon)^{-p}\|u_A + u_{G_\varepsilon}\|^p_{\mathcal{F}_{r,p}} \\ &\leq (1-\varepsilon)^{-p}(\|u_A\|_{\mathcal{F}_{r,p}} + \varepsilon)^p\end{aligned}$$

which yields (3.65) since $\varepsilon > 0$ was arbitrary. □

Now we are in a position to prove that $\operatorname{cap}_{r,p}$ is an outer Choquet capacity.

Theorem 3.1.53. *Let $\mathcal{F}_{r,p}(\mathbb{R}^n;\mathbb{R})$, $1 < p < \infty$, $r > 0$, be weakly regular. Then for any increasing sequence $(A_\nu)_{\nu\in\mathbb{N}}$, $A_\nu \subset \mathbb{R}^n$, it follows that*

$$\operatorname{cap}_{r,p}\Big(\bigcup_{\nu\in\mathbb{N}} A_\nu\Big) = \sup_{\nu\in\mathbb{N}} \operatorname{cap}_{r,p}(A). \tag{3.66}$$

Corollary 3.1.54. *For any weakly regular space $\mathcal{F}_{r,p}(\mathbb{R}^n;\mathbb{R})$, $1 < p < \infty$ and $r > 0$, the capacity $\operatorname{cap}_{r,p}$ is an outer Choquet capacity.*

Proof of Theorem 3.1.53: Obviously we may assume that the right hand side of (3.66) is finite. Set $A := \bigcup_{\nu\in\mathbb{N}} A_\nu$ and denote by $u_\nu = u_{A_\nu}$ the (r,p)-equilibrium potential of the set A_ν. Since $\|u_\nu\|^p_{\mathcal{F}_{r,p}} = \operatorname{cap}_{r,p}(A_\nu) \leq c < \infty$ with a constant c independent of ν, we may use the Banach–Saks theorem and argue as in the proof of Theorem 3.1.31 to find a subsequence $(u_{\nu_l})_{l\in\mathbb{N}}$ such that $\left(\frac{1}{N}\sum_{l=1}^N u_{\nu_l}\right)_{N\in\mathbb{N}}$ converges strongly in $\mathcal{F}_{r,p}(\mathbb{R}^n;\mathbb{R})$ to some $u \in \mathcal{F}_{r,p}(\mathbb{R}^n;\mathbb{R})$. Let us denote for a moment by v_N the function $\frac{1}{N}\sum_{l=1}^N u_{\nu_l}$, and take an

(r,p)-quasi-continuous modification $\tilde{v}_N$ of v_N. It follows that $\liminf_{N\to\infty} \tilde{v}_N(x) \geq 1$ (r,p)-q.e. on A_ν for each $\nu \in \mathbb{N}$, hence on A. From Corollary 3.1.50.A we may deduce that a subsequence of $(\tilde{v}_N)_{N\in\mathbb{N}}$ converges (r,p)-q.e. to an (r,p)-quasi-continuous modification $\tilde{u}$ of u. This implies that $u \in \{w \in \mathcal{F}_{r,p}(\mathbb{R}^n;\mathbb{R}); \tilde{w} \geq 1$ (r,p)-q.e. on $A\}$ and further

$$\begin{aligned} \operatorname{cap}_{r,p}(A) &\leq \|u\|^p_{\mathcal{F}_{r,p}} = \lim_{l\to\infty} \|v_{N_l}\|^p_{\mathcal{F}_{r,p}} \\ &\leq \sup_{\substack{\nu_k \\ k=1,\dots,N_l}} \|u_{\nu_k}\|^p_{\mathcal{F}_{r,p}} \leq \sup_{\nu\in\mathbb{N}} \|u_\nu\|^p_{\mathcal{F}_{r,p}} \\ &= \sup_{\nu\in\mathbb{N}} \operatorname{cap}_{r,p}(A_\nu). \quad \square \end{aligned}$$

3.2 First Results on L^p-Sub-Markovian Semigroups in their Associated Bessel Potential Spaces

In Section I.4.8 we extended a Feller semigroup $(T_t^{(\infty)})_{t\geq 0}$ to the space $B_b(\mathbb{R}^n;\mathbb{R})$ by constructing the sub-Markovian kernels $p_t(x,dy)$. For the extended semigroup $(\tilde{T}_t^{(\infty)})_{t\geq 0}$ we had the representation (I.408), i.e.

$$\tilde{T}_t^{(\infty)} u(x) = \int_{\mathbb{R}^n} u(y) p_t(x,dy). \tag{3.67}$$

From (3.67) there is a straightforward way (as we will see in volume III) to construct an associated Feller process $((X_t)_{t\geq 0}, P^x)_{x\in\mathbb{R}^n}$ by defining the transition function

$$P^x\{X_t \in A\} = p_t(x,A) = \tilde{T}_t^{(\infty)} \chi_A(x) \tag{3.68}$$

for a (bounded) Borel set $A \subset \mathbb{R}^n$. A major theme in the modern theory of stochastic processes is to construct and to study a process starting with an L^p-sub-Markovian semigroup. In particular, for $p = 2$ this is the subject of the theory of Dirichlet forms, see the fundamental paper [97] of M.Fukushima as well as the monographs [99], [102], [251], [35], [204]. Given an L^p sub-Markovian semigroup $(T_t^{(p)})_{t\geq 0}$ on $L^p(\mathbb{R}^n;\mathbb{R})$ one is tempted to define a transition function by

$$p_t(x,A) = T_t^{(p)} \chi_A(x), \quad \lambda^{(n)}(A) < \infty. \tag{3.69}$$

However, the left hand side in (3.69) is only defined as an L^p-function, i.e. it is only unique up to a set $N(A,t)$ of Lebesgue measure zero $\lambda^{(n)}(N(A,t)) = 0$. Since an application of the Kolmogorov theorem requires a simultaneous control of all the sets $N(A,t)$, we cannot construct the process along this line. The idea of M.Fukushima was (for $p = 2$ and $(T_t^{(2)})_{t\geq 0}$ symmetric) to use quasi-continuous modifications of $T_t^{(2)}u$ for sufficiently many u in order to overcome this problem.

In this section we will discuss first in the case of analytic sub-Markovian semigroups how it is possible to obtain suitable refinements of $T_t^{(p)}u$ which are eventually good enough for an application of the Kolmogorov theorem. In this first part of the section we will assume the semigroup $(T_t^{(p)})_{t\geq 0}$ to be analytic, often we will also require that $(T_t^{(p)})_{t\geq 0}$ is symmetric (on $L^p(\mathbb{R}^n;\mathbb{R})\cap L^2(\mathbb{R};\mathbb{R})$). One should recall that symmetric sub-Markovian semigroups are always analytic, compare E.M.Stein's result [255], or Theorem I.4.2.12. Further, by the result of A.Carasso and T.Kato [46], or Theorem I.4.3.26, many subordinate semigroups are analytic even if the original semigroup was not. One of the fundamental properties of an analytic semigroup $(T_t^{(p)})_{t\geq 0}$ with generator $(A^{(p)}, D(A^{(p)}))$ is

$$T_t^{(p)}u \in \bigcap_{k\in\mathbb{N}} D((-A^{(p)})^k), \quad u \in L^p(\mathbb{R}^n;\mathbb{R}), \tag{3.70}$$

compare Proposition I.4.2.5, in particular (I.4.92). Combining (3.70) with Corollary 3.1.21 and Theorem 3.1.26 yields

Theorem 3.2.1. *Let $(T_t^{(p)})_{t\geq 0}$ be an analytic L^p-sub-Markovian semigroup with generator $(A^{(p)}, D(A^{(p)}))$. Then we have for all $u \in L^p(\mathbb{R}^n;\mathbb{R})$*

$$T_t^{(p)}u \in \bigcap_{r\geq 1} \mathcal{F}_{r,p}(\mathbb{R}^n;\mathbb{R}) = \bigcap_{k\in\mathbb{N}} D((-A^{(p)})^k). \tag{3.71}$$

Note that Theorem 3.2.1 has the important

Corollary 3.2.2. *Let $(T_t^{(p)})_{t\geq 0}$ be an analytic L^p-sub-Markovian semigroup with generator $(A^{(p)}, D(A^{(p)}))$ and assume that for some $k_0 \in \mathbb{N}$ we have $D((-A^{(p)})^{k_0}) \subset C(\mathbb{R}^n;\mathbb{R}) \cap L^p(\mathbb{R}^n;\mathbb{R})$. Then it follows for all $t > 0$ that*

$$T_t^{(p)} : L^p(\mathbb{R}^n;\mathbb{R}) \to C(\mathbb{R}^n;\mathbb{R})\cap L^p(\mathbb{R}^n;\mathbb{R}). \tag{3.72}$$

In particular, for any bounded Borel set $A \subset \mathbb{R}^n$ the function $x \mapsto T_t^{(p)}\chi_A(x)$ is continuous in the sense that the equivalence class of $T_t^{(p)}\chi_A$ admits a unique continuous representative.

Definition 3.2.3. Let $(T_t^{(p)})_{t\geq 0}$ be an L^p-sub-Markovian semigroup. We call $(T_t^{(p)})_{t\geq 0}$ a *strong L^p-sub-Markovian semigroup* if (3.72) holds for $t > 0$.

Remark 3.2.4. A. Note that Definition 3.2.3 has a close analogy to the definition of a strong Feller semigroup, compare Definition I.4.8.6.

B. From Corollary 3.2.2 it is obvious that one of the most important problems in the theory is to characterize the spaces $\mathcal{F}_{r,p}(\mathbb{R}^n;\mathbb{R})$ in terms of function spaces. We will turn to this problem in the Section 3.3

An analytic L^p-sub-Markovian semigroup is in general not a strongly L^p-sub-Markovian semigroup and in order to get a "nice" representative for $T_t\chi_A$ we will use the capacities $\text{cap}_{r,p}$ introduced in the previous section.

We start with a few auxiliary results.

Lemma 3.2.5. *Let $(T_t^{(p)})_{t\geq 0}$ be an analytic L^p-sub-Markovian semigroup. Further let $(u_\nu)_{\nu\in\mathbb{N}}$, $u_\nu \in L^p(\mathbb{R}^n;\mathbb{R})$, be a sequence converging in $L^p(\mathbb{R}^n;\mathbb{R})$ to $u \in L^p(\mathbb{R}^n;\mathbb{R})$. For all $0 < r \leq 2$ it follows that $(T_t^{(p)}u_\nu)_{\nu\in\mathbb{N}}$ converges to $T_t^{(p)}u$ in $\mathcal{F}_{r,p}(\mathbb{R}^n;\mathbb{R})$. Moreover $T_t^{(p)}u \to u$ as $t \to 0$ in $\|\cdot\|_{\mathcal{F}_{r,p}}$, provided $u \in \mathcal{F}_{r,p}(\mathbb{R}^n;\mathbb{R})$.*

Proof: The analyticity of $T_t^{(p)})_{t\geq 0}$ yields that $T_t^{(p)}u_\nu$, $T_t^{(p)}u \in \mathcal{F}_{r,p}(\mathbb{R}^n;\mathbb{R})$. Further we find for $S_t^{(p)} := e^{-t}T_t^{(p)}$

$$\begin{aligned}\|S_t^{(p)}(u_\nu - u)\|_{\mathcal{F}_{r,p}} &= \|(\mathrm{id} - A^{(p)})^{r/2}S_t^{(p)}(u_\nu - u)\|_{L^p}\\ &= \|(\mathrm{id} - A^{(p)})^{\frac{(r-2)}{2}}(\mathrm{id} - A^{(p)})S_t^{(p)}(u_\nu - u)\|_{L^p}\\ &\leq \|(\mathrm{id} - A^{(p)})S_t^{(p)}(u_\nu - u)\|_{L^p} \leq \frac{c}{t}\|u_\nu - u\|_{L^p},\end{aligned}$$

where we used the fact that the semigroup $(e^{-t}S_t^{(p)})_{t\geq 0}$ is also analytic, $(\mathrm{id} - A^{(p)})^{-s/2}$, $s > 0$, is an L^p-contraction, see Theorem 3.1.19 and (3.31) and for the last line we used (I.4.101). Thus we arrive at

$$\|T_t^{(p)}(u_\nu - u)\|_{\mathcal{F}_{r,p}} \leq c\frac{e^t}{t}\|u_\nu - u\|_{L^p},$$

implying that $T_t^{(p)}u_\nu \to T_t^{(p)}u$ in $\mathcal{F}_{r,p}(\mathbb{R}^n;\mathbb{R})$ as $\nu \to \infty$. The last statement follows from

$$\begin{aligned}\|T_t^{(p)}u - u\|_{\mathcal{F}_{r,p}} &= \|(\mathrm{id} - A^{(p)})^{r/2}(T_t^{(p)}u - u)\|_{L^p}\\ &= \|T_t^{(p)}(\mathrm{id} - A^{(p)})^{r/2}u - (\mathrm{id} - A^{(p)})^{r/2}u\|_{L^p}\end{aligned}$$

and the continuity of $(T_t^{(p)})_{t\geq 0}$. □

Lemma 3.2.6. *Let $u \in C_0(\mathbb{R}^n;\mathbb{R})$, $u \geq 0$, and let $(u_k)_{k\in\mathbb{N}}$, $u_k \in C_0(\mathbb{R}^n;\mathbb{R}) \cap L^p(\mathbb{R}^n;\mathbb{R})$, be a sequence such that $u_k \to u$ in $L^p(\mathbb{R}^n;\mathbb{R})$ and $u_k \to u$ uniformly on $\mathbb{R}^n$. Then there exists a sequence $(v_k)_{k\in\mathbb{N}}$, $v_k \in C_0(\mathbb{R}^n;\mathbb{R}) \cap L^p(\mathbb{R}^n;\mathbb{R})$, such that $0 \leq v_k \leq v_{k+1} \leq u$, $v_k \to u$ uniformly and in $L^p(\mathbb{R}^n;\mathbb{R})$.*

Proof: Define $K := \operatorname{supp} u$ and $w_k := u_k\chi_K$ and $s_k := w_k \vee 0$. It follows that

$$|u - s_k| \leq |u - w_k| \leq |u - u_k|,$$

hence $s_k \to u$ in $L^p(\mathbb{R}^n;\mathbb{R})$ and uniformly. Now set

$$v_k := (s_1 \vee \ldots \vee s_k) \wedge u,$$

thus $v_k \geq 0$ and $v_k \leq v_{k+1} \leq u$. In addition we have

$$0 \leq u - v_k = u - (s_1 \wedge u) \vee \ldots \vee (s_k \wedge u)$$

and further

$$\begin{aligned} |u - v_k| &= u - v_k \leq u - (s_k \wedge u) = |(s_k \wedge u) - u| \\ &= \left|\frac{1}{2}(s_k + u - |s_k - u| - 2u)\right| = \frac{1}{2}|(s_k - u) - |s_k - u|| \\ &\leq |s_k - u|, \end{aligned}$$

which implies the lemma. □

Further we need a variant of the monotone class argument.

Lemma 3.2.7. *Let H be a family of non-negative functions on $\mathbb{R}^n$ such that*

$$u_1, u_2 \in H \text{ and } \lambda_1 u_1 + \lambda_2 u_2 \geq 0, \text{ then } \lambda_1 u_1 + \lambda_2 u_2 \in H; \tag{3.73}$$

$$\text{if } (u_\nu)_{\nu\in\mathbb{N}} \text{ is a sequence in } H \text{ increasing to } u \in L^p(\mathbb{R}^n;\mathbb{R}), \text{ then } u \text{ belongs to } H; \tag{3.74}$$

$$\text{for every open set } G \subset \mathbb{R}^n \text{ there is an increasing sequence } (u_\nu)_{\nu\in\mathbb{N}} \text{ of functions } u_\nu \in H \text{ which converges pointwise to } \chi_G. \tag{3.75}$$

Then H contains all non-negative Borel functions in $L^p(\mathbb{R}^n;\mathbb{R})$, $1 < p < \infty$.

Proof: For a bounded open set $F \subset \mathbb{R}^n$ we set

$$D_G := \{A \subset G; \quad A \text{ is a Borel set and } \chi_A \in H\}.$$

It follows that D_G is a Dynkin system that contains because of (3.74) and (3.75) all open subsets of G, thus $D_G = \mathcal{B}^{(n)}(G)$ for a bounded open set. Now let $u \in L^p(\mathbb{R}^n; \mathbb{R})$ be any non-negative Borel function and take a sequence of bounded open set G_ν, $\nu \in \mathbb{N}$, increasing to $\mathbb{R}^n$. It follows that $u\chi_{G_\nu} \in H$ for all $\nu \in \mathbb{N}$ and further that $u\chi_{G_\nu}$ increases to u. Hence $u \in H$. □

The following lemma is taken from [102].

Lemma 3.2.8. *Let F be a set and $G \subset F$ be a countable subset. Further let S be a countable family of mappings $s : F \times F \to F$. Then there exists a countable set H such that $G \subset H \subset F$ and $s(H \times H) \subset H$ for all $s \in S$.*

Proof: Let $(s_l)_{l \in \mathbb{N}}$ be a sequence of mappings $s_l : F \times F \to F$ such that each $s \in S$ appears infinitely often in this sequence. Further set $G_0 = G$ and for $k \in \mathbb{N}$ set $G_{k+1} = G_k \cup s_{k+1}(G_k \times G_k)$. The set $H := \bigcup_{k=1}^{\infty} G_k$ has the desired properties. In fact, for $x, y \in H$ and $s \in S$ we find $k \in \mathbb{N}$ such that $x, y \in G_k$ and $s = s_{k+1}$. □

Proposition 3.2.9. *Let $(T_t^{(p)})_{t \geq 0}$ be an analytic L^p-sub-Markovian semigroup on $L^p(\mathbb{R}^n; \mathbb{R})$ and for some $0 < r \leq 2$ suppose that $\mathcal{F}_{r,p}(\mathbb{R}^n; \mathbb{R})$ is contraction regular, see Definition 3.1.35.C. Then for the countable set $\mathbb{Q}^+$ there exists a regular nest $(F_\nu^0)_{\nu \in \mathbb{N}}$ and sub-Markovian kernel $(\tilde{p}_t)_{t \in \mathbb{Q}^+}$ on $(\mathbb{R}^n; \mathcal{B}^n)$ such that*

$$\tilde{T}_t^{(p)} u(\cdot) := \int_{\mathbb{R}^n} u(y) \tilde{p}_t(\cdot, dy) \in C_\infty(\{F_\nu^0\}) \quad u \in C_\infty(\mathbb{R}^n; \mathbb{R}). \tag{3.76}$$

Moreover, $\tilde{T}_t^{(p)} u$ is an (r,p)-quasi-continuous modification of $T_t^{(p)} u$ for all $u \in L^p(\mathbb{R}^n; \mathbb{R}) \cap B(\mathbb{R}^n; \mathbb{R})$.

Proof: Let $W \subset \mathcal{F}_{r,p}(\mathbb{R}^n; \mathbb{R}) \cap C_0(\mathbb{R}^n; \mathbb{R})$ be the subspace required in the definition of contraction regularity, see Definition 3.1.35.C. In particular, $u \in W$ implies $|u| \in W$. First we prove the existence of a countable set $V \subset W$ such that V is dense in $(C_0(\mathbb{R}^n; \mathbb{R}); \|\cdot\|_\infty)$ and further, for $u, v \in V$ and $a \in \mathbb{Q}$ it follows that $|u|, u + v, av \in V$. To construct V let $(u_k)_{k \in \mathbb{N}}$ be a dense sequence in $(C_0(\mathbb{R}^n; \mathbb{R}); \|\cdot\|_\infty)$. Since $W \subset C_0(\mathbb{R}^n; \mathbb{R})$ is dense, we can choose for all $k, \nu \in \mathbb{N}$ some $u_{k,\nu} \in W$ such that $\|u_k - u_{k,\nu}\|_\infty \leq \frac{1}{\nu}$. The set $(u_{k,\nu})_{\nu \in \mathbb{N}}$ is dense in $(C_0(\mathbb{R}^n;\mathbb{R}); \|\cdot\|_\infty)$. Now we define the mapping s_0', s_1' and s_a, $a \in \mathbb{Q}$, from $W \times W$ into W by

$$s_0'(u,v) = |u|, \quad s_1'(u,v) = u+v, \quad s_a(u,v) = a\,u.$$

Applying Lemma 3.2.8 with $F = W$, $G = \{u_{k,\nu}; k,\nu \in \mathbb{N}\}$ and $S = \{s_0', s_1'\} \cup \{s_a; a \in \mathbb{Q}\}$ we obtain a countable set with the desired properties. Set

$$H_0 := \bigcup_{t\in\mathbb{Q}^+} T_t^{(p)}(V). \tag{3.77}$$

Since $V \subset W \subset \mathcal{F}_{r,p}(\mathbb{R}^n;\mathbb{R})$ and $\mathcal{F}_{r,p}(\mathbb{R}^n;\mathbb{R})$ is invariant under $T_t^{(p)}$, recall that the semigroup is analytic, it follows that $H_0 \subset \mathcal{F}_{r,p}(\mathbb{R}^n;\mathbb{R})$. According to Theorem 3.1.47 we may choose for each $u \in H_0$ an (r,p)-quasi continuous modification $\tilde{u}$ in the restricted sense. We put $\tilde{H}_0 := \{\tilde{u}; u \in H_0\}$. From Theorem 3.1.45.A we deduce the existence of a regular nest $(F_k^0)_{k\in\mathbb{N}}$ such that $\tilde{H}_0 \subset C_\infty(\{F_k^0\})$. Let $Y_0 := \bigcup_{k=1}^\infty F_k^0$ and fix $t \in \mathbb{Q}^+$. By Theorem 3.1.45.B it follows that

$$(T_t^{(p)}(u+v))^\sim(x) = (T_t^{(p)})^\sim(x) + (T_t^{(p)}v)^\sim(x) \quad \text{for all } x \in Y_0,\ u,v \in V,$$

$$(T_t^{(p)}(au))^\sim(x) = a(T_t^{(p)})^\sim(x) \quad \text{for all } x \in Y_0,$$

and

$$0 \le u \le 1, \quad u \in V, \quad \text{implies } 0 \le (T_t^{(p)}u)^\sim(x) \le 1 \text{ for all } x_0 \in Y_0.$$

Thus we find

$$|(T_t^{(p)}u)^\sim(x)| \le \|u\|_\infty \quad \text{for all } x \in Y_0 \text{ and } u \in V. \tag{3.78}$$

We claim now that for all $x \in Y_0$ there exists a linear positive functional l_x on $C_\infty(\mathbb{R}^n;\mathbb{R})$ satisfying

$$l_x(u) = (T_t^{(p)}u)^\sim(x) \quad \text{for all } u \in V,\ x \in Y_0, \tag{3.79}$$

and

$$|l_x(u)| \le \|u\|_\infty \quad \text{for all } u \in C_\infty(\mathbb{R}^n;\mathbb{R}),\ x \in Y_0. \tag{3.80}$$

For $u \in C_\infty(\mathbb{R}^n;\mathbb{R})$ choose $(u_\nu)_{\nu\in\mathbb{N}}$, $u_\nu \in V$, such that $\lim\limits_{\nu\to\infty} \|u_\nu - u\|_\infty = 0$. It follows that $(T_t^{(p)}u_\nu)^\sim(x)$ converges uniformly on Y_0 to a limit $l_x(u)$. Since

$l_x(|u|) = \lim_{\nu\to\infty} l_x(|u_\nu|) \geq 0$, it follows that l_x is positivity preserving and linear, as well as (3.79) and (3.80). Observe that $l_.(u) : Y_0 \to \mathbb{R}$, $x \mapsto l_x(u)$ belongs to $C_\infty(\{F_k^0\})$ since $(T_t^{(p)} u_\nu)^\sim \in C_\infty(\{F_k^0\})$. Thus by a variant of the Riesz representation theorem, l_x admits a representing measure $\tilde{p}_t(x,\cdot)$ such that

$$l_x(u) = \int_{\mathbb{R}^n} u(y)\tilde{p}_t(x,dy), \quad x \in Y_0, \tag{3.81}$$

and

$$\tilde{p}_t(x,\mathbb{R}^n) \leq 1 \quad \text{for } x \in Y_0. \tag{3.82}$$

We extend $\tilde{p}_t(x,\cdot)$ by defining $\tilde{p}_t(x,A) = 0$ for all $x \in Y_0^c$ and $A \in \mathcal{B}^{(n)}$. It follows that this extension, which we denote once again by $\tilde{p}_t(x,A)$, is a sub-Markovian kernel on $(\mathbb{R}^n;\mathcal{B}^{(n)})$ satisfying (3.76). We put $\tilde{T}_t^{(p)} u := \int_{\mathbb{R}^n} u(y)\tilde{p}_t(\cdot,dy)$ for $u \in L^p(\mathbb{R}^n;\mathbb{R})$. Now, let $u \in C_0(\mathbb{R}^n;\mathbb{R})$, $u \geq 0$. We choose a sequence $(u_k)_{k\in\mathbb{N}}$, $u_k \in V$, converging to u uniformly. For any $w \in V$, $u \leq w$, it follows that $v_k := (0 \vee u_k) \wedge w$ belongs to V and converges to u uniformly and in $L^p(\mathbb{R}^n;\mathbb{R})$, recall that $u \vee v = \frac{1}{2}(u+v+|u-v|)$ and $u \wedge v = \frac{1}{2}(u+v-|u-v|)$. We know that $l_x(v_k)$ is an (r,p)-quasi-continuous modification of $T_t^{(p)} v_k$ and that $l_x(v_k) \to l_x(u)$ uniformly on Y_0, i.e. in particular

$$\int_{\mathbb{R}^n} u_k(y)\tilde{p}_t(x,dy) \to \int_{\mathbb{R}^n} u(y)\tilde{p}_t(x,dy) \quad (r,p)-q.e.$$

On the other hand it follows that $T_t^{(p)} v_k \to T_t^{(p)} u$ in $\mathcal{F}_{r,p}(\mathbb{R}^n;\mathbb{R})$ due to Lemma 3.2.5. Thus Corollary 3.1.50 implies that $x \mapsto \int_{\mathbb{R}^n} u(y)\tilde{p}_t(x,dy)$ is an (r,p)-quasi-continuous modification of $u \in C_0(\mathbb{R}^n;\mathbb{R})$, $u \geq 0$. Now, set

$$H := \{u \in L^p(\mathbb{R}^n;\mathbb{R}) \cap B(\mathbb{R}^n;\mathbb{R}); \quad u \geq 0 \text{ and } \tilde{T}_t^{(p)} u = T_t^{(p)} u \ (r,p)\text{-q.e.}\}.$$

We know that $\{u \in C_0(\mathbb{R}^n;\mathbb{R}); u \geq 0\} \subset H$, implying that H fulfills (3.73) and (3.75) in Lemma 3.2.7. Thus it remains to prove (3.74) in order to see that $\tilde{T}_t^{(p)} u$ is an (r,p)-quasi continuous modification of $T_t^{(p)} u$ for all $u \in L^p(\mathbb{R}^n;\mathbb{R}) \cap B(\mathbb{R}^n;\mathbb{R})$, $u \geq 0$. For this let $(u_\nu)_{\nu\in\mathbb{N}}$ be a sequence in H increasing to some $u \in L^p(\mathbb{R}^n;\mathbb{R})$. In particular we have u_ν, $u \geq 0$. It follows by the dominated convergence theorem that $u_\nu \to u$ in $L^p(\mathbb{R}^n;\mathbb{R})$ which yields by Lemma 3.2.5 that $T_t^{(p)} u_\nu \to T_t^{(p)} u$ in $\|\cdot\|_{\mathcal{F}_{r,p}}$. Therefore, by Proposition 3.1.49 a subsequence $(T_t^{(p)} u_{\nu_k})_{k\in\mathbb{N}}$ has a modification $(\tilde{T}_t^{(p)} u_{\nu_k})_{k\in\mathbb{N}}$ that converges

(r,p)-quasi-uniformly, hence (r,p)-quasi-everywhere, to an (r,p)-quasi continuous modification of $T_t^{(p)}u$. But

$$\tilde{T}_t^{(p)} u_{\nu_k}(x) = \int_{\mathbb{R}^n} u_{\nu_k}(y)\tilde{p}_t(x,dy) \to \int_{\mathbb{R}^n} u(y)\tilde{p}_t(x,dy)$$

on Y_0. Hence $\tilde{T}_t^{(p)}u$ is by Corollary 3.1.50.B an (r,p)-quasi-continuous modification of $T_t^{(p)}u$.

Given $u \in L^p(\mathbb{R}^n;\mathbb{R}) \cap B(\mathbb{R}^n;\mathbb{R})$ and decompose u according to $u = u^+ - u^-$. Each of the functions u^+ and u^- belongs to $L^p(\mathbb{R}^n;\mathbb{R}) \cap B(\mathbb{R}^n;\mathbb{R})$. It follows that $\tilde{T}_t^{(p)}u^+$ and $\tilde{T}_t^{(p)}u^-$ are (r,p)-quasi-continuous modification of $T_t^{(p)}u^+$ and $T_t^{(p)}u^-$, respectively. Hence, $\tilde{T}_t^{(p)}u^+ - \tilde{T}_t^{(p)}u^-$ is an (r,p)-quasi continuous modification of $T_t^{(p)}u$, and clearly we have $\tilde{T}_t^{(p)}u = \tilde{T}_t^{(p)}u^+ - \tilde{T}_t^{(p)}u^-$. Hence the proposition is proved. □

By Proposition 3.2.9 we have constructed for each $t \in \mathbb{Q}^+$ a kernel $\tilde{p}_t(x,dy)$ on $(\mathbb{R}^n;\mathcal{B}^n)$ such that for every $u \in L^p(\mathbb{R}^n;\mathbb{R}) \cap B(\mathbb{R}^n;\mathbb{R})$ an (r,p)-quasi continuous modification of $T_t^{(p)}u$ is given by $\tilde{T}_t^{(p)}u$. Our next aim is to construct kernels $p_t(x,dy)$ on $(\mathbb{R}^n;\mathcal{B}^n)$ such that $(p_t(x,dy))_{t\in\mathbb{Q}^+}$ form a family of sub-Markovian kernels satisfying the Chapman–Kolmogorov equations and leading to a quasi-continuous representation of $T_t^{(p)}u$. We set

$$\tilde{\mathcal{F}}_{r,p}(\mathbb{R}^n;\mathbb{R}) := \{\tilde{u}; \tilde{u} = u \quad \text{q.e. and } \mathcal{F}_{r,p}(\mathbb{R}^n;\mathbb{R})\}.$$

Note that there is a small abuse of notation. $\tilde{\mathcal{F}}_{r,p}(\mathbb{R}^n;\mathbb{R})$ consists of equivalence classes of functions which coincide in the complement of a set of (r,p)-capacity zero.

Lemma 3.2.10. *Let $\tilde{H} \subset \tilde{\mathcal{F}}_{r,p}(\mathbb{R}^n;\mathbb{R}) \cap B_b(\mathbb{R}^n;\mathbb{R})$ being the smallest algebra over $\mathbb{Q}$ such that $V \subset \tilde{H}$, V taken from the proof of Proposition 3.2.9, and for all $t \in \mathbb{Q}^+$ and $u \in \tilde{H}$ it follows that $\tilde{T}_t^{(p)}u \in \tilde{H}$. Then $\tilde{H}$ is countable.*

Proof: We consider the family of mappings from $(\mathcal{F}_{r,p}(\mathbb{R}^n;\mathbb{R}) \cap B_b(\mathbb{R}^n;\mathbb{R}))^2$ into $\mathcal{F}_{r,p}(\mathbb{R}^n;\mathbb{R}) \cap B_b(\mathbb{R}^n;\mathbb{R})$ consisting of

$$\begin{aligned} s_1(u,v) &= u+v, \\ s_{2,a}(u,v) &= au, \quad a \in \mathbb{Q}, \\ s_3(u,v) &= u \cdot v, \\ s_{4,t}(u,v) &= \tilde{T}_t^{(p)}u \quad t \in \mathbb{Q}^+, \end{aligned}$$

and apply Lemma 3.2.8 with $G = V$ and $S = \{s_1, s_3\} \cup \{s_{2,a}; a \in \mathbb{Q}\} \cup \{s_{4,t}; t \in \mathbb{Q}^+\}$. □

Lemma 3.2.11. *There exists a regular nest $(F_k)_{k\in\mathbb{N}}$ with the following properties*

$$\tilde{H} \subset C_\infty(\{F_k\}), \quad F_k \subset F_k^0, \quad k \in \mathbb{N}, \quad \text{where } F_k^0 \text{ is taken from the proof of Proposition 3.2.9}; \tag{3.83}$$

$$\text{for all } x \in Y_1 := \bigcup_{k=1}^{\infty} F_k \text{ and for all } u \in \tilde{H} \text{ there exists a sequence } (t_k)_{k\in\mathbb{N}},\ t_k \in \mathbb{Q}^+, \text{ decreasing to 0 such that } \tilde{T}_{t_k}^{(p)} u(x) \to u(x) \text{ for all } x \in Y_1; \tag{3.84}$$

$$\text{for all } x \in Y_1,\ t, s \in \mathbb{Q}_+, \text{ and } u \in \tilde{H} \text{ we have } (\tilde{T}_t^{(p)} \circ \tilde{T}_s^{(p)})u(x) = \tilde{T}_{t+s}^{(p)} u(x). \tag{3.85}$$

Proof: For each $u \in \tilde{H}$ there exists a sequence $(t_k)_{k\in\mathbb{N}}$, $t_k \in \mathbb{Q}^+$, decreasing to zero such that (3.84) holds (r,p)-quasi everywhere, compare Corollary 3.1.50 and recall Lemma 3.2.5. Since $\tilde{H}$ is countable, we can using a diagonal argument choose $(t_k)_{k\in\mathbb{N}}$ independent of $u \in \tilde{H}$, i.e. there is a set $N \subset \mathbb{R}^n$, $\mathrm{cap}_{r,p}(N) = 0$, such that (3.84) holds for all $x \in N^c$. Next we apply Theorem 3.1.45 to find a nest $(F_k')_{k\in\mathbb{N}}$ such that (3.83) holds for $(F_k')_{k\in\mathbb{N}}$ and $N \subset \mathbb{R}^n \setminus \bigcup_{k=1}^{\infty} F_k'$. Taking the regularization $(F_k)_{k\in\mathbb{N}}$ of $(F_k')_{k\in\mathbb{N}}$ in the sense of Lemma 3.1.44, we obtain (3.83) and (3.84), whereas (3.85) follows from Theorem 3.1.45 B. In the sense of a.e. equality we have

$$\tilde{T}_t \circ \tilde{T}_s u = T_t \circ T_s u = T_{t+s} u = \tilde{T}_{t+s} u,$$

i.e.

$$\tilde{T}_t \circ \tilde{T}_s u = \tilde{T}_{s+t} u \quad a.e.,$$

which implies that $\tilde{T}_t \circ \tilde{T}_s u = \tilde{T}_{s+t} u$ q.e. on $Y_1 := \bigcup_{\nu=1}^{\infty} F_\nu$. □

Now we may prove

Theorem 3.2.12. *Let $(T_t^{(p)})_{t\geq 0}$ be an analytic L^p-sub-Markovian semigroup and suppose that $\mathcal{F}_{r,p}(\mathbb{R}^n;\mathbb{R})$, $0 < r \leq 2$, is contraction regular. Then there* *exists a family of sub-Markovian kernels $p_t(x, dy)$, $t \in \mathbb{Q}^+$, on $(\mathbb{R}^n, \mathcal{B}^n)$ and a Borel set $Y_2 \subset Y_1$, Y_1 as in Lemma 3.2.11, with the following properties ($\tilde{p}_t(x, dy)$ from Proposition 3.2.9.)*

$$\mathrm{cap}_{r,p}(Y_2^c) = 0 \quad \text{and } \tilde{p}_t(x, Y_2^c) = 0 \text{ for all } x \in Y_2,\ t \in \mathbb{Q}^+, \tag{3.86}$$

$$p_{t+s}(x, A) = \int_{\mathbb{R}^n} p_t(y, A) p_s(x, dy), \quad x \in \mathbb{R}^n, \quad A \in \mathcal{B}^n, \quad s, t \in \mathbb{Q}^+, \tag{3.87}$$

and

$$x \mapsto \int_{\mathbb{R}^n} u(y) p_t(x, dy) \quad t \in \mathbb{Q}^+, \quad u \in L^p(\mathbb{R}^n; \mathbb{R}) \cap B(\mathbb{R}^n; \mathbb{R}) \tag{3.88}$$

is a quasi-continuous modification of $T_t^{(p)} u$.

Proof: Since $\lambda^{(p)}(Y_1^c) = 0$ it follows that $T_t^{(p)} \chi_{Y_1^c} = 0$ a.e. But $\tilde{T}_t^{(p)} \chi_{Y_1^c} = \tilde{p}_t(x, Y_1^c)$ is an (r,p)-quasi continuous modification of $T_t^{(p)} \chi_{Y_1^c}$ by Proposition 3.2.9. Thus there exists a Borel set $Y_1^{(1)} \subset Y_1$, $\mathrm{cap}_{r,p}((Y_1^{(1)})^c) = 0$ and $\tilde{p}_t(x, Y_1^c) = 0$ for all $x \in Y_1^{(1)}$ and $t \in \mathbb{Q}^+$. By induction we find a sequence $(Y_1^{(k)})_{k \in \mathbb{N}}$ of Borel sets such that $Y_1^{(k+1)} \subset Y_1^{(k)} \subset Y_1$, $\mathrm{cap}_{r,p}(Y_1^{(k+1)}) = 0$ and $\tilde{p}_t(x, (Y_1^{(k)})^c) = 0$ for all $x \in Y_1^{(k+1)}$ and $t \in \mathbb{Q}^+$. The set $Y_2 = \bigcap_{k=1}^{\infty} Y_1^{(k)}$ fulfills (3.86). We define now

$$p_t(x, A) := \begin{cases} \tilde{p}_t(x, A) & , x \in Y_2, \quad A \in \mathcal{B}^{(n)} \\ 0 & , x \in Y_2^c \quad A \in \mathcal{B}^{(n)}. \end{cases} \tag{3.89}$$

In order to prove (3.86) it is sufficient to show

$$P_t(P_s u)(x) = P_{t+s} u(x) \quad \text{for all } x \in \mathbb{R}^n,\ t, s \in \mathbb{Q}^+,\ u \in V, \tag{3.90}$$

where

$$P_t u(x) = \int_{\mathbb{R}^n} u(y) p_t(x, dy).$$

Note that both sides in (3.90) (or (3.87)) do vanish for $x \in Y_2^c$. For $x \in Y_2$ it follows from (3.86) that

$$P_t(P_s u)(x) = \tilde{T}_t^{(p)}(P_s u)(x) = \tilde{T}_t^{(p)}(\tilde{T}_s^{(p)} u)(x),$$

but from (3.85) it follows for these x and t, s that

$$\tilde{T}_t^{(p)}(\tilde{T}_s^{(p)} u)(x) = (\tilde{T}_{t+s}^{(p)} u)(x) = P_{t+s} u(x),$$

implying (3.90), i.e. (3.87). Finally, (3.88) follows from Proposition 3.2.9. □

Note that still some work is required to pass in Theorem 3.2.12 from $t \in \mathbb{Q}^+$ to $t \in \mathbb{R}^+$.

Let us return to a general, not necessarily analytic L^p-sub-Markovian semigroup $(T_t^{(p)})_{t\geq 0}$. Our aim is to find some additional characterization of the equilibrium potential. For this some duality theory is needed. Thus we assume that $1 < p < \infty$ and that the dual semigroup $(T_t^{(p)*})_{t\geq 0}$ of $(T_t^{(p)})_{t\geq 0}$ which is a contraction semigroup on $L^{p'}(\mathbb{R}^n;\mathbb{R})$, $p' = \frac{p}{p-1}$, is itself $L^{p'}$-sub-Markovian. This is indeed a serious restriction. There are L^p-sub-Markovian semigroups with a dual semigroup not being sub-Markovian, compare E.M.Ouhabaz [229], or [203].

First we need a characterization of the dual space $\mathcal{F}_{r,p}^*(\mathbb{R}^n;\mathbb{R})$ of $\mathcal{F}_{r,p}(\mathbb{R}^n;\mathbb{R})$. For $(T_t^{(p)})_{t\geq 0}$ and $(T_t^{(p)*})_{t\geq 0}$ as above we know that

$$((\mathrm{id} - A^{(p)})^{-r/2})^* = (\mathrm{id} - A^{(p)*})^{-r/2} \tag{3.91}$$

holds. Writing L^p instead of $L^p(\mathbb{R}^n;\mathbb{R})$, $\mathcal{F}_{r,p}$ instead of $\mathcal{F}_{r,p}(\mathbb{R}^n;\mathbb{R})$ etc. we set as before $\mathcal{F}_{r,p} = (\mathrm{id} - A^{(p)})^{-r/2}(L^p)$ and in addition we write

$$\mathcal{F}_{r,p^*} := (\mathrm{id} - A^{(p)*})^{-r/2}(L^{p'}), \quad p' = \frac{p}{p-1}. \tag{3.92}$$

Clearly, $\mathcal{F}_{r,p^*}$, $r \geq 0$, is the scale of abstract Bessel potential spaces associated with $(T_t^{(p)*})_{t\geq 0}$. In order to determine $\mathcal{F}_{r,p^*}$ we use the method of negative norms due to P.Lax [191], see also K.Yosida [288]. Thus on $L^{p'}$, $p' = \frac{p}{p-1}$, we introduce the family of norms

$$\|g\|_{\mathcal{H}_{-r,p'}} := \sup_{\substack{u\in\mathcal{F}_{r,p}\\ u\neq 0}} \frac{\left|\int g\,u\,dx\right|}{\|u\|_{\mathcal{F}_{r,p}}}. \tag{3.93}$$

Clearly, $\|\cdot\|_{\mathcal{H}_{-r,p'}}$ is a norm for $r \geq 0$ and $1 < p < \infty$. Define the space $\mathcal{H}_{-r,p'}$ as completion of $L^{p'}$ with respect to the norm $\|\cdot\|_{\mathcal{H}_{-r,p'}}$, i.e.

$$\mathcal{H}_{-r,p'} := \overline{L^{p'}}^{\|\cdot\|_{\mathcal{H}_{-r,p'}}}. \tag{3.94}$$

Let $g \in \mathcal{H}_{-r,p'}$ and $(g_\nu)_{\nu\in\mathbb{N}}$, $g_\nu \in L^{p'}$, be a sequence converging in $\|\cdot\|_{-r,p}$ to g. We define the linear mapping $T_g : \mathcal{F}_{r,p} \to \mathbb{R}$ by

$$T_g(\varphi) := \lim_{\nu\to\infty} \int_{\mathbb{R}^n} g_\nu \varphi\,dx, \qquad \varphi \in \mathcal{F}_{r,p}. \tag{3.95}$$

Lemma 3.2.13. *The mapping T_g is independent of the choice of the approximating sequence $(g_\nu)_{\nu\in\mathbb{N}}$.*

Proof: Suppose that $(g_\nu)_{\nu\in\mathbb{N}}$ and $(h_\nu)_{\nu\in\mathbb{N}}$, $g_\nu, h_\nu \in L^{p'}$, are two sequences converging in $\mathcal{H}_{-r,p'}$ to g. For all $\varphi \in \mathcal{F}_{r,p}$ it follows that

$$\Big|\int_{\mathbb{R}^n} (g_\nu - h_\nu)\varphi\, dx\Big| \leq \|g_\nu - h_\nu\|_{\mathcal{H}_{-r,p}} \|\varphi\|_{\mathcal{F}_{r,p}},$$

implying that

$$\lim_{\nu\to\infty} \int_{\mathbb{R}^n} (g_\nu - h_\nu)\varphi\, dx = 0.$$

Now we find

$$\begin{aligned} T_g(\varphi) &= \lim_{\nu\to\infty} \int_{\mathbb{R}^n} g_\nu\varphi\, dx \\ &= \lim_{\nu\to\infty} \Big(\int_{\mathbb{R}^n} h_\nu\varphi\, dx + \int_{\mathbb{R}^n} (g_\nu - h_\nu)\varphi\, dx\Big) \\ &= \lim_{\nu\to\infty} \int_{\mathbb{R}^n} h_\nu\varphi\, dx = T_h(\varphi) \end{aligned}$$

for all $\varphi \in \mathcal{F}_{r,p}$.

Lemma 3.2.14. *The mapping T_g is a continuous linear functional on $\mathcal{F}_{r,p}$, i.e. $T_g \in \mathcal{F}^*_{r,p}$.*

Proof: Let $(g_\nu)_{\nu\in\mathbb{N}}$, $g_\nu \in L^{p'}$, be a sequence converging in $\mathcal{H}_{-r,p'}$ to g. Given $\varphi = (\mathrm{id} - A^{(p)})^{-r/2}\psi \in \mathcal{F}_{r,p}$, $\psi \in L^p$, $\psi \neq 0$, we find

$$\begin{aligned} \frac{\big|\int_{\mathbb{R}^n} g_\nu\varphi\, dx\big|}{\|\varphi\|_{\mathcal{F}_{r,p}}} &= \frac{\big|\int_{\mathbb{R}^n} (\mathrm{id} - A^{(p)*})^{-r/2} g_\nu (\mathrm{id} - A^{(p)})^{r/2}\varphi\, dx\big|}{\|\varphi\|_{\mathcal{F}_{r,p}}} \\ &= \frac{\big|\int_{\mathbb{R}^n} (\mathrm{id} - A^{(p)*})^{-r/2} g_\nu \cdot \psi\, dx\big|}{\|\psi\|_{L^p}}. \end{aligned}$$

Since $\|g_\nu\|_{\mathcal{H}_{-r,p'}} \leq c$ we find

$$\|(\mathrm{id} - A^{(p)*})^{-r/2} g_\nu\|_{L^{p'}} = \sup_{\substack{\psi\in L^p \\ \psi\neq 0}} \frac{\big|\int_{\mathbb{R}^n} (\mathrm{id} - A^{(p)*})^{-r/2} g_\nu\psi\, dx\big|}{\|\psi\|_{L^p}} \leq c. \tag{3.96}$$

Furthermore it follows that

$$\begin{aligned}|T_g(\varphi)| &= \lim_{\nu\to\infty}\left|\int_{\mathbb{R}^n} g_\nu\varphi\,dx\right| \\ &= \lim_{\nu\to\infty}\left|\int_{\mathbb{R}^n}(\mathrm{id}-A^{(p)*})^{-r/2}g_\nu(\mathrm{id}-A^{(p)})^{r/2}\varphi\,dx\right| \\ &\le \lim_{\nu\to\infty}\sup\|(\mathrm{id}-A^{(p)*})^{-r/2}g_\nu\|_{L^{p'}}\|(\mathrm{id}-A^{(p)})^{r/2}\varphi\|_{L^p} \\ &\le c\|\varphi\|_{\mathcal{F}_{r,p}},\end{aligned}$$

which proves the lemma. □

Theorem 3.2.15. *With $\mathcal{F}_{r,p}$ and $\mathcal{H}_{-r,p}$ as before we have*

$$\mathcal{F}^*_{r,p} = \mathcal{H}_{-r,p}. \tag{3.97}$$

Proof: We know already that $\mathcal{H}_{-r,p'} \subset \mathcal{F}^*_{r,p}$. Let $g \in \mathcal{F}^*_{r,p}$. It follows that

$$|\langle g,\varphi\rangle| \le c\|\varphi\|_{\mathcal{F}_{r,p}} = c\|\psi\|_{L^p}$$

for all $\varphi \in \mathcal{F}_{r,p}$, $\varphi = (\mathrm{id}-A^{(p)})^{-r/2}\psi$ with some $\psi \in L^p$. Thus we have

$$|\langle g,(\mathrm{id}-A^{(p)})^{-r/2}\psi\rangle,| \le c\|\psi\|_{L^p} \tag{3.98}$$

for all $\psi \in L^p$. Define $H_g : L^p \to \mathbb{R}$ by

$$H_g(\psi) := \langle g,(\mathrm{id}-A^{(p)})^{-r/2}\psi\rangle, \tag{3.99}$$

i.e. $H_g \in (L^p)^*$ by (3.98). Hence, there exists $h_g \in L^{p'}$, $p' = \frac{p}{p-1}$, such that

$$H_g(\psi) = \int_{\mathbb{R}^n} h_g\psi\,dx$$

for all $\psi \in L^p$, or,

$$\langle g,(\mathrm{id}-A^{(p)})^{-r/2}\psi\rangle = \int_{\mathbb{R}^n} h_g\psi\,dx. \tag{3.100}$$

Now, take a sequence $(h_\nu)_{\nu\in\mathbb{N}}$, $h_\nu \in \mathcal{F}_{r,p^*}$, such that $h_\nu \to h$ in $L^{p'}$, recall that $\mathcal{F}_{r,p^*} \subset L^{p'}$ is a dense subspace. With $g_\nu \in L^{p'}$ defined by $(\mathrm{id}-A^{(p)*})^{-r/2}g_\nu =$

h_ν we arrive at

$$\begin{aligned}\langle g, (\mathrm{id} - A^{(p)})^{-r/2}\psi\rangle &= \lim_{\nu\to\infty} \int_{\mathbb{R}^n} h_\nu \psi \, dx \\ &= \lim_{\nu\to\infty} \int_{\mathbb{R}^n} (\mathrm{id} - A^{(p)*})^{-r/2} g_\nu \cdot \psi \, dx \\ &= \lim_{\nu\to\infty} \int_{\mathbb{R}^n} g_\nu (\mathrm{id} - A^{(p)})^{-r/2} \psi \, dx \\ &= \lim_{\nu\to\infty} \langle g_\nu, \varphi\rangle,\end{aligned}$$

i.e.

$$\lim_{\nu\to\infty} \langle g_\nu, \varphi\rangle = \langle g, \varphi\rangle. \tag{3.101}$$

On the other hand it follows that

$$\begin{aligned}\frac{\left|\int_{\mathbb{R}^n} (g_\nu - g_\mu)\varphi \, dx\right|}{\|\varphi\|_{\mathcal{F}_{r,p}}} &= \frac{\left|\int_{\mathbb{R}^n} (h_\nu - h_\mu)(\mathrm{id} - A^{(p)})^{-r/2}\varphi \, dx\right|}{\|\varphi\|_{\mathcal{F}_{r,p}}} \\ &\leq \|h_\nu - h_\mu\|_{L^{p'}},\end{aligned}$$

implying that $(g_\nu)_{\nu\in\mathbb{N}}$ is a Cauchy sequence in $\mathcal{H}_{-r,p'}$. Thus there exists $\tilde{g} \in \mathcal{H}_{-r,p}$ such that $g_\nu \to \tilde{g}$ in $\mathcal{H}_{-r,p}$ which yields for $\varphi \in \mathcal{F}_{r,p}$

$$\langle \tilde{g}, \varphi\rangle = \lim_{\nu\to\infty} \langle g_\nu, \varphi\rangle = \langle g, \varphi\rangle,$$

i.e. $\tilde{g} = g$ implying $\mathcal{F}^*_{r,p} \subset \mathcal{H}_{-r,p'}$ and the theorem is proved. □

We are looking for a different characterization of $\mathcal{F}^*_{r,p}$. Since $\|\varphi\|_{\mathcal{F}_{r,p}} \leq \|\varphi\|_{\mathcal{F}_{r+t,p}}$ for all $\varphi \in \mathcal{F}_{r+t,p}$, $r, t \geq 0$, it follows that

$$\|g\|_{\mathcal{H}_{-(r+t),p'}} \leq \|g\|_{\mathcal{H}_{-r,p'}}$$

for all $g \in \mathcal{H}_{-r,p'}$. For $h \in L^{p'}$ we define

$$(\mathrm{id} - A^{(p)*})^{r/2} h = g \in \mathcal{H}_{-r,p'} \tag{3.102}$$

by

$$\langle g, \varphi\rangle = \int_{\mathbb{R}^n} h (\mathrm{id} - A^{(p)})^{r/2} \varphi \, dx \tag{3.103}$$

for all $\varphi \in \mathcal{F}_{r,p}$. Thus given $h \in L^{p'}$, it follows that $(\mathrm{id} - A^{(p)*})^{r/2} h \in \mathcal{H}_{-r,p'}$. On the other hand, given $g \in \mathcal{H}_{-r,p'}$ there exists a sequence $(g_\nu)_{\nu \in \mathbb{N}}$, $g_\nu \in L^{p'}$, such that $g_\nu \to g$ in $\mathcal{H}_{-r,p'}$. Defining $h_\nu := (\mathrm{id} - A^{(p)*})^{-r/2} g_\nu$ we find

$$\int_{\mathbb{R}^n} g_\nu \varphi \, dx = \int_{\mathbb{R}^n} h_\nu (\mathrm{id} - A^{(p)})^{r/2} \varphi \, dx.$$

As in the proof of Theorem 3.2.15 it follows that $(h_\nu)_{\nu \in \mathbb{N}}$ is a Cauchy sequence in $L^{p'}$ and for its limit $h \in L^{p'}$ we find

$$\langle g, \varphi \rangle = \int_{\mathbb{R}^n} h (\mathrm{id} - A^{(p)})^{r/2} \varphi \, dx. \tag{3.104}$$

Thus we have proved

Theorem 3.2.16. *For the dual space $\mathcal{F}^*_{r,p}$ of $\mathcal{F}_{r,p}$ it holds*

$$\mathcal{F}^*_{r,p} = (\mathrm{id} - A^{(p)*})^{r/2} L^{p'} \tag{3.105}$$

with norm

$$\|g\|_{\mathcal{F}^*_{r,p}} = \|g\|_{\mathcal{H}_{-r,p'}} = \|h\|_{L^{p'}}, \tag{3.106}$$

for $g = (\mathrm{id} - A^{(p)*})^{r/2} h$, $h \in L^{p'}$.

From (3.106) it is clear that $(\mathrm{id} - A^{(p)*})^{r/2}$ is an isometry from $L^{p'}$ onto $\mathcal{H}_{-r,p'} = \mathcal{F}^*_{r,p}$.

We will also use a straightforward extension of the definition of the *support* of u, $\operatorname{supp} u$, for $u \in \mathcal{F}^*_{r,p} = \mathcal{H}_{-r,p'}$. For $G \subset \mathbb{R}^n$ and $u \in \mathcal{F}^*_{r,p}$ we say $u|_G = 0$ if $\langle u, \varphi \rangle = 0$ for all $\psi \in \mathcal{F}_{r,p} \cap C_b$ such that $\operatorname{supp} \varphi \subset G$. The support of $u \in \mathcal{F}^*_{r,p}$ is by definition the complement of the largest open set $\tilde{G}$ such that $u|_{\tilde{G}} = 0$.

Let $(T_t^{(p)})_{t \geq 0}$ and $(T_t^{(p)*})_{t \geq 0}$ be as before with corresponding generators $(A^{(p)}, D(A^{(p)}))$ and $(A^{(p)*}, D(A^{(p)*}))$, respectively. We want to study the (r,p)-equilibrium potential u_G for an open set G more closely. In order to be conform with the theory of convex minimization problems and monotone operators it is convenient to introduce on $\mathcal{F}_{r,p}(\mathbb{R}^n; \mathbb{R})$ the functionals

$$E_{r,p}(u) := \frac{1}{p} \|u\|^p_{\mathcal{F}_{r,p}} = \frac{1}{p} \int_{\mathbb{R}^n} |(\mathrm{id} - A^{(p)})^{r/2} u|^p \, dx \tag{3.107}$$

and for $f \in L^{p'}(\mathbb{R}^n; \mathbb{R})$, $p' = \frac{p}{p-1}$,

$$E^f_{r,p}(u) := E_{r,p}(u) - \int_{\mathbb{R}^n} f \, u \, dx. \tag{3.108}$$

We assume that $\mathcal{F}_{r,p}(\mathbb{R}^n;\mathbb{R})$ is regular, see Definition 3.1.35.B, and $p \geq 2$. Note that for $p \geq 2$ the function $g : \mathbb{R} \to \mathbb{R}$, $s \mapsto g(s) := |s|^p$ is a convex function with strictly monotone increasing derivative, i.e. it is a *strictly convex function.* Moreover, g has a continuous second derivative and we have

$$g'(s) = p\, s|s|^{p-1}, \tag{3.109}$$

$$g''(s) = p(p-1)|s|^{p-2}, \tag{3.110}$$

and further the estimate

$$(s|s|^{p-2} - r|r|^{p-2})(s-r) \geq 2^{2-p}|s-r|^p \tag{3.111}$$

holds for all $r, s \in \mathbb{R}$, see E.Zeidler [289], p.503. As a shorthand we use sometimes

$$T_{r,p} := (\mathrm{id} - A^{(p)})^{r/2} \quad \text{and} \quad T^*_{r,p} := (\mathrm{id} - A^{(p)*})^{r/2}. \tag{3.112}$$

Definition 3.2.17. Let X be a Banach space.

A. For a convex set $K \subset X$ a functional $E : K \to \mathbb{R}$ is called *convex* if

$$E((1-\mu)u + \mu v) \leq (1-\mu)E(u) + \mu E(v) \tag{3.113}$$

holds for all $\mu \in [0,1]$ and $u, v \in K$. It is called *strictly convex* if

$$E((1-\mu)u + \mu v) < (1-\mu)E(u) + \mu E(v) \tag{3.114}$$

for all $\mu \in (0,1)$ and $u, v \in K$, $u \neq v$.

B. We call a functional $E : G \to \mathbb{R}$, $G \subset X$ open, *Gâteaux differentiable* if for each $u \in G$ there exists some $a_u \in X^*$ such that

$$\lim_{t \to 0} \frac{E(u+th) - E(u)}{t} = \langle a_u, h \rangle \tag{3.115}$$

for all $h \in X$.

C. The functional $E : H \to \mathbb{R}$, $H \subset X$, is called *coercive* if

$$\frac{E(u)}{\|u\|_X} \to \infty \quad \text{as} \quad \|u\|_X \to \infty, \quad u \in H. \tag{3.116}$$

Lemma 3.2.18. *The functional $E_{r,p} : \mathcal{F}_{r,p}(\mathbb{R}^n;\mathbb{R}) \to \mathbb{R}$, $p \geq 2$, is strictly convex, coercive and Gâteaux differentiable. Its Gâteaux derivative at $u \in \mathcal{F}_{r,p}(\mathbb{R}^n;\mathbb{R})$ is given by*

$$\mathcal{A}^{(p)}_r u := (\mathrm{id} - A^{(p)*})^{r/2}\left(|(\mathrm{id} - A^{(p)})^{r/2} u|^{p-2} (\mathrm{id} - A^{(p)})^{r/2} u\right). \tag{3.117}$$

Proof: The strict convexity is obvious and the coercivity follows from

$$\frac{E_{r,p}(u)}{\|u\|_{\mathcal{F}_{r,p}}} = \frac{1}{p}\|u\|^{p-1}_{\mathcal{F}_{r,p}}.$$

Now, fix $u, h \in \mathcal{F}_{r,p}(\mathbb{R}^n;\mathbb{R})$ and consider on $[0,1]$ the quotient

$$\begin{aligned}
&\frac{E_{r,p}(u+th) - E_{r,p}(u)}{t} \\
&= \frac{1}{p}\int_{\mathbb{R}^n} \frac{|(\mathrm{id} - A^{(p)})^{r/2}(u+th)|^p - |(\mathrm{id} - A^{(p)})^{r/2}u|^p}{t}\,dx \\
&= \int_{\mathbb{R}^n} ((1-\alpha)T_{r,p}(u+th) + \alpha T_{r,p}u) \\
&\times |(1-\alpha)T_{r,p}(u+th) + \alpha T_{r,p}(u)|^{p-2} T_{r,p}h\,dx,
\end{aligned}$$

where we used for the last line the mean value theorem, $\alpha \in (0,1)$. For $t \to 0$ we arrive at

$$\lim_{t\to 0}\frac{E_{r,p}(u+th) - E_{r,p}(u)}{t} = \int_{\mathbb{R}^n} (T_{r,p}(u))|T_{r,p}(u)|^{p-2}T_{r,p}h\,dx,$$

which yields

$$\begin{aligned}
\lim_{t\to 0}\frac{E_{r,p}(u+th) - E_{r,p}(u)}{t} &= \langle T^*_{r,p}(|T_{r,p}(u)|^{p-2}T_{r,p}(u)), h\rangle \\
&= \langle \mathcal{A}^{(p)}_r u, h\rangle,
\end{aligned}$$

and the lemma is proved. □

Remark 3.2.19. For $p = 2$ we find for the Gâteaux derivative of $E_{r,2}(u)$ at u

$$\mathcal{A}^{(2)}_r u = (\mathrm{id} - A^{(2)*})^{r/2} \circ (\mathrm{id} - A^{(2)})^{r/2} u \tag{3.118}$$

where $(\mathrm{id} - A^{(2)})^{r/2} : \mathcal{F}_{r,2} \to L^2$ and according to Theorem 3.2.16 we have $(\mathrm{id} - A^{(2)})^{r/2} : L^2 \to \mathcal{F}^*_{r,2} = \mathcal{H}_{-r,2}$.

Corollary 3.2.20. *The function $E^f_{r,p}$ is on $\mathcal{F}_{r,p}(\mathbb{R}^n;\mathbb{R})$ strictly convex, coercive and Gâteaux differentiable with Gâteaux derivative at $u \in \mathcal{F}_{r,p}(\mathbb{R}^n;\mathbb{R})$ given by*

$$\mathcal{A}^{(p)}_{r,f} u := \mathcal{A}^{(p)}_r u - f. \tag{3.119}$$

Proof: The strict convexity of $E^f_{r,p}$ follows since $E_{r,p}$ is strictly convex and $u \mapsto -\int_{\mathbb{R}^n} f\, u\, dx$ is linear. Furthermore we find

$$\begin{aligned}\frac{E^f_{r,p}(u)}{\|u\|_{\mathcal{F}_{r,p}}} &= \frac{1}{p}\|u\|^{p-1}_{\mathcal{F}_{r,p}} - \frac{1}{\|u\|_{\mathcal{F}_{r,p}}}\int_{\mathbb{R}^n} f\, u\, dx \\ &\geq \frac{1}{p}\|u\|^{p-1}_{\mathcal{F}_{r,p}} - \frac{\|f\|_{L^{p'}}\|u\|_{L^p}}{\|u\|_{\mathcal{F}_{r,p}}} \\ &\geq \frac{1}{p}\|u\|^{p-1}_{\mathcal{F}_{r,p}} - \|f\|_{L^{p'}},\end{aligned}$$

implying the coercivity of $E^f_{r,p}$. Since

$$\frac{-\int_{\mathbb{R}^n} f(u+th)\, dx + \int_{\mathbb{R}^n} f\, u\, dx}{t} = -\int_{\mathbb{R}^n} f\, h\, dx,$$

it follows from the preceding lemma that $E^f_{r,p}$ is Gâteaux differentiable with Gâteaux derivative (3.119). □

We want to study the operators $\mathcal{A}^{(p)}_r$, $\mathcal{A}^{(p)}_{r,f} : \mathcal{F}_{r,p} \to \mathcal{F}^*_{r,p}$ more closely and start with some definitions.

Definition 3.2.21. Let Y be a reflexive, separable real Banach space and $K \subset Y$ a closed convex set. Further let $T : K \to Y^*$ be an operator.

A. We call T *monotone* if

$$\langle Tu - Tv, u - v\rangle \geq 0 \tag{3.120}$$

for all $u, v \in K$.

B. The operator is called *strictly monotone* if

$$\langle Tu - Tv, u - v\rangle > 0 \tag{3.121}$$

for all $u, v \in K$ and $u \neq v$.

C. If there is a strictly increasing continuous function $\gamma : \mathbb{R}_+ \to \mathbb{R}$, $\gamma(0) = 0$ and $\lim\limits_{t\to\infty} \gamma(t) = \infty$, such that for all $u, v \in K$

$$\langle Tu - Tv, u - v\rangle \geq \gamma(\|u - v\|_Y)\|u - v\|_Y \tag{3.122}$$

holds, then T is called *uniformly monotone.*

D. We say that T is *coercive* with respect to K if there is an element $\varphi \in K$ such that

$$\lim_{\|u\|_Y\to\infty} \frac{\langle Tu - T\varphi, u - \varphi\rangle}{\|u - \varphi\|_Y} = \infty, \quad u \in K. \tag{3.123}$$

Clearly, for a linear space K we may pose the condition

$$\lim_{\|u\|_Y\to\infty} \frac{\langle Tu, u\rangle}{\|u\|_Y} = \infty, \quad u \in K, \tag{3.124}$$

instead of (3.123). If K is an unbounded set and T uniformly monotone, then T is coercive as well as strictly monotone, hence monotone.

Proposition 3.2.22. *Let $\mathcal{A}_r^{(p)}$ be as in* (3.117). *Then for any closed convex set $K \subset \mathcal{F}_{r,p}(\mathbb{R}^n;\mathbb{R})$, $p \geq 2$, we have the estimate*

$$\langle \mathcal{A}_r^{(p)}u - \mathcal{A}_r^{(p)}v, u - v\rangle \geq 2^{2-p}\|u-v\|^p_{\mathcal{F}_{r,p}}, \tag{3.125}$$

implying that $\mathcal{A}_r^{(p)}$ is uniformly monotone and coercive on every unbounded closed convex set K.

Proof: From the very definition of $\mathcal{A}_r^{(p)}$ we find

$$\begin{aligned}\langle \mathcal{A}_r^{(p)}u - \mathcal{A}_r^{(p)}v, u - v\rangle = \int_{\mathbb{R}^n} \Big\{ &\big(|(\mathrm{id} - A^{(p)})^{r/2}u|^{p-2}(\mathrm{id} - A^{(p)})^{r/2}u \\ &- |(\mathrm{id} - A^{(p)})^{r/2}v|^{p-2}(\mathrm{id} - A^{(p)})^{r/2}v\big) \\ &\times \big((\mathrm{id} - A^{(p)})^{r/2}u - (\mathrm{id} - A^{(p)})^{r/2}v\big)\Big\}\, dx\end{aligned}$$

and (3.111) yields

$$\begin{aligned}\langle \mathcal{A}_r^{(p)}u - \mathcal{A}_r^{(p)}v, u - v\rangle &\geq 2^{2-p}\int_{\mathbb{R}^n} |(\mathrm{id} - A^{(p)})^{r/2}(u-v)|^p\, dx \\ &= 2^{2-p}\|u-v\|^p_{\mathcal{F}_{r,p}}. \quad \square\end{aligned}$$

Corollary 3.2.23. *For every $f \in L^{p'}(\mathbb{R}^n;\mathbb{R})$, $p' = \frac{p}{p-1}$, $p \geq 2$ it follows that*

$$\langle \mathcal{A}_{r,f}^{(p)}u - \mathcal{A}_{r,f}^{(p)}v, u - v\rangle \geq 2^{2-p}\|u-v\|^p_{\mathcal{F}_{r,p}} \tag{3.126}$$

holds.

Definition 3.2.24. Let Y be a reflexive, separable real Banach space and $K \subset Y$ a closed convex set. Further let $T : K \to Y^*$ be an operator.

A. We call T *radially continuous* if for all $u, v \in K$ the function

$$s \mapsto \langle T(u + sv), v\rangle$$

is continuous on $[0, 1]$.

B. The operator T is called *hemicontinuous* if for all $u, v \in K$ and $h \in Y$ the function

$$s \mapsto \langle T(u+sv), h\rangle$$

is continuous on $[0,1]$.

C. If for every sequence $(u_\nu)_{\nu\in\mathbb{N}}$, $u_\nu \in K$, which converges strongly to $u \in K$ it follows that $(Tu_\nu)_{\nu\in\mathbb{N}}$ converges weakly (in Y^*) to Tu, then T is said to be *demicontinuous*.

D. We say that T satisfies *condition* (S) if for every sequence $(u_\nu)_{\nu\in\mathbb{N}}$, $u_\nu \in K$, converging weakly to $u \in K$ with the property that

$$\lim_{\nu\to\infty} \langle Tu_\nu - Tu, u_\nu - u\rangle = 0,$$

then u_ν converges strongly to u.

E. The operator T is *bounded* if it maps bounded sets into bounded sets.

Proposition 3.2.25. *Every uniformly monotone operator $T : Y \to Y^*$ satisfies condition (S).*

Proof: Taking γ from Definition 3.2.21.C we find for $(u_\nu)_{\nu\in\mathbb{N}}$, $u_\nu \in Y$, and $u \in Y$ that

$$\gamma(\|u_\nu - u\|_Y)\|u_\nu - u\|_Y \leq \langle Tu_\nu - Tu, u_\nu - u\rangle.$$

Hence, if $u_\nu \to u$ weakly and $\lim_{\nu\to\infty} \langle Tu_\nu - Tu, u_\nu - u\rangle = 0$ the properties of γ imply $u_\nu \to u$ in Y. □

Corollary 3.2.26. *The operator $\mathcal{A}_r^{(p)}$ and $\mathcal{A}_{r,f}^{(p)}$ satisfy condition (S).*

Proposition 3.2.27. *The operator $\mathcal{A}_r^{(p)} : \mathcal{F}_{r,p}(\mathbb{R}^n;\mathbb{R}) \to \mathcal{F}_{r,p}^*(\mathbb{R}^n;\mathbb{R})$, $p \geq 2$, is hemicontinuous, hence radially continuous.*

Proof: The operator $(\mathrm{id} - A^{(p)})^{r/2} : \mathcal{F}_{r,p} \to L^p$ is continuous and in addition for $u, v \in \mathcal{F}_{r,p}$ fixed and for $0 \leq s \leq 1$ it follows for $s \to s_0$ that

$$\begin{aligned}|(\mathrm{id} - A^{(p)})^{r/2}(u+sv)|^{p-2}(\mathrm{id} - A^{(p)})^{r/2}(u+sv)\\ \to |(\mathrm{id} - A^{(p)})^{r/2}(u+s_0v)|^{p-2}(\mathrm{id} - A^{(p)})^{r/2}(u+s_0v)\end{aligned}$$

almost everywhere. In addition we find for all $s \in [0,1]$

$$\begin{aligned} &\big|\,|(\mathrm{id}-A^{(p)})^{r/2}(u+sv)|^{p-2}(\mathrm{id}-A^{(p)})^{r/2}(u+sv)\big| \\ &\qquad = |(\mathrm{id}-A^{(p)})^{r/2}(u+sv)|^{p-1} \\ &\qquad \le c_p\Big(|(\mathrm{id}-A^{(p)})^{r/2}u|^{p-1} + |(\mathrm{id}-A^{(p)})^{r/2}v|^{p-1}\Big). \end{aligned}$$

Thus the dominated convergence theorem implies

$$\begin{aligned} &|(\mathrm{id}-A^{(p)})^{r/2}(u+sv)|^{p-2}(\mathrm{id}-A^{(p)})^{r/2}(u+sv) \\ &\qquad \to |(\mathrm{id}-A^{(p)})^{r/2}(u+s_0v)|^{p-2}(\mathrm{id}-A^{(p)})^{r/2}(u+s_0v) \end{aligned}$$

in $L^{p'}$, which yields the hemicontinuity of $\mathcal{A}_r^{(p)}$. □

Corollary 3.2.28. *A. For every closed convex set $K \subset \mathcal{F}_{r,p}(\mathbb{R}^n;\mathbb{R})$ the operator $\mathcal{A}_r^{(p)} : K \to \mathcal{F}_{r,p}^*(\mathbb{R}^n;\mathbb{R})$ is hemicontinuous.*

B. For every closed convex set $K \subset \mathcal{F}_{r,p}(\mathbb{R}^n;\mathbb{R})$ and $f \in L^{p'}(\mathbb{R}^n;\mathbb{R})$, $p' = \frac{p}{p-1}$, the operator $\mathcal{A}_{r,f}^{(p)} : K \to \mathcal{F}_{r,p}^*(\mathbb{R}^n;\mathbb{R})$ is hemicontinuous.

The following two fundamental results from F.Browder's and G.Minty's theory of monotone operators are quoted without proof. A proof can be found in E.Zeidler [289],p. 557, and of course in the original papers of F.Browder [38] and G.Minty [218].

Theorem 3.2.29. *A. Let $T : Y \to Y^*$ be a radially continuous, monotone and coercive operator. Then for every $f \in Y^*$ the set of solutions of*

$$Tu = f \tag{3.127}$$

is non-empty, weakly closed and convex.

B. Let $T : Y \to Y^*$ be radially continuous, strictly monotone and coercive. Then the inverse operator $T^{-1} : Y^* \to Y$ exists, is strictly monotone, bounded and demicontinuous. If in addition T satisfies condition (S), then T^{-1} is continuous.

Theorem 3.2.30. *For $f \in L^{p'}(\mathbb{R}^n;\mathbb{R}) \subset \mathcal{F}_{r,p}^*(\mathbb{R}^n;\mathbb{R})$, $p \ge 2$, there exists a unique solution $u \in \mathcal{F}_{r,p}(\mathbb{R}^n;\mathbb{R})$ of the equation*

$$\mathcal{A}_r^{(p)}u = f. \tag{3.128}$$

Moreover, $\mathcal{A}_r^{(p)} : \mathcal{F}_{r,p}(\mathbb{R}^n;\mathbb{R}) \to \mathcal{F}_{r,p}^*(\mathbb{R}^n;\mathbb{R})$ has a continuous inverse which is given on $L^{p'}(\mathbb{R}^n;\mathbb{R}) \subset \mathcal{F}_{r,p}^*(\mathbb{R}^n;\mathbb{R})$ by

$$f \mapsto (\mathrm{id}-A^{(p)})^{-r/2}\Big(|(\mathrm{id}-A^{(p)*})^{-r/2}f|^{p'-2}(\mathrm{id}-A^{(p)*})^{-r/2}f\Big). \tag{3.129}$$

Proof: It remains to prove (3.129). For this we use that for $j(x) = x|x|^{p-2}$ the associated superposition operator $J : L^p \to L^{p'}$, $p' = p/(p-1)$, $u \mapsto J(u) = u|u|^{p-2}$, is invertible and its inverse is the superposition operator J^{-1}, where $j^{-1}(y) = y|y|^{p'-2}$, and $J^{-1} : L^{p'} \to L^p$, $v \mapsto J(v) = v|v|^{p'-2}$. Thus for $g \in L^p$ it follows that $h = g|g|^{p-2} \in L^{p'}$ if and only if $g = h|h|^{p-2} \in L^p$. Now, given $f \in L^{p'}(\mathbb{R}^n; \mathbb{R})$ we find a unique $u \in \mathcal{F}_{r,p}$ such that

$$(\mathrm{id} - A^{(p)*})^{r/2}(|(\mathrm{id} - A^{(p)})^{r/2}u|^{p-2}(\mathrm{id} - A^{(p)})^{r/2}u) = f,$$

hence

$$|(\mathrm{id} - A^{(p)})^{r/2}u|^{p-2}(\mathrm{id} - A^{(p)})^{r/2}u = (\mathrm{id} - A^{(p)*})^{-r/2}f,$$

i.e.

$$(\mathrm{id} - A^{(p)})^{r/2}u = |(\mathrm{id} - A^{(p)*})^{-r/2}f|^{p'-2}(\mathrm{id} - A^{(p)*})^{-r/2}f,$$

which leads to (3.129). □

Let $K \subset \mathcal{F}_{r,p}(\mathbb{R}^n; \mathbb{R})$, $p \geq 2$, be a closed set. We want to study the minimization problem

$$\inf\{E^f_{r,p}(u); u \in K\}, \tag{3.130}$$

where $f \in L^{p'}(\mathbb{R}^n; \mathbb{R})$, $p' = \frac{p}{p-1}$, is a given function. For an open set $G \subset \mathbb{R}^n$

$$K := \{u \in \mathcal{F}_{r,p}(\mathbb{R}^n; \mathbb{R}); u \geq 1 \quad \text{a.e. on } G\}$$

is a closed convex set. Indeed, the convexity is obvious. Now let $(u_\nu)_{\nu \in \mathbb{N}}$, $u_\nu \in K$, be a sequence converging in $\mathcal{F}_{r,p}(\mathbb{R}^n; \mathbb{R})$ to $u \in \mathcal{F}_{r,p}(\mathbb{R}^n; \mathbb{R})$. It follows that $(u_\nu)_{\nu \in \mathbb{N}}$ converges in $L^p(\mathbb{R}^n; \mathbb{R})$, hence a subsequence converges almost everywhere to u, implying $u \geq 1$ a.e. on G. Thus determining $\mathrm{cap}_{r,p}(G)$ is the minimization problem (3.130). We know that $E^f_{r,p}$ is a strictly convex, coercive and Gâteaux differentiable function. Thus it follows that (3.130) has always a minimizer on K, compare E.Zeidler [289], Theorem 25.D. However, for us it is not necessary to apply this general result. The proof of Theorem 3.1.31 applies word for word to the new situation. Hence we have a unique minimizer u_K to problem (3.130).

Now for $0 \leq t \leq 1$ and $\varphi \in K$ consider the function $(1-t)u_K + t\varphi \in K$. It follows that

$$\begin{aligned}0 &\leq \frac{d}{dt}\Big|_{t=0} E^f_{r,p}((1-t)u_K + t\varphi)\\ &= \frac{d}{dt}\Big|_{t=0} \Big\{ \frac{1}{p}\int_{\mathbb{R}^n} |(\mathrm{id} - A^{(p)})^{r/2}((1-t)u_K + t\varphi)|^p\,dx \\ &\quad - \int_{\mathbb{R}^n} f\cdot((1-t)u_K + t\varphi)\,dx\Big\}\\ &= \langle \mathcal{A}^{(p)}_r u_K, \varphi - u_K\rangle - \int_{\mathbb{R}^n} f\cdot(\varphi - u_K)\,dx\\ &= \langle \mathcal{A}^{(p)}_{r,f} u_K, \varphi - u_K\rangle.\end{aligned}$$

Thus, as it is expected from the general theory of variational inequalities, see for example the monograph [175] of D.Kinderlehrer and G.Stampacchia, we obtain

Proposition 3.2.31. *Necessary for $u_K \in K$ to be minimizer of* (3.130) *is that*

$$\langle \mathcal{A}^{(p)}_r u_K, \varphi - u_K\rangle \geq 0 \tag{3.131}$$

holds for all $\varphi \in K$, i.e.

$$\begin{aligned}&\int_{\mathbb{R}^n} |(\mathrm{id} - A^{(p)})^{r/2} u_K|^{p-2}(\mathrm{id} - A^{(p)})^{r/2} u_K \cdot (\mathrm{id} - A^{(p)})^{r/2}(\varphi - u_K)\,dx\\ &\geq \int_{\mathbb{R}^n} f\cdot(\varphi - u_K)\,dx.\end{aligned} \tag{3.132}$$

Remark 3.2.32. In later sections, especially in volume III, when handling the balayage problem, we will work with closed sets of the form

$$\{u \in \mathcal{F}_{r,p}(\mathbb{R}^n;\mathbb{R});\, u|_G = 0 \quad a.e.\}, \qquad G \subset \mathbb{R}^n \text{ open},$$

or

$$\overline{\mathcal{F}_{r,p}(\mathbb{R}^n;\mathbb{R}) \cap C_0(G;\mathbb{R})}^{\|\cdot\|_{\mathcal{F}_{r,p}}}, \quad G \subset \mathbb{R}^n \quad \text{open}.$$

These sets are obviously closed subspaces of $\mathcal{F}_{r,p}(\mathbb{R}^n;\mathbb{R})$. Therefore we may test in (3.131) with $\varphi = \pm\psi + u_K$, $\psi \in K$, to find

$$\pm\langle \mathcal{A}^{(p)}_{r,f} u_K, \psi\rangle \geq 0,$$

which yields for all $\psi \in K$

$$\langle \mathcal{A}_r^{(p)} u_K, \psi \rangle = \int_{\mathbb{R}^n} f\psi \, dx, \tag{3.133}$$

or

$$\int_{\mathbb{R}^n} |(\mathrm{id} - A^{(p)})^{r/2} u|^{p-2} (\mathrm{id} - A^{(p)})^{r/2} u \cdot (\mathrm{id} - A^{(p)})^{r/2} \psi \, dx = \int_{\mathbb{R}^n} f\psi \, dx. \tag{3.134}$$

In order to have a closer analogy to the theory of Dirichlet forms let us introduce the form

$$\mathcal{E}_r^{(p)}(u,v) := \int_{\mathbb{R}^n} |(\mathrm{id} - A^{(p)})^{r/2} u|^{p-2} (\mathrm{id} - A^{(p)})^{r/2} u \cdot (\mathrm{id} - A^{(p)})^{r/2} v \, dx \tag{3.135}$$

defined on $\mathcal{F}_{r,p}(\mathbb{R}^n;\mathbb{R})$. It follows that

$$\mathcal{E}_r^{(p)}(u,u) = \|u\|^p_{\mathcal{F}_{r,p}} = pE_{r,p}(u) \geq 0, \tag{3.136}$$

and

$$\begin{aligned} |\mathcal{E}_r^{(p)}(u,v)| &\leq \| |(\mathrm{id} - A^{(p)})^{r/2} u|^{p-1} \|_{L^{p'}} \|(\mathrm{id} - A^{(p)})^{r/2} v\|_{L^p} \\ &= \|u\|^{p/p'}_{\mathcal{F}_{r,p}} \|v\|_{\mathcal{F}_{r,p}}, \end{aligned} \tag{3.137}$$

thus $\mathcal{E}_r^{(p)} : \mathcal{F}_{r,p} \times \mathcal{F}_{r,p} \to \mathbb{R}$ and it is linear and continuous in the second variable for fixed first variable. Further, from the proof of Proposition 3.2.27 it follows that the mapping

$$s \mapsto \mathcal{E}_r^{(p)}(u+s\omega, v), \quad u, v, \omega \in \mathcal{F}_{r,p}(\mathbb{R}^n;\mathbb{R}) \tag{3.138}$$

is continuous from $\mathbb{R}$ to $\mathbb{R}$. Using $\mathcal{E}_r^{(p)}$ we find for $u, v \in \mathcal{F}_{r,p}(\mathbb{R}^n;\mathbb{R})$

$$\langle \mathcal{A}_r^{(p)} u, v \rangle = \mathcal{E}_r^{(p)}(u,v), \tag{3.139}$$

compare also Remark 3.2.19.

Remark 3.2.33. Let $p = 2$ and $r = 1$ and suppose that $(T_t^{(2)})_{t\geq 0}$ is a symmetric sub-Markovian semigroup on $L^2(\mathbb{R}^n;\mathbb{R})$. It follows that

$$\mathcal{E}_1^{(2)}(u,v) = \int_{\mathbb{R}^n} (\mathrm{id} - A^{(2)})^{1/2} u \cdot (\mathrm{id} - A^{(2)})^{1/2} v \, dx,$$

i.e. $\mathcal{E}_1^{(2)}$ is the symmetric Dirichlet form generating by $A^{(2)} - \mathrm{id}$ on the Dirichlet space $\mathcal{F}_{1,2}(\mathbb{R}^n;\mathbb{R})$.

Now let K in (3.130) be the set $\{u \in \mathcal{F}_{r,p}(\mathbb{R}^n;\mathbb{R}); u \geq 1$ a.e. on $G\}$ for an open set G and consider $\|u\|^p_{\mathcal{F}_{r,p}} = pE_{r,p}(u)$ instead of $E^f_{r,p}$, in particular $f = 0$. Then the minimizer becomes the (r,p)-equilibrium potential u_G provided that $\text{cap}_{r,p}(G) < \infty$, compare Theorem 3.1.31 and Definition 3.1.32. From Proposition 3.2.31 together with the notation (3.135) we derive

Corollary 3.2.34. *Let $G \subset \mathbb{R}^n$ be an open set, $p \geq 2$ and $\text{cap}_{r,p}(G) < \infty$. Then the (r,p)-equilibrium potential $u_G \in \mathcal{F}_{r,p}(\mathbb{R}^n;\mathbb{R})$ of G satisfies*

$$\langle \mathcal{A}^{(p)}_r u_G, \varphi - u_G \rangle \geq 0 \tag{3.140}$$

or

$$\mathcal{E}^{(p)}_r (u_G, \varphi - u_G) \geq 0 \tag{3.141}$$

for all $\varphi \in \{v \in \mathcal{F}_{r,p}(\mathbb{R}^n;\mathbb{R}); v \geq 1$ a.e. on $G\}$.

Proof: First we arrive using Proposition 3.2.31 at

$$\langle p\mathcal{A}^{(p)}_r u_G, \varphi - u_G \rangle \geq 0$$

for all $\varphi \in \{v \in \mathcal{F}_{r,p}(\mathbb{R}^n;\mathbb{R}); v \geq 1$ a.e. on $G\}$ which implies the corollary. □

Suppose that $\psi \in \mathcal{F}_{r,p}(\mathbb{R}^n;\mathbb{R})$ is such that $\psi|_G \geq 0$ a.e. It follows that $\varphi := u_G + \psi \in \{v \in \mathcal{F}_{r,p}(\mathbb{R}^n;\mathbb{R}); v \geq 1$ a.e. on $G\}$ implying

$$\langle \mathcal{A}^{(p)}_r u_G, \psi \rangle = \mathcal{E}^{(p)}_r (u_G, \psi) \geq 0 \tag{3.142}$$

for all $\psi \in \mathcal{F}_{r,p}(\mathbb{R}^n;\mathbb{R})$, $\psi|_G \geq 0$ a.e.

Note that (3.142) is part of assertion (iii) of Lemma 2.1.1 in the monograph [102] of M.Fukushima et.al. in case $p = 2$.

In order to improve properties of the (r,p)-equilibrium potential u_G we have to use a truncation property.

Definition 3.2.35. Let $(\mathcal{H}(\mathbb{R}^n;\mathbb{R}); \|\cdot\|_{\mathcal{H}})$ be a subspace of $L^p(\mathbb{R}^n;\mathbb{R})$.

A. We say that the *Lipschitz functions* T with constant 1 *operate* on $\mathcal{H}(\mathbb{R}^n;\mathbb{R})$ if

$$u \in \mathcal{H}(\mathbb{R}^n;\mathbb{R}) \quad \text{implies } T \circ u \in \mathcal{H}(\mathbb{R}^n;\mathbb{R}) \tag{3.143}$$

and

$$\|T \circ u\|_{\mathcal{H}} \leq c\|u\|_{\mathcal{H}} \tag{3.144}$$

holds for all $u \in \mathcal{H}(\mathbb{R}^n;\mathbb{R})$ and all $T : \mathbb{R} \to \mathbb{R}$ such that $|T(x) - T(y)| \leq |x - y|$ and $T(0) = 0$.

B. If the Lipschitz functions operate on $\mathcal{H}(\mathbb{R}^n;\mathbb{R})$ and (3.144) holds with constant $c = 1$, then we say that the *truncation property* holds for $\mathcal{H}(\mathbb{R}^n;\mathbb{R})$.

In the next section we will discuss the truncation property more detailed. But we want to emphasize already now that the validity of the truncation property is a strong restriction for the space $\mathcal{F}_{r,p}(\mathbb{R}^n;\mathbb{R})$.

Now suppose that $\mathcal{F}_{r,p}(\mathbb{R}^n;\mathbb{R})$ satisfies the truncation property and let $G \subset \mathbb{R}^n$ be an open set, $\operatorname{cap}_{r,p}(G) < \infty$ and let u_G be its (r,p)-equilibrium potential. We want to show that in this situation $u_G\big|_G = 1$ a.e. holds and give a further characterization of u_G. First observe that $(0 \vee u_G) \wedge 1 \in \{u \in \mathcal{F}_{r,p}(\mathbb{R}^n;\mathbb{R}); u \geq 1$ a.e. on $G\}$ and

$$\|((0 \vee u_G) \wedge 1)\|^p_{\mathcal{F}_{r,p}} \leq \|u_G\|^p_{\mathcal{F}_{r,p}},$$

implying that $u_G = (0 \vee u_G) \wedge 1$, i.e. $0 \leq u_G \leq 1$ a.e. and $u_G|_G = 1$ a.e. Further we claim that u_G is the unique element in $\mathcal{F}_{r,p}(\mathbb{R}^n;\mathbb{R})$ satisfying $u_G = 1$ a.e. on G such that

$$\mathcal{E}_r^{(p)}(u_G, v) \geq 0 \tag{3.145}$$

for all $v \in \mathcal{F}_{r,p}(\mathbb{R}^n;\mathbb{R})$, $v \geq 0$ a.e. on G. From (3.142) we know already that for u_G inequality (3.145) holds. Now let $u_G \in \mathcal{F}_{r,p}(\mathbb{R}^n;\mathbb{R})$, $u_G = 1$ a.e. on G and $\mathcal{E}_r^{(p)}(u_G, v) \geq 0$ for all $v \in \mathcal{F}_{r,p}(\mathbb{R}^n;\mathbb{R})$, $v \geq 0$ a.e. on G. It follows that $w - u_G \geq 0$ a.e. on G for all $w \in \mathcal{F}_{r,p}(\mathbb{R}^n;\mathbb{R})$, $w \geq 1$ a.e. on G, thus we have

$$\mathcal{E}_r^{(p)}(u_G, w - u_G) \geq 0$$

and the variational inequality characterization of the (r,p)-equilibrium potential yields the result, compare Proposition 3.2.31. In particular, it follows that for $v \in \mathcal{F}_{r,p}(\mathbb{R}^n;\mathbb{R})$ such that $v = 1$ a.e. on G

$$\mathcal{E}_r^{(p)}(u_G, v) = \operatorname{cap}_{r,p}(G). \tag{3.146}$$

Indeed, for $v = 1$ a.e. on G we find that $\pm(u_G - v) = 0$ a.e. on G implying

$$0 \leq \pm\mathcal{E}_r^{(p)}(u_G, u_G - v) = \pm(\mathcal{E}_r^{(p)}(u_G, u_G) - \mathcal{E}_r^{(p)}(u_G, v)),$$

i.e.

$$\operatorname{cap}_{r,p}(G) \leq \mathcal{E}_r^{(p)}(u_G, v) \leq \operatorname{cap}_{r,p}(G).$$

Hence we have proved the following

Lemma 3.2.36. *Let $\mathcal{F}_{r,p}(\mathbb{R}^n;\mathbb{R})$, $p \geq 2$, satisfy the truncation property and let $G \subset \mathbb{R}^n$ be an open set with* $\text{cap}_{r,p}(G) < \infty$. *The (r,p)-equilibrium potential u_G fulfills $0 \leq u_G \leq 1$ a.e., $u_G|_G = 1$ a.e. and* (3.141), (3.145) *as well as* (3.146) *holds.*

Before we will continue to develop the potential theory of L^p-sub-Markovian semigroups in their associated Bessel potential spaces we want to prepare the ground to give concrete examples. For this we will introduce and study in the next section Bessel potential spaces associated with a real-valued negative definite function.

Remark 3.2.37. For a metric space X and a finite, fully supported measure μ on X in [172] T.Kazumi and I.Shigekawa studied very systematically the spaces $\mathcal{F}_{r,p}$ and (r,p)-capacities for an L^p-sub-Markovian semigroup $(T_t^{(p)})_{t\geq 0}$ with the property that $(T_t^{(p)*})_{t\geq 0}$ is also an $L^{p'}$-sub-Markovian semigroup, $p' = \frac{p}{p-1}$. In particular the dual spaces $\mathcal{F}^*_{r,p}$ are investigated and positive "distributions", i.e. elements in $\mathcal{F}^*_{r,p}$ satisfy $\langle u, \varphi \rangle \geq 0$ for all $\varphi \geq 0$, $\varphi \in \mathcal{F}_{r,p}$, are characterized as measures provided $1 \in \mathcal{F}_{r,p}$. Further some elements of an L^p-potential theory are presented. Moreover, following D.Feyel and de La Pradelle [92], (r,p)-capacities of functions are studied. We will come back to parts of these considerations in later sections.

3.3 Bessel Potential Spaces Associated with a Continuous Negative Definite Function

Let $(T_t)_{t\geq 0}$ be the operator semigroup (defined on $S(\mathbb{R}^n)$) associated to the continuous real-valued negative definite function $\psi : \mathbb{R}^n \to \mathbb{R}$, i.e.

$$T_t u(x) = (2\pi)^{-n/2} \int_{\mathbb{R}^n} e^{ix\cdot\xi} e^{-t\psi(\xi)} \hat{u}(\xi)\, d\xi. \tag{3.147}$$

As we know from Chapter I.4, for each $1 \leq p < \infty$ there is an extension of $(T_t)_{t\geq 0}$ to an L^p-sub-Markovian semigroup $(T_t^{(p)})_{t\geq 0}$. Our aim in this section is to determine in terms of function spaces and to study for these semigroups the spaces $\mathcal{F}_{r,p}(\mathbb{R}^n;\mathbb{R})$, $1 < p < \infty$. It is clear that these spaces should also be of greater relevance when studying L^p-semigroups generated by a pseudo-differential operator $-p(x,D)$ with a symbol comparable to ψ, i.e. $\text{Re}\, p(x,\xi) \sim \psi(\xi)$ and $|\text{Im}\, p(x,\xi)| \leq \text{Re}\, p(x,\xi)$. For $p = 2$ such results had been considered in Chapter 2. Our presentation in this section follows closely the joint papers [86] and [87] with W.Farkas and R.L.Schilling.

In Example 3.1.28 we identified already the space $\mathcal{F}_{r,p}(\mathbb{R}^n)$ with the space $H^{\psi,s}(\mathbb{R}^n)$ which was introduces in Section I.3.10 and intensively used in later sections. However, in the Hilbert space case the spaces $H^{\psi,s}(\mathbb{R}^n)$ and $B^s_{\psi,2}(\mathbb{R}^n)$, see Section I.3.10, do coincide since by Plancherel's theorem

$$\|F^{-1}((1+\psi(\cdot))^{s/2}\hat{u})\|_{L^2} = \|(1+\psi(\cdot))^{s/2}\hat{u}\|_{L^2} \tag{3.148}$$

holds first for $u \in S(\mathbb{R}^n)$ and then by extension for general u. For $p \neq 2$ the result is quite different, the spaces $B^s_{\psi,p}(\mathbb{R}^n)$ are distinct from the spaces $\mathcal{F}_{r,p}(\mathbb{R}^n)$. Note that $\mathcal{F}_{2,p}(\mathbb{R}^n)$ must be the domain of $\mathrm{id} - A^{(p)}$, hence $A^{(p)}$, $A^{(p)}$ being the generator of $(T_t^{(p)})_{t\geq 0}$, and on $S(\mathbb{R}^n)$ the operator $A^{(p)}$ is given by

$$A^{(p)}u(x) = -\psi(D)u(x) = -(2\pi)^{-n/2}\int_{\mathbb{R}^n} e^{ix\cdot\xi}\psi(\xi)\hat{u}(\xi)\,d\xi, \tag{3.149}$$

compare Example I.4.6.29 (and note that the arguments given there apply also for $S(\mathbb{R}^n)$ instead of $C_0^\infty(\mathbb{R}^n)$). Thus the norm $\|\cdot\|_{\mathcal{F}_{2,p}}$ should essentially be the graph norm of $A^{(p)}$; having in mind Theorem 3.1.25, we want to identify the norm $\|\cdot\|_{\mathcal{F}_{r,p}}$ with

$$\|(\mathrm{id}+\psi(D))u\|_{L^p} = \|F^{-1}((1+\psi(\cdot))\hat{u}(\cdot))\|_{L^p}.$$

However the definition of $F^{-1}((1+\psi(\cdot))\hat{u}(\cdot))$ for non-smooth u leads to a serious problem: For example let $u \in L^p(\mathbb{R}^n)$ and $p > 2$. In general we know only $\hat{u} \in S'(\mathbb{R}^n)$. A typical continuous negative definite function $\psi : \mathbb{R}^n \to \mathbb{R}$ is not differentiable, hence $\psi(\cdot)\hat{u}(\cdot)$ is in general not defined! This is the main problem we have to overcome to define spaces $H_p^{\psi,s}(\mathbb{R}^n)$. Before attacking this question we have to make a further remark. A general continuous negative definite function $\psi : \mathbb{R}^n \to \mathbb{R}$ has the Lévy–Khinchin representation

$$\psi(\xi) = c + q(\xi) + \int_{\mathbb{R}^n\setminus\{0\}} (1-\cos(y\cdot\xi))\,\nu(dy) \tag{3.150}$$

with $c \geq 0$, a non-negative definite quadratic form q and the Lévy measure ν integrating the function $y \mapsto |y|^2 \wedge 1$, compare Corollary I.3.7.9 and the remark following Definition I.3.7.11. Further we know that the function $\varphi(\xi) := \int_{\mathbb{R}^n\setminus\{0\}}(1-\cos(y\cdot\xi))\,\nu(dy)$ has at most quadratic growth. Hence in any direction ξ_0 where $q(\xi_0)$ is non-degenerate, i.e. $q(\mu\xi_0) = \mu^2 q(\xi_0) \neq 0$ for $\mu \neq 0$, $\varphi(\mu\xi_0)$ is dominated by $q(\mu\xi_0)$ for $\mu \to \infty$. We may decompose $\mathbb{R}^n = V_1 \oplus V_2$ where $q|_{V_1}$ is strictly positive definite and $q|_{V_2}$ is degenerate. Clearly on V_1 the quadratic form $q|_{V_1}$ dominates the asymptotic behaviour of $\psi|_{V_1}$ and therefore in the definition of norms such as (3.148) we could substitute $\psi|_{V_1}$ by $q|_{V_1}$. Thus a change of variable would lead to function spaces associated with ψ, $\psi = \psi(\eta, \xi)$, consisting of functions $u : \mathbb{R}^{m_1} \times \mathbb{R}^{m_2} \to \mathbb{R}$ (or $\mathbb{C}$) and with respect to the variables $y \in \mathbb{R}^{m_1}$ the behaviour of the Fourier transform $\hat{u}(\eta, \xi)$ is controlled by $|\eta|^2$. Therefore, in these directions we may work with classical Sobolev or Bessel potential spaces. For this reason we will restrict our attention to continuous negative definite functions having the representation

$$\psi(\xi) = \int_{\mathbb{R}^n\setminus\{0\}} (1-\cos(y\cdot\xi))\, \nu(dy), \tag{3.151}$$

where the Lévy measure ν integrates as before the function $y \mapsto (|y|^2 \wedge 1)$.

In order to overcome the first mentioned difficulty we use the observation of W.Hoh [128] discussed in Theorem I.3.7.13 of our treatise. Let $\varphi : \mathbb{R}^n \to \mathbb{R}$ be a continuous negative definite function with representation (3.151) and consider its decomposition

$$\psi(\xi) = \psi_R(\xi) + \tilde{\psi}_R(\xi) \tag{3.152}$$

where for $R > 0$ the function ψ_R is defined by

$$\psi_R(\xi) = \int_{\mathbb{R}^n\setminus\{0\}} (1-\cos(y\cdot\xi))\chi_{B_R(0)}(y)\, \nu(dy). \tag{3.153}$$

It follows that the Lévy measure ν_R associated with the continuous negative definite function ψ_R is given by $\chi_{B_R(0)}(\cdot)\nu$, hence it has absolute moments of all orders, i.e.

$$\int_{\mathbb{R}^n\setminus\{0\}} |y|^l \nu_R(dy) = \int_{B_R(0)\setminus\{0\}} |y|^l \nu_R(dy) = M_{R,l} < \infty, \quad l \geq 2.$$

Therefore it follows by Theorem I.3.7.13 that ψ_R is for any $R > 0$ an arbitrary often differentiable function with partial derivatives having at most quadratic growth. In particular it follows that for each $u \in S'(\mathbb{R}^n)$ the expression $\psi_R u$ is well-defined and belongs to $S'(\mathbb{R}^n)$. Thus for $u \in L^p(\mathbb{R}^n)$ the expression $\psi_R \widehat{u}$ is well-defined in $S'(\mathbb{R}^n)$. On the other hand the continuous negative definite function $\tilde{\psi}_R$ is a bounded function

$$|\tilde{\psi}_R(\xi)| \leq 2\nu(B_R^c(0)) < \infty. \tag{3.154}$$

As we noted before $S(\mathbb{R}^n) \subset D(A^{(p)})$ for all $1 < p < \infty$ and on $S(\mathbb{R}^n)$ we have $A^{(p)} = -\psi(D)$. We decompose $\psi(D)$ according to the decomposition (3.152), i.e.

$$\psi(D)u = \psi_R(D)u + \tilde{\psi}_R(D)u, \quad u \in S(\mathbb{R}^n). \tag{3.155}$$

Lemma 3.3.1. *The operator $\tilde{\psi}_R(D)$ extends to a bounded operator in $L^p(\mathbb{R}^n)$, $1 < p < \infty$. We will denote the extension again by $\tilde{\psi}_R(D)$.*

Proof: From (3.153) it follows that

$$\tilde{\psi}_R(\xi) = \int_{\mathbb{R}\setminus\{0\}} (1 - \cos(y \cdot \xi)) \chi_{B_R^c(0)}(y) \nu(dy)$$

implying for $u \in S(\mathbb{R}^n)$

$$\tilde{\psi}_R(D)u(x) = \int_{B_R^c(0)} (u(x) - u(x-y)) \nu(dy)$$

which yields

$$\begin{aligned} \|\tilde{\psi}_R(D)u\|_{L^p} &\le \left\| \int_{|y|\ge R} (u - u(\cdot - y)) \nu(dy) \right\|_{L^p} \\ &\le \int_{|y|\ge R} \|u - u(\cdot - y)\|_{L^p} \nu(dy) \le 2\nu(B_R^c(0)) \|u\|_{L^p}. \quad \square \end{aligned}$$

Next consider the operator $\psi_R(D)$ which is defined on $S(\mathbb{R}^n)$ by

$$\psi_R(D)u = F^{-1}(\psi_R \hat{u}). \tag{3.156}$$

But (3.156) makes sense for $u \in S'(\mathbb{R}^n)$, i.e we have a natural extension of $\psi_R(D)$ to $S'(\mathbb{R}^n)$. Now we are in a position to extend $\psi(D)$ to $L^p(\mathbb{R}^n)$ by

$$\psi(D)u := F^{-1}(\psi_R \hat{u}) + \int_{B_R^c(0)} (u(\cdot) - u(\cdot - y)) \, \nu(dy). \tag{3.157}$$

This extension is independent of R: Let $R_1 > R_2 > 0$ and denote for a moment by $\psi^{R_1}(D)$ and $\psi^{R_2}(D)$ the respective extension of $\psi(D)$ using the decompositions belonging to R_1 and R_2. Since

$$F^{-1}(\psi_{R_1} \hat{u}) - F^{-1}(\psi_{R_2} \hat{u}) = \int_{B_{R_2}^c(0) \setminus B_{R_1}^c(0)} (u(\cdot) - u(\cdot - y)) \, \nu(dy) \tag{3.158}$$

it follows that $\psi^{R_1}(D) = \psi^{R_2}(D)$.

For $R > 0$ let us define the family of norms

$$\|u\|_{\psi,R,p} := \|(\mathrm{id} - \psi_R(D))u\|_{L^p} \tag{3.159}$$

for those $u \in L^p(\mathbb{R}^n)$ for which (3.159) is finite. To understand this norm better observe that ψ_R is associated with an L^p-sub-Markovian semigroup $(T_t^{(p,R)})_{t\geq 0}$ and that $S(\mathbb{R}^n) \subset D(A^{(p,R)})$ where $A^{(p,R)}$ denotes the generator of $(T_t^{(p,R)})_{t\geq 0}$ which restricted to $S(\mathbb{R}^n)$ is just given by $\psi_R(D)u = F^{-1}(\psi_R(\cdot)\hat{u})$. It follows from Lemma 3.1.22 that

$$\|u\|_{L^p} \leq \|u\|_{\psi,R,p} \tag{3.160}$$

holds for all $u \in S(\mathbb{R}^n)$, $R > 0$ and $1 < p < \infty$. In particular we conclude that $\|\cdot\|_{\psi,R,p}$ is a norm.

Lemma 3.3.2. *For $0 < R_2 < R_1$ the norms $\|\cdot\|_{\psi,R_1,p}$ and $\|\cdot\|_{\psi,R_2,p}$ are equivalent on $S(\mathbb{R}^n)$.*

Proof: Let us consider the measure $\nu_{R_1R_2} = \chi_{B_{R_1}(0)}(\cdot)\nu - \chi_{B_{R_2}(0)}(\cdot)\nu = \chi_{B_{R_1}(0)\setminus B_{R_2}(0)}(\cdot)\nu$, where ν is the Lévy measure associated to ψ. It follows that $\nu_{R_1R_2}$ is bounded measure and further for $u \in S(\mathbb{R}^n)$ we find

$$\begin{aligned}\|\psi_{R_1}(D)u - \psi_{R_2}(D)u\|_{L^p} &= \left\| \int_{\mathbb{R}^n\setminus\{0\}} (u(\cdot) - u(\cdot - y))\nu_{R_1R_2}(dy) \right\|_{L^p} \\ &\leq \int_{\mathbb{R}^n\setminus\{0\}} \|(u(\cdot) - u(\cdot - y)\|_{L^p}\nu_{R_1R_2}(dy) \leq 2\nu(B_{R_1}(0)\setminus B_{R_2}(0))\|u\|_{L^p}.\end{aligned}$$

Thus we find using (3.160)

$$\begin{aligned}\|(\mathrm{id} + \psi_{R_2}(D)u\|_{L^p} &\leq \|(\mathrm{id} + \psi_{R_1}(D)u\|_{L^p} + \|\psi_{R_2}(D)u - \psi_{R_1}(D)u\|_{L^p} \\ &\leq c_0\|(\mathrm{id} + \psi_{R_1}(D)u\|_{L^p}\end{aligned}$$

as well as

$$\begin{aligned}\|(\mathrm{id} + \psi_{R_1}(D)u\|_{L^p} &\leq \|(\mathrm{id} + \psi_{R_2}(D)u\|_{L^p} + \|\psi_{R_1}(D)u - \psi_{R_2}(D)u\|_{L^p} \\ &\leq c_0\|(\mathrm{id} + \psi_{R_2}(D)u\|_{L^p},\end{aligned}$$

proving the lemma. □

Now let $u \in S(\mathbb{R}^n)$, $R > 0$, $1 < p < \infty$, and observe the estimates

$$\begin{aligned}\|(\mathrm{id} + \psi(D))u\|_{L^p} &\leq \|(\mathrm{id} + \psi_R(D))u\|_{L^p} + \|\tilde{\psi}_R(D)u\|_{L^p} \\ &\leq (1 + 2\nu(B_R^c(0)))\|u\|_{\psi,R,p}\end{aligned}$$

and

$$\begin{aligned}\|u\|_{\psi,R,p} &\leq \|(\mathrm{id}+\psi_R(D)+\tilde{\psi}_R(D))u\|_{L^p}+\|\tilde{\psi}_R(D)u\|_{L^p}\\ &\leq (1+2\nu(B_R^c(0))\|(\mathrm{id}+\psi(D))u\|_{L^p},\end{aligned}$$

where we used in particular Lemma 3.3.1.

These considerations justify to start with

Definition 3.3.3. Let $\psi : \mathbb{R}^n \to \mathbb{R}$ be a real-valued continuous negative definite function with representation (3.151). The *ψ-Bessel potential space of order 2* with respect to $L^p(\mathbb{R}^n)$, $1 < p < \infty$, is the space

$$H_p^{\psi,2}(\mathbb{R}^n) := \{u \in L^p(\mathbb{R}^n); \|u\|_{H_p^{\psi,2}} < \infty\} \tag{3.161}$$

where

$$\|u\|_{H_p^{\psi,2}} := \|(\mathrm{id}+\psi(D))u\|_{L^p}. \tag{3.162}$$

compare (3.157).

Remark 3.3.4. A. From our definition it follows that to every $u \in H_p^{\psi,2}(\mathbb{R}^n)$ there exists a unique $f \in L^p(\mathbb{R}^n)$ such that $(\mathrm{id}+\psi(D))u = f$ and $\|u\|_{H_p^{\psi,2}} = \|f\|_{L^p}$.

B. It is clear that for $R > 0$ the norm $\|\cdot\|_{H_p^{\psi,2}}$ and $\|\cdot\|_{H_p^{\psi_R,2}}$ are equivalent and that $H_p^{\psi,2}(\mathbb{R}^n) = H_p^{\psi_R,2}(\mathbb{R}^n)$. Moreover, for $u \in S(\mathbb{R}^n)$ it follows that $u \in H_p^{\psi,2}(\mathbb{R}^n)$ and since $(1+\psi)\widehat{u} \in \bigcap_{r\geq 1} L^r(\mathbb{R}^n)$

$$\|u\|_{H_p^{\psi,2}} = \|F^{-1}((1+\psi(\cdot))\hat{u})\|_{L^p}. \tag{3.163}$$

Our next aim is to prove that $S(\mathbb{R}^n)$ is dense subspace in $H_p^{\psi,2}(\mathbb{R}^n)$. We need some preparation.

Lemma 3.3.5. *The operator* $\mathrm{id}+\psi(D)$ *maps* $S(\mathbb{R}^n)$ *into* $\bigcap_{t\geq 0} H^t(\mathbb{R}^n) \subset C_\infty^\infty(\mathbb{R}^n)$ *and for any* $R > 0$ *the operator* $\mathrm{id}+\psi_R(D)$ *maps* $S(\mathbb{R}^n)$ *into itself.*

Proof: For $u \in S(\mathbb{R}^n)$ and $\alpha \in \mathbb{N}_0^n$ we find

$$\begin{aligned}&\|D^\alpha(\mathrm{id}+\psi(D))u\|_{L^2}^2 = \int_{\mathbb{R}^n} |\xi^u(1+\psi(\xi))|^2|\hat{u}(\xi)|^2\,d\xi\\ &\leq c\int_{\mathbb{R}^n}(1+|\xi|^2)^{|\alpha|}(1+|\xi|^2)^2|\hat{u}(\xi)|^2\,d\xi = c\|u\|_{H^{2+|\alpha|}}^2,\end{aligned}$$

implying $D^\alpha((\mathrm{id}+\psi(D))u) \in L^2(\mathbb{R}^n)$ which together with Sobolev's embedding theorem proves the first part of the lemma. Now, for $R > 0$ and $\alpha, \beta \in \mathbb{N}_0^n$ we find

$$\begin{aligned}(2\pi)^{n/2} x^\beta D_x^\alpha \psi_R(D)u(x) &= x^\beta D_x^\alpha \int_{\mathbb{R}^n} e^{ix\cdot\xi}\psi_R(\xi)\hat{u}(\xi)\,d\xi \\ &= i^{|\beta|}\int_{\mathbb{R}^n} e^{ix\cdot\xi}\xi^\alpha \partial_\xi^\beta[\psi_R(\xi)\hat{u}(\xi)]\,d\xi\end{aligned}$$

and we arrive at

$$\begin{aligned}&\sup_{x\in\mathbb{R}^n} |x^\beta D_x^\alpha \psi_R(D)u(x)| \\ &\quad \le \tilde{c}\cdot\int_{\mathbb{R}^n} \frac{d\xi}{(1+|\xi|^2)^{\frac{n+1}{2}}} \sup_{\xi\in\mathbb{R}^n} \left|\xi^\alpha (1+|\xi|^2)^{\frac{n+1}{2}} \partial_\xi^\beta[\psi_R(\xi)\hat{u}(\xi)]\right|.\end{aligned}$$

Since $\hat{u} \in S(\mathbb{R}^n)$ and ψ_R is an arbitrary often differentiable function which together with its partial derivative has at most quadratic growth, the second assertion follows from the last estimate, i.e. $(\mathrm{id}+\psi_R(D)) : S(\mathbb{R}^n) \to S(\mathbb{R}^n)$.

Remark 3.3.6. For a general continuous negative definite function it is not possible to prove that $\mathrm{id}+\psi(D)$ maps $S(\mathbb{R}^n)$ into itself since for $u \in S(\mathbb{R}^n)$ the function $\xi \mapsto \psi(\xi)\widehat{u}(\xi)$ will not be smooth, i.e. we can not expect that $F^{-1}(\psi\widehat{u})$ decays at infinity faster than every polynomial.

Lemma 3.3.7. *Let ψ be a continuous negative definite function with representation* (3.151)*, $R > 0$ and ψ_R defined as in* (3.153)*. Further, for $\lambda > 0$ define on $S(\mathbb{R}^n)$*

$$(\lambda\,\mathrm{id}+\psi_R(D))^{-1}u(x) := (2\pi)^{-n/2}\int_{\mathbb{R}^n} e^{ix\cdot\xi}\frac{1}{\lambda+\psi_R(\xi)}\hat{u}(\xi)\,d\xi. \tag{3.164}$$

Then the operator $(\lambda\,\mathrm{id}+\psi_R(D))^{-1}$ maps $S(\mathbb{R}^n)$ continuously into itself and it is on $S(\mathbb{R}^n)$ the inverse operator to $\lambda\,\mathrm{id}+\psi_R(D)$.

Proof: As in the proof of the preceding lemma we find using the fact that $\partial^\alpha\left(\frac{1}{\psi_R(\cdot)+\lambda}\right)$ is polynomially bounded that

$$\begin{aligned}&\sup_{x\in\mathbb{R}^n} |x^\beta D_x^\alpha (\lambda\,\mathrm{id}+\psi_R(D))^{-1}u| \\ &\quad \le c\cdot\int_{\mathbb{R}^n} \frac{1}{(1+|\xi|^2)^{\frac{n+1}{2}}}\,d\xi\cdot\sup_{\xi\in\mathbb{R}^n}\left|\xi^\alpha(1+|\xi|^2)^{\frac{n+1}{2}}\partial_\xi^\beta\left(\frac{\hat{u}(\xi)}{1+\psi_R(\xi)}\right)\right|\end{aligned}$$

which implies that $(\lambda \,\mathrm{id} + \psi_R(D))^{-1}$ maps $S(\mathbb{R}^n)$ continuously into itself. Moreover we find for $u \in S(\mathbb{R}^n)$

$$\begin{aligned}(\lambda \,\mathrm{id} + \psi_R(D))^{-1}(\lambda \,\mathrm{id} + \psi_R(D))u &= F^{-1}\Big(\frac{1}{\lambda + \psi_R(\cdot)} F((\lambda \,\mathrm{id} + \psi_R(D))u)\Big) \\ &= F^{-1}\Big(\frac{1}{(\lambda + \psi_R(\cdot))}(\lambda + \psi_R(\cdot))\hat{u}\Big) = u\end{aligned}$$

and analogously we prove that

$$(\lambda \,\mathrm{id} + \psi_R(D))(\lambda \,\mathrm{id} + \psi_R(D))^{-1} = \mathrm{id} \quad \text{on} \quad S(\mathbb{R}^n). \quad \square$$

Now we may prove

Theorem 3.3.8. *For $1 < p < \infty$ the space $S(\mathbb{R}^n)$ is dense in $H_p^{\psi,2}(\mathbb{R}^n)$.*

Proof: From Remark 3.3.4.A we know for $u \in H_p^{\psi,2}(\mathbb{R}^n)$ and $R > 0$ that $(\lambda \,\mathrm{id} + \psi_R(D))u = f \in L^p(\mathbb{R}^n)$, i.e. $u = (\lambda \,\mathrm{id} + \psi_R(D))^{-1} f$. Let $(\varphi_\nu)_{\nu \in \mathbb{N}}$ be a sequence in $S(\mathbb{R}^n)$ converging in $L^p(\mathbb{R}^n)$ to f. It follows from Lemma 3.3.7 that

$$v_\nu := F^{-1}((\lambda \,\mathrm{id} + \psi_R(\cdot))^{-1}\hat{\varphi}_\nu(\cdot)) = (\lambda \,\mathrm{id} + \psi_R(D))^{-1}\varphi_\nu \in S(\mathbb{R}^n)$$

and that $\|u - v_\nu\|_{\psi,R,p} = \|f - \varphi_\nu\|_{L^p}$ which yields that $\lim_{\nu\to\infty} \|u - v_\nu\|_{\psi,R,p} = 0$. From (3.160) we also conclude that $\lim_{\nu\to\infty} \|u - v_\nu\|_{L^p} = 0$. The L^p-continuity of the operator $\tilde{\psi}_R(D)$ gives $\lim_{\nu\to\infty} \|\tilde{\psi}_R(D)u - \tilde{\psi}_D(D)v_\nu\|_{L^p} = 0$. Consequently we find using Remark 3.3.4.B

$$\begin{aligned}\lim_{\nu\to\infty} \|u - v_\nu\|_{H_p^{\psi,2}} \le{}& \lim_{\nu\to\infty} \|u - v_\nu\|_{\psi,R,p} \\ &+ \lim_{\nu\to\infty} \|\tilde{\psi}_R(D)u - \tilde{\psi}_D(D)v_\nu\|_{L^p} = 0,\end{aligned}$$

which proves the theorem. $\square$

It turns out that $S(\mathbb{R}^n)$ is in fact a core for $A^{(p)}$. For this we need

Lemma 3.3.9. *Let $(T_t)_{t\ge 0}$ be a strongly continuous contraction semigroup with generator $(A, D(A))$ and resolvent $(R_\lambda^A)_{\lambda>0}$ on a Banach space $(X, \|\cdot\|_X)$. If $D \subset D(A)$ is a cone such that D is a dense subset in X and $R_\lambda^A D \subset D$ for all $\lambda > 0$ then D is an operator core for $(A, D(A))$.*

Proof: We write as before $\|u\|_{A,X}$ for the graph norm of $(A, D(A))$, see (I.2.153). We have to prove that $\overline{D}^{\|\cdot\|_{A,X}} = D(A)$. Since $D \subset X$ is dense, we find for every $f \in D(A)$ a sequence $(f_\nu)_{\nu\in\mathbb{N}}$ such that $\lim_{\nu\to\infty} \|f - f_\nu\|_X = 0$. Since R_λ^A leaves D invariant, we have $R_\lambda^A f_\nu \in D$ implying

$$\begin{aligned}\|R_\lambda^A f_\nu - R_\lambda^A f\|_{A,X} &= \|R_\lambda^A f_\nu - R_\lambda^A f\|_X + \|AR_\lambda^A f_\nu - AR_\lambda^A f\|_X \\ &\leq \frac{1}{\lambda}\|f_\nu - f\|_X + 2\|f_\nu - f\|_X\end{aligned}$$

where we used that $\|AR_\lambda^A\| \leq 2$ note that $AR_\lambda^A = (A-\lambda)R_\lambda^A + \lambda R_\lambda^A$. It follows that $(R_\lambda^A f_\nu)_{\nu\in\mathbb{N}}$ is a sequence in D which converges in the graph norm to $R_\lambda^A f$ and $R_\lambda^A f \in \overline{D}^{\|\cdot\|_{A,X}}$. Since D is a cone we find in addition $\lambda R_\lambda^A f \in \overline{D}^{\|\cdot\|_{A,X}}$ and therefore

$$\begin{aligned}\|\lambda R_\lambda^A f - f\|_{A,X} &= \|\lambda R_\lambda^A f - f\|_X + \|\lambda AR_\lambda^A f - Af\|_X \\ &\leq \int_0^\infty \lambda e^{-\lambda t}\|T_t f - f\|_X\,dt + \int_0^\infty \lambda e^{-\lambda t}\|T_t Af - Af\|_X\,dt \\ &= \int_0^\infty e^{-s}(\|T_{s/\lambda} f - f\|_X + \|T_{s/\lambda}Af - Af\|_X)\,ds.\end{aligned}$$

Since $f \in D(A)$ and $T_{s/\lambda}$ is a contraction, an application of the dominated convergence theorem yields for $\lambda \to \infty$ that $\lambda R_\lambda^A f \to f \in \overline{D}^{\|\cdot\|_{A,X}}$ where the convergence is with respect to the graph norm. Hence $D(A) \subset \overline{D}^{\|\cdot\|_{A,X}}$ and the lemma is proved. □

Remark 3.3.10. A semigroup version of Lemma 3.3.9 can be found in E.B.Davies [60], Theorem 1.9.

Theorem 3.3.11. *Let $1 < p < \infty$ and ψ be a continuous negative definite function with representation* (3.151). *Then $D(A^{(p)}) = H_p^{\psi,2}(\mathbb{R}^n)$ and $S(\mathbb{R}^n)$ is an operator core for $(A^{(p)}, D(A^{(p)}))$.*

Proof: Consider the operator $\psi_R(D)$ where $R > 0$ and ψ_R is given by (3.153). We already know that $-\psi_R(D)$ extends to a generator of a strongly continuous L^p-semigroup and we denote this extension by $(A_R^{(p)}, D(A_R^{(p)}))$. From Lemma 3.3.7 it follows for the resolvent of $A_R^{(p)}$

$$R_\lambda^{A_R^{(p)}}|_{S(\mathbb{R}^n)} = (\lambda + \psi_R(D))^{-1}$$

and each of the operators $R_\lambda^{A_R^{(p)}}$ preserves $S(\mathbb{R}^n)$. By Lemma 3.3.9 we conclude that $S(\mathbb{R}^n)$ is a core for $(A_R^{(p)}, D(A_R^{(p)}))$ and because of Theorem 3.3.8 we get using Remark 3.3.4.B

$$D(A_R^{(p)}) = \overline{S(\mathbb{R}^n)}^{\|\cdot\|_{A_R^{(p)}, H_p^{\psi_R,2}}} = H_p^{\psi_R,2}(\mathbb{R}^n) = H_p^{\psi,2}(\mathbb{R}^n).$$

Since $S(\mathbb{R}^n) \subset L^p(\mathbb{R}^n)$ is dense and since

$$(A^{(p)} - A_R^{(p)})|_{S(\mathbb{R}^n)} = \tilde{\psi}_R(D)|_{S(\mathbb{R}^n)}$$

is (uniformly) bounded for $R \geq 1$, we also see that $D(A^{(p)}) = H_p^{\psi,2}(\mathbb{R}^n)$. It remains to prove that $S(\mathbb{R}^n)$ is also a core for $A^{(p)}$. This follows from $A^{(p)} = A_R^{(p)} + (A^{(p)} - A_R^{(p)})$ and the fact that the second operator in this decomposition is an L^p-bounded operator. □

Corollary 3.3.12. *Let ψ be a continuous negative definite function with representation* (3.151). *For the Bessel potential space $\mathcal{F}_{2,p}(\mathbb{R}^n)$ associated with the semigroup $(T_t^{(p)})_{t\geq 0}$, where $(T_t^{(p)})_{t\geq 0}$ is related to ψ by $(T_t^{(p)}u)^\wedge(\xi) = e^{-t\psi(\xi)}\hat{u}(\xi)$, $u \in S(\mathbb{R}^n)$, we have*

$$F_{2,p}(\mathbb{R}^n) = H_p^{\psi,2}(\mathbb{R}^n). \tag{3.165}$$

We want to use the result of the preceding corollary to identify all Bessel potential spaces $\mathcal{F}_{s,p}(\mathbb{R}^n)$ associated with the semigroup $(T_t^{(p)})_{t\geq 0}$, $(T_t^{(p)}u)^\wedge(\xi) = e^{-t\psi(\xi)}\hat{u}(\xi)$, with certain function spaces. Therefore we introduce for $s \geq 0$ and $1 < p < \infty$ the notation

$$H_p^{\psi,s}(\mathbb{R}^n) := \mathcal{F}_{s,p}(\mathbb{R}^n) = (\mathrm{id} - A^{(p)})^{-s/2} L^p(\mathbb{R}^n), \tag{3.166}$$

and hence by Corollary 3.1.21

$$D((\mathrm{id} - A^{(p)})^{s/2}) = H_p^{\psi,s}(\mathbb{R}^n). \tag{3.167}$$

For a moment we take on $H_p^{\psi,s}(\mathbb{R}^n)$ the norm $\|\cdot\|_{\mathcal{F}_{s,p}}$ where as before $\|u\|_{\mathcal{F}_{s,p}} = \|(\mathrm{id} - A^{(p)})^{s/2}u\|_{L^p}$. We want to prove (at least on $S(\mathbb{R}^n)$)

$$\|u\|_{H_p^{\psi,s}} = \|F^{-1}((1+\psi(\cdot))^{s/2}\hat{u}(\cdot))\|_{L^p}. \tag{3.168}$$

We know already from Lemma 3.1.22 that the continuous embeddings

$$H_p^{\psi,s+t}(\mathbb{R}^n) \hookrightarrow H_p^{\psi,s}(\mathbb{R}^n) \hookrightarrow L^p(\mathbb{R}^n) \hookrightarrow S'(\mathbb{R}^n) \tag{3.169}$$

hold for $s, t > 0$. As in Section 3.1 we use the relation

$$(\mathrm{id} - A^{(p)})^{-s/2} u = V_s^{(p)} u = \frac{1}{\Gamma(\frac{s}{2})} \int_0^\infty t^{\frac{s}{2}-1} e^{-t} T_t^{(p)} u \, dt, \tag{3.170}$$

the semigroup property on $L^p(\mathbb{R}^n)$,

$$(\mathrm{id} - A^{(p)})^{-s/2} \circ (\mathrm{id} - A^{(p)})^{-t/2} = (\mathrm{id} - A^{(p)})^{-\frac{s+t}{2}}, \quad s, t \geq 0, \tag{3.171}$$

and the fact that $(\mathrm{id} - A^{(p)})^{-s/2}$, $s \geq 0$, is an L^p-contraction. Further, since the operator $(\mathrm{id} - A^{(p)})^{-s/2} : H_p^{\psi,s+r}(\mathbb{R}^n) \to H_p^{\psi,r}(\mathbb{R}^n)$, $s, r \geq 0$, is continuous (since it is continuous between the spaces $\mathcal{F}_{s+r,p}(\mathbb{R}^n)$ and $\mathcal{F}_{r,p}(\mathbb{R}^n)$) we may apply the calculus for fractional powers of generators of contraction semigroups, compare Section I.3.3, to find

$$(\mathrm{id} - A^{(p)})^{s/2} \circ (\mathrm{id} - A^{(p)})^{r/2} = (\mathrm{id} - A^{(p)})^{\frac{s+r}{2}}, \quad s, r \geq 0, \tag{3.172}$$

and

$$\|u\|_{H_p^{\psi,s+r}} = \|(\mathrm{id} - A^{(p)})^{\frac{s+r}{2}} u\|_{L^p}, \tag{3.173}$$

moreover we have

$$(\mathrm{id} - A^{(p)})^{-s/2} \circ (\mathrm{id} - A^{(p)})^{r/2} = (\mathrm{id} - A^{(p)})^{\frac{r-s}{2}} \quad s, r \geq 0, \tag{3.174}$$

and

$$\|(\mathrm{id} - A^{(p)})^{r/2} u\|_{L^p} \leq \|(\mathrm{id} - A^{(p)})^{\frac{s+r}{2}} u\|_{L^p}. \tag{3.175}$$

For $k \in \mathbb{N}$ and $u \in S(\mathbb{R}^n)$

$$(\mathrm{id} - A^{(p)})^k u(x) = (2\pi)^{-n/2} \int_{\mathbb{R}^n} e^{ix \cdot \xi} (1 + \psi(\xi))^k \hat{u}(\xi) \, d\xi,$$

implying that $S(\mathbb{R}^n) \subset \bigcap_{t \geq 0} H_p^{\psi,t}(\mathbb{R}^n)$. The continuity of $(\mathrm{id} - A^{(p)})^{s/2}$ from $H_p^{\psi,s+r}(\mathbb{R}^n)$ to $H_p^{\psi,r}(\mathbb{R}^n)$ implies

$$(\mathrm{id} - A^{(p)})^{s/2} u \in \bigcap_{t \geq 0} H_p^{\psi,t}(\mathbb{R}^n) \quad \text{for } s \geq 0 \text{ and } u \in S(\mathbb{R}^n). \tag{3.176}$$

Further it follows for all $r, s, t \geq 0$ such that $s - r \geq 0$ the continuity of the operators

$$(\mathrm{id} - A^{(p)})^{s/2} \circ (\mathrm{id} - A^{(p)})^{-r/2} : H_p^{\psi,t+s-r}(\mathbb{R}^n) \to H_p^{\psi,t}(\mathbb{R}^n) \tag{3.177}$$

and

$$(\mathrm{id}-A^{(p)})^{-r/2} \circ (\mathrm{id}-A^{(p)})^{s/2} : H_p^{\psi,t+s-r}(\mathbb{R}^n) \to H_p^{\psi,t}(\mathbb{R}^n). \tag{3.178}$$

Hence, for $s \geq 0$ the operator

$$(\mathrm{id}-A^{(p)})^{s/2} : H_p^{\psi,t+s}(\mathbb{R}^n) \to H_p^{\psi,t}(\mathbb{R}^n) \tag{3.179}$$

is injective. Therefore for $g \in H_p^{\psi,t}(\mathbb{R}^n)$ there exists $f \in L^p(\mathbb{R}^n)$ such that $g = (\mathrm{id}-A^{(p)})^{-t/2} f$ and setting

$$u := (\mathrm{id}-A^{(p)})^{-s/2} g = (\mathrm{id}-A^{(p)})^{-\frac{t+s}{2}} f \in H_p^{\psi,s+t}(\mathbb{R}^n)$$

we find $(\mathrm{id}-A^{(p)})^{s/2} u = g$. Thus we have proved

Corollary 3.3.13. *For $s, t \geq 0$ and $1 < p < \infty$ the operator*

$$(\mathrm{id}-A^{(p)})^{s/2} : H_p^{\psi,t+s}(\mathbb{R}^n) \to H_p^{\psi,t}(\mathbb{R}^n) \tag{3.180}$$

is a bijective continuous operator with continuous inverse.

Consider for $R > 0$ the decomposition of ψ according to (3.152) and the extension of the operator $\psi_R(D)$ from $L^p(\mathbb{R}^n)$ to $S'(\mathbb{R}^n)$ and that from $\tilde{\psi}_R(D)$ as bounded operator on $L^p(\mathbb{R}^n)$, see the results in the beginning of this section. Let us show that $\tilde{\psi}_R(D)$ is bounded from $H_p^{\psi,s}(\mathbb{R}^n)$ to $H_p^{\psi,s}(\mathbb{R}^n)$: we have for $u \in H_p^{\psi,s}(\mathbb{R}^n)$, taking into account the translation invariance of $A^{(p)}$,

$$\begin{aligned}
\|\tilde{\psi}_R(D)u\|_{H_p^{\psi,s}} &= \|\tilde{\psi}_R(D)u\|_{\mathcal{F}_{s,p}} \\
&= \left\|(\mathrm{id}-A^{(p)})^{s/2} \int_{B_R^c(0)} (u(\cdot) - u(\cdot - y))\nu(dy)\right\|_{L^p} \\
&= \left\| \int_{B_R^c(0)} \{((\mathrm{id}-A^{(p)})^{s/2}u)(\cdot) - ((\mathrm{id}-A^{(p)})^{s/2}u)(\cdot - y)\}\nu(dy)\right\|_{L^p} \\
&\leq 2\nu(B_R^c(0))\|(\mathrm{id}-A^{(p)})^{s/2}u\|_{L^p} \\
&= 2\nu(B_R^c(0))\|u\|_{F_{s,p}} = 2\nu(B_R^c(0))\|u\|_{H_p^{\psi,s}}.
\end{aligned}$$

Next we observe that $(\mathrm{id}-A^{(p)})^{s/2}u = (\mathrm{id}+\psi(D))^{s/2}u = ((\mathrm{id}+\psi_R(D)) - \psi_R(D))^{s/2}u$, hence

$$\begin{aligned}
\|(\mathrm{id}-A^{(p)})^{s/2}u\|_{L^p} &\leq c(\|(\mathrm{id}+\psi_R(D))^{s/2}u\|_{L^p} + \|u\|_{L^p}) \\
&\leq c'\|(\mathrm{id}+\psi_R(D))^{s/2}u\|_{L^p}
\end{aligned}$$

as well as

$$\begin{aligned}\|(\mathrm{id}+\psi_R(D))^{s/2}u\|_{L^p} &\leq c''\|(\mathrm{id}+\psi_R(D)+\tilde{\psi}_R(D))^{s/2}u\|_{L^p}\\ &= c''\|(\mathrm{id}-A^{(p)})^{s/2}u\|_{L^p},\end{aligned}$$

which yields that $\|u\|_{H_p^{\psi,s}} = \|(\mathrm{id}-A^{(p)})^{s/2}u\|_{L^p}$ and $\|(\mathrm{id}+\psi_R(D))^{s/2}u\|_{L^p}$ are equivalent norms and $H_p^{\psi,s}(\mathbb{R}^n) = H_p^{\psi_R,s}(\mathbb{R}^n)$ as Banach spaces. Further, the proof of Lemma 3.3.7 yields that $(\mathrm{id}+\psi_R(D))^{-s/2}$ maps $S(\mathbb{R}^n)$ continuously into itself.

Proposition 3.3.14. *For $s \geq 0$ the space $S(\mathbb{R}^n)$ is a dense subspace of $H_p^{\psi,s}(\mathbb{R}^n)$.*

Proof: Take $u \in H_p^{\psi,s}(\mathbb{R}^n)$ and for some $R > 0$ set $f = (\mathrm{id}+\psi_R(D))^{s/2}u$, thus $\|u\|_{H_p^{\psi_R,s}} = \|f\|_{L^p}$. Given $\varepsilon > 0$ there exists $v_\varepsilon \in S(\mathbb{R}^n)$ such that $\|f - v_\varepsilon\|_{L^p} < \varepsilon$. With $w_\varepsilon := (\mathrm{id}+\psi_R(D))^{-s/2}v_\varepsilon$ we find $w_\varepsilon \in S(\mathbb{R}^n)$ by the preceding remark and further

$$\|u - w_\varepsilon\|_{H_p^{\psi_R,s}} = \|f - v_\varepsilon\|_{L^p} < \varepsilon.$$

Since $\|\cdot\|_{H_p^{\psi_R,s}}$ and $\|\cdot\|_{H_p^{\psi,s}}$ are equivalent norms the proposition is proved. □

An obvious consequence of Proposition 3.3.14 is that $H_p^{\psi,s+t}(\mathbb{R}^n)$ is dense in $H_p^{\psi,s}(\mathbb{R}^n)$ for any $t, s \geq 0$. To proceed further we need a result generalizing formula (I.4.151) which can be found in U.Westphal [280], [281].

Proposition 3.3.15. *Let $(A, D(A))$ be the generator of a strongly continuous contraction semigroup $(T_t)_{t\geq 0}$ with resolvent $(R_\lambda^A)_{\lambda>0}$ on the Banach space $(X, \|\cdot\|_X)$. For $k < s < k+1$, $k \in \mathbb{N}$, we have for $u \in D(A^{k+1})$*

$$(-A)^s u = \frac{\sin \pi(s-k)}{\pi} \int_0^\infty \lambda^{s-k-1} R_\lambda^A A^{k+1} u \, d\lambda. \tag{3.181}$$

Theorem 3.3.16. *Let $\psi : \mathbb{R}^n \to \mathbb{R}$ be a continuous negative definite function with representation* (3.151) *and let $s \geq 0$, $1 < p < \infty$. Then for all $u \in S(\mathbb{R}^n)$ we have*

$$(\mathrm{id}-A^{(p)})^s u = F^{-1}((1+\psi(\cdot))^s \hat{u}(\cdot)). \tag{3.182}$$

Proof: For $s = k \in \mathbb{N}_0$ formula (3.182) is already known. Let $u \in S(\mathbb{R}^n)$, $s = k + \sigma$, $k \in \mathbb{N}_0$ and $0 < \sigma < 1$. Using Proposition 3.3.15 we find the representation

$$\begin{aligned}(\mathrm{id} - A^{(p)})^s u(x) &= \frac{\sin \pi\sigma}{\pi} \int_0^\infty \lambda^{\sigma-1} R_\lambda^{(A^{(p)}-\mathrm{id})} (\mathrm{id} - A^{(p)})^{k+1} u(x)\, d\lambda \\ &= (2\pi)^{-n/2} \frac{\sin \pi\sigma}{\pi} \int_0^\infty \int_{\mathbb{R}^n} \lambda^{\sigma-1} (\lambda + (1+\psi(\xi)))^{-1} \\ &\quad (1+\psi(\xi))^{k+1} \hat{u}(\xi)\, d\xi d\lambda,\end{aligned}$$

where we used for $k \in \mathbb{N}_0$ and $u \in S(\mathbb{R}^n)$ the fact that (3.182) holds and for $R_\lambda^{(A^{(p)}-\mathrm{id})}$ the representation

$$R_\lambda^{(A^{(p)}-\mathrm{id})} u = F^{-1}\Big(\frac{1}{\lambda + (1+\psi)} \hat{u}\Big), \quad u \in S(\mathbb{R}^n).$$

Since $\hat{u} \in S(\mathbb{R}^n)$ and since for $\tau > 0$

$$\int_0^\infty \frac{\lambda^{\sigma-1}}{\lambda + \tau}\, d\lambda = \frac{\Gamma(\sigma)\Gamma(1-\sigma)}{\Gamma(1)} \tau^{\sigma-1} = \frac{\pi}{\sin \pi\sigma} \tau^{\sigma-1}$$

holds, we find further

$$\begin{aligned}(\mathrm{id} - A^{(p)})^s u(x) &= (2\pi)^{-n/2} \frac{\sin \pi\sigma}{\pi} \int_{\mathbb{R}^n} e^{ix\cdot\xi} (1+\psi(\xi))^{k+1} \\ &\quad \times \int_0^\infty \frac{\lambda^{\sigma-1}}{(\lambda + (1+\psi(\xi)))}\, d\lambda\, \hat{u}(\xi)\, d\xi \\ &= (2\pi)^{-n/2} \frac{\sin \pi\sigma}{\pi} \frac{\pi}{\sin \pi\sigma} \int_{\mathbb{R}^n} e^{ix\cdot\xi} (1+\psi(\xi))^{k+\sigma} \hat{u}(\xi)\, d\xi \\ &= (2\pi)^{-n/2} \int_{\mathbb{R}^n} e^{ix\cdot\xi} (1+\psi(\xi))^s \hat{u}(\xi)\, d\xi,\end{aligned}$$

which proves the theorem. □

From (3.182) we deduce that for $u \in S(\mathbb{R}^n)$

$$\|u\|_{H_p^{\psi,s}} = \|F^{-1}((1+\psi(\cdot))^{s/2} \hat{u}(\cdot))\|_{L^p}, \tag{3.183}$$

i.e. we have proved (3.168) on $S(\mathbb{R}^n)$. Since $S(\mathbb{R}^n)$ is dense in $H_p^{\psi,s}(\mathbb{R}^n)$ it follows further that $H_p^{\psi,s}(\mathbb{R}^n)$ is the closure of $S(\mathbb{R}^n)$ with respect to the norm $\|F^{-1}((1+\psi(\cdot))^{s/2}\hat{u}\|_{L^p}$. Further, the proof of Theorem 3.3.16 yields

Corollary 3.3.17. *For ψ, s and p as in Theorem* 3.3.16 *the operator* $(-A^{(p)})^s$ *has on* $S(\mathbb{R}^n)$ *the representation*

$$(-A^{(p)})^s u(x) = (2\pi)^{-n/2} \int_{\mathbb{R}^n} e^{ix\cdot\xi} (\psi(\xi))^s \hat{u}(\xi)\, d\xi. \tag{3.184}$$

Applying now Theorem 3.1.26 we arrive at

Theorem 3.3.18. *For $u \in S(\mathbb{R}^n)$ and $1 < p < \infty$ the estimates*

$$\begin{aligned}\gamma_0(\|F^{-1}(\psi^{s/2}\hat{u})\|_{L^p} + \|u\|_{L^p}) &\leq \|u\|_{H_p^{\psi,s}} \\ &\leq \gamma_1(\|F^{-1}(\psi^{s/2}\hat{u})\|_{L^p} + \|u\|_{L^p})\end{aligned} \tag{3.185}$$

hold and by a density argument (3.185) extends to all $u \in H_p^{\psi,s}(\mathbb{R}^n)$ with an appropriate interpretation of $F^{-1}(\psi^{s/2}\hat{u})$ and $F^{-1}((1+\psi)^{s/2}\hat{u})$, respectively (see the consideration in the beginning of this section, where we discussed the problem to define $F^{-1}(\psi^{s/2}\widehat{u})$ and $F^{-1}((1+\psi)^{s/2}\widehat{u})$, respectively).

Remark 3.3.19. We excluded in our considerations the continuous negative definite function $\xi \mapsto |\xi|^2$. However the continuous negative definite function $\xi \mapsto |\xi|^{2r_0}$, $0 < r_0 < 1$, are included in our considerations. The spaces $H_p^{|\cdot|^{2r_0},s}(\mathbb{R}^n)$ are as Banach spaces equivalent to the classical Bessel potential spaces $H_p^{r_0 s}(\mathbb{R}^n)$, compare Definition I.3.11.9, and (3.185) is the well known comparison of the L^p-norm of Riesz and Bessel potentials, i.e.

$$\gamma_0(\|(-\Delta)^{sr_0/2}u\|_{L^p}+\|u\|_{L^p}) \leq \|u\|_{H_p^{sr_0}} \leq \gamma_1(\|(-\Delta)^{sr_0/2}u\|_{L^p}+\|u\|_{L^p}) \tag{3.186}$$

see E.M.Stein [256], Section V.3.2.

In Section 3.2, especially Theorem 3.2.15, we gave a description of the dual space $\mathcal{F}^*_{r,p}(\mathbb{R}^n)$ of a general (abstract) Bessel potential space $\mathcal{F}_{r,p}(\mathbb{R}^n)$. We want to make this description concrete for the spaces $H_p^{\psi,s}(\mathbb{R}^n)$.

Definition 3.3.20. Let $\psi : \mathbb{R}^n \to \mathbb{R}$ be a continuous negative definite function with representation (3.151), let $1 < p < \infty$ and $s < 0$. We define the space $H_p^{\psi,s}(\mathbb{R}^n)$ as the closure of $S(\mathbb{R}^n)$ with respect to the norm

$$\|u\|_{H_p^{\psi,s}} = \|F^{-1}((1+\psi(\cdot))^{s/2}\hat{u}(\cdot))\|_{L^p}, \qquad s < 0. \tag{3.187}$$

This definition gives us a scale of ψ-Bessel potential spaces $H_p^{\psi,s}(\mathbb{R}^n)$ for the range of $s \in \mathbb{R}$ with $S(\mathbb{R}^n)$ being a dense subspace. This implies, in particular that a continuous linear functional on $H_p^{\psi,s}(\mathbb{R}^n)$ can be interpreted as an element of $S'(\mathbb{R}^n)$. More precisely, $l \in S'(\mathbb{R}^n)$ belong to $(H_p^{\psi,s}(\mathbb{R}^n))^*$ if and only if there exists $c > 0$ such that

$$|l(\varphi)| \leq c\|\varphi\|_{H_p^{\psi,s}} \quad \text{for all } \varphi \in S(\mathbb{R}^n). \tag{3.188}$$

Theorem 3.3.21. *Let $\psi : \mathbb{R}^n \to \mathbb{R}$ be a continuous negative definite function with representation* (3.151) *and let $s \in \mathbb{R}$, $1 < p < \infty$ and $p' = \frac{p}{p-1}$. The (topological) dual space of $H_p^{\psi,s}(\mathbb{R}^n)$ is the space $H_{p'}^{\psi,-s}(\mathbb{R}^n)$, i.e. for any $l \in (H_p^{\psi,s}(\mathbb{R}^n))^*$ there exists $v \in H_{p'}^{\psi,-s}(\mathbb{R}^n)$ such that for all $\varphi \subset S(\mathbb{R}^n)$*

$$l(\varphi) = \langle F^{-1}((1+\psi(\cdot))^{-s/2}\widehat{v}, F^{-1}((1+\psi(\cdot))^{s/2}\widehat{\varphi})\rangle. \tag{3.189}$$

The norm of l is given by $\|v\|_{H_{p'}^{\psi,-s}}$. Conversely, for any $v \in H_{p'}^{\psi,-s}(\mathbb{R}^n)$ an element $l \in (H_p^{\psi,s}(\mathbb{R}^n))^*$ is given by (3.189).

Remark 3.3.22. Note that $F^{-1}((1+\psi(\cdot))^{-s/2}\hat{v}) = (\mathrm{id} - A^{(p)})^{-s/2}v \in L^{p'}(\mathbb{R}^n)$ for $v \in H_{p'}^{\psi,-s}(\mathbb{R}^n)$ implying that $F^{-1}((1+\psi(\cdot))^{-s/2}\hat{v}) \in L^{p'}(\mathbb{R}^n)$. Therefore (3.189) can be written as convergent integral

$$l(\varphi) = \int_{\mathbb{R}^n} F^{-1}((1+\psi(\cdot))^{-s/2}\hat{v})(x)F^{-1}((1+\psi(\cdot))^{s/2}\hat{\varphi})(x)\,dx. \tag{3.190}$$

Proof of Theorem 3.3.21: Let $v \in H_{p'}^{\psi,-s}(\mathbb{R}^n)$. Since the integral in (3.190) is convergent, an application of Hölder's inequality yields

$$|l(\varphi)| \leq \|v\|_{H_{p'}^{\psi,-s}}\|\varphi\|_{H_p^{\psi,s}},$$

and a standard argument, compare the proof of Theorem 3.2.15, shows that $\|l\| = \|v\|_{H_{p'}^{\psi,-s}}$.

Conversely, let us assume that l is a continuous linear functional on $H_p^{\psi,s}(\mathbb{R}^n)$; in particular, we have (3.188). On $S(\mathbb{R}^n)$ the norm $|||\varphi||| = \|\varphi\|_{H_p^{\psi,s}}$ is introduced and we consider l as a continuous linear functional on $(S(\mathbb{R}^n), |||\cdot|||)$. As such it has a representation by a function $\omega \in L^{p'}(\mathbb{R}^n)$, i.e.

$$l(\varphi) = \int_{\mathbb{R}^n} F^{-1}((1+\psi(\cdot))^{s/2}\widehat{\varphi}(\cdot))(x)\omega(x)\,dx, \quad \varphi \in S(\mathbb{R}^n). \tag{3.191}$$

Since $\omega \in L^{p'}(\mathbb{R}^n)$ we find a sequence $(\omega_\nu)_{\nu\in\mathbb{N}}$, $\omega_\nu \in S(\mathbb{R}^n)$, which converges in $L^{p'}(\mathbb{R}^n)$ to ω. Clearly $v_\nu := F^{-1}((1+\psi)^{s/2}\hat{\omega}_\nu) \in H_{p'}^{\psi,-s}(\mathbb{R}^n)$ and for $\nu \neq \mu$ we have

$$\|F^{-1}((1+\psi(\cdot))^{s/2}F(\omega_\nu - \omega_\mu))\|_{H_{p'}^{\psi,-s}} = \|\omega_\nu - \omega_\mu\|_{L^{p'}},$$

implying that $(v_\nu)_{\nu\in\mathbb{N}}$ is a Cauchy sequence in $H_{p'}^{\psi,-s}(\mathbb{R}^n)$. Denote by $v \in H_{p'}^{\psi,-s}(\mathbb{R}^n)$ the limit of $(v_\nu)_{\nu\in\mathbb{N}}$ it follows that

$$\omega = F^{-1}((1+\psi(\cdot))^{-s/2}\hat{v}), \tag{3.192}$$

and the theorem is proved. □

Next we want to study some embedding theorems for the spaces $H_p^{\psi,s}(\mathbb{R}^n)$. The first elementary result is

Proposition 3.3.23. *Let $\psi : \mathbb{R}^n \to \mathbb{R}$ be a continuous negative definite function with representation* (3.151), $1 < p < \infty$, *and* $r \geq 0$. *For any $s \in \mathbb{R}$ we have the continuous embedding*

$$H_p^{\psi,s+r}(\mathbb{R}^n) \hookrightarrow H_p^{\psi,s}(\mathbb{R}^n). \tag{3.193}$$

Proof: For $s \geq 0$ the statement is just Lemma 3.1.22. Let $s < 0$, $1 < p < \infty$ and $-s-r \geq 0$. We may use Lemma 3.1.22 to find the continuous embedding

$$H_{p'}^{\psi,-s}(\mathbb{R}^n) \hookrightarrow H_{p'}^{\psi,-s-r}(\mathbb{R}^n), \quad p' = \frac{p}{p-1},$$

and Theorem 3.3.21 yields (3.193). If $s < 0$, $1 < p < \infty$ and $-s-r < 0$, we have the continuous embedding $H_p^{\psi,s+r}(\mathbb{R}^n) \hookrightarrow L^p(\mathbb{R}^n)$. Since $H_{p'}^{\psi,-s}(\mathbb{R}^n) \hookrightarrow L^{p'}(\mathbb{R}^n)$, we find by duality $L^p(\mathbb{R}^n) \hookrightarrow H_p^{\psi,s}(\mathbb{R}^n)$ which proves the proposition. □

More general embedding result can be obtained by using Fourier multipliers of type (p, q) as they were investigated by L.Hörmander in [136].

Definition 3.3.24. Let $1 \leq p, q \leq \infty$ We call a distribution $m \in S'(\mathbb{R}^n)$ a *Fourier multiplier of type (p, q)* if

$$\|m\|_{M_{p,q}} := \sup\left\{\frac{\|F^{-1}(m\hat{\varphi})\|_{L^q}}{\|\varphi\|_{L^p}};\quad 0 \neq \varphi \in S(\mathbb{R}^n)\right\} < \infty. \tag{3.194}$$

The set of all Fourier multipliers of type (p, q) is denoted by $M_{p,q}$.

We need some properties of the set $M_{p,q}$. For $p = q = 2$ we may use Plancherel's theorem to find

$$\begin{aligned}\|m\|_{M_{2,2}} &:= \sup\left\{\frac{\|m\hat{\varphi}\|_{L^2}}{\|\hat{\varphi}\|_{L^2}};\quad 0 \neq \varphi \in S(\mathbb{R}^n)\right\} \\ &= \sup\left\{\frac{\|m\varphi\|_{L^2}}{\|\varphi\|_{L^2}};\quad 0 \neq \varphi \in S(\mathbb{R}^n)\right\}\end{aligned}$$

since the Fourier transform is a bijective on $S(\mathbb{R}^n)$. From the last line we derive immediately that $L^\infty(\mathbb{R}^n) \subset M_{2,2}$. On the other hand, if $m \in M_{2,2}$ it follows that

$$\|m\varphi\|_{L^2} \le c\|\varphi\|_{L^2} \tag{3.195}$$

for all $\varphi \in S(\mathbb{R}^n)$, hence $m \in L^2_{loc}(\mathbb{R}^n)$. Further, the operator $\varphi \mapsto m\varphi$ extends continuously as an operator on $L^2(\mathbb{R}^n)$ such that (3.195) remains valid. Suppose that m is not in $L^\infty(\mathbb{R}^n)$. Then for every $N \in \mathbb{N}$ there exists $A_N \subset \mathbb{R}^n$ such that $0 < \lambda^{(n)}(A_N) < \infty$ and $|m| \ge N$ on A_N. Taking in (3.195) the function $\chi_{A_N} \in L^2(\mathbb{R}^n)$ we arrive at

$$N\lambda^{(n)}(A_N)^{1/2} \le \|m\chi_{A_N}\|_{L^2} \le c\|\chi_{A_N}\|_{L^2} = c\lambda^{(n)}(A_N)^{1/2},$$

which gives contradiction, hence we have

$$M_{2,2} = L^\infty(\mathbb{R}^n). \tag{3.196}$$

Next we observe that

$$F(L^1(\mathbb{R}^n)) \subset M_{p,p} \quad 1 < p < \infty. \tag{3.197}$$

Indeed, for $m \in F(L^1(\mathbb{R}^n))$ the convolution theorem yields

$$F^{-1}(m\hat{\varphi}) = (2\pi)^{-n/2}(F^{-1}m) * \varphi$$

with $F^{-1}m \in L^1(\mathbb{R}^n)$. Now (3.197) follows from Young's inequality, Lemma I.2.3.15.

In Proposition I.2.6.19 it was mentioned that every sequentially continuous and translation invariant operator $A : C_0^\infty(\mathbb{R}^n) \to C^\infty(\mathbb{R}^n)$ is a convolution operator, i.e. $Au = K * u$ for some $K \in \mathcal{D}'(\mathbb{R}^n)$. A result of the same type holds for bounded translation invariant operators $A : L^p(\mathbb{R}^n) \to L^q(\mathbb{R}^n)$. As proved in L.Hörmander [136] for such an operator exists $K \in S'(\mathbb{R}^n)$ such that $Au = K * u$ for all $u \in L^p(\mathbb{R}^n)$. Moreover, in the same paper it was proved that for $q < p < \infty$ a bounded translation invariant operator from $L^p(\mathbb{R}^n)$ to $L^q(\mathbb{R}^n)$ is trivial, i.e. identically zero. If m is a Fourier multiplier of type (p, q) then by $\varphi \mapsto F^{-1}(m\hat{\varphi})$ a bounded translation invariant operator from $L^p(\mathbb{R}^n)$ to $L^q(\mathbb{R}^n)$ is defined and the convolution theorem gives the representation

$$F^{-1}(m\hat{\varphi}) = (2\pi)^{-n/2}\hat{m} * \varphi, \quad \hat{m} \in S'(\mathbb{R}^n).$$

Hence we have a one-to-one correspondence between the set $L_{p,q}$ of all $u \in S'(\mathbb{R}^n)$ such that $\|(u * \varphi\|_{L^q} \leq c\|\varphi\|_{L^p}$ holds for all $\varphi \in S(\mathbb{R}^n)$, and the set $M_{p,q}$ of all Fourier multipliers of type (p,q).

Given $u \in S'(\mathbb{R}^n)$, we are interested in the range of all p,q such that $u \in L_{p,q}$. The following result is taken from L.Hörmander [136].

Theorem 3.3.25. *Let $u \in S'(\mathbb{R}^n)$, $u \neq 0$. The set $U(p,q)$ of points $(\frac{1}{p}, \frac{1}{q}) \in \mathbb{R}^2$ such that $u \in L_{p,q}$ is a convex subset of the triangle $D := \{(\frac{1}{x}, \frac{1}{y}) \in \mathbb{R}^2; 0 \leq \frac{1}{x} \leq 1, 0 \leq \frac{1}{y} \leq 1, \frac{1}{y} \leq \frac{1}{x}\}$, and it is symmetric with respect to the line $\frac{1}{x} + \frac{1}{y} = 1$, i.e. $u(p,q) = u(q',p')$.*

Proof: As noted above, necessary for the operator associated with u to be non-trivial is that $q \geq p$, i.e. $\frac{1}{q} \leq \frac{1}{p}$ proving that $U(p,q)$ is contained in D. Let $u \in L_{p,q}$ and define p', q', by $\frac{1}{p} + \frac{1}{p'} = 1 = \frac{1}{q} + \frac{1}{q'}$. Now, if $u \in L_{p,q}$, i.e. $\|u * \varphi\|_{L^q} \leq c_1\|\varphi\|_{L^p}$ for all $\varphi \in S(\mathbb{R}^n)$ we find for $\varphi, \psi \in S(\mathbb{R}^n)$ by Hölder's inequality

$$\begin{aligned} |[(u * \varphi) * \psi](0)| &= \left| \int_{\mathbb{R}^n} (u * \varphi)(y)\psi(y)\, dy \right| \\ &\leq c\|u * \varphi\|_{L^q} \|\psi\|_{L^{q'}} \leq c\|\varphi\|_{L^p} \|\psi\|_{L^{q'}}. \end{aligned}$$

By basic properties of the convolution we find

$$[(u*\varphi)*\psi](0) = [(u*\psi)*\varphi](0),$$

which leads to

$$\left| \int_{\mathbb{R}^n} (u*\psi)(y)\varphi(y)\, dy \right| = |[(u*\varphi)*\psi](0)| \leq c\|\varphi\|_{L^p} \|\psi\|_{L^{q'}}$$

or

$$\|u*\psi\|_{L^{p'}} = \sup_{\varphi \in L^p} \frac{\left| \int_{\mathbb{R}^n} (u * \psi)(y)\varphi(y)\, dy \right|}{\|\varphi\|_{L^p}} \leq c\|\psi\|_{L^{q'}},$$

which proves the symmetry of $U(p,q)$ stated in the theorem. Finally we may apply the Riesz–Thorin convexity theorem, Theorem I.2.8.1, to obtain the convexity of $U(p,q)$. □

Corollary 3.3.26. *For $1 < p < \infty$ we have $M_{p,p} \subset M_{2,2} = L^\infty(\mathbb{R}^n)$.*

Proof: By the above theorem $\hat{m} \in L_{p,p}$ implies $\hat{m} \in L_{p',p'}$ and therefore by the convexity of $U(p,p)$ it follows $\hat{m} \in L_{2,2}$ which implies the corollary. □

Corollary 3.3.27. *For $1 < p < \infty$ the equality*

$$M_{1,p} = M_{p',\infty} = F(L^p(\mathbb{R}^n)), \quad p' = \frac{p}{p-1}. \tag{3.198}$$

holds.

Proof: It is sufficient to show that $L_{p',\infty} = L_{1,p} = L^p(\mathbb{R}^n)$ holds. The first equality follows from Theorem 3.3.25 and the second equality follows by the same calculation done in Theorem 3.3.25 and the fact that $L^{p'}(\mathbb{R}^n)$ is the dual space of $L^p(\mathbb{R}^n)$. □

Now we may return to the spaces $H_p^{\psi,s}(\mathbb{R}^n)$.

Theorem 3.3.28. *Let $\psi_1, \psi_2 : \mathbb{R}^n \to \mathbb{R}$ be two continuous negative definite functions both with Lévy–Khinchin representation* (3.151). *Further let $s, r \in \mathbb{R}$ and $1 < p, q < \infty$. Then the continuous embedding*

$$H_p^{\psi_1,s}(\mathbb{R}^n) \hookrightarrow H_q^{\psi_2,r}(\mathbb{R}^n) \tag{3.199}$$

holds if and only if

$$m := (1+\psi_2)^{r/2}(1+\psi_1)^{-s/2} \in M_{p,q}. \tag{3.200}$$

Proof: Assume that (3.200) holds. For $u \in S(\mathbb{R}^n) \subset H_p^{\psi_1,s}(\mathbb{R}^n)$ we find $v := F^{-1}((1+\psi_1)^{s/2}\hat{u}) \in L^p(\mathbb{R}^n)$ which yields

$$\begin{aligned}
\|u\|_{H_q^{\psi_2,r}} &= \|F^{-1}((1+\psi_2)^{r/2}\hat{u}\|_{L^q} \\
&= \|F^{-1}(m\hat{v})\|_{L^q} \leq \|m\|_{M_{p,q}}\|v\|_{L^p} \\
&= \|m\|_{M_{p,q}}\|u\|_{H_p^{\psi_1,s}},
\end{aligned}$$

implying (3.199). Now suppose that (3.200) holds and take $\varphi \in S(\mathbb{R}^n)$. It follows that

$$\begin{aligned}
\|F^{-1}(m\hat{\varphi})\|_{L^q} &= \|F^{-1}((1+\psi_1)^{-s/2})\hat{\varphi}\|_{H_q^{\psi_2,r}} \\
&\leq c\|F^{-1}((1+\psi_1)^{-s/2})\hat{\varphi}\|_{H_p^{\psi_1,s}} \leq c\|\varphi\|_{L^p},
\end{aligned}$$

which gives (3.200) and the theorem is proved. □

Corollary 3.3.29. *A. Let $1 < p < \infty$ and let $\psi : \mathbb{R}^n \to \mathbb{R}$ be an unbounded continuous negative definite function with representation* (3.151). *Then $H_p^{\psi,s}(\mathbb{R}^n) \hookrightarrow H_p^{\psi,r}(\mathbb{R}^n)$ if and only if $s \geq r$.*

B. Let $1 < p < \infty$, $s > 0$, and $\psi_1, \psi_2 : \mathbb{R}^n \to \mathbb{R}$ be two continuous negative definite functions with representation (3.151). *Then the embedding $H_p^{\psi_1,s}(\mathbb{R}^n) \hookrightarrow H_p^{\psi_2,s}(\mathbb{R}^n)$ implies that there exists a constant $c > 0$ such that*

$$1+\psi_2(\xi) \leq c(1+\psi_1(\xi)), \quad \xi \in \mathbb{R}^n. \tag{3.201}$$

If, in particular, $H_p^{\psi_1,s}(\mathbb{R}^n) = H_p^{\psi_2,s}(\mathbb{R}^n)$ we have

$$\frac{1}{c}(1+\psi_1(\xi)) \leq 1+\psi_2(\xi) \leq c(1+\psi_1(\xi)). \tag{3.202}$$

The converse assertion holds for $p = 2$.

Proof: A. If $s \geq r$ then the validity of the embedding was proved in Proposition 3.3.23 even for the case that ψ is bounded. Assume now that $H_p^{\psi,s}(\mathbb{R}^n) \hookrightarrow H_p^{\psi,r}(\mathbb{R}^n)$. By Theorem 3.3.25 we find that

$$m := (1+\psi)^{(r-s)/2} \in M_{p,p} \subset L^\infty(\mathbb{R}^n).$$

Since ψ is unbounded, it follows that $s \geq r$.

B. We just have to apply Theorem 3.3.25 and to use $M_{p,p} \subset L^\infty(\mathbb{R}^n)$. □

It is also possible to prove a Sobolev-type embedding theorem for the spaces $H_p^{\psi,s}(\mathbb{R}^n)$.

Theorem 3.3.30. *Let $\psi : \mathbb{R}^n \to \mathbb{R}$ be a continuous negative definite function with representation* (3.151). *Further let $1 < p < \infty$ and $s \in \mathbb{R}$. The embedding*

$$H_p^{\psi,s}(\mathbb{R}^n) \hookrightarrow C_\infty(\mathbb{R}^n) \tag{3.203}$$

holds if and only if

$$F^{-1}((1+\psi)^{-s/2}) \in L^{p'}(\mathbb{R}^n), \quad p' = \frac{p}{p-1}, \tag{3.204}$$

i.e. if and only if $(1+\psi)^{-s/2} \in M_{p,\infty}$.

Proof: Suppose that (3.204) is satisfied and put $m(\cdot) := (1+\psi(\cdot))^{-s/2}$. From the definition of a Fourier multiplier of type (p, ∞) we find for $\varphi \in$

$S(\mathbb{R}^n) \subset H_p^{\psi,s}(\mathbb{R}^n)$

$$\begin{aligned}\|\varphi\|_{L^\infty} &= \|F^{-1}((1+\psi)^{-s/2}F(F^{-1}((1+\psi)^{s/2}\hat{\varphi}))\|_{L^\infty}\\ &\le \|m\|_{M_{p,\infty}}\|F^{-1}((1+\psi)^{s/2}\hat{\varphi})\|_{L^p}\\ &= c\|\varphi\|_{H_p^{\psi,s}}.\end{aligned}$$

Since $S(\mathbb{R}^n)$ is dense in $H_p^{\psi,s}(\mathbb{R}^n)$ each element $u \in H_p^{\psi,s}(\mathbb{R}^n)$ is uniform limit of continuous functions, hence it has a continuous representative and (3.203) follows. Conversely, suppose that (3.203) holds. It follows that

$$|\varphi(0)| \le c\|\varphi\|_{H_p^{\psi,s}}, \quad \varphi \in S(\mathbb{R}^n).$$

This however means that the Dirac measure at zero ε_0 is a continuous linear functional on $H_p^{\psi,s}(\mathbb{R}^n)$. By the duality theorem, Theorem 3.3.21, $\varepsilon_0 \in H_{p'}^{\psi,-s}(\mathbb{R}^n)$. But now the definition of $H_{p'}^{\psi,-s}(\mathbb{R}^n)$ and the fact that $\hat{\varepsilon}_0 = (2\pi)^{-n/2}$ yields (3.204). □

We want to find some sufficient criterion directly posed on ψ in order that (3.203) or (3.200) with $\psi_1(\xi) = |\xi|^2$ hold. For this let $\psi : \mathbb{R}^n \to \mathbb{R}$ be a continuous negative definite function with representation (3.151) and assume in addition

$$1+\psi(\xi) \ge c_0(1+|\xi|^2)^{r_0}, \quad \xi \in \mathbb{R}^n, \tag{3.205}$$

for some $r_0 > 0$. Since for every $R > 0$ the function $\tilde{\psi}_R$ is bounded it follows that (3.205) holds also for ψ_R with $c_0 = c_0(R)$, where ψ_R and $\tilde{\psi}_R$ are given by (3.152) and (3.153).

Lemma 3.3.31. *Suppose that the continuous negative definite function* $\psi : \mathbb{R}^n \to \mathbb{R}$ *has the representation* (3.151) *and satisfies* (3.205). *Then for all* $R > 0$, $t > 0$, $1 < p < \infty$, *and* $s > \frac{t+n}{r_0}$ *the function*

$$\frac{(1+|\cdot|^2)^{t/2}}{(1+\psi_R(\cdot))^{s/2}} \tag{3.206}$$

is a Fourier multiplier of type (p,p).

Proof: Since (3.205) holds also (maybe with a different constant) for ψ_R we derive immediately that

$$\frac{(1+|\cdot|^2)^{t/2}}{(1+\psi_R(\cdot))^{s/2}} \in L^1(\mathbb{R}^n) \quad \text{for } s > \frac{t+n}{r_0}. \tag{3.207}$$

Further, using Theorem I.3.7.13 we find for $k \in \mathbb{N}$

$$\Big|(1-\Delta_\xi)^k\Big(\frac{(1+|\xi|^2)^{t/2}}{(1+\psi_R(\xi))^{s/2}}\Big)\Big| \leq c_k \frac{(1+|\xi|^2)^{t/2}}{(1+\psi_R(\xi))^{s/2}},$$

implying

$$(1-\Delta)^k\Big(\frac{(1+|\cdot|^2)^{t/2}}{(1+\psi_R(\cdot))^{s/2}}\Big) \in L^1(\mathbb{R}^n) \quad \text{for } s > \frac{t+n}{r_0}. \tag{3.208}$$

Combining (3.207) and (3.208) we conclude that

$$F\Big(\frac{(1+|\cdot|^2)^{t/2}}{(1+\psi_R(\cdot))^{s/2}}\Big) \in L^1(\mathbb{R}^n) \quad \text{for } s > \frac{t+n}{r_0}. \tag{3.209}$$

But (3.209) means by (3.197) that for $s > \frac{t+n}{r_0}$ the function $\xi \mapsto \dfrac{(1+|\cdot|^2)^{t/2}}{(1+\psi_R(\cdot))^{s/2}}$ is a Fourier multiplier of type (p,p). □

Corollary 3.3.32. *A. Suppose that the continuous negative definite function $\psi : \mathbb{R}^n \to \mathbb{R}$ with representation* (3.151) *satisfies* (3.205). *Then for $s > \frac{t+n}{r_0}$ the space $H_p^{\psi,s}(\mathbb{R}^n)$ is continuously embedded into the space $H_p^t(\mathbb{R}^n)$.*

B. Suppose that ψ is as in part A and that $s > \frac{n}{r_0}\frac{1+p}{p}$. Then the space $H_p^{\psi,s}(\mathbb{R}^n)$ is continuously embedded into $C_\infty(\mathbb{R}^n)$.

Proof: Part A is clear from Lemma 3.3.31 and Theorem 3.3.28. For part B observe that $H_p^t(\mathbb{R}^n) \subset C_\infty(\mathbb{R}^n)$ for $t > \frac{n}{p}$, see Section I.3.11. Hence combining this embedding with the result of part A we obtain part B. □

Remark 3.3.33. A. A further, more implicit sufficient criterion for (3.203) to hold is the assumption that there is an integer $N \in \mathbb{N}$ such that with $\kappa := n(\frac{1}{p'} - \frac{1}{2}) > N$, $p' = \frac{p}{p-1}$, the function $(1+\psi_R)^{-s/2}$ has (distributional) derivatives $D^\alpha((1+\psi_k)^{-s/2} \in L^2(\mathbb{R}^n)$ for all $|\alpha| \leq N$. Indeed, in this case the following *Carlson–Beurling inequality*

$$\|F^{-1}((1+\psi_R(\cdot))^{-s/2})\|_{L^{p'}} \leq c\|(1+\psi_R(\cdot))^{-s/2}\|_{L^2}^{1-\kappa/N} \sum_{|\alpha|=N} \|D^\alpha((1+\psi_R(\cdot))^{-s/2}\|_{L^2}^{\kappa/N} \tag{3.210}$$

holds. We refer to D.Brenner et.al [36], p.18, or to H.Dappa and W.Trebels [59], Lemma 1.1.

B. In [87], Theorem 2.3.16, it was proved that if $\psi : \mathbb{R}^n \to \mathbb{R}$ has the representation (3.151) and satisfies (3.205) then if $0 < \varepsilon < 1$ is given and $\theta \in (\frac{1}{2-\varepsilon}, 1)$ then for $p = p_{\varepsilon,\theta} := \frac{1+\theta_\varepsilon}{1+(\varepsilon-1)\theta}$ the space $H_p^{\psi,\theta_n/r_0}(\mathbb{R}^n)$ is continuously embedded into $C_\infty(\mathbb{R}^n)$. Note that for $\theta \to \frac{1}{2-\varepsilon}$, i.e. $p \to 2$, we get for $\varepsilon \to 0$ the continuous embedding $H_2^{\psi,s}(\mathbb{R}^n) \hookrightarrow C_\infty(\mathbb{R}^n)$ for $s > \frac{n}{2r_0}$. This limiting case follows of course from a direct calculation using Plancherel's theorem. Lemma 3.3.31 is taken from the joint paper [166] with R.Schilling.

Corollary 3.3.34. . *Suppose that the continuous negative definite function $\psi : \mathbb{R}^n \to \mathbb{R}$ has the representation* (3.151) *and satisfies* (3.205). *Then $C_0^\infty(\mathbb{R}^n)$ is a dense subspace of $H_p^{\psi,s}(\mathbb{R}^n)$, $s \geq 0$.*

Proof: As in the proof of Lemma 3.3.31 we find that for $t > s + n$

$$(1-\Delta)^k \left(\frac{(1+\psi_R(\cdot))^{s/2}}{(1+|\cdot|^2)^{t/2}} \right) \in L^1(\mathbb{R}^n)$$

for all $k \in \mathbb{N}_0$. Hence $H_p^t(\mathbb{R}^n)$ is continuously embedded into $H_p^{\psi,s}(\mathbb{R}^n)$ implying the corollary. □

An immediate consequence of Corollary 3.3.34 is

Corollary 3.3.35. *For all $s \in \mathbb{R}$ and $1 < p < \infty$*

$$H_p^{\psi,s}(\mathbb{R}^n) \subset D'(\mathbb{R}^n) \tag{3.211}$$

holds.

Next we want to determine the complex interpolation spaces for two spaces $H_{p_0}^{\psi,s_0}(\mathbb{R}^n)$ and $H_{p_1}^{\psi,s_1}(\mathbb{R}^n)$. For doing so we need first some additional considerations to Section I.2.8. Again we set

$$G := \{z \in \mathbb{C}; 0 < \operatorname{Re} z < 1\}. \tag{3.212}$$

For two (complex) Banach spaces $(X_0, \|\cdot\|_{X_0})$ and $(X_1, \|\cdot\|_{X_1})$ both embedded into some common Hausdorff space $\mathcal{H}$ we set $X := X_0 + X_1$ equipped with the norm $\|\cdot\|_X := \max(\|\cdot\|_{X_0}, \|\cdot\|_{X_1})$, which is obviously equivalent to $\|\cdot\|_{X_0} + \|\cdot\|_{X_1}$, and which turns X into a Banach space. Denote by $W(G;X)$ the space of all continuous functions $\omega : \overline{G} \to X$ with the following properties:

i) $\omega|_G$ is analytic and $\sup_{z \in \overline{G}} \|\omega(z)\|_X < \infty$;

ii) $\omega(iy) \in X_0$ and $\omega(1+iy) \in X_1$ for $y \in \mathbb{R}$ with continuous maps $y \mapsto \omega(iy)$ and $y \mapsto \omega(1+iy)$, respectively;

iii) $\|\omega\|_{W(G;X)} := \max(\sup_{y\in\mathbb{R}} \|\omega(iy)\|_{X_0}, \sup_{y\in\mathbb{R}} \|\omega(1+iy)\|_{X_1}) < \infty$.

By the maximum principle for analytic functions $(W(G;X), \|\cdot\|_{W(G;X)})$ is a Banach space. In the situation just described we call the Banach spaces $\{X_0, X_1\}$ an *interpolation couple.* For any interpolation couple and $0 < \theta < 1$ we define its *complex interpolation spaces*, compare also Definition I.2.8.5, by

$$[X_0, X_1]_\theta := \{u \in X; \quad \text{there exists } \omega \in W(G,X) \text{ such that } \omega(\theta) = u\} \tag{3.213}$$

and on $[X_0, X_1]_\theta$ we introduce the norm

$$\|u\|_{[X_0,X_1]_\theta} := \inf\{\|\omega\|_{W(G;X)}; \quad \omega \in W(G;X) \text{ and } \omega(\theta) = u\}. \tag{3.214}$$

Clearly, $([X_0, X_1]_\theta, \|\cdot\|_{[X_0,X_1]_\theta})$ is a Banach space. The following lemma, taken from [87] is a modification of Lemma 24.6.3 in H.Triebel [268], and it gives a useful equivalent norm on $[X_0, X_1]_\theta$.

Lemma 3.3.36. *Let* $\{X_0, X_1\}$ *be an interpolation couple and let* $0 < \theta < 1$. *Then we have*

$$\|u\|_{[X_0,X_1]_\theta} = \inf\{(\sup_{y\in\mathbb{R}} \|\omega(iy)\|_{X_0})^{1-\theta} \cdot \sup_{y\in\mathbb{R}} \|\omega(1+iy)\|^\theta_{X_1}\}, \tag{3.215}$$

where the infimum ranges over all $\omega \in W(G;X)$ *such that* $\omega(\theta) = u$.

Proof: Since by iii) of the definition of $W(G;X)$ we have

$$\sup_{y\in\mathbb{R}} \|\omega(iy)\|_{X_0}, \quad \sup_{y\in\mathbb{R}} \|\omega(1+iy)\|_{X_1} \le \|\omega\|_{W(G;X)},$$

it follows that

$$(\sup_{y\in\mathbb{R}} \|\omega(iy)\|_{X_0})^{1-\theta} (\sup_{y\in\mathbb{R}} \|\omega(1+iy)\|_{X_1})^\theta \le \|u\|_{[X_0,X_1]_\theta}.$$

On the other hand, the function $z \mapsto a^{z-\theta}\omega(z)$, $a > 0$ and $z \mapsto \omega(z)$ as in (3.214), is admissible in (3.214) too, which yields by using iii)

$$\|u\|_{[X_0,X_1]_\theta} \le \max(a^{-\theta} \sup_{y\in\mathbb{R}} \|\omega(iy)\|_{X_0}, a^{1-\theta} \sup_{y\in\mathbb{R}} \|\omega(1+iy)\|_{X_1}). \tag{3.216}$$

If $\|\omega(1+iy)\|_{X_1} \neq 0$ we may choose

$$a = \frac{\sup_{y\in\mathbb{R}} \|\omega(iy)\|_{X_0}}{\sup_{y\in\mathbb{R}} \|\omega(1+iy)\|_{X_1}}$$

to get the estimate

$$\|u\|_{[X_0,X_1]_\theta} \le (\sup_{y\in\mathbb{R}} \|\omega(iy)\|_{X_0})^{1-\theta} \cdot \sup_{y\in\mathbb{R}} \|\omega(1+iy)\|_{X_1})^{\theta}.$$

If however $\|\omega(1+iy)\|_X = 0$, then we let a tend to infinity to arrive at (3.215). □

An application at this lemma to $L^{p_0}(\mathbb{R}^n)$ and $L^{p_1}(\mathbb{R}^n)$, $1 < p_0, p_1 < \infty$, for $0 < \theta < 1$ and $\frac{1}{p} = \frac{1-\theta}{p_0} + \frac{\theta}{p_1}$ yields

$$[L^{p_0}(\mathbb{R}^n), L^{p_1}(\mathbb{R}^n)]_\theta = L^p(\mathbb{R}^n), \tag{3.217}$$

and

$$\|u\|_{L^p} = \inf\{(\sup_{y\in\mathbb{R}} \|g(iy)\|_{L^{p_0}})^{1-\theta}, \sup_{y\in\mathbb{R}} \|g(1+iy)\|_{L^{p_1}})^{\theta}\}, \tag{3.218}$$

where the infimum is taken over all $g \in W(G; S'(\mathbb{R}^n))$ such that $g(\theta) = u$.

In order to apply complex interpolation to the spaces $H^{\psi,s_0}_{p_0}(\mathbb{R}^n)$ and $H^{\psi,s_1}_{p_1}(\mathbb{R}^n)$ we need bounds for the imaginary powers of $\mathrm{id}+\psi(D)$ on $L^p(\mathbb{R}^n)$. It turns out that this is a rather deep and involved result. We will state it here without a proof, the proof will be discussed in the next section.

Theorem 3.3.37. *Let $\psi : \mathbb{R}^n \to \mathbb{R}$ be a continuous negative definite function and $1 < p < \infty$. Then there exists constants $\gamma \ge 0$ and $c_\gamma \ge 1$ such that*

$$\|(\mathrm{id}+\psi(D))^{iy} u\|_{L^p} \le c_\gamma e^{\gamma|y|} \|u\|_{L^p} \tag{3.219}$$

holds for all $y \in \mathbb{R}$.

Next we can prove the announced in interpolation theorem which is again taken from [87].

Theorem 3.3.38. *Let $\psi : \mathbb{R} \to \mathbb{R}$ be a continuous negative definite function. Further let $0 < p_0, p_1 < \infty$, $s_0, s_1 \in \mathbb{R}$ and $0 < \theta < 1$. For*

$$s = (1-\theta)s_0 + \theta s_1 \quad \textit{and} \quad \frac{1}{p} = \frac{1-\theta}{p_0} + \frac{\theta}{p_1}$$

it follows that

$$[H^{\psi,s_0}_{p_0}(\mathbb{R}^n); H^{\psi,s_1}_{p_1}(\mathbb{R}^n)]_\theta = H^{\psi,s}_p(\mathbb{R}^n) \tag{3.220}$$

holds.

Proof: Let $X := H^{\psi,s_0}_{p_0}(\mathbb{R}^n) + H^{\psi,s_1}_{p_1}(\mathbb{R}^n) \hookrightarrow S'(\mathbb{R}^n)$ and $G = \{z \in \mathbb{C}; 0 < \operatorname{Re} z < 1\}$. Further denote for a moment $H_\theta(\mathbb{R}^n) := [H^{\psi,s_0}_{p_0}(\mathbb{R}^n); H^{\psi,s_1}_{p_1}(\mathbb{R}^n)]_\theta$. First let $u \in H_\theta(\mathbb{R}^n)$ and choose any $\omega \in W(G;X)$ with $\omega(\theta) = u$. We define on G

$$g_\omega(z) = F^{-1}\Big(e^{(z-\theta)^2}(1+\psi)^{((1-z)s_0+zs_1)/2}F(\omega(z))\Big). \tag{3.221}$$

First we prove that g_ω satisfies the conditions i)–iii) for the interpolation couple $\{L^{p_0}(\mathbb{R}^n), L^{p_1}(\mathbb{R}^n)\}$. Clearly g_ω is analytic in G, continuous in $\overline{G}$, and the boundedness condition i) follows from the definition of the norms in $H^{\psi,s_j}_p(\mathbb{R}^n)$, $j = 0,1$. In addition we have for $y \in \mathbb{R}$

$$\begin{aligned} g_\omega(iy) = e^{(iy-\theta)^2}&(1+\psi(D))^{iy(s_1-s_0)/2} \\ &F^{-1}((1+\psi)^{s_0/2}F(\omega(iy))) \end{aligned}$$

and

$$\begin{aligned} g_\omega(1+iy) = e^{(iy+1-\theta)^2}&(1+\psi(D))^{iy(s_1-s_0)/2} \\ &F^{-1}((1+\psi)^{s_1/2}F(\omega(1+iy))). \end{aligned}$$

By Theorem 3.3.37 we find with some $\gamma = \gamma(s_0, s_1) \geq 0$ and $c_\gamma \geq 1$ that

$$\begin{aligned} \|g_\omega(iy)\|_{L^{p_0}} &\leq |e^{(iy-\theta)^2}| \|(1+\psi(D))^{iy(s_1-s_0)/2}\| \\ &\quad \|F^{-1}((1+\psi)^{s_0/2}F(\omega(iy))\|_{L^{p_0}} \\ &\leq c_\gamma e^{-y^2+\theta^2}e^{\gamma|y|}\|F^{-1}((1+\psi)^{s_0/2}F(\omega(iy))\|_{L^{p_0}} \\ &\leq M_\gamma \|F^{-1}((1+\psi)^{s_0/2}F(\omega(iy))\|_{L^{p_0}}, \end{aligned}$$

or

$$\|g_\omega(iy)\|_{L^p} \leq M_\gamma \|\omega(iy)\|_{H^{\psi,s_0}_{p_0}} \tag{3.222}$$

with

$$M_\gamma = \sup_{y\in\mathbb{R}}(c_\gamma e^{-y^2+\theta^2}e^{\gamma|y|}) < \infty.$$

In particular we have $g_\omega(iy) \in L^{p_0}(\mathbb{R}^n)$. An analogous calculation yields

$$\|g_\omega(1+iy)\|_{L^{p_1}} \leq \tilde{M}_\gamma \|\omega(1+iy)\|_{H^{\psi,s_1}_{p_1}}. \tag{3.223}$$

Moreover we find

$$g_\omega(\theta) = F^{-1}((1+\psi)^{s/2}F(\omega(\theta))),$$

and in the following we set

$$v := g_\omega(\theta).$$

Now, using (3.217) we arrive at

$$\begin{aligned}\|u\|_{H^{\psi,s_1}_{p_1}} &= \|F^{-1}((1+\psi)^{s/2}F(\omega(\theta))\|_{L^p} = \|v\|_{L^p} \\ &= \inf_{\substack{g\in W(G;S'(\mathbb{R}^n)) \\ g(\theta)=v}} \{\sup_{y\in\mathbb{R}} \|g(iy)\|^{1-\theta}_{L^{p_0}} \cdot \sup_{y\in\mathbb{R}} \|g(1+iy)\|^{\theta}_{L^{p_1}}\} \\ &\leq \sup_{y\in\mathbb{R}} \|g_\omega(iy)\|^{1-\theta}_{L^{p_0}} \cdot \sup_{y\in\mathbb{R}} \|g_\omega(1+iy)\|^{\theta}_{L^{p_1}}\end{aligned}$$

with g_ω given by (3.221). Using (3.222) and (3.223) we find

$$\|u\|_{H^{\psi,s}_p} \leq M^{1-\theta}_\gamma \tilde{M}^\theta_\gamma \sup_{y\in\mathbb{R}} \|\omega(iy)\|^{1-\theta}_{H^{\psi,s_0}_{p_0}} \cdot \sup_{y\in\mathbb{R}} \|\omega(1+iy)\|^{\theta}_{H^{\psi,s_1}_{p_1}}.$$

Since $\omega \in W(G;X)$, $\omega(\theta) = u$, was arbitrary we may pass to the infimum and conclude from Lemma 3.3.36 that

$$\|u\|_{H^{\psi,s}_p} \leq M^{1-\theta}_\gamma \tilde{M}^\theta_\gamma \|u\|_{H_\theta}$$

implying

$$[H^{\psi,s_0}_{p_0}(\mathbb{R}^n), H^{\psi,s_1}_{p_1}(\mathbb{R}^n)]_\theta = H_\theta \subset H^{\psi,s}_p(\mathbb{R}^n).$$

Now we will prove the converse inclusion. Note that

$$(\mathrm{id} + \psi(D))^{s/2} H^{\psi,s}_p(\mathbb{R}^n) = L^p(\mathbb{R}^n),$$

i.e.

$$v := F^{-1}((1+\psi)^{s/2}\hat{u}) \in L^p(\mathbb{R}^n)$$

for any $u \in H^{\psi,s}_p(\mathbb{R}^n)$. We choose now an arbitrary $g \in W(G;S'(\mathbb{R}^n))$ with $g(\theta) = v$ and $g(k+iy) \in L^{p_k}(\mathbb{R}^n)$, $k = 0,1$, recall $[L^{p_0}(\mathbb{R}^n), L^{p_1}(\mathbb{R}^n)]_\theta = L^p(\mathbb{R}^n)$, $\frac{1}{p} = \frac{1-\theta}{p_0} + \frac{\theta}{p_1}$. Analogously to the previous considerations, compare (3.221), we define now

$$\omega_g(z) := F^{-1}(e^{(z-\theta)^2}(1+\psi)^{(z-1)s_0/2 - zs_1/2}F(g(z)))$$

and calculations analogous to those done before show that ω_g satisfies i)-iii) for the interpolation couple $\{H^{\psi,s_0}_{p_0}(\mathbb{R}^n), H^{\psi,s_1}_{p_1}\}$. Further, arguing as in the previous case we will find that $H^{\psi,s}_p(\mathbb{R}^n) \subset H_\theta$, recall $H_\theta = [H^{\psi,s_0}_{p_0}(\mathbb{R}^n), H^{\psi,s_1}_{p_1}]_\theta$, proving the theorem. □

Finally we will use the (r,p)-capacity in order to examine elements in $H_p^{\psi,r}(\mathbb{R}^n)$ more detailed. Once again the function ψ is assumed to have the representation (3.151), i.e.

$$\psi(\xi) = \int_{\mathbb{R}^n\setminus\{0\}} (1-\cos y\cdot\xi)\,\nu(dy).$$

Since by (3.167) we have $H_p^{\psi,r}(\mathbb{R}^n) = \mathcal{F}_{r,p}$, where $\mathcal{F}_{r,p}$ is the (r,p)-Bessel potential space associated with the semigroup $(T_t^{(p)})_{t\geq 0}$,

$$T_t^{(p)}u(x) = (2\pi)^{-n/2}\int_{\mathbb{R}^n} e^{ix\cdot\xi}e^{-t\psi(\xi)}\hat{u}(\xi)\,d\xi,$$

we can use all results from Section 3.1 and 3.2. In particular, the capacity $\mathrm{cap}_{r,p}^{\psi}$ in $H_p^{\psi,r}(\mathbb{R}^n)$ is defined as in Definition 3.1.30 and it is a Choquet capacity. Moreover, for $r\geq 0$, $1<p<\infty$, all the spaces $H_p^{\psi,r}(\mathbb{R}^n)$ are regular in the sense of Definition 3.1.35.B which follows from Proposition 3.3.14 (or Corollary 3.3.34). In the following we will speak shortly of (r,p)-quasi-everywhere properties, when $\mathrm{cap}_{r,p}^{\psi}$-quasi-everywhere properties are meant.

Theorem 3.3.39. *Let $1<p<\infty$, $r>0$ and $\psi:\mathbb{R}^n\to\mathbb{R}$ be a continuous negative definite function with representation* (3.151). *The following statements hold.*

A. If $u\in H_p^{\psi,r}(\mathbb{R}^n;\mathbb{R})$ is (r,p)-quasi-everywhere and $u\geq 0$ a.e. on an open set G, then $u\geq 0$ (r,p)-quasi-everywhere on G.

B. Each $u\in H_p^{\psi,r}(\mathbb{R}^n;\mathbb{R})$ admits an (r,p)-quasi-continuous modification $\tilde{u}$ which is unique up to (r,p)-quasi-everywhere equality. Moreover, for $\rho>0$ the inequality

$$\mathrm{cap}_{r,p}^{\psi}\{|\tilde{u}|>\rho\} \leq \frac{c}{\rho^p}\|u\|^p_{H_p^{\psi,r}}$$

holds.

C. let $(u_k)_{k\in\mathbb{N}}$, $u_k\in H_p^{\psi,r}(\mathbb{R}^n;\mathbb{R})$, be a sequence converging in $H_p^{\psi,r}(\mathbb{R}^n;\mathbb{R})$ to $u\in H_p^{\psi,r}(\mathbb{R}^n;\mathbb{R})$. Then there exists a subsequence $(u_{k_l})_{l\in\mathbb{N}}$ such that $(\tilde{u}_{k_l})_{l\in\mathbb{N}}$ converges (r,p)-quasi-uniformly to $\tilde{u}$, which implies also the (r,p)-quasi-everywhere convergence of a subsequence of $(\tilde{u}_k)_{k\in\mathbb{N}}$ to $\tilde{u}$.

D. For all $A\subset\mathbb{R}^n$ with $\mathrm{cap}_{r,p}^{\psi}(A)<\infty$ there exists an (r,p)-equilibrium potential *u_A satisfying*

$$u_A\in\{u\in H_p^{\psi,r}(\mathbb{R}^n;\mathbb{R});\quad \tilde{u}\geq 1\ (r,p)\text{-}q.e.\ on\ A\} \tag{3.224}$$

and

$$\|u_A\|^p_{H_p^{\psi,r}} = \mathrm{cap}^{\psi}_{r,p}(A). \tag{3.225}$$

E. If a sequence $(u_k)_{k\in\mathbb{N}}$ *of* (r,p)*-quasi-continuous functions* u_k *belonging to* $H_p^{\psi,r}(\mathbb{R}^n;\mathbb{R})$ *converges to* $u \in H_p^{\psi,r}(\mathbb{R}^n;\mathbb{R})$ *in the norm* $\|\cdot\|_{H_p^{\psi,r}}$*, then a subsequence of* $(u_k)_{k\in\mathbb{N}}$ *converges* (r,p)*-quasi-everywhere to an* (r,p)*-quasi-continuous modification* $\tilde{u}$ *of* u*.*

Proof: For A we have to apply Proposition 3.1.39 and B follows from Theorem 3.1.41. Further, C is a consequence of Proposition 3.1.49 and Corollary 3.1.50. The existence of the (r,p)-equilibrium potential, part D, is implied by Theorem 3.1.51, and finally, E follows from D and the result proved before. □

Without proof we state for our situation the following result due to S.Albeverio and Z.-M.Ma [4]:

Theorem 3.3.40. *There exists a kernel* $v_r^{(p)}(x,dy)$ *such that for all* $u \in H_p^{\psi,r}(\mathbb{R}^n;\mathbb{R})$ *an* (r,p)*-quasi-continuous version of* u *is given by*

$$x \mapsto \tilde{u}(x) = \int_{\mathbb{R}^n} f(y) v_r^{(p)}(x,dy)$$

where $u = (\mathrm{id} + \psi(D))^{-r/2} f$*,* $f \in L^p(\mathbb{R}^n;\mathbb{R})$*.*

Remark 3.3.41. From the formula

$$(\mathrm{id} - A^{(p)})^{-r/2} u = \frac{1}{\Gamma(\frac{r}{2})} \int_0^\infty t^{r/2-1} e^{-t} T_t u \, dt$$

it is clear that $(\mathrm{id} - A^{(p)})^{-r/2}$ has a kernel representation. The importance of Theorem 3.3.40 is that we may take a single kernel to get for each $u \in H_p^{\psi,r}(\mathbb{R}^n)$ an (r,p)-quasi-continuous representation.

To complete our discussion of the spaces $H_p^{\psi,r}(\mathbb{R}^n;\mathbb{R})$ we ask when the space $H_p^{\psi,r}(\mathbb{R}^n;\mathbb{R})$ will have the contraction property, compare Definition 3.2.35.B. For $p = 2$ this is equivalent to the question when $H_2^{\psi,r}(\mathbb{R}^n;\mathbb{R})$ will be a Dirichlet space and we get the first result:

Lemma 3.3.42. *For any continuous negative definite function* ψ *with representation* (3.151) *the space* $H_2^{\psi,r}(\mathbb{R}^n;\mathbb{R})$*,* $r \le 1$*, has the contraction property.*

Remark 3.3.43. Note that for a continuous negative definite function ψ also ψ^α, $0 < \alpha < 1$, is a continuous negative definite function and that $H_2^{\psi,r} = H_2^{\psi^\alpha, r/\alpha}$. This shows that there is a principal problem consisting of the trade-off between the "order" of ψ and the order of "differentiability" that makes a sharp formulation of results like Lemma 3.3.42 difficult.

An L^p-version of Lemma 3.3.42 is much more complicated and requires a more delicate analysis which we will develop in the next sections and volume III. As a result we will obtain the following theorem due to F.Hirsch [119].

Theorem 3.3.44. *For a continuous negative definite function ψ satisfying* (3.151) *the spaces $H_p^{\psi,r}(\mathbb{R}^n;\mathbb{R})$, $2 \le p$ and $0 < r < 1$, have the contraction property. In this case the (r,p)-equilibrium potential u_G for an open set G has the property that $u_G|_G = 1$ a.e.*

In the paper [270] H.Triebel discussed the problem for which function spaces $B^s_{pq}(\mathbb{R}^n;\mathbb{R})$ and $F^s_{pq}(\mathbb{R}^n;\mathbb{R})$ the Lipshitz function operate on these spaces, compare Definition 3.2.35. Extensions are given in his monograph [271]. It is noteworthy that he could show that the Lipschitz functions operate on these spaces also in some situations where we are out of the scope of Dirichlet spaces. In particular this applies to the spaces $H^s_p(\mathbb{R}^n;\mathbb{R})$ for $1 < p < \infty$, $1 < s < 1 + \frac{1}{p}$.

3.4 Stein's Littlewood–Paley Theory for Sub-Markovian Semigroups

We remind that reader to our introduction to understand and appreciate the role of E.M.Stein's work [255] for this part of our monograph.

Often students in probability theory will hear that the important distinction of measure theory and probability theory lies in the concept of conditional expectation, and the theory of martingales invented by J.L.Doob [69] as a major tool in probability theory dependent from its very definition on the notion of conditional expectation. In volume III of this treatise we will come back to "probabilistic" martingale theory. However, once established in probability theory, the concept of martingales can be seen and used in a plain analytic context. This point of view will be taken in this section in order to discuss certain maximal inequalities for sub-Markovian semigroups.

In the following $(\Omega, \mathcal{A}, P)$ is a probability space and for a measurable function $f : \Omega \to \mathbb{R}$ (or $\overline{R}$) we denote its integral by

$$E(f) := \int_{\Omega} f(\omega) P(d\omega) \tag{3.226}$$

and call it the *expectation* of f provided it exists, i.e. provided $f \geq 0$ or $f \in L^1(\Omega, P)$. The following result is a consequence of the Radon–Nykodym theorem, see H.Bauer [19], Theorem 15.1.

Theorem 3.4.1. *For every sub-σ-field $\mathcal{F} \subset \mathcal{A}$ and every measurable function $f : \Omega \to \mathbb{R}$ with existing integral, i.e. either $f \geq 0$ or $f \in L^1(\Omega, P)$, there exists a measurable function f_0 which is in addition $\mathcal{F}$-measurable and satisfies*

$$\int_C f_0 \, dP = \int_C f \, dP \tag{3.227}$$

for all $C \in \mathcal{F}$. For $f \in L^1(\Omega, P)$ and $f \geq 0$ a.e. it follows in addition that $f_0 \geq 0$ a.e.

Definition 3.4.2. In the situation of Theorem 3.4.1 we call f_0 the *conditional expectation* of f under $\mathcal{F}$ and denote it by

$$E(f|\mathcal{F}). \tag{3.228}$$

Remark 3.4.3. In probability theory a measurable mapping is called a random variable and the integral of a real-valued random variable is called its expectation. In this context the name conditional expectation makes more sense, but we think that it does not make sense to invent a new name in the analytic context.

The following two lemmata summarize properties of conditional expectation. A proof of the first lemma can be found in H.Bauer [19], §15, the proof of the second lemma is taken from E.M.Stein [255], p.90.

Lemma 3.4.4. *Let $(\Omega, \mathcal{A}, P)$ be a probability space, f and g real-valued measurable functions on Ω such that the expectations of f, g and $f + g$ exist (if needed). Further suppose that $\mathcal{F}, \mathcal{G}$ are sub-σ-fields of $\mathcal{A}$ and let $\alpha \in \mathbb{R}$. Then we have*

$$E(E(f|\mathcal{F})) = E(f); \tag{3.229}$$

$$f \geq 0 \quad a.e. \text{ implies } E(f|\mathcal{F}) \geq 0 \ a.e. \tag{3.230}$$

$$\textit{if } f \textit{ is } \mathcal{F}\textit{-measurable, then } E(f|\mathcal{F}) = f \textit{ a.e.}; \tag{3.231}$$

$$E(f+g|\mathcal{F}) = E(f|\mathcal{F})+E(g|\mathcal{F}) \textit{ a.e. and } E(\alpha f|\mathcal{F}) = \alpha E(f|\mathcal{F}) \textit{ a.e.}; \tag{3.232}$$

$$E(E(f|\mathcal{F})|\mathcal{F}) = E(f|\mathcal{F}), \quad \textit{a.e.}; \tag{3.233}$$

$$\textit{if } g \textit{ is bounded and } \mathcal{F} \textit{ measurable, then } E(gf|\mathcal{F}) = gE(f|\mathcal{F}) \textit{ a.e.}; \tag{3.234}$$

$$\textit{if } \mathcal{G} \subset \mathcal{F} \textit{ then } E(E(f|\mathcal{F})|\mathcal{G}) = E(f|\mathcal{G}) \textit{ a.e.} \tag{3.235}$$

Remark 3.4.5. Clearly (3.230) implies for suitable f and g that $f \geq g$ a.e. yields $E(f|\mathcal{F}) \geq E(g|\mathcal{F})$ a.e., and further, from (3.231) it follows that

$$E(1|\mathcal{F}) = 1 \quad \text{a.e. for all } \mathcal{F} \subset \mathcal{A}. \tag{3.236}$$

Lemma 3.4.6. *A. If $f, g \in L^2(\Omega, P)$ then*

$$\int_\Omega E(f|\mathcal{F})g \, dP = \int_\Omega fE(g|\mathcal{F}) \, dP. \tag{3.237}$$

B. For $f \in L^p(\Omega, P)$ it follows that

$$\|E(f|\mathcal{F})\|_{L^p} \leq \|f\|_{L^p}. \tag{3.238}$$

Proof: A. Since by (3.229)

$$\begin{aligned}\int_\Omega E(f|\mathcal{F})g \, dP &= E(E(f|\mathcal{F})g) = E(E(E(f|\mathcal{F})g|\mathcal{F})) \\ &= E(E(f|\mathcal{F})E(g|\mathcal{F}))\end{aligned}$$

and on the other hand

$$\begin{aligned}\int_\Omega fE(g|\mathcal{F}) \, dP &= E(fE(g|\mathcal{F})) = E(E(fE(g|\mathcal{F})|\mathcal{F})) \\ &= E(E(f|\mathcal{F})E(g|\mathcal{F})),\end{aligned}$$

equality (3.237) is proved.

B. First note that for $\mathcal{F} \subset \mathcal{A}$ it follows that $L^p(\Omega, \mathcal{F}, P) \subset L^p(\Omega, \mathcal{A}, P)$.

Now, since $E(f|\mathcal{F})$ is $\mathcal{F}$-measurable we find with $p' = \frac{p}{p-1}$

$$\begin{aligned}
\|E(f|\mathcal{F})\|_{L^p} &= \sup\{\int_\Omega E(f|\mathcal{F})g\,dP; g \in L^{p'}(\Omega, \mathcal{F}, P) \quad \text{and } \|g\|_{L^{p'}} \leq 1\} \\
&= \sup\{\int_\Omega f(\omega)g(\omega)dP; g \in L^{p'}(\Omega, \mathcal{F}, P) \quad \text{and } \|g\|_{L^{p'}} \leq 1\} \\
&\leq \sup\{\int_\Omega f(\omega)g(\omega)dP; g \in L^{p'}(\Omega, \mathcal{A}, P) \quad \text{and } \|g\|_{L^{p'}} \leq 1\} \\
&= \|f\|_{L^p},
\end{aligned}$$

and the lemma is proved. □

Remark 3.4.7. We may consider the taking of conditional expectation with respect to $\mathcal{F}$ as an operator. Since for $\mathcal{F} \subset \mathcal{A}$ we have $L^p(\Omega, \mathcal{F}, P) \subset L^p(\Omega, \mathcal{A}, P)$ it follows from the preceding lemmata that

$$E(\cdot|\mathcal{F}) : L^p(\Omega, \mathcal{A}, P) \to L^p(\Omega, \mathcal{F}, P) \subset L^p(\Omega, \mathcal{A}, P)$$

is a linear, positivity preserving operator which is idempotent, i.e. a projection. In case $p = 2$ it is even selfadjoint by (3.237).

Remark 3.4.8. We will need a variant of the *Marcinkiewicz interpolation theorem*, Theorem I.2.8.9, which is proved in E.M.Stein [255], p.92-93. Let $(\Omega, \mathcal{A}, P)$ be a probability space and suppose that $T : L^1(\Omega, P) + L^\infty(\Omega, P) \to L^1(\Omega, P) + L^\infty(\Omega, P)$ is a sublinear operator which is of weak type (1,1) and bounded from $L^\infty(\Omega, P)$ into itself. Then for $1 < p < \infty$ the operator T maps $L^p(\Omega, P)$ into itself and is bounded, i.e.

$$\|Tu\|_{L^p} \leq c_p \|u\|_{L^p}. \tag{3.239}$$

In the following let $(\mathcal{F}_\nu)_{\nu\in\mathbb{N}}$ be an increasing sequence of sub-σ-fields of $\mathcal{A}$, i.e. $\mathcal{F}_\nu \subset \mathcal{F}_{\nu+1}$ and $\mathcal{F}_\nu \subset \mathcal{A}$ for all $\nu \in \mathbb{N}$. For f with existing expectation we define

$$E_m(f) := E(f|\mathcal{F}_m) \tag{3.240}$$

and the *maximal function*

$$f^*(\omega) := \sup_{m\geq 1} |E_m(f)(\omega)|. \tag{3.241}$$

Lemma 3.4.9. *Let $(f_\nu)_{\nu=1,\dots,m}$ be a finite family of functions in $L^1(\Omega,P)+L^\infty(\Omega,P)$ such that f_ν is measurable with respect to $\mathcal{F}_\nu$. Suppose that for $\nu=1,\dots,m$*

$$f_\nu = E(f_m|\mathcal{F}_\nu) \quad a.e.$$

holds and define

$$f_m^*(\omega) := \sup_{1\le\nu\le m} f_\nu(\omega).$$

Then we have the following inequalities

$$P\{\omega; f_m^*(\omega)>\alpha\} \le \frac{c_0}{\alpha}\|f_m\|_{L_1}, \quad \alpha>0; \tag{3.242}$$

and

$$\|f_m^*\|_{L^p} \le c_p\|f_m\|_{L^p}, \tag{3.243}$$

where c_0 and c_p are independent of m.

Proof: Without loss of generality we may assume $f_m \ge 0$, otherwise use the representation $f_m = f_m^+ - f_m^-$ and handle f_m^+ and f_m^- separately. Hence we may also assume that $f_\nu \ge 0$. For

$$S_\alpha := \{\omega; f_m^*(\omega)>\alpha\} = \{\omega; f_\nu(\omega)>\alpha \quad \text{for some } 1\le\nu\le m\}$$

we have the decomposition

$$S_\alpha = \bigcup_{\nu=1}^m \{\omega; f_\nu(\omega)>\alpha \quad \text{but } f_\mu(\omega)\le\alpha \text{ for } \mu<\nu\} = \bigcup_{\nu=1}^m S_\alpha^{(\nu)}.$$

Note that $S_\alpha^{(\nu_1)} \cap S_\alpha^{(\nu_2)} = \emptyset$ for $\nu_1 \ne \nu_2$ and $S_\alpha^{(\nu)} \in \mathcal{F}_\nu$. Now it follows that

$$\begin{aligned}\int_{S_\alpha} f_m(\omega)P(d\omega) &= \sum_{\nu=1}^m \int_{S_\alpha^{(\nu)}} f_m(\omega)P(d\omega)\\ &= \sum_{\nu=1}^m \int_{S_\alpha^{(\nu)}} E(f_m|\mathcal{F}_j)(\omega)P(d\omega)\\ &= \sum_{\nu-1}^m \int_{S_\alpha^{(\nu)}} f_\nu(\omega)P(d\omega) \ge \sum_{\nu=1}^m \alpha P(S_\alpha^{(\nu)}) = \alpha P(S_\alpha),\end{aligned}$$

or

$$P\{\omega \in \Omega; f_m^*(\omega) > \alpha\} \leq \frac{1}{\alpha} \int_{\{\omega\in\Omega; f_m^*(\omega)>\alpha\}} f_m(\omega) P(d\omega)$$

which yields (3.242). Obviously we have

$$\begin{aligned} \|f_m^*\|_{L^\infty} &= \sup_{\omega\in\Omega} \sup_{1\leq\nu\leq m} |E(f_m|\mathcal{F}_\nu)| \\ &\leq (\sup_{\omega\in\Omega} |f_m(\omega)|) \sup_{1\leq\nu\leq m} E(1|\mathcal{F}_\nu) = \|f_m\|_{L^\infty}, \end{aligned}$$

and the Marcinkiewicz interpolation theorem in the variant of Remark 3.4.8 implies the lemma. □

Now we turn to our first maximal function estimates:

Theorem 3.4.10 (Martingale Maximal Theorem). *Let f, $(\Omega, \mathcal{A}, P)$ and $(\mathcal{F}_\nu)_{\nu\in\mathbb{N}}$ be as before. With $E_m(f)$ given by (3.240) and f^* given by (3.241) we have*

$$P\{\omega; f^*(\omega) > \alpha\} \leq \frac{c_0}{\alpha}\|f\|_{L^1}, \tag{3.244}$$

i.e. $f \mapsto f^*$ is of weak-type (1,1), and

$$\|f^*\|_{L^p} \leq c_p\|f\|_{L^p}, \quad 1 < p \leq \infty. \tag{3.245}$$

Proof: By Lemma 3.4.9 we know for the sequence $(f_m)_{m\in\mathbb{N}}$, $f_m = E_m(f)$ the estimate

$$\|\sup_{\nu\leq m} f_\nu\|_{L^p} \leq c_p\|f_m\|_{L^p} \leq c_p\|f\|_{L^p}, \quad 1 < p \leq \infty,$$

by (3.243) with c_p independent of m. Since $(\sup_{\nu\leq m} f_\nu)_{m\in\mathbb{N}}$ increases to f^* we obtain (3.245) by monotone convergence. Analogously we have with c_0 independent of m

$$P\{\omega; f_m^*(\omega) \geq \alpha\} \leq \frac{c_0}{\alpha}\|f_m\|_{L^1} \leq \frac{c}{\alpha}\|f\|_{L^1}$$

where we used (3.242). Again we may let m tend to infinity to get (3.245) since $P\{\omega; f_m^*(\omega) \geq \alpha\} \uparrow P\{\omega; f^*(\omega) \geq \alpha\}$, proving the theorem. □

Later on in this section we will discuss a maximal theorem for semigroups involving analogues of the Hardy–Littlewood maximal function or the

Littlewood–Paley function. Before doing so we need more results on martingales. We continue with an auxiliary result due to R.Gundy [114] which has a certain analogy to the Calderón–Zygmund decomposition, Lemma I.3.12.2. Our formulation and proof is taken from E.M.Stein [255]. We need some preparations.

Let $f \in L^1(\Omega, P)$, $f \geq 0$ and $\alpha > 0$. As before we set $f_\nu = E_\nu(f) = E(f|\mathcal{F}_\nu)$.

Definition 3.4.11. A mapping $\tau : \Omega \to \mathbb{N} \cup \{\infty\}$ is called a *stopping time* with respect to $(\mathcal{F}_\nu)_{\nu\in\mathbb{N}}$ if for each $\nu \in \mathbb{N}$ the set $\{\omega \in \Omega; \tau(\omega) = \nu\}$ belong to $\mathcal{F}_\nu$.

Remark 3.4.12. Again, the name stopping time will become clear when we turn to probability theory in volume III.

Lemma 3.4.13. *For a stopping time τ we find with $f_\infty(\omega) := f(\omega)$*

$$\int_\Omega f(\omega)P(d\omega) = \int_\Omega f_{\tau(\omega)}(\omega)P(d\omega). \tag{3.246}$$

Proof: For f and τ as above we have

$$\begin{aligned}\int_\Omega f_{\tau(\omega)}(\omega)P(d\omega) &= \sum_{\nu=1}^{\infty} \int_{\{\omega;\tau(\omega)=\nu\}} f_\nu(\omega)P(d\omega) \\ &= \sum_{\nu=1}^{\infty} \int_{\{\omega;\tau(\omega)=\nu\}} E(f|\mathcal{F}_\nu)P(d\omega) \\ &= \sum_{\nu=1}^{\infty} \int_{\{\omega;\tau(\omega)=\nu\}} f(\omega)P(d\omega) = \int_\Omega f(\omega)P(d\omega). \quad \square\end{aligned}$$

Lemma 3.4.14. *Every $k \in \mathbb{N}$ is a stopping time with respect to $(\mathcal{F}_\nu)_{\nu\in\mathbb{N}}$ and if τ_ν, $\nu \in \mathbb{N}$, are stopping times with respect to $(\mathcal{F}_\nu)_{\nu\in\mathbb{N}}$, the same holds for $\tau_1 + k$, $k \in \mathbb{N}$, $\sup_{\nu\in\mathbb{N}} \tau_\nu$ and $\min_{\nu\leq k} \tau_\nu$.*

A proof of Lemma 3.4.14 can be found for example in K.L.Chung [52]. Recall that $f_\nu = E(f|\mathcal{F}_\nu)$ implying

$$f_\nu = E(f_{\nu+1}|\mathcal{F}_\nu) \tag{3.247}$$

since by (3.235) we have $E(E(f|\mathcal{F}_{\nu+1})|\mathcal{F}_\nu) = E(f|\mathcal{F}_\nu)$. Defining for a stopping time τ

$$f_{\nu\wedge\tau}(\omega) = f_{\nu\wedge\tau(\omega)}(\omega) \tag{3.248}$$

we claim

$$f_{\nu\wedge\tau} = E(f_{(\nu+1)\wedge\tau}|\mathcal{F}_\nu). \tag{3.249}$$

For a stopping time τ we find

$$\begin{aligned} E(f_{(\nu+1)\wedge\tau}|\mathcal{F}_\nu) &= E(f_{\nu+1}\chi_{\{\tau\geq\nu+1\}}|\mathcal{F}_\nu) + E(f_\tau\chi_{\{\tau<\nu+1\}}|\mathcal{F}_\nu) \\ &= E(f_{\nu+1}\chi_{\{\tau>\nu\}}|\mathcal{F}_\nu) + E(f_\tau\chi_{\{\tau\leq\nu\}}|\mathcal{F}_\nu) \\ &= \chi_{\{\tau>\nu\}}E(f_{\nu+1}|\mathcal{F}_\nu) + f_\tau\chi_{\{\tau\leq\nu\}} \\ &= \chi_{\{\tau>\nu\}}f_\nu + f_\tau\chi_{\{\tau\leq\nu\}} \\ &= f_{\tau\wedge\nu}, \end{aligned}$$

proving (3.249).

Remark 3.4.15. The property (3.247) is called the *martingale property* and $(f_\nu)_{\nu\in\mathbb{N}}$ is a *martingale* with respect to $(\mathcal{F}_\nu)_{\nu\in\mathbb{N}}$. The sequence $(f_{\nu\wedge\tau})_{\nu\in\mathbb{N}}$ is called the *stopped martingale.* with respect to the stopping time τ. The interpretation of (3.249) is therefore that a stopped martingale is once again a martingale with respect to $(\mathcal{F}_\nu)_{\nu\in\mathbb{N}}$.

Now we prove the result due to R.Gundy [114] which gives a type of Calderón–Zygmund decomposition for L^1-martingale, compare Lemma 4.12.2. Our proof follows once again E.M.Stein [255].

Lemma 3.4.16. *Let $f \in L^1(\Omega, P)$, $f \geq 0$, be given. For $\lambda > 0$ we may decompose $f = g + h + k$ such that*

$$P\{\omega\in\Omega; \sup_{\nu\in\mathbb{N}}|E_\nu(g)(\omega)| > 0\} \leq \frac{c}{\lambda}\|f\|_{L^1} \quad \textit{and} \quad \|g\|_{L^1} \leq c\|f\|_{L^1}; \tag{3.250}$$

$$\Big\|\sum_{\nu=1}^{\infty}|E_\nu(h)-E_{\nu-1}(h)|\Big\|_{L^1} \leq c\|f\|_{L^1} \quad \textit{implying also } \|h\|_{L^1} \leq c\|f\|_{L^1}; \tag{3.251}$$

$$\|k\|_{L^\infty} \leq c\lambda \quad \textit{and } \|k\|_{L^1} \leq c\|f\|_{L^1}. \tag{3.252}$$

Here we used throughout the notion $E_\nu(g) = E(g|\mathcal{F}_\nu)$ etc., and the definition $E_0(g) = 0$.

Proof: A. Construction of g: Given $f \geq 0$, $f \in L^1(\Omega, P)$ and $\lambda > 0$. We claim that

$$\rho(\omega) := \inf\{n; f_n(\omega) > \lambda\} \tag{3.253}$$

is a stopping time. Indeed we have

$$\{\omega; \rho(\omega) = n\} = \{\omega; f_1(\omega), \ldots, f_{n-1}(\omega) \leq \lambda \quad \text{but} \quad f_n(\omega) > \lambda\} \in \mathcal{F}_n.$$

With $\varphi_k := f_k - f_{k-1}$ and $f_0 = 0$ we find

$$f_\nu = \sum_{k=1}^{\nu} \varphi_k,$$

and define further

$$\varepsilon_\nu(\omega) := \varphi_\nu(\omega)\chi_{\{\omega'; \rho(\omega')=\nu\}}(\omega).$$

The definition of ρ implies that $\varepsilon_\nu \geq 0$. Next we introduce

$$\sigma(\omega) := \inf\{\nu; \sum_{k=0}^{\nu} E(\varepsilon_{k+1}|\mathcal{F}_k)(\omega) > \lambda\}.$$

Since $(\mathcal{F}_\nu)_{\nu \in \mathbb{N}}$ is increasing and $E(\varepsilon_{k+1}|\mathcal{F}_k)$ is $\mathcal{F}_k$-measurable, it follows that σ is a stopping time. Finally we set

$$\tau := \sigma \wedge \rho$$

which is by Lemma 3.4.14 a stopping time. We prove now

$$P\{\omega \in \Omega;\ \tau(\omega) < \infty\} \leq \frac{c}{\lambda}\|f\|_{L^1}. \tag{3.254}$$

For this note that

$$\{\omega \in \Omega;\ \rho(\omega) \neq +\infty\} = \{\omega \in \Omega;\ \sup_\nu f_\nu(\omega) > \lambda\} = \{\omega \in \Omega;\ f^*(\omega) > \lambda\},$$

or by the martingale maximal theorem, Theorem 3.4.10,

$$P\{\omega \in \Omega;\ \rho(\omega) < \infty\} \leq \frac{c}{\lambda}\|f\|_{L^1}. \tag{3.255}$$

Further, since $f \geq 0$

$$\begin{aligned}
\int_\Omega \sum_{k=0}^{\infty} E(\varepsilon_{k+1}|\mathcal{F}_k)(\omega)P(d\omega) &= \int_\Omega \sum_{k=0}^{\infty} \varepsilon_{k+1}(\omega)P(d\omega) \\
&= \sum_{k=0}^{\infty} \int_{\{\omega;\rho(\omega)=k+1\}} (f_{k+1}(\omega) - f_k(\omega))P(d\omega) \\
&\leq \sum_{k=0}^{\infty} \int_{\{\omega;\rho(\omega)=k+1\}} f_{k+1}(\omega)P(d\omega) = \int_{\{\omega;\rho(\omega)<\infty\}} f_{\rho(\omega)}(\omega)P(d\omega) \\
&\leq \int_\Omega f_{\rho(\omega)}(\omega)P(d\omega) = \int_\Omega f(\omega)P(d\omega) = \|f\|_{L^1},
\end{aligned}$$

where we used in the last step (3.246). From Chebyshev's inequality we deduce

$$\begin{aligned} P\{\omega;\ \sum_{k=0}^{\infty} E(\varepsilon_{k+1}|\mathcal{F}_k)(\omega) > \lambda\} &\le \frac{1}{\lambda}\int_{\Omega}\Big|\sum_{k=0}^{\infty} E(\varepsilon_{k+1}|\mathcal{F}_k)(\omega)\Big|\, P(d\omega) \\ &\le \frac{1}{\lambda}\|f\|_{L^1}. \end{aligned}$$

But

$$\{\omega \in \Omega; \sigma(\omega) < \infty\} = \{\omega \in \Omega; \sum_{k=0}^{\infty} E(\varepsilon_{k+1}|\mathcal{F}_k)(\omega) > \lambda\},$$

which leads to

$$P\{\omega; \sigma(\omega) < \infty\} \le \frac{1}{\lambda}\|f\|_{L^1}. \tag{3.256}$$

Thus we find

$$\begin{aligned} P\{\omega \in \Omega;\ \tau(\omega) < \infty\} &\le P\{\omega \in \Omega;\ \rho(\omega) < \infty\} + P\{\omega \in \Omega;\ \sigma(\omega) < \infty\} \\ &\le \frac{c}{\lambda}\|f\|_{L^1}, \end{aligned}$$

and (3.254) is proved. Now we define

$$g(\omega) := f(\omega) - f_{\tau(\omega)}(\omega). \tag{3.257}$$

It follows immediately that

$$\{\omega \in \Omega; g(\omega) \ne 0)\} \subset \{\omega \in \Omega; \tau(\omega) \ne \infty\}$$

which gives by (3.254)

$$P\{\omega \in \Omega; g(\omega) \ne 0\} \le \frac{c}{\lambda}\|f\|_{L^1}. \tag{3.258}$$

Since $E_\nu(g) = f_\nu - f_{\nu\wedge\tau}$ we arrive at

$$P\{\omega \in \Omega; \sup_{\nu\in\mathbb{N}} E_\nu(g)(\omega) \ne 0\} \subset P\{\omega \in \Omega, \tau(\omega \ne \infty\}$$

proving

$$P\{\omega \in \Omega; \sup_{\nu\in\mathbb{N}} E_\nu(g)(\omega) \ne 0\} \le \frac{c}{\lambda}\|f\|_{L^1},$$

and further we have

$$\|g\|_{L^1} \leq \|f\|_{L^1} + \|f_{\tau(\cdot)}\|_{L^1} \leq 2\|f\|_{L^1},$$

where we used once again (3.246).

B. Preparations to construct h and k: From the definition

$$g(\omega) = f(\omega) - f_{\tau(\omega)}(\omega)$$

is remains to find h and k such that

$$h(\omega) + k(\omega) = f_{\tau(\omega)}(\omega).$$

We shall construct h and k as L^1-limits of sequences $(h_\nu)_{\nu\in\mathbb{N}}$ and $(k_\nu)_{\nu\in\mathbb{N}}$. For this we look for another representation of $f_{\nu\wedge\tau}$, namely recalling $\varphi_j = f_j - f_{j-1}$ we start with

$$f_{\nu\wedge\tau} = \sum_{j=1}^{\nu} \varphi_j \chi_{\{\omega';\tau(\omega')\geq j\}}$$

and using $\{\omega' \in \Omega; \tau(\omega') \geq j\} = \{\omega' \in \Omega; \rho(\omega') \geq j\} \cap \{\omega' \in \Omega; \sigma(\omega') \geq j\}$ we find with

$$\varepsilon_j := \varphi_j \chi_{\{\omega';\rho(\omega')=j\}} \quad \text{and} \quad \gamma_j := \varphi_j \chi_{\{\omega';\rho(\omega')>j\}}$$

that

$$f_{\nu\wedge\tau} = \sum_{j=1}^{\nu} (\gamma_j + \varepsilon_j) \chi_{\{\omega';\sigma(\omega')\geq j\}}.$$

We set

$$h_\nu = \sum_{j=1}^{\nu} (\varepsilon_j - E(\varepsilon_j|\mathcal{F}_{j-1})) \chi_{\{\omega';\sigma(\omega')\geq j\}} =: \sum_{j=1}^{\nu} \psi_j \tag{3.259}$$

and

$$k_\nu = \sum_{j=1}^{\nu} (\gamma_j + E(\varepsilon_j|\mathcal{F}_{j-1})) \chi_{\{\omega';\sigma(\omega')\geq j\}}. \tag{3.260}$$

Since $f_{\nu\wedge\tau} = h_\nu + k_\nu$ by construction, for the L^1-limit, if it exists, we will find $h + k = f_\tau$, or $g + h + k = f$. Let us prove that $(h_\nu)_{\nu\in\mathbb{N}}$ and $(k_\nu)_{\nu\in\mathbb{N}}$ are martingales with respect to $(\mathcal{F}_\nu)_{\nu\in\mathbb{N}}$, i.e.

$$h_\nu = E(h_{\nu+1}|\mathcal{F}_\nu) \quad \text{and} \quad k_\nu = E(k_{\nu+1}|\mathcal{F}_\nu).$$

From $k_\nu = f_{\nu\wedge\tau} - h_\tau$ and the fact that $(f_{\nu\wedge\tau})_{\nu\in\mathbb{N}}$ is an $(\mathcal{F}_\nu)$-martingale, see (3.249), it suffices to prove that $(h_\nu)_{\nu\in\mathbb{N}}$ is a martingale with respect to $(\mathcal{F}_\nu)_{\nu\in\mathbb{N}}$. Now, using the notation $E_\nu(l) = E(l|\mathcal{F}_\nu)$ we find

$$E_\nu(h_{\nu+1}) = E_\nu\Big(\sum_{j-1}^{\nu+1}\psi_j\Big) = E_\nu(\psi_{\nu+1})+\sum_{j=1}^{\nu}\psi_j,$$

where we used the $\mathcal{F}_j$-measurability of ψ_j. Thus

$$E_\nu(h_{\nu+1}) = E_\nu(\psi_{\nu+1})+h_\nu,$$

and it remains to show that $E_\nu(\psi_{\nu+1}) = 0$. By the $\mathcal{F}_\nu$-measurability of $\{\omega'\in\Omega;\sigma(\omega')\geq n+1\} = \Big(\bigcup_{j=1}^{n}\{\omega'\in\Omega;\sigma(\omega')=j\}\Big)^c$ and our definition we find

$$\begin{aligned}
E_\nu(\psi_{\nu+1}) &= E_\nu((\varepsilon_{\nu+1} - E_\nu(\varepsilon_{\nu+1}))\chi_{\{\omega'\in\Omega;\sigma(\omega')\geq\nu+1\}})\\
&= (E_\nu(\varepsilon_{\nu+1} - E_\nu(\varepsilon_{\nu+1})))\chi_{\{\omega'\in\Omega;\sigma(\omega')\geq\nu+1\}}\\
&= (E_\nu(\varepsilon_{\nu+1}) - E_\nu(\varepsilon_{\nu+1}))\chi_{\{\omega'\in\Omega;\sigma(\omega')\geq\nu+1\}} = 0,
\end{aligned}$$

implying that $(h_\nu)_{\nu\in\mathbb{N}}$ and $(k_\nu)_{\nu\in\mathbb{N}}$ are $(\mathcal{F})_{\nu\in\mathbb{N}}$ -martingales.

C. Prove of (3.251): We claim that the estimate

$$\sum_{j=1}^{\infty}\|\psi_j\|_{L^1} \leq c\|f\|_{L^1} \tag{3.261}$$

will imply (3.251). Since by (3.259) $h_\nu = \sum_{j=1}^{\nu}\psi_j$, a first consequence of (3.261) is the L^1-convergence of $(h_\nu)_{\nu\in\mathbb{N}}$ to some limit $h\in L^1(\Omega,P)$. The martingale property of $(f_\nu)_{\nu\in\mathbb{N}}$ yields $E_\nu(h) = E(h|\mathcal{F}_\nu) = h_\nu$, thus $E_\nu(h) - E_{\nu-1}(h) = \psi_\nu$ implying

$$\sum_{\nu=1}^{\infty}\|E_\nu(h)-E_{\nu-1}(h)\|_{L^1} \leq c\|f\|_{L^1}$$

which is (3.251) and $\|h\|_{L^1}\leq c\|f\|_{L^1}$ is clear from the above considerations. Thus we have to prove (3.261), i.e.

$$\sum_{j=1}^{\infty}\int_\Omega|\varepsilon_j-E(\varepsilon_j|\mathcal{F}_{j-1})(\omega)|\chi_{\{\omega'\in\Omega;\sigma(\omega')\geq j\}}(\omega)P(d\omega) \leq c\|f\|_{L^1}.$$

Now, $\varepsilon_j \geq 0$ and therefore using the measurability properties of ε_j we get, recall $f \geq 0$,

$$\begin{aligned}
&\sum_{j=1}^{\infty} \int_{\Omega} |\varepsilon_j - E(\varepsilon_j|\mathcal{F}_{j-1})(\omega)| \chi_{\{\omega';\sigma(\omega')>j\}} dP \\
&\leq \sum_{j=1}^{\infty} \int_{\Omega} \varepsilon_j \chi_{\{\omega';\sigma(\omega')\geq j\}} dP + \sum_{j=1}^{\infty} \int_{\Omega} E(\varepsilon_j|\mathcal{F}_{j-1})(\omega)| \chi_{\{\omega';\sigma(\omega')\geq j\}} dP \\
&= 2\sum_{j=1}^{\infty} \int_{\Omega} \varepsilon_j \chi_{\{\omega';\sigma(\omega')\geq j\}} dP \\
&\leq 2\int_{\Omega} \sum_{j=1}^{\infty} \varepsilon_j(\omega) dP(d\omega) \\
&= 2\int_{\Omega} \sum_{j=1}^{\infty} (f_j(\omega) - f_{j-1}(\omega)) \chi_{\{\omega';\rho(\omega')=j\}}(\omega) dP(d\omega) \\
&\leq 2\int_{\Omega} \sum_{j=1}^{\infty} f_j(\omega) \chi_{\{\omega';\rho(\omega')=j\}}(\omega) dP(d\omega) \\
&= 2\int_{\{\omega';\rho(\omega')<\infty\}} f_{\rho(\omega)}(\omega) P(d\omega) \leq 2\int_{\Omega} f_{\rho(\omega)}(\omega) P(d\omega) = 2\|f\|_{L^1},
\end{aligned}$$

where we used in the last step (3.246). Thus (3.251) is proved.

Note that the proof of (3.259) yields also, due to the L^1-bound for $f_{\nu\wedge\tau}$, that $(k_\nu)_{\nu\in\mathbb{N}}$ converges in L^1. In fact the argument goes as Doob's convergence theorem for L^1-martingales which states that L^1-bounded martingales are almost surely convergent, see H.Bauer [19], § 19.

D. Proof of (3.252): It remains to show (3.252) and for this we exploit the properties of the stopping time σ. Since $k = f - g - h$ we know already that

$$\|k\|_{L^1} \leq c\|f\|_{L^1}.$$

Further we have the representation

$$k = \sum_{j=1}^{\infty} (\gamma_j + E(\varepsilon_j|\mathcal{F}_{j-1})) \cdot \chi_{\{\omega';\sigma(\omega')\geq j\}} \quad \text{a.e.},$$

and we have to show that

$$|k(\omega)| \leq c\lambda \quad \text{a.e.}$$

We proceed by proving

$$\left\| \sum_{j=1}^{\infty} \gamma_j \chi_{\{\omega';\sigma(\omega')\geq j\}} \right\|_{\infty} \leq c\lambda \tag{3.262}$$

and

$$\left\| \sum_{j=1}^{\infty} E(\varepsilon_j|\mathcal{F}_{j-1}) \chi_{\{\omega';\sigma(\omega')\geq j\}} \right\|_{\infty} \leq c\lambda. \tag{3.263}$$

Now, since

$$\begin{aligned} \sum_{j=1}^{\infty} \gamma_j \chi_{\{\omega';\sigma(\omega')\geq j\}}(\omega) &= \sum_{j=1}^{\infty} \varphi_j(\omega) \chi_{\{\omega';\rho(\omega')>j\}}(\omega) \chi_{\{\omega;\sigma(\omega')\geq j\}}(\omega) \\ &= \sum_{j=1}^{(\rho(\omega)-1)\wedge\sigma(\omega)} \varphi_j(\omega) = f_{(\rho(\omega)-1)\wedge\sigma(\omega)}(\omega), \end{aligned}$$

and the definition of ρ yields (3.262). On the other hand

$$\begin{aligned} 0 &\leq \sum_{j=1}^{\infty} E(\varepsilon_j|\mathcal{F}_{j-1})(\omega) \chi_{\{\omega';\sigma(\omega')\geq j\}}(\omega) \\ &= \sum_{j=1}^{\sigma(\omega)} E(\varepsilon_j|\mathcal{F}_{j-1})(\omega) = \sum_{l=1}^{\sigma(\omega)-1} E(\varepsilon_{l+1}|\mathcal{F}_l)(\omega) \leq \lambda, \end{aligned}$$

by the definition of σ, thus (3.263) and therefore lemma is proved. □

As a further tool we will need the following result on Rademacher functions. For $\nu \in \mathbb{N}_0$ we define on $[0,1]$ the *Rademacher functions* by

$$r_\nu(t) := \begin{cases} 1; & \frac{j}{2^\nu} \leq t < \frac{j+1}{2^\nu} \text{ and } j \text{ even} \\ -1; & \frac{j}{2^\nu} \leq t < \frac{j+1}{2^\nu} \text{ and } j \text{ odd}. \end{cases} \tag{3.264}$$

The Rademacher functions form an orthogonal family in $L^2([0,1])$ but they are not complete. Thus for a sequence $(b_\nu)_{\nu\in\mathbb{N}_0}$ such that $\sum_{\nu=0}^{\infty} |b_\nu|^2 < \infty$, an element in $L^2([0,1])$ is given by

$$F(t) := \sum_{\nu=0}^{\infty} b_\nu r_\nu(t)$$

and

$$\|F\|_{L^2} = \Big(\sum_{\nu=0}^{\infty} |b_\nu|^2\Big)^{1/2}.$$

A proof of the following result is given in A.Zygmund [292], Theorem 8.4, p.213.

Theorem 3.4.17. *Let $(b_\nu)_{\nu\in\mathbb{N}}$ be a sequence such that $\sum_{\nu=0}^{\infty} |b_\nu|^2 < \infty$. Then the function $F(t) := \sum_{\nu=0}^{\infty} b_\nu r_\nu(t)$ belongs to $L^p([0,1])$, $1 \le p < \infty$, and with constants depending only on p we have*

$$c_{0,p}\Big(\sum_{\nu=0}^{\infty} |b_\nu|^2\Big)^{1/2} \le \|F\|_{L^p} \le c_{1,p}\Big(\sum_{\nu=0}^{\infty} |b_\nu|^2\Big)^{1/2}. \tag{3.265}$$

With notations of Lemma 3.4.16 we find that $(E_\nu f - E_{\nu-1} f)_{\nu\in\mathbb{N}}$ is an orthogonal family in $L^2(\Omega, P)$ and therefore, for any sequence $(a_\nu)_{\nu\in\mathbb{N}}$ such that $\sum_{\nu=0}^{\infty} |a_\nu|^2 < \infty$, we may define for $f \in L^2(\Omega, P)$

$$T_a f := \sum_{k=1}^{\infty} a_\nu (E_\nu - E_{\nu-1}) f \in L^2(\Omega, P). \tag{3.266}$$

In addition we introduce the *Littlewood–Paley function*

$$G(f)(\omega) := \Big(\sum_{\nu=1}^{\infty} |E_\nu(f(\omega)) - E_{\nu-1}(f(\omega))|^2\Big)^{1/2} \tag{3.267}$$

Due to the orthogonality of the family $(E_\nu f - E_{\nu-1} f)_{\nu\in\mathbb{N}}$ it follows that

$$\begin{aligned}\|G(f)\|_{L^2}^2 &= \int_\Omega \sum_{\nu=1}^{\infty} |E_\nu f - E_{\nu-1} f|^2 \, dP \\ &= \sum_{\nu=1}^{\infty} \int_\Omega ((E_{\nu-1} f)^2 - (E_\nu f)^2) \, dP \\ &= \int_\Omega (E_0 f)^2 \, dP = \int_\Omega |f|^2 \, dP = \|f\|_{L^2}^2,\end{aligned}$$

i.e. the Littlewood–Paley function give rise to an isometry on $L^2(\Omega, P)$.

The following theorem is a martingale version of the Littlewood–Paley inequality (see also below). Again, we follow the representation of E.M.Stein [255] which is based on Lemma 3.4.16 and R.Gundy's work [114]. We use the notations of Lemma 3.4.16.

Theorem 3.4.18. *Let* $(a_\nu)_{\nu\in\mathbb{N}}$, $|a_\nu| \leq 1$, *be a sequence and for* $f \in L^2(\Omega, P)$ *consider* $T_a f$ *as given by* (3.266). *Then we have*

$$\|T_a f\|_{L^p} \leq c_p \|f\|_{L^p} \quad \text{for } f \in L^p(\Omega,P) \cap L^2(\Omega,P),\ 1 < p < \infty; \tag{3.268}$$

$$P\{\omega \in \Omega; T_a f(\omega) > \lambda\} \leq \frac{c_1}{\lambda}\|f\|_{L^1}, \quad \text{i.e. } T_a \text{ is weak type (1,1)}; \tag{3.269}$$

$$\|G(f)\|_{L^p} \leq c_p \|f\|_{L^p} \quad \text{for } f \in L^p(\Omega,P) \cap L^2(\Omega,P),\ 1 < p < \infty; \tag{3.270}$$

$$\|E_\infty(f)\|_{L^p} \leq c_p \|G(f)\|_{L^p} \quad \text{if} \quad E_\infty f := \lim_{\nu\to\infty} E_\nu f. \tag{3.271}$$

In all inequalities c_p, $1 \leq p < \infty$, *is independent of the particular choice of the sequence* $(a_\nu)_{\nu\in\mathbb{N}}$.

Proof: Let $f \in L^1(\Omega,P) \cap L^2(\omega,P)$, $f \geq 0$, and decompose $f = g + h + k$ according to Lemma 3.4.16. We find

$$P\{\omega \in \Omega; T_a f(\omega) > \lambda\} \leq P\Big\{\omega \in \Omega; T_a g(\omega) > \frac{\lambda}{3}\Big\} + P\Big\{\omega \in \Omega; T_a h(\omega) > \frac{\lambda}{3}\Big\} + P\Big\{\omega \in \Omega; T_a k(\omega) > \frac{\lambda}{3}\Big\},$$

and estimate each term separately. Obviously we have

$$\{\omega \in \Omega; T_a g(\omega) > \frac{\lambda}{3}\} \subset \{\omega \in \Omega; \sup_{\nu\in\mathbb{N}} |E_\nu(g)(\omega)| \neq 0\},$$

which leads to

$$P\{\omega \in \Omega; T_a g(\omega) > \frac{\lambda}{3}\} \subset P\{\omega \in \Omega; \sup_{\nu\in\mathbb{N}} |E_\nu(g)(\omega)| \neq 0\} \leq \frac{c}{\lambda}\|f\|_{L^1}.$$

Since $T_a h(\omega) \leq \sum_{\nu=1}^\infty |(E_\nu - E_{\nu-1})h(\omega)|$ we find using Chebyshev's inequality

$$P\Big\{\omega \in \Omega; T_a h(\omega) > \frac{\lambda}{3}\Big\} \leq P\Big\{\omega \in \Omega; \sum_{\nu=1}^\infty |(E_\nu - E_{\nu-1})h(\omega)| > \frac{\lambda}{3}\Big\}$$
$$\leq \frac{c'}{\lambda}\Big\| \sum_{\nu=1}^\infty |(E_\nu - E_{\nu-1})h| \Big\|_{L^1} \leq \frac{c''}{\lambda}\|f\|_{L^1}.$$

Finally, from (3.252) we deduce

$$\|k\|_{L^2}^2 \leq \|k\|_\infty \|k\|_{L^1} \leq \lambda\tilde{c}\|f\|_{L^1},$$

and since T_a is a bounded operator on $L^2(\Omega, P)$ we get further, again by using Chebyshev's inequality,

$$P\{\omega \in \Omega; T_a k(\omega) \geq \frac{\lambda}{3}\} \leq \frac{9}{\lambda^2}\|T_a k\|_{L^2}^2 \leq \frac{9}{\lambda^2}\|k\|_{L^2}^2 \leq \frac{\tilde{c}'}{\lambda}\|f\|_{L^1},$$

and we arrive at (3.269) for $f \geq 0$. For general f we have to use the decomposition $f = f^+ - f^-$, $f^+, f^- \geq 0$, to obtain (3.269). So far we know that T_a is of weak-type (1,1), and it is continuous on $L^2(\Omega, P)$. Thus by the Marcinkiewicz interpolation theorem, we know that T_a is bounded from $L^p(\Omega, P)$ to $L^p(\Omega, P)$ for $1 < p \leq 2$, and by duality we find that T_a is bounded on $L^p(\Omega, P)$, $1 < p < \infty$, proving (3.268).

In order to see (3.270) note that by (3.268) we have

$$\int_\Omega \Big|\sum_{\nu=1}^\infty a_\nu(E_\nu f - E_{\nu-1}f)\Big|^p dP \leq c_p\|f\|_{L^p}^p$$

for any sequence $(a_\nu)_{\nu\in\mathbb{N}}$, $|a_\nu| \leq 1$, with c_p independent of the sequence. For $t \in [0,1]$ we may take $a_\nu = r_\nu(t)$, the νth Rademacher function, to get

$$\int_\Omega \Big|\sum_{\nu=1}^\infty (E_\nu f - E_{\nu-1}f) r_\nu(t)\Big|^p dP \leq c_p\|f\|_{L^p}^p,$$

with c_p independent of t. Integrating with respect to t and interchanging the order of integration yields

$$\int_\Omega \Big(\int_0^1 \Big|\Big(\sum_{\nu=1}^\infty E_\nu(f(\omega)) - E_{\nu-1}(f(\omega))\Big) r_\nu(t)\Big|^p dt\Big) dP \leq c_p\|f\|_{L^p}^p.$$

Now we apply Theorem 3.4.17 with $b_\nu = b_\nu(\omega) = E_\nu(f)(\omega) - E_{\nu-1}(f)(\omega)$ and find that

$$\int_0^1 \Big|\sum_{\nu=1}^\infty (E_\nu f(\omega) - E_{\nu-1}f(\omega)) r_\nu(t)\Big|^p dt$$
$$\sim \Big(\sum_{\nu=1}^\infty |E_\nu f(\omega) - E_{\nu-1}f(\omega)|^2\Big)^{p/2},$$

or

$$\int_0^1 \Big|\sum_{\nu=1}^\infty (E_\nu f(\omega) - E_{\nu-1}f(\omega)) r_\nu(t)\Big|^p dt \sim |G(f)(\omega)|^p,$$

where as usual $A \sim B$ means $c_0 \leq \frac{A}{B} \leq c_1$ for some $0 < c_0 \leq c_1$. Thus we find with a suitable constant $\tilde{c}_p$

$$\tilde{c}_p \int_\Omega |G(f)|^p \, dP \leq \int_\Omega \Big(\int_0^1 \Big(\Big| \sum_{\nu=1}^{\infty} (E_\nu f(\omega) - E_{\nu-1} f(\omega)) r_\nu(t) \Big|^p \, dt \Big) dP$$

$$\leq c_p \|f\|_{L^p}^p,$$

proving (3.270). □

We continue to follow E.M.Stein [255]:

Theorem 3.4.19. *Let $\mathcal{F}_1 \subset \mathcal{F}_2 \subset \ldots$ be an increasing family of sub-σ-fields of $\mathcal{A}$ and denote by E_ν the operator $E(\cdot|\mathcal{F}_\nu)$. Suppose that $(f_\nu)_{\nu\in\mathbb{N}}$ is a sequence of functions on $(\Omega, \mathcal{A}, P)$, f_ν is not necessary $\mathcal{F}_\nu$-measurable, and let $(n_\nu)_{\nu\in\mathbb{N}}$ be a sequence of positive integers. Then we have*

$$\Big\| \Big(\sum_{\nu=1}^{\infty} |E_{n_\nu}(f_\nu)|^2 \Big)^{1/2} \Big\|_{L^p} \leq c_p \Big\| \Big(\sum_{\nu=1}^{\infty} |f_\nu|^2 \Big)^{1/2} \Big\|_{L^p}, \quad 1 < p < \infty, \tag{3.272}$$

with c_p depending only on p.

Proof: Denote by $L^p(l_q)$ the Banach space of all sequences of functions $(f_\nu)_{\nu\in\mathbb{N}}$ for which

$$\|(f_\nu)\|_{L^p(l_q)} := \Big(\int_\Omega \Big(\sum_{\nu=1}^{\infty} |f_\nu(\omega)|^q \Big)^{p/q} dP \Big)^{1/p}$$

is finite, with the usual modification for $q = \infty$, i.e.

$$\|(f_\nu)\|_{L^p(l_\infty)} = \Big(\int_\Omega (\sup_{\nu\in\mathbb{N}} |f_\nu(\omega)|)^p \, dP \Big)^{1/p}.$$

For a linear operator T mapping both $L^{p_0}(l_{q_0})$ and $L^{p_1}(l_{q_1})$ boundedly into themselves a variant of the Riesz–Thorin theorem, compare Theorem I.2.8.1, holds stating that T maps also $L^{p_t}(l_{q_t})$ continuously into itself with

$$\frac{1}{p_t} = \frac{1-t}{p_0} + \frac{t}{p_1}, \quad \frac{1}{q_t} = \frac{1-t}{q_0} + \frac{t}{q_1}, \quad t \in [0,1],$$

see A.Benedeck and R.Panzone [23]. Now, consider T defined for a sequence $(f_\nu)_{\nu\in\mathbb{N}}$ by

$$(f_\nu)_{\nu\in\mathbb{N}} \mapsto (E_{n_\nu} f_\nu)_{\nu\in\mathbb{N}}.$$

Since

$$\begin{aligned}\int_\Omega \Big(\sum_{\nu=1}^{\infty} |E_{n_\nu} f_\nu(\omega)|^p\Big)^{p/p} dP(\omega) &= \sum_{\nu=1}^{\infty} \int_\Omega |E_{n_\nu} f_\nu(\omega)|^p dP(\omega)\\ &\leq \sum_{\nu=1}^{\infty} \int_\Omega |f_\nu(\omega)|^p dP(\omega) = \int_\Omega \Big(\sum_{\nu=1}^{\infty} |f_\nu(\omega)|^p\Big)^{p/p} dP(\omega),\end{aligned}$$

T maps $L^p(l_p)$ continuously into itself. Next we claim that T maps $L^p(l_\infty)$, $1 < p \leq \infty$, boundedly into itself. Using the maximal function of $\varphi^*(\omega) := \sup_{\nu\in\mathbb{N}} |f_\nu(\omega)|$, i.e.

$$\varphi^*(\omega) = \sup_{k\geq 1} |E_k(\varphi)(\omega)|$$

we find

$$\begin{aligned}\int_\Omega |\sup_{\nu\in\mathbb{N}} E_{n_\nu} f_\nu(\omega)|^p \, P(d\omega) &\leq \int_\Omega |\sup_{\nu,k\in\mathbb{N}} E_k f_\nu(\omega)|^p \, P(d\omega)\\ &\leq \int_\Omega |\varphi^*(\omega)|^p \, P(d\omega) \leq c_p \int_\Omega (\sup_{\nu\in\mathbb{N}} |f_\nu(\omega)|)^p \, P(d\omega),\end{aligned}$$

where we used in the last step the martingale maximal theorem, Theorem 3.4.10. The variant of the Riesz–Thorin theorem mentioned above leads to the boundedness of T on $L^p(l_q)$, $1 < p \leq q \leq +\infty$, in particular, if $1 < p \leq 2$, T is bounded on $L^p(l_2)$, proving the theorem. □

Note that for $2 \leq p < \infty$ we get by duality the result for $L^p(l_q)$.

Remark 3.4.20. Instead of a family $\mathcal{F}_1 \subset \mathcal{F}_2 \subset \cdots$ of sub-σ-fields one may also work with sub-σ-fields $\cdots \subset \mathcal{F}_{n+1} \subset \mathcal{F}_n \subset \cdots \subset \mathcal{F}_1$ and the sequence $E_\nu(f|\mathcal{F}_\nu)$, $\nu \in \mathbb{N}$. This will be a *reversed martingale*, see H.Bauer [19], p.163, or R.M.Dudley [72], 10.6, for details, especially for the equivalent approach by martingales indexed by $-\mathbb{N}$. For us it is important to note: All estimates done so far are valid for reversed martingales too.

Next we want to establish some connections between martingales $(f_\nu)_{\nu\in\mathbb{N}}$ and one parameter semigroup $(T_t)_{t\geq 0}$. For this we will need a result due to G.C.Rota [237] on powers of selfadjoint positivity preserving contractions on L^p-space.

Theorem 3.4.21. *Let Q be a linear operator having the following properties:*

$$\|Qu\|_{L^p} \leq \|u\|_{L^p}, \quad 1 \leq p \leq \infty; \tag{3.273}$$

$$Q \quad \text{is selfadjoint on } L^2(\Omega, P); \tag{3.274}$$

$$Qu \geq 0 \quad \text{for } u \geq 0; \tag{3.275}$$

$$Q1 = 1. \tag{3.276}$$

Then there exists a measure space $(M, \mathcal{M}, \mu)$ and a family of sub-σ-fields of $\mathcal{M}$ $\cdots \subset \mathcal{F}_{\nu+1} \subset \mathcal{F}_\nu \subset \cdots \subset \mathcal{F}_1 \subset \mathcal{M}$ and $\tilde{\mathcal{F}}_0 \subset \mathcal{M}$ with the following properties:

i) There is an isomorphism j from the measure space $(\Omega, \mathcal{A}, P)$ into the measure space $(M, \tilde{\mathcal{F}}_0, \mu)$ inducing also an isomorphism between the spaces $L^p(M, \tilde{\mathcal{F}}_0, \mu)$ and $L^p(\Omega, \mathcal{A}, P)$, which we denote again by j.

ii) If $f \in L^p(M, \tilde{\mathcal{F}}_0, \mu)$ then $Q^{2\nu}(j(f)) = \tilde{E}E_\nu f$, where $\tilde{E}$ and E_ν denote the conditional expectation with respect to $\mathcal{F}_0$ and $\mathcal{F}_\nu$, respectively.

The proof of this theorem is very much inspired by the Kolmogorov construction of a canonical stochastic process associated with a transition function and these ideas will be discussed in detail in volume III of these treaties, thus we might be a bit sketchy for some readers now.

Proof of Theorem 3.4.21: Set $M = \Omega^{\mathbb{N}_0}$ and let $\mathcal{M}$ be the σ-field generated by the cylinder sets

$$A_1 \times A_2 \times \cdots \times A_N \times \Omega \times \Omega \times \ldots, \quad A_j \in \mathcal{A}. \tag{3.277}$$

For a cylinder set $A = A_0 \times A_1 \times \cdots \times A_N \times \Omega \times \Omega \times \ldots$ we define

$$\tilde{\mu}(A) = \int_\Omega \chi_{A_0} Q(\chi_{A_1} \cdot Q(\ldots(\chi_{A_{N-1}} \cdot Q\chi_{A_N}))\ldots)\, dP, \tag{3.278}$$

and $\tilde{\mu}$ extends to a measure μ on $(M, \mathcal{M})$ (by the usual arguments, see volume III). Define now

$$\tilde{\mathcal{F}}_0 := \{A_0 \times \Omega^{\mathbb{N}};\ A_0 \in \mathcal{A}\} \tag{3.279}$$

and

$$\mathcal{F}_\nu := \{\Omega^\nu \times A;\ A \in \mathcal{M}\}. \tag{3.280}$$

It follows that

$$\cdots \mathcal{F}_{\nu+1} \subset \mathcal{F}_\nu \subset \cdots \subset \mathcal{F}_1 \subset \tilde{\mathcal{F}}_0 \subset \mathcal{M}$$

and by $j : M \to \Omega$, $j(x_0, x_1, \ldots) = x_0$ an isomorphism of the measure spaces $(M, \tilde{\mathcal{F}}_0, \mu)$ and $(\Omega, \mathcal{A}, P)$ is given, note that for this $Q1 = 1$ is needed. Thus we have proved i).

In order to prove ii) we show

a) if $g : M \to \mathbb{R}$ depends only on x_ν, i.e. $g(x_0, x_1, \ldots) = g_\nu(x_\nu)$ then we have

$$\tilde{E}(g)(x_0, x_1, \ldots) = (Q^\nu g_\nu)(x_0)$$

b) if $g : M \to \mathbb{R}$ depends only on x_0, hence is $\tilde{\mathcal{F}}_0$-measurable and $g(x_0, x_1, \ldots) = g_0(x_0)$, then it follows that

$$E_\nu(g)(x_0, x_1, \ldots) = Q^\nu g_0(x_\nu).$$

Suppose that a) and b) hold. It follows for $f \in L^p(\Omega, \mathcal{A}, P)$ that

$$\tilde{E}E_\nu(j^{-1}f) = j^{-1}(Q^{2\nu} f)$$

which implies already ii).

We will show now a): We claim that $Q^\nu g_\nu(x_0)$ is the conditional expectation of g with respect to $\tilde{\mathcal{F}}_0$. Clearly, $Q^\nu g(\cdot)$ is $\tilde{\mathcal{F}}_0$-measurable, thus we have to prove

$$\int_A g_\nu(x_\nu)\, \mu(dx) = \int_A Q^\nu g_\nu(x_0)\, \mu(dx). \tag{3.281}$$

for $A \in \tilde{\mathcal{F}}_0$, i.e. $A = A_0 \times \Omega^{\mathbb{N}}$. If g_ν is a characteristic function of $B \subset \Omega$, then both sides in (3.281) are equal to $\int_{A_0} Q^\nu \chi_B(\omega)\, P(d\omega)$ by the definition of μ, see (3.278). Hence a standard limit procedure (see again volume III) yields (3.281).

Finally we prove b): As in a) this can be reduces to show

$$\int_A g_0(x_0)\, \mu(dx) = \int_A (Q^\nu g_0)(x_\nu)\, \mu(dx), \tag{3.282}$$

where $A = \Omega^{\nu-1} \times A_\nu \times A_{\nu+1} \times \cdots \times A_N \times \Omega \times \ldots$ and g_0 is the characteristic function $g_0 = \chi_B$, $B \subset \Omega$ measurable. Again we use (3.278) to find

$$\begin{aligned} \int_A g_0(x_0)\, \mu(dx) &= \int_M g_0(x_0) \chi_{A_\nu} \cdot \chi_{A_{\nu+1}} \cdot \cdots \cdot \chi_{A_N}\, \mu(dx) \\ &= \int_\Omega g_0(\omega) Q^\nu (\chi_{A_\nu} Q(\chi_{A_{\nu+1}} Q(\ldots)\ldots)\, P(d\omega). \end{aligned} \tag{3.283}$$

Note that the properties of Q imply

$$\int_\Omega (Qh)(\omega)\, P(d\omega) = \int_\Omega h(\omega)\, P(d\omega)$$

for $h \in L^1(\Omega, P) \cap L^\infty(\Omega, P)$ (for example). Now we find for the right-hand side in (3.282)

$$\begin{aligned}\int_A (Q^\nu g_0)(x_\nu)\, \mu(dx) &= \int_M (Q^\nu g_0)(x_0)\chi_{A_\nu} \cdot \chi_{A_{\nu+1}} \cdot \dots \cdot \chi_{A_N}\, \mu(dx) \\ &= \int_\Omega Q^\nu(Q^\nu g_0(x)\chi_{A_\nu} Q(\chi_{A_{\nu+1}} Q(\dots)\dots))\, dP \\ &= \int_\Omega (Q^\nu g_0(\cdot)\chi_{A_\nu} Q(\chi_{A_{\nu+1}} Q(\dots)\dots))\, dP.\end{aligned} \tag{3.284}$$

Setting $\varphi := \chi_{A_\nu} Q(\chi_{A_{\nu+1}} Q(\dots)\dots)$ we obtain from (3.283) that

$$\int_A g_0(x_0)\mu(dx) = \int_\Omega g_0(\omega) Q^\nu \varphi(\omega)\, P(d\omega),$$

while (3.284) implies

$$\int_A (Q^\nu g_0)(x_\nu)\, \mu(dx) = \int_\Omega (Q^\nu g_0)(\omega)\varphi(\omega)\, P(d\omega).$$

Since Q^ν is selfadjoint, it follows that

$$\begin{aligned}\int_A g_0(x_0)\mu(dx) &= \int_\Omega g_0(\omega) Q^\nu \varphi(\omega)\, P(d\omega) \\ &= \int_\Omega (Q^\nu g_0)(\omega)\varphi(\omega)\, P(d\omega) = \int_A (Q^\nu g_0)(x_\nu)\, \mu(dx),\end{aligned}$$

and the theorem is proved. □

Let us discuss how to use Theorem 3.4.21 for examining sub-Markovian semigroups. Let $(T_t)_{t\geq 0}$ be a symmetric sub-Markovian semigroup which is conservative, i.e. $T_t 1 = 1$ for all $t \geq 0$. Thus for each $t \geq 0$ and $1 \leq p < \infty$ the operator $T_t = T_t^{(p)}$ fulfills (3.273)-(3.276). We may define for $k \in \mathbb{N}$

$$Q := T_{1/(2^{k+1})}$$

and find that Q satisfies (3.273)–(3.276). Now, the martingale maximal theorem, Theorem 3.4.10 and part ii) of Theorem 3.4.21 yields

$$\begin{aligned}\|\sup_{\nu\in\mathbb{N}} Q^{2\nu}(j(f))(\cdot)\|_{L^p(\Omega,P)} &= \|j^{-1}(\sup_{\nu\in\mathbb{N}} Q^{2\nu}(j(f)))\|_{L^p(M,\tilde{\mathcal{F}}_0,\mu)}\\ &= \|\sup_{\nu\in\mathbb{N}} \tilde{E}E_\nu(f(\cdot))\|_{L^p(M,\tilde{\mathcal{F}}_0,\mu)}\\ &\leq \|\tilde{E}(\sup_{\nu\in\mathbb{N}} E_\nu f)\|_{L^p(M,\tilde{\mathcal{F}}_0,\mu)} \leq \|\sup_{\nu\in\mathbb{N}} E_\nu f\|_{L^p(M,\tilde{\mathcal{F}}_0,\mu)}\\ &\leq c_p\|f\|_{L^p(M,\tilde{\mathcal{F}}_0,\mu)} = c_p\|j(f)\|_{L^p(\Omega,P)},\end{aligned}$$

or for $u \in L^p(\Omega, P)$

$$\|\sup_{\nu\in\mathbb{N}} Q^{2\nu}u\|_{L^p} \leq c_p\|u\|_{L^p},$$

i.e.

$$\|\sup_{\nu\in\mathbb{N}} T_{\nu/2^k}u\|_{L^p} \leq c_p\|u\|_{L^p} \tag{3.285}$$

with c_p independent of k. For $k \to \infty$ we arrive at

Proposition 3.4.22. *Let $(T_t)_{t\geq 0}$ be a conservative, symmetric sub-Markovian semigroup. Then the maximal estimate*

$$\|\sup_{t>0} T_t u\|_{L^p} \leq c_p\|u\|_{L^p}. \tag{3.286}$$

holds.

Now we come to this part of our discussion where we shall follow Stein's work to get better estimates for maximal and Littlewood–Paley functions associated with a symmetric sub-Markovian semigroup, however, from now on we will omit larger parts of the proofs.

We need the following ergodic theorem due to N.Dunford and J.T.Schwartz [73], generalizing an earlier result of E.Hopf, which we state without proof.

Theorem 3.4.23. *Let $(T_t^{(p)})_{t\geq 0}$ be a strongly continuous contraction semigroup on $L^p(\mathbb{R}^n;\mathbb{R})$, $1 \leq p < \infty$, and a contraction semigroup on $L^\infty(\mathbb{R}^n;\mathbb{R})$. For $u \in L^p(\mathbb{R}^n;\mathbb{R})$ there is a null set $N(u) \subset \mathbb{R}^n$ such that for $x \notin N(u)$*

$$M_t^{(p)}u(x) := \frac{1}{t}\int_0^t T_s^{(p)}u(x)\,ds \tag{3.287}$$

and

$$M^*_{(p)}u(x) := \sup_{t>0} |M_t^{(p)}u(x)| \tag{3.288}$$

are defined and we have

$$|M_t^{(p)}u(x)| \leq M^*_{(p)}u(x), \quad x \notin N(u); \tag{3.289}$$

$$\|M^*_{(p)}u\|_{L^p} \leq 2\Big(\frac{p}{p-1}\Big)^{1/p}\|u\|_{L^p}, \quad 1 < p < \infty, \tag{3.290}$$

hence $\|M_t^{(p)}u\|_{L^p} \leq c_p\|u\|_{L^p}$ *for* $1 < p < \infty$, *and* $M_t^{(p)}u$ *converges almost everywhere to* u *as* $t \to 0$.

Note that $t \mapsto \frac{1}{t}\int_0^t T_s^{(p)}u(x)\,ds$ is for all $x \notin N(u)$, $\lambda^{(n)}(N(u)) = 0$, a continuous function. In order to prove maximal inequalities for $t \mapsto T_t^{(p)}u(x)$ we need more regularity for this mapping which we may derive from the analyticity of $(T_t^{(p)})_{t\geq 0}$.

Lemma 3.4.24. *Let* $(T_t^{(p)})_{t\geq 0}$ *be an analytic* L^p*-sub-Markovian semigroup and let* $u \in L^p(\mathbb{R}^n;\mathbb{R})$, $1 < p < \infty$. *For each* $t > 0$ *we can redefine* $T_tu(x)$ *on a null set such that for every fixed* x *the mapping* $t \mapsto T_tu(x)$ *is on* $(0,\infty)$ *a real-analytic function.*

Proof: A function $t \mapsto f(t) \in X$, $(X, \|\cdot\|_X)$ being a Banach space, is analytic in $(0,\infty)$ if and only if for every $t_0 \in (0,\infty)$ there is a neighbourhood $U_{2\varepsilon}(t_0)$ and elements $f_k \in X$, $k \in \mathbb{N}_0$, such that $f(t) = \sum_{k=0}^\infty f_k \cdot (t-t_0)^k$ and $\sum_{k=0}^\infty \|f_k\|_X \varepsilon^k < \infty$. Thus we may find elements $\tilde{u}_k \in L^p(\mathbb{R}^n;\mathbb{R})$ each having a representative u_k (a pointwise defined function) such that

$$T_t^{(p)}u(x) = \sum_{k=0}^\infty u_k(x)(t-t_0)^k \quad \text{a.e.} \quad x \in \mathbb{R}^n, \quad t \in U_{2\varepsilon}(t_0). \tag{3.291}$$

Now we may redefine $T_t^{(p)}u$ on a null set such that (3.291) holds for every $x \in \mathbb{R}^n$ and $\sum_{k=0}^\infty \|u_k\|_{L^p}\varepsilon^k < \infty$. Next we may cover $(0,\infty)$ with countable many neighbourhoods $U_\varepsilon(t_l)$ and using the analyticity of $(T_t^{(p)})_{t\geq 0}$ and the fact that the countable union of null sets is again a null set we are done. □

Let $(T_t^{(p)})_{t\geq 0}$, $1 < p < \infty$, be a symmetric, hence analytic, L^p-sub-Markovian semigroup. For each p we define the maximal function

$$u^*_{(p)}(x) := \sup_{t>0} |T_t^{(p)}u(x)| \tag{3.292}$$

which makes even pointwise sense by Lemma 3.4.24.

Lemma 3.4.25. *For* $(T_t^{(2)})_{t\geq 0}$ *as above we have the estimate*

$$\|u^*_{(2)}\|_{L^2} \leq c_2 \|u\|_{L^2}. \tag{3.293}$$

Proof: Let us introduce the *Littlewood–Paley function* for a semigroup by

$$g_1(u)(x) := \Big(\int_0^\infty t \Big| \frac{\partial}{\partial t} T_t u(x) \Big|^2 \, dt \Big)^{1/2}. \tag{3.294}$$

By the spectral theorem there exists a resolution of identity such that

$$T_t^{(2)} u = \Big(\int_{0+}^\infty e^{-\lambda t} \, dE_\lambda \Big) u. \tag{3.295}$$

(Note that we have taken now the resolution of identity corresponding to $-A^{(2)}$, $A^{(2)}$ being the generator of $(T_t^{(2)})_{t\geq 0}$, whereas in the proof of Theorem I.4.2.12 we worked with that corresponding to $T_1^{(2)}$.) From (3.295) it follows that

$$\frac{\partial}{\partial t} T_t^{(2)} u = \Big(- \int_0^\infty \lambda e^{-\lambda t} \, dE_\lambda \Big) u,$$

which yields by a further application of the spectral theorem

$$\begin{aligned} \int_{\mathbb{R}^n} |g_1(u)(x)|^2 \, dx &= \int_{\mathbb{R}^n} \int_0^\infty t \Big| \frac{\partial}{\partial t} T_t^{(2)} u(x) \Big|^2 \, dt \, dx \\ &= \int_0^\infty t \Big\| \frac{\partial}{\partial t} T_t^{(2)} u \Big\|_{L^2}^2 \, dt = \int_0^\infty t \int_{0+}^\infty \lambda^2 e^{-2\lambda t} (dE_\lambda u, u)_0 \, dt \\ &= \int_{0+}^\infty \lambda^2 \int_0^\infty t e^{-2\lambda t} \, dt (dE_\lambda u, u)_0 = \frac{1}{4} \int_{0+}^\infty 1 (dE_\lambda u, u)_0, \end{aligned}$$

and we arrive at

$$\|g_1(u)\|_{L^2} \leq \frac{1}{2} \|u\|_{L^2}. \tag{3.296}$$

An integration by parts yields

$$\int_0^t s \Big(\frac{\partial}{\partial s} T_s^{(2)} u \Big) \, ds = t T_t^{(2)} u - \int_0^t T_s^{(2)} u \, ds, \tag{3.297}$$

from which we deduce

$$\begin{aligned}|T_t^{(2)}u(x)| &\leq \Big|\frac{1}{t}\int_0^t T_s^{(2)}u(x)\,ds\Big| + \Big|\frac{1}{t}\int_0^t s\Big(\frac{\partial}{\partial s}T_s^{(2)}u(x)\Big)\,ds\Big| \\ &\leq \Big|\frac{1}{t}\int_0^t T_s^{(2)}u(x)\,ds\Big| \\ &\quad + \frac{1}{t}\Big(\int_0^t s\,ds\Big)^{1/2}\Big(\int_0^t s\Big|\frac{\partial}{\partial s}T_s^{(2)}u(x)\Big|^2\,ds\Big)^{1/2} \\ &\leq M_{(2)}^*u(x) + g_1(u)(x).\end{aligned}$$

Now, applying the Dunford–Schwartz–Hopf ergodic theorem, Theorem 3.4.23, especially (3.290), and (3.296) we find

$$\begin{aligned}\|u_{(2)}^*\|_{L^2} &= \|\sup_{t>0}|T_t^{(2)}u(\cdot)|\|_{L^2} \\ &\leq \|M_{(2)}^*u\|_{L^2} + \|g_1(u)\|_{L^2} \leq c_2\|u\|_{L^2},\end{aligned}$$

and the lemma is proved. □

As already announced, let us now state partly without proof further auxiliary results. They all are taken from E.M.Stein [255], where also proofs are given.

Lemma 3.4.26. *For* $(T_t^{(2)})_{t\geq 0}$ *as above we introduce the* Paley–Littlewood functions

$$g_k(u) := \int_0^\infty s^{2k-1}\Big|\frac{\partial^k T_s^{(p)}u}{\partial s^k}\Big|^2\,ds, \quad k\in\mathbb{N}, \tag{3.298}$$

and we claim

$$\|g_k(u)\|_{L^2} \leq \tilde{c}_k\|u\|_{L^2} \tag{3.299}$$

for all $u\in L^2(\mathbb{R}^n;\mathbb{R})$.

Lemma 3.4.27. *For* $(T_t^{(2)})_{t\geq 2}$ *as in Lemma* 3.4.25 *suppose that*

$$\Big\|\sup_{t>0} t^{k-1}\frac{\partial^{k-1}}{\partial t^{k-1}}T_t^{(2)}u\Big\|_{L^2} \leq a_{k-1}\|u\|_{L^2} \tag{3.300}$$

and

$$\Big\|\sup_{t>0}\frac{1}{t}\int_0^t s^{k-1}\frac{\partial^{k-1}T_s^{(2)}u}{\partial s^{k-1}}\,ds\Big\|_{L^2} \leq b_{k-1}\|u\|_{L^2} \tag{3.301}$$

hold. Then it follows that

$$\left\| \sup_{t>0} \frac{1}{t} \int_0^t s^k \frac{\partial^k}{\partial s^k} T_s^{(2)} u \, ds \right\|_{L^2} \leq c_k \|u\|_{L^2}. \tag{3.302}$$

Proposition 3.4.28. *Let $(T_t^{(p)})_{t\geq 0}$, $1 < p < \infty$, be an analytic symmetric L^p-sub-Markovian semigroup. For the maximal function*

$$u_{k,(p)}^* := \sup_{t>0} \left| t^k \frac{\partial^k T_t^{(p)} u(x)}{\partial t^k} \right|, \quad k \in \mathbb{N}_0, \tag{3.303}$$

the estimates

$$\|u_{k,(2)}^*\| \leq c_k \|u\|_{L^2} \tag{3.304}$$

hold.

Combining the Dunford–Schwartz–Hopf ergodic theorem, Theorem 3.4.23, with Lemma 3.4.27 we find the estimates

$$\left\| \sup_{t>0} \left| t^{-1} \int_0^t T_s^{(p)} u(\cdot) \, ds \right| \right\|_{L^p} \leq c'_p \|u\|_{L^p}, \quad p > 1, \tag{3.305}$$

and

$$\left\| \sup_{t>0} \left| t^k \frac{\partial^k}{\partial t^k} T_t^{(2)} u \right| \right\|_{L^2} \leq \tilde{c}'_p \|u\|_{L^p}, \quad k \geq 0. \tag{3.306}$$

The aim is to interpolate (3.305) and (3.306) to obtain

$$\left\| \sup_{t>0} T_t^{(p)} u \right\|_{L^p} \leq c_p \|u\|_{L^p}, \quad 1 \leq p < \infty. \tag{3.307}$$

It turns out that fractional derivatives and integrals which we encountered already in Section I.4.7 will provide us with the analytic family of operators we are longing for to interpolate (3.305) and (3.306). As before we mention the monograph [243] of S.Samko et.al. as our standard reference for *fractional integrals* and *derivatives.* A further source is the book [240] of B.Rubin. For $\alpha \in \mathbb{C}$, $\operatorname{Re}\alpha > 0$, we define the fractional integral on $L^1((0,\infty))$ by

$$I^{(\alpha)} u(t) = \frac{1}{\Gamma(\alpha)} \int_0^\infty (t-s)^{\alpha-1} u(s) \, ds. \tag{3.308}$$

Proposition 3.4.29. *For* $u \in C_0^\infty((0,\infty)) \subset S(\mathbb{R})$ *the function* $\alpha \mapsto I^{(\alpha)}u(t)$ *defined on* $\operatorname{Re}\alpha > 0$ *has an analytic continuation onto* $\mathbb{C}$. *Further, we have*

$$I^{(-k)}u = \frac{d^k\, u(t)}{d\, t^k} \qquad k \in \mathbb{N}, \tag{3.309}$$

and

$$I^{(\alpha)} \circ I^{(\beta)} = I^{(\alpha+\beta)}, \quad I^{(0)} = \mathrm{id}\,. \tag{3.310}$$

Since fractional derivatives are becoming more and more important when dealing with certain sub-Markovian semigroups and Markov processes, we prefer to provide a detailed proof of this proposition, again, following closely E.M.Stein [255].

Proof of Proposition 3.4.29: It is clear that $\alpha \mapsto I^{(\alpha)}u(t)$ is for each $u \in C_0^\infty((0,\infty))$ an analytic function in $\operatorname{Re}\alpha > 0$. Now, for $\operatorname{Re}\alpha \leq 0$ the singularity which may trouble is at point $s = t$, hence it is sufficient to prove that the operators

$$\begin{aligned} M^{(\alpha)}u(t) :&= t^{-\alpha}\frac{1}{\Gamma(\alpha)}\int_0^t (t-s)^{\alpha-1}u(s)\,ds \\ &= \frac{1}{\Gamma(\alpha)}\int_0^1 (1-s)^{\alpha-1}u(st)\,ds \end{aligned} \tag{3.311}$$

have analytic continuation to $\mathbb{C}$. Note that we have now a fixed singularity at $s = 1$. Therefore we split the integral into two parts

$$M^{(\alpha)}u(t) = M_1^{(\alpha)}u(t) + M_2^{(\alpha)}u(t) \tag{3.312}$$

where

$$M_1^{(\alpha)}u(t) = \frac{1}{\Gamma(\alpha)}\int_0^{1/2} (1-s)^{\alpha-1}u(st)\,ds$$

and

$$M_2^{(\alpha)}u(t) = \frac{1}{\Gamma(\alpha)}\int_{1/2}^1 (1-s)^{\alpha-1}u(st)\,ds.$$

Clearly, $\alpha \mapsto M_1^{(\alpha)}u(t)$ has an analytic continuation to $\mathbb{C}$ since there is no singularity at all for $\operatorname{Re}\alpha \leq 0$, recall that $\alpha \mapsto \frac{1}{\Gamma(\alpha)}$ is an entire function. To handle $M_2^{(\alpha)}$ we note that an integration by parts yields

$$M_2^{(\alpha)}u(t) = \frac{1}{\Gamma(\alpha+1)}\int_{1/2}^{1}(1-s)^{\alpha}\frac{d}{ds}u(st)\,ds + \frac{1}{\Gamma(\alpha+1)}\Big(\frac{1}{2}\Big)^{\alpha}u(\frac{1}{2}t), \tag{3.313}$$

which we may take to get an analytic continuation for $\alpha \mapsto M_2^{(\alpha)}u(t)$ to the half plane $\operatorname{Re}\alpha > -1$. Further integrations by part, say l-times, lead to an expression of $M_2^{(\alpha)}u(t)$ which allows us to arrive at an analytic continuation for $\alpha \mapsto M_2^{(\alpha)}u(t)$ into the half plane $\operatorname{Re}\alpha > -l$ implying the possibility to continue $\alpha \mapsto M_2^{(\alpha)}u(t)$, hence $I^{(\alpha)}u(t)$, analytically to $\mathbb{C}$. For $\operatorname{Re}\alpha, \operatorname{Re}\beta > 0$ we find using Fubini's theorem that

$$\begin{aligned} I^{(\alpha)}(I^{(\beta)}u)(r) &= \frac{1}{\Gamma(\alpha)\Gamma(\beta)}\int_0^r (r-t)^{\alpha-1}\int_0^t (t-s)^{\beta-1}u(s)\,ds \\ &= \frac{B(\alpha,\beta)}{\Gamma(\alpha)\Gamma(\beta)}\int_0^r (r-s)^{\alpha+\beta-1}u(s)\,ds, \end{aligned}$$

where $B(\alpha,\beta) = \frac{\Gamma(\alpha)\Gamma(\beta)}{\Gamma(\alpha+\beta)}$ is the Beta-function. Thus the semigroup property (3.309) holds for $\operatorname{Re}\alpha, \operatorname{Re}\beta > 0$ and extends by analytic continuation to all $\alpha, \beta \in \mathbb{C}$. Next, recall the definition of a fractional derivative D_+^{α}, $0 < \alpha < 1$, given in Section I.4.7. We may apply (I.4.382) to $u \in C_0^{\infty}((0,\infty)) \subset S(\mathbb{R})$, and comparing with (3.308) we find

$$D_+^{\alpha}u(t) = \frac{d}{dt}I^{(1-\alpha)}u(t) \tag{3.314}$$

as well as

$$D_+^{\alpha}u(t) = F^{-1}_{\xi\mapsto t}((-i\xi)^{\alpha}\hat{u}(\xi))(t), \tag{3.315}$$

which yields for $0 < \alpha, \beta,\ \alpha+\beta < 1$

$$D_+^{\alpha}D_+^{\beta} = D_+^{\alpha+\beta}. \tag{3.316}$$

Furthermore we have for $0 < \alpha < 1$ and $u \in C_0^{\infty}((0,\infty))$, note that $u(0) = 0$,

$$\begin{aligned} \frac{d}{dt}I^{(1-\alpha)}u(t) &= \frac{d}{dt}\frac{1}{\Gamma(1-\alpha)}\int_0^t (t-s)^{-\alpha}u(s)\,ds \\ &= \frac{d}{dt}\frac{1}{\Gamma(1-\alpha)}\Big(-\frac{1}{1-\alpha}\Big)\int_0^t \Big(\frac{d}{ds}(t-s)^{1-\alpha}\Big)u(s)\,ds \end{aligned}$$

$$
\begin{aligned}
&= \frac{d}{dt}\frac{1}{\Gamma(1-\alpha)}\frac{1}{1-\alpha}\int_0^t (t-s)^{1-\alpha}\frac{d\,u(s)}{ds}\,ds \\
&= \frac{1}{(1-\alpha)\Gamma(1-\alpha)}\int_0^t \frac{d}{dt}(t-s)^{1-\alpha}\frac{d\,u(s)}{ds}\,ds \\
&= \frac{1}{\Gamma(1-\alpha)}\int_0^t (t-s)^{-\alpha}\frac{d\,u(s)}{ds}\,ds = I^{(1-\alpha)}\Big(\frac{d\,u(s)}{ds}\Big)t.
\end{aligned}
$$

Thus we find

$$
D_+^\alpha u(t) = \frac{d}{dt}I^{(1-\alpha)}u(t) = I^{(1-\alpha)}\Big(\frac{d\,u(s)}{ds}\Big)(t).
$$

In particular we have

$$
\begin{aligned}
I^{(\alpha)}D_+^\alpha u(t) &= I^{(\alpha)}\Big(\frac{d}{ds}I^{(1-\alpha)}u\Big)(t) \\
&= I^{(\alpha)}I^{(1-\alpha)}\Big(\frac{du}{ds}\Big)(t) = I^{(1)}\Big(\frac{du}{ds}\Big)(t) \\
&= \int_0^t \frac{du(s)}{ds}\,ds = u(t),
\end{aligned}
$$

and

$$
\begin{aligned}
D_+^\alpha I^{(\alpha)}u(t) &= \Big(\frac{d}{dt}I^{(1-\alpha)}\Big)(I^{(\alpha)}u)(t) \\
&= \frac{d}{dt}(I^{(1)}u)(t) = \frac{d}{dt}\int_0^t u(s)\,ds = u(t).
\end{aligned}
$$

Thus for $0 < \alpha \leq 1$ we have on $C_0^\infty((0,\infty))$

$$
I^{(\alpha)}D_+^\alpha u = D_+^\alpha I^\alpha u = u, \tag{3.317}
$$

which implies by the first part of (3.309) (which is already shown) that

$$
I^{(-\alpha)}u = D_+^\alpha u. \tag{3.318}
$$

Since (3.314) implies obviously that $\lim\limits_{\alpha\to 0} D_+^\alpha u = u$ we arrive at

$$
\lim_{\alpha\to 0} I^{(-\alpha)}u = u, \quad \text{i.e.} \quad I^{(0)} = \mathrm{id}. \tag{3.319}
$$

Now, (3.309) follows easily from $I^{(k)} \circ I^{(-k)} = \mathrm{id}$ and

$$
I^{(k)}u(t) = \frac{1}{(k-1)!}\int_0^t (t-s)^{k-1}u(s)\,ds = \int_0^t\int_0^s \dots \int_0^r u(r)\,dr\dots ds,
$$

proving the proposition. □

With the notation (3.311) we have

$$\|\sup_{t>0} M^1(T_t u)(\cdot)\|_{L^p} \le A_p^{(1)} \|u\|_{L^p}, \quad 1 < p < \infty, \tag{3.320}$$

and

$$\|\sup_{t>0} M^{-k}(T_t u)(\cdot)\|_{L^2} \le A^{(-k)} \|u\|_{L^2}, \quad k \in \mathbb{N}, \tag{3.321}$$

and we want to prove

$$\|\sup_{t>0} M^0(T_t u)(\cdot)\|_{L^p} \le A_p^0 \|u\|_{L^p}, \quad 1 < p < \infty, \tag{3.322}$$

by interpolating the operator M_k^* defined by

$$M_k^* v := \sup_{t>0} |M^k(v)(t)|, \quad v = T_t u. \tag{3.323}$$

In order to apply Stein's interpolation theorem, Theorem I.2.8.3, (or variants of it) we have to overcome two difficulties: We need estimates for M_{1+iy}^* and M_{-k+iy}^* for all $y \in \mathbb{R}$, and we have to find reasonable substitutes for M_k^* by linear operators.

Lemma 3.4.30. *Let $(T_t^{(p)})_{t\ge 0}$, $1 < p < \infty$, be an analytic and symmetric L^p-sub-Markovian semigroup. Then there are constants K_p and A_k such that*

$$\|M_{1+iy}^*(T_t^{(p)} u)\|_{L^p} \le K_p e^{\pi|y|} \|u\|_{L^p}, \quad u \in L^p(\mathbb{R}^n;\mathbb{R}), \tag{3.324}$$

and

$$\|M_{-k+iy}^*(T_t^{(2)} u)\|_{L^p} \le A_k e^{\pi|y|} \|u\|_{L^p}, \quad u \in L^2(\mathbb{R}^n;\mathbb{R}), \tag{3.325}$$

hold.

Lemma 3.4.31. *Let $(T_t^{(p)})_{t\ge 0}$ be as in Lemma 3.4.30 and let $\tau : \mathbb{R}^n \to (0,\infty)$ be a measurable mapping. Further define for $\alpha \in \{l + iy; l \in \{1\} \cup -\mathbb{N}_0$ and $y \in \mathbb{R}\}$*

$$S_{\alpha,\tau}^{(p)} u(x) := \frac{1}{\Gamma(\alpha)} \tau(x)^{-\alpha} \int_0^{\tau(x)} T_s^{(p)} u(x)\, ds.$$

Then there exists a constant κ_p such that

$$\|S_{0,\tau}^{(p)} u\|_{L^p} = \|T_{\tau(\cdot)}^{(p)} u(\cdot)\|_{L^p} \le \kappa_p \|u\|_{L^p} \tag{3.326}$$

holds.

Finally we are in position to discuss *Stein's maximal theorem*, see [255].

Theorem 3.4.32. *Let* $(T_t^{(p)})_{t\geq 0}$, $1 < p < \infty$, *be an analytic and symmetric* L^p*-sub-Markovian semigroup and denote as in* (3.292) *by*

$$u^*_{(p)}(x) = \sup_{t>0} |T_t^{(p)} u(x)|$$

its maximal function (at $u \in L^p(\mathbb{R}^n;\mathbb{R})$*). Then there exists a constant such that*

$$\|u^*\|_{L^p} \leq A_p \|u\|_{L^p}, \quad 1 < p < \infty, \tag{3.327}$$

holds and for $u \in L^p(\mathbb{R}^n;\mathbb{R})$ *we have*

$$\lim_{t\searrow 0} T_t^{(p)} u = u \quad a.e. \tag{3.328}$$

Proof (sketch): In order to get (3.327) one determines a function τ such that

$$|T_{\tau(x)}^{(p)} u(x)| \geq \frac{1}{2} \sup_{t>0} |T_t^{(p)} u(x)|$$

for each $x \in \mathbb{R}^n$. Now, (3.326) yields

$$\| \sup_{t>0} T_t^{(p)} u\|_{L^p} \leq 2\kappa_p \|u\|_{L^p}$$

which is (3.327). It remains to prove that $\lim_{t\searrow 0} T_t^{(p)} u = u$ a.e. Since $t \mapsto T_t^{(p)} u(x)$ is real-analytic by Lemma 3.4.24, it follows that $\lim_{t\searrow 0} T_t^{(p)}(T_s^{(p)} u) = T_s^{(p)} u$ for almost all $x \in \mathbb{R}^n$ and $s > 0$. Thus we find for $u \in L^2(\mathbb{R}^n;\mathbb{R})$

$$\begin{aligned}
&\limsup_{t\searrow 0} |T_t^{(2)} u(x) - u(x)| \\
&\qquad \leq \limsup_{t\searrow 0} |T_t^{(2)}(u - T_s^{(2)})(x)| + \limsup_{t\searrow 0} |T_t^{(2)} T_s^{(2)} u(x) - T_s^{(2)} u(x)| \\
&\qquad\quad + |T_s^{(2)} u(x) - u(x)| \\
&\qquad \leq \sup_{t>0} |T_t^{(2)}(u - T_s^{(2)})(x)| + |T_s^{(2)} u(x) - u(x)| \\
&\qquad = (u - T_s^{(2)} u)^*(x) + |T_s^{(2)} u(x) - u(x)|,
\end{aligned}$$

or since the maximal function is L^2-bounded

$$\|\limsup_{t\searrow 0}|T_t^{(2)}u(\cdot)-u(\cdot)|\|_{L^2}\le 2\|(u-T_s^{(2)})^*\|_{L^2}$$
$$\le c\|u-T_s^{(2)}u\|_{L^2}.$$

But $(T_t^{(2)})_{t\ge 0}$ is strongly continuous, implying as $s\to 0$ that

$$\limsup_{t\searrow 0}|T_t^{(2)}u(x)-u(x)|=0\quad\text{a.e.},$$

or $T_t^{(2)}u(x)\to u(x)$ a.e. for $u\in L^2(\mathbb{R}^n;\mathbb{R})$. Now, let $u\in L^p(\mathbb{R}^n;\mathbb{R})$, $1<p<\infty$, and for $\varepsilon>0$ given, take $v\in L^2(\mathbb{R}^n;\mathbb{R})\cap L^p(\mathbb{R}^n;\mathbb{R})$ such that $\|u-v\|_{L^p}<\varepsilon$. As before we find using that $T_t^{(2)}v\to v$ a.e.

$$\limsup_{t\searrow 0}|T_t^{(p)}u(x)-u(x)|\le$$
$$\le |T_t^{(p)}u(x)-T_t^{(p)}v(x)|+\limsup_{t\searrow 0}|T_t^{(2)}v(x)-v(x)|+|v(x)-u(x)|$$
$$\le \limsup_{t\searrow 0}|T_t^{(p)}(u-v)(x)|+|(u-v)(x)|$$
$$\le (u-v)^*(x)+|(u-v)(x)|.$$

The L^p-boundedness of $(u-v)^*$ yields further

$$\|\limsup_{t\searrow 0}|T_t^{(p)}u(\cdot)-u(\cdot)|\|_{L^p}\le K\|u-v\|_{L^p}\le K\varepsilon.$$

For $\varepsilon\to 0$ we get finally

$$\limsup_{t\searrow 0}|T_t^{(p)}u(x)-u(x)|=0\quad\text{a.e.},$$

which means that $T_t^{(p)}u\to u$ a.e. as $t\to 0$. □

We will state now *E.M.Stein's Littlewood–Paley inequality* in the general case. For this let $(T_t)_{t\ge 0}$ be a symmetric and conservative sub-Markovian semigroup defined consistently on all spaces $L^p(\mathbb{R}^n;\mathbb{R})$, $1\le p<\infty$. We denote the corresponding *Littlewood–Paley function* once again by

$$g_1(u)(x):=\left(\int_0^\infty t\left|\frac{\partial}{\partial t}T_tu(x)\right|^2dt\right)^{1/2}. \tag{3.329}$$

Theorem 3.4.33 (E.M.Stein). *If $u \in L^p(\mathbb{R}^n;\mathbb{R})$ then $g_1(u) \in L^p(\mathbb{R}^n;\mathbb{R})$ and*

$$\|g_1(u)\|_{L^p} \le c_p \|u\|_{L^p} \tag{3.330}$$

holds for $1 < p < \infty$.

Corollary 3.4.34. *For all $k \in \mathbb{N}$ the estimate*

$$\|g_k(u)\|_{L^p} \le c_p \|u\|_{L^p}, \quad 1 < p < \infty, \tag{3.331}$$

holds.

Corollary 3.4.35. *Denote by E_0 the orthogonal projection $E_0 : L^2(\mathbb{R}^n) \to \{h \in L^2(\mathbb{R}^n); T_t h = h \text{ for all } t > 0\}$. Then the estimate*

$$\|u\|_{L^p} \le c_p \|g_1(u)\|_{L^p} + \|E_0(u)\|_{L^p} \tag{3.332}$$

holds.

Proof (sketch): First we have to show that $\|E_0(u)\|_{L^p}$ makes sense which will follow from

$$\|E_0(u)\|_{L^p} \le c_p \|u\|_{L^p}, \quad 1 < p < \infty. \tag{3.333}$$

Since in addition $(T_t^{(2)})_{t\ge 0}$ is symmetric its generator is selfadjoint and the spectral theorem yields

$$T_t u = \int_0^\infty e^{-\lambda t}\, dE_\lambda u = E_0(u) + \int_{0+}^\infty e^{-\lambda t}\, dE_\lambda u$$

and further

$$\frac{\partial}{\partial t} T_t u = \int_{0+}^\infty (-\lambda) e^{-\lambda t}\, dE_\lambda u$$

which gives

$$\int_{\mathbb{R}^n} \left| \frac{\partial}{\partial t} T_t u(x) \right|^2 dx = \int_{0+}^\infty \lambda^2 e^{-2\lambda t}\, (dE_\lambda u, u).$$

Therefore we find

$$\begin{aligned} \int_{\mathbb{R}^n} |g_1(u)|^2\, dx &= \int_0^\infty t \int_{\mathbb{R}^n} \left| \frac{\partial T_t}{\partial t} u(x) \right|^2 dx\, dt \\ &= \int_{0+}^\infty \lambda^2 \int_0^\infty e^{\ 2\lambda t} t\, dt\, (dE_\lambda u, u) \\ &= \frac{1}{4} \int_{0+}^\infty 1\, d(E_\lambda u, u), \end{aligned}$$

or

$$\|u\|_{L^2}^2 = 4\|g_1(u)\|_{L^2}^2 + \|E_0(u)\|_{L^2}. \tag{3.334}$$

We may polarize (3.334) if $E_0(u_j) = 0$ to get

$$\int_{\mathbb{R}^n} u_1(x)u_2(x)\,dx = 4\int_0^\infty \int_{\mathbb{R}^n} t\,\frac{\partial T_t u_1}{\partial t}\frac{\partial T_t u_2}{\partial t}\,dx\,dt. \tag{3.335}$$

Now, for $u \in L^2(\mathbb{R}^n) \cap L^p(\mathbb{R}^n)$ and $v \in L^2(\mathbb{R}^2) \cap L^{p'}(\mathbb{R}^n)$, $p' = \frac{p}{p-1}$, $E_0(u) = E_0(v) = 0$, we find

$$\begin{aligned}
\|u\|_{L^p} &= \sup_{\substack{v\in L^2\cap L^{p'} \\ \|v\|_{L^{p'}}\le 1}} \int_{\mathbb{R}^n} u(x)v(x)\,dx \\
&\le 4 \sup_{\substack{v\in L^2\cap L^{p'} \\ \|v\|_{L^{p'}}\le 1}} \int_0^\infty \int_{\mathbb{R}^n} t\,\frac{\partial T_t u}{\partial t}\frac{\partial T_t v}{\partial t}\,dx\,dt \\
&\le 4 \sup_{\substack{v\in L^2\cap L^{p'} \\ \|v\|_{L^{p'}}\le 1}} \int_{\mathbb{R}^n} g_1(u)g_2(v)\,dx,
\end{aligned}$$

where we used the Cauchy–Schwarz inequality in the last step.

Now, Hölder inequality leads to

$$\|u\|_{L^p} \le \Big(4 \sup_{\substack{v\in L^2\cap L^{p'} \\ \|v\|_{L^{p'}}\le 1}} \|g_1(v)\|_{L^{p'}}\Big)\|g_1(u)\|_{L^p},$$

or since $\|g_1(v)\|_{L^{p'}} \le c_{p'}\|v\|_{L^p}$ we find

$$\|u\|_{L^p} \le c_p\|g_1(u)\|_{L^p}$$

for $u \in L^p(\mathbb{R}^n) \cap L^2(\mathbb{R}^n)$, $E_0(u) = 0$. The general case follows when substituting u by $u - E_0 u$ and the observation that $\frac{\partial}{\partial t} T_t E_0 u = \frac{\partial}{\partial t} E_0 u = 0$:

$$\begin{aligned}
\|u\|_{L^p} - \|E_0(u)\|_{L^p} \le \|u - E_0(u)\|_{L^p} &\le c_p\|g_1(u - E_0(u))\|_{L^p} \\
&= c_p\|g_1(u)\|_{L^p},
\end{aligned}$$

which is (3.332). □

We continue to use the spectral decomposition of T_t, i.e. the representation

$$T_t u = \int_0^\infty e^{-\lambda t}\,dE_\lambda u.$$

More generally, we define operators for bounded measurable function on $(0,\infty)$ by

$$S_m u := \int_{0+}^{\infty} m(\lambda)\, dE_\lambda u \tag{3.336}$$

and we call m of *Laplace transform type* if

$$m(\lambda) = \lambda \int_0^\infty e^{-\lambda t} M(t)\, dt, \quad \lambda > 0, \tag{3.337}$$

holds with a bounded measurable function M on $(0,\infty)$. Again we follow E.M.Stein [255] to get

Proposition 3.4.36. *Let m be of Laplace transform type. Then S_m is a bounded operator on $L^p(\mathbb{R}^n)$, $1 < p < \infty$.*

Proof: We will prove

$$g_1(S_m u) \le c\, g_2(u). \tag{3.338}$$

for all $u \in L^p(\mathbb{R}^n)$ with c independent of u. Once the estimate (3.338) is proved the proposition follows from the above corollaries:

$$\begin{aligned} \|S_m(u)\|_{L^p} &\le c_p \|g_1(S_m u)\| + \|E_0(S_m(u))\|_{L^p} \\ &\le \tilde{c}\|g_2(u)\|_{L^p} \le c_p' \|u\|_{L^p}, \end{aligned}$$

where we also used the fact that $E_0(S_m u) = 0$ which is due to the representation (3.336). To prove (3.337) we need some preparation.

For u such that $E_0(u) = 0$ it follows that

$$\begin{aligned} \Big|\frac{\partial}{\partial t} T_t u\Big| &\le \int_0^\infty \Big|\Big(\frac{\partial^2}{\partial s^2} T_s u\Big)\Big|\, ds \\ &= \int_t^\infty \Big|\frac{\partial^2}{\partial s^2} T_s u\Big| s \frac{ds}{s} \le \Big(\int_t^\infty \Big|\frac{\partial^2}{\partial s^2} T_s u\Big|^2 s^2\, ds\Big)^{1/2} \cdot t^{-1/2}, \end{aligned}$$

where in the last step we used the Cauchy–Schwarz inequality. Now the last estimate and Fubini's theorem yields

$$\begin{aligned} (g_1(u))^2 &= \int_0^\infty t \Big|\frac{\partial}{\partial t} T_t u\Big|^2 dt \\ &\le \int_0^\infty \Big(\int_t^\infty \Big|\frac{\partial^2 T_s u}{\partial s^2}\Big|^2 s^2\, ds\Big)\, dt \\ &= \int_0^\infty s^3 \Big|\frac{\partial^2}{\partial s^2} T_s u\Big|^2 ds = g_2(u)^2, \end{aligned}$$

or

$$g_1(u) \le g_2(u) \quad \text{for } E_0(u) = 0. \tag{3.339}$$

Next observe that

$$\begin{aligned}\int_0^\infty \frac{\partial}{\partial t} T_t u\, M(t)\, dt &= \int_0^\infty \frac{\partial}{\partial t} \int_{0+}^\infty e^{-\lambda t}\, dE_\lambda u\, M(t)\, dt \\ &= \int_0^\infty \int_{0+}^\infty -\lambda e^{-\lambda t}\, dE_\lambda u\, M(t)\, dt = \int_{0+}^\infty \int_0^\infty -\lambda e^{-\lambda t} M(t)\, dt\, dE_\lambda u \\ &= -\int_{0+}^\infty m(\lambda)\, dE_\lambda u = -S_m u,\end{aligned}$$

i.e.

$$S_m(u) = -\int_0^\infty \frac{\partial}{\partial t} T_t u\, M(t)\, dt. \tag{3.340}$$

Now we may use exactly the same calculation leading to (3.339) to find

$$g_1(S_m u)^2 = \int_0^\infty s \left| \frac{\partial}{\partial s} T_s(S_m u) \right|^2 ds \le c \int_0^\infty t^3 \left| \frac{\partial^2}{\partial t^2} T_t u \right|^2 dt = c g_2(u)^2,$$

and the proposition is proved. □

This proposition has an important corollary:

Corollary 3.4.37. *Let $(T_t)_{t \ge 0}$ be as above, in particular let $A^{(2)}$ be a self-adjoint operator. Then for every real number $r \in \mathbb{R}$ the operator $(-A)^{ir}$ is a bounded operator from $L^p(\mathbb{R}^n)$ into itself, $1 < p < \infty$.*

Proof: We only have to observe that λ^{ir} is of Laplace type. □

Corollary 3.4.38. *Let $\psi : \mathbb{R}^n \to \mathbb{R}$ be a continuous negative definite function satisfying $\psi(0) = 0$. Then for all $r \in \mathbb{R}$ the operator $(\psi(D))^{ir}$ is bounded on $L^p(\mathbb{R}^n)$, $1 < p < \infty$.*

3.5 Global Properties of L^p-Sub-Markovian Semigroups

Let us suppose that we may associate to a given L^p-sub-Markovian semigroup $(T_t^{(p)})_{t \ge 0}$, $1 < p < \infty$, a Markov process $(\Omega, \mathcal{A}, (X_t)_{t \ge 0}, P^x)_{x \in \mathbb{R}^n}$. We may ask for properties which are common for all the measures P^x, $x \in \mathbb{R}^n$, or for almost all paths $t \mapsto X_t(\omega)$, $X_0(\omega) = x \in \mathbb{R}^n$. Following M.Fukushima, Y.Oshima and M.Takeda [102] we will call these properties global properties of the process. In this section we will study properties of $(T_t^{(p)})_{t\geq 0}$ which will imply global property of the associated process. Therefore we call these properties *global properties* of the semigroup.

Some of our considerations depend on whether a given L^p-sub-Markovian semigroup $(T_t^{(p)})_{t\geq 0}$ has an extension to a positivity preserving L^1-contraction. This however is equivalent to the fact that $(T_t^{(p)*})_{t\geq 0}$ is an $L^{p'}$-sub-Markovian semigroup as we may deduce from the next theorem which is taken from our joint paper [166] with R.L.Schilling.

Theorem 3.5.1. *Let $1 < p < \infty$ and $T : L^p(\mathbb{R}^n;\mathbb{R}) \to L^p(\mathbb{R}^n;\mathbb{R})$ be a linear (bounded) operator. Then $T^* : L^{p'}(\mathbb{R}^n;\mathbb{R}) \to L^{p'}(\mathbb{R}^n;\mathbb{R})$ is a (bounded) $L^{p'}$-sub-Markovian operator, $p' = p/(p-1)$, if and only if T is positivity preserving and extends to a contraction $T : L^1(\mathbb{R}^n;\mathbb{R}) \to L^1(\mathbb{R}^n;\mathbb{R})$. Hence T and T^* are sub-Markovian if and only if both operators extend to positivity preserving contractions onto all spaces $L^q(\mathbb{R};\mathbb{R})$, $1 \leq q \leq \infty$.*

Proof: Throughout the proof $(B_k)_{k\in\mathbb{N}}$ denotes a sequence of Borel sets in $\mathbb{R}^n$ increasing to $\mathbb{R}^n$, i.e. $B_k \uparrow \mathbb{R}^n$, each having finite measure, i.e. $\lambda^{(n)}(B_k) < \infty$, and we set as before $p' = \frac{p}{p-1}$. Assume that $T : L^p(\mathbb{R}^n;\mathbb{R}) \to L^p(\mathbb{R}^n;\mathbb{R})$ is positivity preserving. For $v \in L^{p'}(\mathbb{R}^n;\mathbb{R})$, $v \geq 0$ a.e., we set $A_k := B_k \cap \{T^*v < 0\}$. It follows that $\chi_{A_k} \in L^p(\mathbb{R}^n;\mathbb{R})$ and

$$\begin{aligned}\int_{B_k\cap\{T^*v<0\}} (T^*v)(x)\,dx &= \int_{\mathbb{R}^n} (T^*v)(x)\chi_{A_k}(x)\,dx \\ &= \int_{\mathbb{R}^n} v(x)T(\chi_{A_k})(x)\,dx \geq 0\end{aligned}$$

implying that $\lambda^{(n)}(B_k \cap \{T^*v < 0\}) = 0$ for all $k \in \mathbb{N}$, i.e. T^* is positivity preserving if T is positivity preserving. Interchanging the role of T and T^* we find that T^* is positivity preserving if and only if T is positivity preserving. Now assume that $T : L^1(\mathbb{R}^n;\mathbb{R}) \to L^1(\mathbb{R}^n;\mathbb{R})$ is positivity preserving and an L^1-contraction. For $v \in L^{p'}(\mathbb{R}^n;\mathbb{R})$ such that $0 \leq v \leq 1$ a.e. we set $C_k := B_k \cap \{T^*v > 1\}$. We find that

$$\int_{B_k\cap\{T^*v>1\}} (T^*v)(x)\,dx = \int_{\mathbb{R}^n} (T^*v)(x)\chi_{C_k}(x)\,dx$$

$$= \int_{\mathbb{R}^n} v(x) T(\chi_{C_k})(x)\, dx \leq \|v\|_\infty \|T\chi_{C_k}\|_{L^1} \leq \|\chi_{C_k}\|_{L^1}$$
$$= \lambda^{(n)}(B_k \cap \{T^* v > 1\})$$

which yields $\lambda^{(n)}(\{T^*v > 1\}) = 0$. Since T is by assumption positivity preserving it follows that for $0 \leq v \leq 1$ a.e., we find $0 \leq T^*v \leq 1$ a.e., i.e. T^* is sub-Markovian.

Suppose now that $T^* : L^{p'}(\mathbb{R}^n; \mathbb{R}) \to L^{p'}(\mathbb{R}^*; \mathbb{R})$ is sub-Markovian. In particular it follows from the first part of this proof that T is positivity preserving. Therefore, for $u \in L^p(\mathbb{R}^n; \mathbb{R}) \cap L^1(\mathbb{R}^n; \mathbb{R})$ it follows that

$$\int_{\mathbb{R}^n} |(Tu)(x)| \chi_{B_k}(x)\, dx \leq \int_{\mathbb{R}^n} T|u|(x) \chi_{B_k}(x)\, dx$$
$$= \int_{\mathbb{R}^n} |u(x)| (T^* \chi_{B_k})(x)\, dx \leq \int_{\mathbb{R}} |u(x)|\, dx.$$

By monotone convergence we find as $k \to \infty$ that

$$\|Tu\|_{L^1} \leq \|u\|_{L^1} \quad \text{for all } u \in L^p(\mathbb{R}^n; \mathbb{R}) \cap L^1(\mathbb{R}^n; \mathbb{R}),$$

which clearly implies that T extends to an L^1-contraction. Finally, if T and T^* are both sub-Markovian operators, the above proof and the consequences of Remark 2.6.15.B give that both T and T^* extend to sub-Markovian contractions on all the spaces $L^q(\mathbb{R}^n; \mathbb{R})$, $1 \leq q \leq \infty$. □

Corollary 3.5.2. *Suppose that for some $1 < p < \infty$ an L^p-sub-Markovian semigroup $(T_t^{(p)})_{t \geq 0}$ is given such that its dual semigroup $(T_t^{(p)*})_{t \geq 0}$ is $L^{p'}$-sub-Markovian. Then $(T_t^{(p)})_{t \geq 0}$ and $(T_t^{(p)*})_{t \geq 0}$ extend both to L^q-sub-Markovian semigroups $(T_t^{(q)})_{t \geq 0}$ and $(\tilde{T}_t^{(q)})_{t \geq 0}$ for all $1 < q < \infty$. In addition we have $T_t^{(q)*} = \tilde{T}_t^{(q')}$ and for all $u \in L^1(\mathbb{R}^n; \mathbb{R}) \cap L^\infty(\mathbb{R}^n; \mathbb{R})$ it holds*

$$T_t^{(q)} u = T_t^{(p)} u \quad \text{and} \quad T_t^{(q)*} u = T_t^{(p)*} u, \quad t > 0. \tag{3.341}$$

Moreover, $(T_t^{(1)})_{t \geq 0}$ is a semigroup of sub-Markovian operators.

Remark 3.5.3. A. The relations (3.341) justify that in the following we will write just $T_t u$ or $T_t^* v$ for $T_t^{(q)} u$ or $T_t^{(q)*} v$, respectively.

B. In the situation of Corollary 3.5.2 we may also consider for all $1 \leq q < \infty$ the corresponding resolvent operators $R_\lambda = \int_0^\infty e^{-\lambda t} T_t\, dt$ which form for $1 < q < \infty$ strongly continuous contraction resolvents, whereas for $q = 1$ a contraction resolvent is given.

C. Note that in general $(T_t^{(1)})_{t\geq 0}$ will not be strongly continuous, see J.Voigt [278].

Let $T : L^p(\mathbb{R}^n;\mathbb{R}) \to L^p(\mathbb{R}^n;\mathbb{R})$ be a bounded linear and positivity preserving operator. The operator norm $\|T\|$ of T is given by

$$\|T\| = \sup_{u\neq 0} \frac{\|Tu\|_{L^p}}{\|u\|_{L^p}}. \tag{3.342}$$

The next lemma taken from E.B.Davies [60] shows that we may restrict the supremum to non-negative functions $u \in L^p(\mathbb{R}^n;\mathbb{R})$.

Lemma 3.5.4. *For a linear, bounded and positivity preserving operator T : $L^p(\mathbb{R}^n;\mathbb{R}) \to L^p(\mathbb{R}^n;\mathbb{R})$ the operator norm equals*

$$\|T\| = \sup\left\{\frac{\|Tu\|_{L^p}}{\|u\|_{L^p}};\quad u \in L^p(\mathbb{R}^n;\mathbb{R}) \text{ and } u \geq 0 \text{ a.e.}\right\}. \tag{3.343}$$

Proof: Clearly we have that

$$C_T := \sup\left\{\frac{\|Tu\|_{L^p}}{\|u\|_{L^p}};\quad u \in L^p(\mathbb{R}^n;\mathbb{R}) \text{ and } u \geq 0 \text{ a.e.}\right\} \leq \|T\|.$$

Now, for $u \in L^p(\mathbb{R}^n;\mathbb{R})$ it follows that $-|u| \leq u \leq |u|$ and therefore $-T|u| \leq Tu \leq T|u|$ implying

$$\|Tu\|_{L^p} \leq \|T|u|\|_{L^p} \leq C_T\||u|\|_{L^p} = C_T\|u\|_{L^p},$$

i.e. $\|T\| \leq C_T$. □

We arrive now at a first central definition in this section.

Definition 3.5.5. Let $T : L^p(\mathbb{R}^n;\mathbb{R}) \to L^p(\mathbb{R}^n;\mathbb{R})$, $1 \leq p \leq \infty$, be a linear bounded and positivity preserving operator.

A. A Borel set $G \subset \mathbb{R}^n$ is called *T-invariant* (or invariant with respect to T) if

$$T\chi_G u = \chi_G T\chi_G u \quad \text{a.e.} \tag{3.344}$$

holds for all $u \in L^p(\mathbb{R}^n;\mathbb{R})$.

B. A Borel set $G \subset \mathbb{R}^n$ is called *strongly T-invariant* if

$$T\chi_G u = \chi_G Tu \quad \text{a.e.} \tag{3.345}$$

holds for all $u \in L^p(\mathbb{R}^n;\mathbb{R})$. Using $\chi_G u$ instead u shows that every strongly T-invariant set is T-invariant.

Lemma 3.5.6. *Let $T : L^p(\mathbb{R}^n;\mathbb{R}) \to L^p(\mathbb{R}^n;\mathbb{R})$, $1 \leq p < \infty$, be a linear bounded positivity preserving operator.*

A. If G is T-invariant then G^c is T^*-invariant. Further, arbitrary intersections and countable unions of T-invariant sets and a countable intersection of T^*-invariant sets are T-invariant. Thus an arbitrary union of T^*-invariant sets and a countable intersection of T^*-invariant sets are T-invariant.

B. If G is strongly T-invariant then G^c is strongly T-invariant. Further, G is strongly T-invariant if and only if G is T-and T^*-invariant.

C. An arbitrary intersection and union of strongly T-invariant sets are strongly T-invariant.

Proof: A. For a T-invariant set G we find for $u \in L^{p'}(\mathbb{R}^n;\mathbb{R})$ and $v \in L^p(\mathbb{R}^n;\mathbb{R})$ that

$$\int_{\mathbb{R}^n} \chi_G v T^*(\chi_{G^c} u)\,dx = \int_{\mathbb{R}^n} \chi_{G^c} u T \chi_G\,dx = \int_{\mathbb{R}^n} (\chi_{G^c} u)\chi_G T \chi_G v\,dx = 0$$

which means that

$$\int_{\mathbb{R}^n} (1-\chi_{G^c})(T^*\chi_{G^c} u) v\,dx = 0, \quad v \in L^p(\mathbb{R}^n;\mathbb{R}),$$

so $T^*\chi_{G^c}u = \chi_{G^c}T^*\chi_{G^c}u$. Next let $(G_j)_{j\in I}$ be an arbitrary family of T-invariant sets. Since T is positivity preserving it is monotone (in the sense that $u_1 \leq u_2$ implies $Tu_1 \leq Tu_2$) implying

$$T\chi_{\cap G_l} u \leq T\chi_{G_j} u = \chi_{G_j} T \chi_{G_j} u \quad \text{for all } j \in I.$$

Hence $(T\chi_{\cap G_l}u)(x) = 0$ for $x \notin \bigcap_{j\in I} G_j$ or $T\chi_{\cap G_l}u = \chi_{\cap G_l}T\chi_{\cap G_l}$ for all $u \in L^p(\mathbb{R}^n;\mathbb{R})$. For a countable family of mutually disjoint T-invariant sets $(G_j)_{j\in\mathbb{N}}$ we find

$$T(\chi_{\cup G_j} u) = \sum_{j=1}^{\infty} T\chi_{G_j} u,$$

but $\sum_{j=1}^{\infty} T\chi_{G_j}u(x) = 0$ for $x \in (\cup G_j)^c$, implying that

$$T(\chi_{\cup G_j} u) = \sum_{j=1}^{\infty} T(\chi_{G_j} u) = \chi_{\cup G_j}\Big(\sum_{j=1}^{\infty} T(\chi_{G_j} u)\Big) = \chi_{\cup G_j} T \chi_{\cup G_j} u.$$

If G_1, G_2 are two arbitrary T-invariant sets we have trivially

$$\chi_{G_1\cup G_2} T \chi_{G_1\cup G_2} u \leq T\chi_{G_1\cup G_2} u \quad \text{if} \quad u \geq 0,$$

but in addition we find

$$\begin{aligned} T\chi_{G_1\cup G_2}u &= T\chi_{G_1}u + T\chi_{G_2}u - T\chi_{G_1\cap G_2}u \\ &= \chi_{G_1}T\chi_{G_1}u + \chi_{G_2}T\chi_{G_2}u - \chi_{G_1\cap G_2}T\chi_{G_1\cap G_2}u. \end{aligned}$$

Further, since $T\chi_{G_1\cap G_2}u = \chi_{G_1\cap G_2}T\chi_{G_1\cap G_2}u + \chi_{G_1\cup G_2}T\chi_{G_1\cap G_2}u$ we find

$$\begin{aligned} T\chi_{G_1\cup G_2}u &= \chi_{G_1\cup G_2}(T\chi_{G_1}u + T\chi_{G_2}u - T\chi_{G_1\cap G_2}u) \\ &= \chi_{G_1\cup G_2}T\chi_{G_1\cup G_2}u, \end{aligned}$$

where we used for $u \geq 0$ the trivial fact $\chi_{G_1\cup G_2}T\chi_{G_1\cup G_2}u \leq T\chi_{G_1\cup G_2}u$. In order to prove the general case we just have to pass from an arbitrary countable family of T-invariant sets to an appropriate family with mutually disjoint T-invariant sets.

B. For a strongly T-invariant set G we deduce from (3.345) that

$$T\chi_{G^c}u = Tu - T\chi_G u = Tu - \chi_G Tu = \chi_{G^c}Tu,$$

i.e. G^c is strongly T-invariant. We prove now that the T-invariance and T^*-invariance of G imply the strong T-invariance of G. For G we have $T\chi_G u = \chi_G T\chi_G u$ or $\chi_{G^c}T\chi_G u = 0$. Further, since G is also T^*-invariant, G^c is T-invariant and it follows that $T\chi_{G^c}u = \chi_{G^c}T\chi_{G^c}u$ or $\chi_G T\chi_{G^c}u = 0$. Thus we find

$$\begin{aligned} \chi_G Tu &= \chi_G T\chi_G u + \chi_G T\chi_{G^c}u = \chi_G T\chi_G u \\ &= \chi_G T\chi_G u + \chi_{G^c}T\chi_G u = T\chi_G u. \end{aligned}$$

On the other hand, if G is strongly T-invariant it follows that G^c is strongly invariant, hence G^c is T-invariant implying that G is T^*-invariant. (Note for $p = 1$ we have to use the fact that $T^{**}|_{L^1} = T$).

C. Part B of the lemma says that the family of strong T-invariant sets is the intersection of the two families of T-and T^*-invariant sets. Therefore part A and B yields part C.

Corollary 3.5.7. *Suppose that $T|_{L^p\cap L^2}$ is symmetric on $L^2(\mathbb{R}^n;\mathbb{R})$. Then G is T-invariant if and only if it is strongly T-invariant.*

Definition 3.5.8. We say that a Borel set $G \subset \mathbb{R}^n$ is *$(T_t)_{t\geq 0}$-invariant* or *invariant* with respect to an L^p-sub-Markovian semigroup $(T_t)_{t\geq 0}$ if it is T_t-invariant for all $t > 0$. Analogously *strongly $(T_t)_{t\geq 0}$-invariant* sets are defined.

Lemma 3.5.9. *Let $(T_t)_{t\geq 0}$ be an L^p-sub-Markovian semigroup with resolvent $(R_\lambda)_{\lambda>0}$. Then G is $(T_t)_{t\geq 0}$-invariant (strongly $(T_t)_{t\geq 0}$-invariant) if and only if it is $(R_\lambda)_{\lambda>0}$-invariant (strongly $(R_\lambda)_{\lambda>0}$-invariant), i.e. invariant (strongly invariant) for all R_λ, $\lambda > 0$.*

Proof: Since $R_\lambda u = \int_0^\infty e^{-\lambda t} T_t u \, dt$ we find for a $(T_t)_{t>0}$-invariant (strongly invariant) set $G \subset \mathbb{R}^n$ and $u \in L^p(\mathbb{R}^n; \mathbb{R})$ immediately

$$R_\lambda \chi_G u = \int_0^\infty e^{-\lambda t} T_t \chi_G u \, dt = \chi_G \int_0^\infty e^{-\lambda t} T_t \chi_G u \, dt = \chi_G R_\lambda \chi_G u$$

and

$$R_\lambda \chi_G u = \int_0^\infty e^{-\lambda t} T_t \chi_G u \, dt = \chi_G \int_0^\infty e^{-\lambda t} T_t u \, dt = \chi_G R_\lambda u,$$

respectively. On the other hand, since the Yosida approximation of the generator A of $(T_t)_{t\geq 0}$ is given by $A_\lambda = \lambda A R_\lambda = \lambda(\lambda R_\lambda - \mathrm{id})$ it follows that a $(R_\lambda)_{\lambda>0}$-invariant (strongly invariant) set G, is also invariant (strongly invariant) for all operators A_λ implying the invariance (strong invariance) of G with respect to the operators

$$e^{\lambda A R_\lambda t} = \sum_{k=0}^{\infty} \frac{t^k}{k!} (\lambda A R_\lambda)^k, \quad t > 0. \tag{3.346}$$

But $e^{\lambda A R_\lambda t} \to T_t$ strongly in $L^p(\mathbb{R}^n; \mathbb{R})$ implying the invariance (strong invariance) of G with respect to T_t for all $t > 0$. □

Remark 3.5.10. Clearly the results of Lemma 3.5.6 carry over to semigroups and their resolvents.

The following result relates strongly invariant sets to the Dirichlet form associated with an L^2-sub-Markovian semigroup and it is taken from the work [226] of Y.Oshima.

Theorem 3.5.11. *Let $(T_t)_{t\geq 0}$ be an L^2-sub-Markovian semigroup defined on $L^2(\mathbb{R}^n; \mathbb{R})$ with dual-semigroup $(T_t^*)_{t\geq 0}$ being also L^2-sub-Markovian. Denote by $(\mathcal{E}, D(\mathcal{E}))$ the corresponding non-symmetric Dirichlet form, compare Definition I.4.7.21 and Theorem I.4.7.23. Then a Borel set $G \subset \mathbb{R}^n$ is strongly $(T_t)_{t\geq 0}$-invariant if and only if for all $u, v \in D(\mathcal{E})$ it follows that $\chi_G u, \chi_G v \in D(\mathcal{E})$ and*

$$\mathcal{E}(u, v) = \mathcal{E}(\chi_G u, \chi_G v) + \mathcal{E}(\chi_{G^c} u, \chi_{G^c} v). \tag{3.347}$$

Proof: Suppose that $G \subset \mathbb{R}^n$ is an $(T_t)_{t\geq 0}$-invariant set. It follows from the $(R_\lambda)_{\lambda>0}$-invariance of G for $u \in D(\mathcal{E})$ that

$$((\mathrm{id}-\lambda R_\lambda)u, u)_0 = ((\mathrm{id}-\lambda R_\lambda)\chi_G u, \chi_G u)_0 + ((\mathrm{id}-\lambda R_\lambda)\chi_{G^c} u, \chi_{G^c} u)_0.$$

Using Lemma I.4.7.18 we deduce that $\chi_G u, \chi_{G^c} u \in D(\mathcal{E})$ and (3.347). Conversely, suppose that $\chi_G u, \chi_G v \in D(\mathcal{E})$. Then of course $\chi_{G^c} u = u - \chi_G u \in D(\mathcal{E})$, hence $\chi_{G^c} u \in D(\mathcal{E})$ too and we claim

$$\mathcal{E}(\chi_G u, \chi_{G^c} v) = 0. \tag{3.348}$$

Indeed, now (3.347) implies by a straightforward calculation

$$\mathcal{E}(\chi_G u, \chi_{G^c} v) + \mathcal{E}(\chi_{G^c} u, \chi_G v) = 0 \tag{3.349}$$

for all $u, v \in D(\mathcal{E})$ and taking $\chi_G u$ instead of u the second term in (3.349) vanished which yields (3.348). Since for $\lambda > 0$ and $v \in D(\mathcal{E})$

$$\mathcal{E}_\lambda(\chi_{G^c} u, R_\lambda v) = (v, \chi_{G^c} u)_0 = \mathcal{E}_\lambda(\chi_{G^c} u, R_\lambda \chi_{G^c} v)$$

we arrive at

$$\mathcal{E}_\lambda(\chi_{G^c} u, R_\lambda v - R_\lambda(\chi_{G^c} v)) = 0$$

and with $u := R_\lambda v - R_\lambda(\chi_{G^c} v))$ we have

$$\mathcal{E}_\lambda(\chi_{G^c} u, u) = 0.$$

Now (3.348) yields

$$\mathcal{E}_\lambda(\chi_{G^c} u, \chi_{G^c} u) = \mathcal{E}_\lambda(\chi_{G^c} u, u) = 0,$$

and therefore $\chi_{G^c} u = 0$, or $\chi_{G^c} R_\lambda v = \chi_{G^c} R_\lambda(\chi_{G^c} v)$. Taking G instead of G^c we have $\chi_G R_\lambda v = \chi_G R_\lambda \chi_G v$.

Thus we find

$$\begin{aligned} R_\lambda \chi_G v &= \chi_G R_\lambda \chi_G v + \chi_{G^c} R_\lambda \chi_G v \\ &= \chi_G R_\lambda v + \chi_{G^c} R_\lambda (v - \chi_{G^c} v) \\ &= \chi_G R_\lambda v + \chi_{G^c} R_\lambda v - \chi_{G^c} R_\lambda \chi_{G^c} v = \chi_G R_\lambda v \end{aligned}$$

proving the $(R_\lambda)_{\lambda>0}$-invariance of G which implies the theorem. □

Remark 3.5.12. Since (3.347) is symmetric in G and G^c, accordingly to Lemma 3.5.6 we should not expect (3.347) to be equivalent to the $(T_t)_{t\geq 0}$-invariance of G.

Definition 3.5.13. An L^p-sub-Markovian semigroup $(T_t)_{t\geq 0}$ is called *irreducible* if for any $(T_t)_{t\geq 0}$-invariant set G either $\lambda^{(n)}(G) = 0$ or $\lambda^{(n)}(G^c) = 0$ holds. Since for $G \subset \mathbb{R}^n$ such that $\lambda^{(n)}(G) = 0$ or $\lambda^{(n)}(G^c) = 0$ it follows that

$$T_t\chi_G u = \chi_G T u \quad \text{a.e.} \quad \text{and} \quad T_t\chi_{G^c} u = \chi_{G^c} T u \quad \text{a.e.},$$

a set G of measure zero as well as its complement are always $(T_t)_{t\geq 0}$-invariant.

Suppose that a Dirichlet form $(\mathcal{E}, D(\mathcal{E}))$ has the property

$$u, v \in D(\mathcal{E}) \text{ and } \operatorname{supp} u \cap \operatorname{supp} v = \varnothing \text{ implies } \mathcal{E}(u, v) = 0. \tag{3.350}$$

Then (3.347) holds for all $u, v \in D(\mathcal{E})$. Dirichlet forms satisfying (3.350) are called *local Dirichlet forms.* Typically they are associated with a differential operator, i.e. they have the (typical) structure

$$\mathcal{E}(u, v) = \sum_{k,l=1}^{n} \int_{\mathbb{R}^n} \frac{\partial u}{\partial x_k} \frac{\partial v}{\partial x_l} \nu_{kl}(dx) \tag{3.351}$$

with some suitable measures $\nu_{kl}(dx)$. In case that $\nu_{kl}(dx) = a_{kl}(x)\lambda^{(n)}(dx)$ with $\mu|\xi|^2 \leq \sum_{k,l=1}^{n} a_{kl}(x)\xi_k\xi_l \leq \frac{1}{\mu}|\xi|^2$ and $a_{kl} = a_{lk} \in C_b^1(\mathbb{R}^n)$ (for simplicity), then $D(\mathcal{E}) = H^1(\mathbb{R}^n; \mathbb{R})$ and typically $\chi_G u \notin H^1(\mathbb{R}^n; \mathbb{R})$ for a Borel set such that $\lambda^{(n)}(G)$ or $\lambda^{(n)}(G^c)$ is strictly positive. This observation is just a special case of

Lemma 3.5.14. *Suppose that the L^p-sub-Markovian semigroup $(T_t)_{t\geq 0}$ has a kernel representation*

$$T_t u(x) = \int_{\mathbb{R}^n} p_t(x, y)u(y)\, dy \tag{3.352}$$

with a kernel function $p_t(x, y) > 0$ for almost all $x, y \in \mathbb{R}^n$ and $t > 0$. Then $(T_t)_{t\geq 0}$ is irreducible.

Proof: Suppose that $G \subset \mathbb{R}^n$ is an invariant Borel set with $\lambda^{(n)}(G) > 0$. Take $u \in L^p(\mathbb{R}^n; \mathbb{R})$. It follows that

$$\chi_G(x)(T_t\chi_G u)(x) = (T_t\chi_G u)(x) = \int_{\mathbb{R}^n} p_t(x, y)\chi_G(y)u(y)\, dy,$$

i.e. $\int_{\mathbb{R}^n} p_t(x,y)\chi_G(y)u(y)\,dy = 0$ for all $x \in G^c$ implying that $\int_G p_t(x,y)\,dy = 0$ for all $x \in G^c$ which is a contradiction to $p_t(x,y) > 0$ unless where $\lambda^{(n)}(G) = 0$ or $\lambda^{(n)}(G^c) = 0$. $\square$

Example 3.5.15. Since the Gauss kernel is strictly positive we find that the Gaussian semigroup is irreducible. Further, if $L(x,D)$ is a second order differential operator generating an L^p-sub-Markovian semigroup $(T_t)_{t\geq 0}$ and satisfying all the assumptions needed to prove Aronson's estimates (2.442) then the semigroup $(T_t)_{t\geq 0}$ is irreducible. Moreover, all semigroups subordinate to $(T_t)_{t\geq 0}$ with respect to a convolution semigroup $(\eta_t)_{t\geq 0}$, $\operatorname{supp}\eta_t \subset [0,\infty)$, admitting for each η_t a strictly positive and continuous density on $(0,\infty)$ are irreducible too. In particular, since it is possible to prove with some effort, compare V.M.Zolotarev [291], Section 2.7, that the one-sided stable semigroup $(\eta_t^{(\alpha)})_{t\geq 0}$, $0 < \alpha < 1$, has smooth and strictly positive densities, the semigroups generated by fractional powers of $L(x,D)$ are all irreducible.

Let $(T_t)_{t\geq 0}$ be an L^p-sub-Markovian semigroup with generator $(A, D(A))$. The elements in the kernel of A, $\operatorname{Ker}(A) := \{u \in D(A); Au = 0\}$ are called *A-harmonic functions* (belonging to $D(A)$). It is interesting to note that for an irreducible semigroup we can obtain easily information about $\operatorname{Ker} A$. The following results are borrowed from E.B.Davies [60].

Proposition 3.5.16. *Let $(T_t)_{t\geq 0}$ be an L^p-sub-Markovian semigroup.*

A. The space $\operatorname{Ker} A$ *is a closed sub-lattice of $L^p(\mathbb{R}^n;\mathbb{R})$, i.e. $u \in \operatorname{Ker}(A)$ implies $|u| \in \operatorname{Ker}(A)$.*

B. If $(T_t)_{t\geq 0}$ is irreducible then $\dim \operatorname{Ker}(A) \leq 1$. *Further if* $\dim \operatorname{Ker}(A) = 1$ *then there exists a strictly positive element $u_0 \in \operatorname{Ker}(A)$, i.e. $u_0 > 0$ a.e.*

Proof: A. If $u \in D(A)$ and $T_t u = u$ for all $t \geq 0$ then $u \in \operatorname{Ker}(A)$. But conversely, since by Lemma I.4.1.14

$$T_t u - u = \int_0^t T_s A u\,ds, \quad u \in D(A),$$

it follows for $u \in \operatorname{Ker}(A)$ that $T_t u = u$ for all $t \geq 0$. Thus we find

$$\operatorname{Ker}(A) = \bigcap_{t\geq 0}\{u \in L^p(\mathbb{R}^n;\mathbb{R}); T_t u = u\}$$

and it remains to show that each of the sets $\{u \in L^p(\mathbb{R}^n;\mathbb{R}); T_t u = u\}$ is a closed sub-lattice of $L^p(\mathbb{R}^n;\mathbb{R})$. Clearly $\{u \in L^p(\mathbb{R}^n;\mathbb{R}); T_t u = u\}$ is a closed sub-space of $L^p(\mathbb{R}^n;\mathbb{R})$. Now, for $u \in \{u \in L^p(\mathbb{R}^n;\mathbb{R}); T_t u = u\}$ we find

$$0 \leq |u| = |T_t u| \leq T_t|u|,$$

and if $|u| \neq T_t|u|$ a.e. then it follows that $\||u|\|_{L^p} < \|T_t|u|\|_{L^p}$ contradicting the contraction property of $(T_t)_{t\geq 0}$.

B. That $0 \in \operatorname{Ker}(A)$, hence $\dim \operatorname{Ker}(A) \geq 0$ is trivial. Hence suppose that $\dim \operatorname{Ker}(A) > 0$. Since $\operatorname{Ker}(A)$ is a sub-lattice, $u \in \operatorname{Ker}(A)$ yields $|u| \in \operatorname{Ker}(A)$, thus $\operatorname{Ker}(A)$ contains at least one non-trivial non-negative element $v \geq 0$ a.e. For $\omega \in L^p(\mathbb{R}^n;\mathbb{R})$ such that $|\omega| \leq \alpha v$ a.e. for some $\alpha \geq 0$ it follows that

$$|T_t\omega| \leq T_t|\omega| \leq T_t(\alpha v) = \alpha v \quad \text{a.e..}$$

Thus we find that $M := \{\omega \in L^p(\mathbb{R}^n;\mathbb{R}); |\omega| \leq \alpha v \text{ a.e. for some } \alpha \geq 0\}$ is invariant under the action of T_t for all $t \geq 0$, i.e. $T_t M \subset M$. Since the closure of M is the set of all $\omega \in L^p(\mathbb{R}^n;\mathbb{R})$ such that $\{x \in \mathbb{R}^n;\ \omega(x) \neq 0 \text{ a.e.}\} \subset \{x \in \mathbb{R}^n;\ v(x) \neq 0 \text{ a.e.}\}$, we deduce that $\{x \in \mathbb{R}^n;\ v(x) \neq 0 \text{ a.e.}\}$ is $(T_t)_{t\geq 0}$-invariant implying by the irreducibility assumption that $\{x \in \mathbb{R}^n;\ v(x) \neq 0 \text{ a.e.}\} = \mathbb{R}^n$. Therefore, if $v \in \operatorname{Ker}(A)$, $v \geq 0$ a.e. and $v \neq 0$, then $\{x \in \mathbb{R}^n;\ v(x) \neq 0 \text{ a.e.}\} = \mathbb{R}^n$. The sub-lattice property implies further that $v^+ = \frac{1}{2}(|v| + v)$ a.e. and $v^- = \frac{1}{2}(|v| - v)$ a.e. belong also to $\operatorname{Ker}(A)$, but $\{x \in \mathbb{R}^n;\ v(x) \geq 0 \text{ a.e.}\} \cap \{x \in \mathbb{R}^n;\ v(x) \leq 0 \text{ a.e.}\} \subset \{x \in \mathbb{R}^n;\ v(x) = 0 \text{ a.e.}\}$ which means that either v^+ or v^- must vanish a.e. Therefore either $v \geq 0$ a.e. or $-v \geq 0$ a.e. Finally take two positive functions $u, v \in \operatorname{Ker}(A)$. Then for all $\lambda \in \mathbb{Q}$ the function $u + \lambda v$ must be either almost everywhere positive or negative and the exeptional set is independent of $\lambda \in \mathbb{Q}$. Hence we may construct some $\lambda_0 \in \mathbb{R}$ such that $u + \lambda_0 v = 0$ a.e. proving that $\operatorname{Ker}(A)$ is one dimensional. $\square$

Next we will discuss transient and recurrent semigroups. In order to get an idea of the probabilistic relevance of these notions and some intuition for analytic formulations we start with a rather heuristic discussion. Suppose our semigroup has an extension to $L^1(\mathbb{R}^n;\mathbb{R})$ as a semigroup $(T_t)_{t\geq 0}$ of positivity preserving contractions which is not necessarily strongly continuous. For a (bounded) Borel set $G \in \mathbb{R}^n$ the term

$$p_t(x, G) := T_t\chi_G(x)$$

is the probabilistic interesting quantity, giving the probability for being at time t in the set A if the process associated with $(T_t)_{t\geq 0}$ starts at $t = 0$ in the point

x. Thus the sum

$$\sum_{k=0}^{M} p_{tk}(x,G) = \sum_{k=0}^{M} T_{tk}\chi_G(x) = \sum_{k=0}^{M} T_t^k \chi_G(x)$$

gives information to be at the time instants $0, t, 2t, \dots, Mt$ in the set G. Fix t and $M \in \mathbb{N}$, divide $[0, Mt]$ into N equally long disjoint intervals, i.e. consider the partition

$$\{0, 1\cdot\frac{Mt}{N}, \dots, k\cdot\frac{Mt}{N}, \dots, Mt\}$$

and consider the sum

$$\sum_{k=0}^{N-1} T_{\frac{kMt}{N}}\chi_G(x) = \sum_{k=0}^{N-1} T^k_{\frac{Mt}{N}}\chi_G(x).$$

It follows that

$$(\mathrm{id} - T_{\frac{Mt}{N}}) \sum_{k=0}^{N-1} T^k_{\frac{Mt}{N}}\chi_G(x) = (\mathrm{id} - T^N_{\frac{Mt}{N}})\chi_G(x) = (\mathrm{id} - T_{Mt})\chi_G(x)$$

or

$$\Big(\frac{\mathrm{id} - T_{\frac{Mt}{N}}}{\frac{Mt}{N}}\Big) \sum_{k=0}^{N-1} T^k_{\frac{Mt}{N}}\chi_G(x)\frac{Mt}{N} = (\mathrm{id} - T_{Mt})\chi_G(x).$$

With the substitution $\frac{Mt}{N} = s$ we find

$$\Big(\frac{\mathrm{id} - T_s}{s}\Big) \sum_{k=0}^{N-1} sT_s^k\chi_G(x) = \Big(\frac{\mathrm{id} - T_s}{s}\Big) \sum_{k=0}^{N-1} sT_{ks}\chi_G(x) = (\mathrm{id} - T_{Mt})\chi_G(x).$$

But $\sum_{k=0}^{N-1} sT_{ks}\chi_G(x)$ is a Riemann sum for $\int_0^N T_{t'}\chi_G(x)\,dt'$. Suppose now that $\lim_{r\to\infty} p_r(x,G) = \lim_{r\to\infty} T_r\chi_G(x) = 0$ and let N tend to infinity such that $Mt/N = s$ remains constant, so $Mt \to \infty$ too. Assuming that $\sum_{k=0}^{N-1} sT_{ks}\chi_G(x)$ will tend to the improper integral $\int_0^\infty T_{t'}\chi_G(x)\,dt'$ we will arrive at

$$\Big(\frac{\mathrm{id} - T_s}{s}\Big) \int_0^\infty T_{t'}\chi_G(x)\,dt' = \chi_G(x).$$

But now we may also pass to the limit $s \to 0$ to find

$$-A\Big(\int_0^\infty T_{t'}\chi_G(x)\,dt'\Big) = \chi_G(x), \tag{3.353}$$

where A is the "generator" of $(T_t)_{t\geq 0}$ (but note that there is of course a problem since we did not assume $(T_t)_{t\geq 0}$ to be strongly continuous on $L^1(\mathbb{R}^n;\mathbb{R})$). Assuming further that we may pass from (3.353) to all $f \in L^1(\mathbb{R}^n;\mathbb{R})$, $f \geq 0$, we arrive at

$$-A\Big(\int_0^\infty T_{t'} f\, dt'\Big) = f. \tag{3.354}$$

Now let us try to interpret some of our heuristic arguments. First, note that $\lim\limits_{r\to\infty} p_r(x,G) = 0$ means that for large r there is "no" probability to be in the set G. Suppose that this condition yields both, the existence of $\int_0^\infty T_t\chi_G(x)\,dt < \infty$ and the convergence of $\sum_{k=0}^\infty T_t^k\chi_G(x) < \infty$. The latter has the interpretation that the process may visit the set G with some positive probability finitely may often, but for large k visits to G are very unlikely. This phenomenon is called *transience of a process* . In addition (3.354) has also an interesting analytic interpretation. Since

$$\int_0^\infty T_t f(x)\, dt = \lim_{\lambda\to 0} R_\lambda f(x) = \lim_{\lambda\to 0}\int_0^\infty e^{-\lambda t} T_t f(x)\, dx,$$

(3.354) means that in the "transient case" for $\lambda \to 0$ the resolvent of the semigroups $(T_t)_{t\geq 0}$ should tend to an inverse of $-A$. Because the inverse of $-\Delta$ is the classical Newton potential, $\int_0^\infty T_t\, dt$ should be considered as an abstract potential operator.

It is also worth to have a look when $p_{ks}(x,G) \geq \eta_0 > 0$ for infinitely many k. Clearly, in this case the series $\sum_{k=0}^\infty T_s^k\chi_G(x)$ will diverge and we should expect the same for the improper integral $\int_0^\infty T_t\chi_G(x)\,dt$. But $p_{ks}(x,G) \geq \eta_0 > 0$ for infinitely many k has the interpretation that the process associated with $(T_t)_{t\geq 0}$ will visit the set G infinitely often, such a phenomenon will be called *recurrence of the process.*

In volume III of this treatise we will be more precise with these interpretations and arguments. But now we are prepared and motivated to develop the analytic theory of transient and recurrent semigroups. For this we assume that the L^p-sub-Markovian semigroup $(T_t)_{t\geq 0}$ has a sub-Markovian dual semigroup. From Theorem 3.5.1 we know that in this case T_t, $t \geq 0$, extends to a sub-Markovian contraction $T_t : L^1(\mathbb{R}^n;\mathbb{R}) \to L^1(\mathbb{R}^n;\mathbb{R})$. Therefore the operators

$$S_t u := \int_0^t T_s u\, ds, \quad t > 0, \tag{3.355}$$

satisfy for $1 \leq p < \infty$ the estimates

$$\|S_t u\|_{L^p} \leq t\|u\|_{L^p}, \quad u \in L^p(\mathbb{R}^n;\mathbb{R}). \tag{3.356}$$

In addition S_t is positivity preserving and since

$$S_{t_1+t_2}u = \int_0^{t_1+t_2} T_s u\, ds = \int_0^{t_1} T_s u\, ds + \int_{t_1}^{t_2} T_s u\, ds,$$

the fact that T_s, $s \geq 0$, is positivity preserving yields for $1 \leq p < \infty$ and $u \in L^p(\mathbb{R}^n;\mathbb{R})$, $u \geq 0$ a.e.,

$$S_t u \leq S_{t'} u \quad \text{for } t \leq t'.$$

According to Remark 3.5.3.B the resolvent $R_\lambda = \int_0^\infty e^{-\lambda t} T_t\, dt$ is also well defined on $L^p(\mathbb{R}^n;\mathbb{R})$ for $1 \leq p < \infty$.

Definition 3.5.17. Let $(T_t)_{t\geq 0}$ be an L^p-sub-Markovian semigroup, $1 < p < \infty$, with dual semigroup being sub-Markovian. For $u \in L^1(\mathbb{R}^n;\mathbb{R})$, $u \geq 0$ a.e., we define the *potential operator* associated with $(T_t)_{t\geq 0}$ by

$$Gu(x) = \lim_{N\to\infty} S_N u(x) = \sup_{N\in\mathbb{N}} S_N u(x). \tag{3.357}$$

Remark 3.5.18. A. Note that the limit in (3.357) exists but might be infinite, i.e. in general we only have $Gu(x) \leq +\infty$ a.e.

B. Since

$$\begin{aligned}\sup_{N\in\mathbb{N}} S_N u(x) &= \sup_{N\in\mathbb{N}} \sup_{\lambda>0} \int_0^N e^{-\lambda t} T_t u\, dt \\ &= \sup_{\lambda>0} \sup_{N\in\mathbb{N}} \int_0^N e^{-\lambda t} T_t u\, dt = \sup_{\lambda>0} R_\lambda u,\end{aligned}$$

we have

$$\lim_{N\to\infty} S_N u(x) = \lim_{\lambda\to 0} R_\lambda u(x) \quad \text{a.e.}$$

C. Often G is also called the *Green operator* associated with $(T_t)_{t\geq 0}$.

Definition 3.5.19. Let $(T_t)_{t\geq 0}$ be an L^p-sub-Markovian semigroup, $1 < p < \infty$, with dual semigroup being sub-Markovian too.

A. We call $(T_t)_{t\geq 0}$ to be *transient* if $Gu(x) < \infty$ a.e. for all $u \in L^1(\mathbb{R}^n;\mathbb{R})$, $u \geq 0$ a.e.

B. The semigroup $(T_t)_{t\geq 0}$ is called *recurrent* if for all $u \in L^1(\mathbb{R}^n;\mathbb{R})$, $u \geq 0$ a.e., $Gu(x) \in \{0, \infty\}$ for almost all $x \in \mathbb{R}^n$.

Remark 3.5.20. A. Note that for a general semigroup $(T_t)_{t\geq 0}$ as in definition 3.5.19 transience–recurrence does not form a dichotomy, see however Example 3.5.23 where translation invariant sub-Markovian semigroups are discussed.

B. A semigroup as in Definition 3.5.19 extends by Theorem 3.5.1 to an L^q-sub-Markovian semigroup for all $1 < q < \infty$. The definition of transience and recurrence however depends on the L^1 extension, hence if $(T_t)_{t\geq 0}$ is transient (or recurrent) for some $1 < p < \infty$ it is transient (or recurrent) for all $1 \leq p < \infty$.

To proceed further we need *Hopf's maximal ergodic inequality* the proof of which we have adapted from M.Fukushima et.al. [102].

Proposition 3.5.21. *Let $(T_t)_{t\geq 0}$ be an L^p-sub-Markovian semigroup with dual semigroup being sub-Markovian too and let S_t, $t \geq 0$, be defined by* (3.355). *For $u \in L^1(\mathbb{R}^n;\mathbb{R})$ and $h > 0$ define $E_h := \{x \in \mathbb{R}^n; \sup_{k\in\mathbb{N}} S_{kh}u(x) > 0\}$. Then we have*

$$\int_{E_h} S_h u(x)\,dx \geq 0. \tag{3.358}$$

Proof: We define

$$E_h^k := \{x \in \mathbb{R}^n;\ \max_{1\leq\nu\leq k} S_{\nu h}u(x) > 0\} = \{x \in \mathbb{R}^n;\ \max_{1\leq\nu\leq k} (S_{\nu h}u)^+(x) > 0\}$$

where as usual f^+ denotes the positive part $f \vee 0$ of f. For $x \in E_h^k$ and $\nu = 1, \dots, k$ we have

$$\begin{aligned} S_{\nu h}u - S_h u = T_h S_{(\nu-1)h}u &\leq (T_h S_{(\nu-1)h}u)^+ \\ &\leq \max_{1\leq\nu\leq k} (T_h S_{\nu h})^+ = \max_{1\leq\nu\leq k} (S_{(\nu+1)h}u - S_h u)^+, \end{aligned}$$

hence for all $\nu = 1, \dots, k$ we get

$$S_{\nu h}u \leq \max_{1\leq\nu\leq k} (S_{(\nu+1)h}u - S_h u)^+ + S_h u,$$

or

$$\max_{1\leq\nu\leq k} S_{\nu h}u \leq \max_{1\leq\nu\leq k} (S_{(\nu+1)h}u - S_h u)^+ + S_h u,$$

implying

$$S_h u(x) + \max_{1\leq\nu\leq k} (S_{(\nu+1)h}u - S_h u)^+(x) \geq \max_{1\leq\nu\leq k} (S_{\nu h}u)^+(x).$$

Since T_t is positivity preserving we get further

$$\begin{aligned}
&\max_{1\leq\nu\leq k} (S_{(\nu+1)}u - S_h u)^+(x) \\
&\quad = \max_{1\leq\nu\leq k} \Big(\int_0^{(\nu+1)h} T_s u(x)\, ds - \int_0^h T_s u(x)\, ds \Big)^+ \\
&\quad \leq T_h(\max_{1\leq\nu\leq k} (S_{\nu h}u)^+)(x),
\end{aligned}$$

where we used calculus results for semigroups.

Hence

$$\begin{aligned}
\int_{E_h^k} S_h u(x)\, dx &\geq \int_{E_h^k} \Big(\max_{1\leq\nu\leq k} (S_{\nu h}u)^+(x) - T_h(\max_{1\leq\nu\leq k} (S_{\nu h}u)^+)(x) \Big)\, dx \\
&\geq \| \max_{1\leq\nu\leq k} (S_{\nu h}u)^+ \|_{L^1} - \| T_h(\max_{1\leq\nu\leq k} (S_{\nu h}u)^+) \|_{L^1} \geq 0,
\end{aligned}$$

where we used that T_h is an L^1-contraction. Now we let k tend to infinity and arrive at (3.358). □

Lemma 3.5.22. *Let $(T_t)_{t\geq 0}$ be an L^p-sub-Markovian semigroup with dual semigroup being sub-Markovian too. The semigroup $(T_t)_{t\geq 0}$ is transient if and only if there exists $g \in L^1(\mathbb{R}^n;\mathbb{R})$ such that $g > 0$ a.e. and $Gg(x) < \infty$. a.e.*

Proof: Let $g \in L^1(\mathbb{R}^n;\mathbb{R})$ such that $Gg(x) < \infty$ a.e. and $g > 0$ a.e. If $u \in L^1(\mathbb{R}^n;\mathbb{R})$, $u \geq 0$ a.e., and $a > 0$, $h > 0$ we consider the sets

$$A := \{x \in \mathbb{R}^n; \sup_{k\in\mathbb{N}} S_{kh}(u - ag)(x) > 0\} \quad \text{and } B = \{x \in \mathbb{R}^n; Gu(x) = \infty\}.$$

With a set C of measure zero we clearly have $B \subset A \cup C$. By Proposition 3.5.21 it follows that

$$\int_A S_h(u - ag)(x)\, dx \geq 0,$$

and therefore

$$h \int_{\mathbb{R}^n} u\, dx \geq \int_{\mathbb{R}^n} S_h u\, dx \geq a \int_A S_h g\, dx \geq a \int_B S_h g\, dx.$$

Thus for all $N \in \mathbb{N}$ and any compact Borel set $K_\nu \subset \mathbb{R}^n$ we find

$$\frac{1}{a} \int_{\mathbb{R}^n} u\, dx \geq \frac{1}{h} \int_{B\cap K_\nu} S_h(g \wedge N)\, dx.$$

Now $\frac{1}{h}S_h(g \wedge N)$ converge in $L^p(\mathbb{R}^n;\mathbb{R})$ (or any $L^q(\mathbb{R}^n;\mathbb{R})$, $1 < q < \infty$) to $g \wedge N$ as $h \downarrow 0$ and this implies

$$\frac{1}{a}\int_{\mathbb{R}^n} u\,dx \geq \int_{B\cap K_\nu} (g \wedge N)\,dx.$$

Finally, if $K_\nu \uparrow \mathbb{R}^n$, $N \uparrow \infty$ and then $a \uparrow \infty$, we arrive at

$$\int_B g(x)\,dx = 0$$

implying $\lambda^{(n)}(B) = 0$, i.e. $Gu(x) < \infty$ a.e. The converse is trivial. □

Example 3.5.23. Let $(\mu_t)_{t\geq 0}$ be a convolution semigroup on $\mathbb{R}^n$ with a corresponding continuous negative definite function $\psi : \mathbb{R}^n \to \mathbb{R}$. Hence there corresponds a symmetric L^2-sub-Markovian semigroup to $(\mu_t)_{t\geq 0}$ given by $T_t u = \mu_t * u$. Further let us assume that $\psi(\xi) > 0$ for $\xi \neq 0$ and that $\psi(\xi) \geq c_0$ in $B_1^c(0)$. We claim that in this situation $(T_t)_{t\geq 0}$ is transient if $\int_{B_1(0)} \frac{1}{\psi(\xi)}\,d\xi < \infty$. In this case we find for the resolvent of $(T_t)_{t\geq 0}$

$$R_\lambda u = (2\pi)^{-n/2}\int_{\mathbb{R}^n} e^{ix\cdot\xi}\frac{1}{\lambda+\psi(\xi)}\hat{u}(\xi)\,d\xi.$$

By Lemma I.3.1.4 we know that $g(x) = e^{-|x|^2/2}$ is a fixed point of the Fourier transform and we need to prove

$$\lim_{\lambda\to 0}(2\pi)^{-n/2}\int_{\mathbb{R}^n} e^{ix\cdot\xi}\frac{1}{\lambda+\psi(\xi)}e^{-|\xi|^2/2}\,d\xi < \infty$$

if $\int_{B_1(0)} \frac{1}{\psi(\xi)}\,d\xi < \infty$. Since

$$\int_{\mathbb{R}^n} e^{ix\cdot\xi}\frac{1}{\lambda+\psi(\xi)}e^{-|\xi|^2/2}\,d\xi = \int_{\mathbb{R}^n\setminus B_1(0)} e^{ix\cdot\xi}\frac{1}{\lambda+\psi(\xi)}e^{-|\xi|^2/2}\,d\xi + \int_{B_1(0)} e^{ix\cdot\xi}\frac{1}{\lambda+\psi(\xi)}e^{-|\xi|^2/2}\,d\xi$$

we may examine each term separately. Now for $\lambda \to 0$

$$\chi_{\mathbb{R}^n\setminus B_1(0)}(\xi)e^{ix\cdot\xi}\frac{1}{\lambda+\psi(\xi)}e^{-|\xi|^2/2} \to \chi_{\mathbb{R}^n\setminus B_1(0)}(\xi)e^{ix\cdot\xi}\frac{1}{\psi(\xi)}e^{-|\xi|^2/2}$$

and

$$\left|\chi_{\mathbb{R}^n\setminus B_1(0)}(\xi)e^{ix\cdot\xi}\frac{1}{\lambda+\psi(\xi)}e^{-|\xi|^2/2}\right| \leq \frac{1}{c_0}e^{-|\xi|^2/2},$$

and further

$$\chi_{B_1(0)}(\xi)e^{ix\cdot\xi}\frac{1}{\lambda+\psi(\xi)}e^{-|\xi|^2/2} \to \chi_{B_1(0)}(\xi)e^{ix\cdot\xi}\frac{1}{\psi(\xi)}e^{-|\xi|^2/2}$$

almost everywhere on $B_1(0)$ as $\lambda \to 0$ and in addition

$$\left|\chi_{B_1(0)}(\xi)e^{ix\cdot\xi}\frac{1}{\lambda+\psi(\xi)}e^{-|\xi|^2/2}\right| \leq \chi_{B_1(0)}(\xi)\frac{1}{\psi(\xi)}.$$

Thus Lebesgue's theorem of dominated convergence implies the result.

Note that a much deeper result is valid: A convolution semigroup $(\mu_t)_{t\geq 0}$ with associated continuous negative definite function $\psi : \mathbb{R}^n \to \mathbb{C}$ give rise to a transient sub-Markovian semigroup if and only if $\operatorname{Re}\frac{1}{\psi} \in L^1_{loc}(\mathbb{R}^n)$, compare [27] or [244].

Let $(T_t)_{t\geq 0}$ be as in Lemma 3.5.22 and suppose that $(T_t)_{t\geq 0}$ is transient. It follows that $\{Gg(x) = \infty\}$, $g \in L^1(\mathbb{R}^n;\mathbb{R})$ and $g(x) \geq 0$ a.e., has measure zero. Hence it is an invariant set with respect to $(T_t)_{t\geq 0}$, and the same holds true for $\{Gg(x) < \infty\} = \{Gg(x) = \infty\}^c$ since $\lambda^{(n)}\{Gg(x) = \infty\} = 0$. Our aim is to determine further invariant sets and for this we need some preparation.

Let $(T_t)_{t\geq 0}$ be an L^p-sub-Markovian semigroup such that its dual semigroup $(T_t^*)_{t\geq 0}$ is sub-Markovian too. As before we may associate with $(T_t)_{t\geq 0}$ its potential operator G, but of course we may also associate with $(T_t^*)_{t\geq 0}$ the corresponding potential operator which we will denote by $\tilde{G}$, i.e.

$$\tilde{G}g(x) := \lim_{N\to\infty}\int_0^N T_t^* g(x)\,dt, \quad g \in L^1(\mathbb{R}^n;\mathbb{R}), \quad g \geq 0 \quad \text{a.e.} \tag{3.359}$$

Clearly $\tilde{G}$ is not the adjoint or conjugate operator to G. However for $u, v \in L^1(\mathbb{R}^n;\mathbb{R}) \cap L^\infty(\mathbb{R}^n;\mathbb{R})$, $u, v \geq 0$ a.e. we have

$$\begin{aligned}\int_{\mathbb{R}^n}(Gu)(x)v(x)\,dx &= \lim_{N\to\infty}\int_0^N\int_{\mathbb{R}^n}(T_t u)(x)v(x)\,dx\,dt \\ &= \lim_{N\to\infty}\int_0^N\int_{\mathbb{R}^n}u(x)T_t^* v(x)\,dx\,dt = \int_{\mathbb{R}^n}u(x)(\tilde{G}v)(x)\,dx,\end{aligned}$$

i.e.

$$\int_{\mathbb{R}^n} (Gu)(x)v(x)\,dx = \int_{\mathbb{R}^n} u(x)\tilde{G}v(x)\,dx. \tag{3.360}$$

Moreover, if $(T_t)_{t\geq 0}$ is transient, then $(T_t^*)_{t\geq 0}$ is transient too. For this note that $Gu(x) < +\infty$ a.e. for all $u \in L^1(\mathbb{R}^n;\mathbb{R})$, $u \geq 0$ a.e., implies by Lemma 3.5.22 the existence of $g \in L^1(\mathbb{R}^n;\mathbb{R})$, $g > 0$ a.e., such that $\|Gg\|_{L^\infty} \leq M$. (Note added in the proofs: We follow in these considerations the lecture notes [226] of Y.Oshima, where the existence of g is taken for granted. R.Schilling quite recently pointed out that this might not be the case and provided a proof in a forthcoming paper dealing with global properties of L^p-sup-Markovian semigroups.) Therefore we can find $h \in L^1(\mathbb{R}^n;\mathbb{R})$, $h > 0$ a.e., such that

$$\int_{\mathbb{R}^n} (Gg)(x)h(x)\,dx = \int_{\mathbb{R}^n} g(x)\tilde{G}h(x)\,dx < \infty$$

implying $\tilde{G}h(x) < \infty$ a.e. and again Lemma 3.5.22 yields the transience of $\tilde{G}$.

Lemma 3.5.24. *Let $(T_t)_{t\geq 0}$ be an L^p-sub-Markovian semigroup such that $(T_t^*)_{t\geq 0}$ is sub-Markovian too. For any $g \in L^1(\mathbb{R}^n;\mathbb{R})$, $g \geq 0$ a.e., the sets $\{x \in \mathbb{R}^n; Gg(x) = \infty\}$, $\{x \in \mathbb{R}^n; \tilde{G}g(x) < \infty\}$, $\{x \in \mathbb{R}^n; Gg(x) > 0\}$ and $\{x \in \mathbb{R}^n; \tilde{G}g(x) = 0\}$ are invariant with respect to $(T_t)_{t\geq 0}$.*

Proof: For $g \in L^1(\mathbb{R}^n;\mathbb{R})$, $g \geq 0$ a.e., and $B_k := \{x \in \mathbb{R}^n; \tilde{G}g(x) \leq k\}$ we find for any $u \in L^1(\mathbb{R}^n;\mathbb{R})$, $u \geq 0$ a.e.,

$$\int_{\mathbb{R}^n} T_t(\chi_{B_k}u)\tilde{G}g\,dx = \int_{\mathbb{R}^n} \chi_{B_k}uT_t^*\tilde{G}g\,dx \leq \int_{\mathbb{R}^n} \chi_{B_k}u\tilde{G}g\,dx \leq k\int_{\mathbb{R}^n} u\,dx \tag{3.361}$$

where we used that for $g \geq 0$

$$T_t^*\tilde{G}g = T_t^*\int_0^\infty T_s^*g\,ds = \int_t^\infty T_r^*g\,dr \leq \int_0^\infty T_r^*g\,dr = \tilde{G}g.$$

From (3.361) it follows that $\int_{\mathbb{R}^n} T_t(\chi_{B_k}u)\tilde{G}g\,dx$ is finite, which implies that $T_t\chi_{B_k}u = 0$ on $\{x \in \mathbb{R}^n; \tilde{G}g(x) = \infty\}$, or $\chi_{\{\tilde{G}g<\infty\}}T_t\chi_{B_k}u = T_t\chi_{B_k}u$. For k tending to infinity it follows that $\{x \in \mathbb{R}^n; \tilde{G}g(x) < \infty\}$ is $(T_t)_{t\geq 0}$-invariant. By the same type of argument we see that $\{x \in \mathbb{R}^n; Gg(x) < \infty\}$ is $(T_t^*)_{t\geq 0}$-invariant which yields that its complement is $(T_t)_{t\geq 0}$-invariant, i.e.

$\{x \in \mathbb{R}^n; Gg(x) = \infty\}$ is $(T_t)_{t\geq 0}$-invariant. Next observe that

$$\int_{\mathbb{R}^n} T_t(\chi_{\{\tilde{G}g=0\}} u)\tilde{G}g\,dx = \int_{\mathbb{R}^n} \chi_{\{\tilde{G}g=0\}} u\, T_t^* \tilde{G}g\,dx \leq \int_{\mathbb{R}^n} \chi_{\{\tilde{G}g=0\}} u\, \tilde{G}g\,dx = 0.$$

Thus, $T_t\chi_{\{\tilde{G}g=0\}}u = 0$ on $\{\tilde{G}g > 0\}$ or $T_t\chi_{\{\tilde{G}g=0\}}u = \chi_{\{\tilde{G}g=0\}}T_t\chi_{\{\tilde{G}g=0\}}u$ which is the invariance of $\{x \in \mathbb{R}^n; \tilde{G}g(x) = 0\}$ with respect to $(T_t)_{t\geq 0}$. As before we conclude that $\{x \in \mathbb{R}^n; Gg(x) = 0\}$ is invariant with respect to $(T_t^*)_{t\geq 0}$. Thus its complement $\{x \in \mathbb{R}^n; Gg(x) > 0\}$ is $(T_t)_{t\geq 0}$. □

Corollary 3.5.25. *A. If $(T_t)_{t\geq 0}$ is as in Lemma 3.5.24 but in addition symmetric, then for $g \in L^1(\mathbb{R}^n;\mathbb{R})$, $g \geq 0$ a.e., the sets $\{x \in \mathbb{R}^n; Gg(x) < \infty\}$, $\{x \in \mathbb{R}^n; Gg(x) = \infty\}$, $\{x \in \mathbb{R}^n; Gg(x) > 0\}$ and $\{x \in \mathbb{R}^n; Gg(x) = 0\}$ are invariant with respect to $(T_t)_{t\geq 0}$.*

B. If $(T_t)_{t\geq 0}$ is as in Lemma 3.5.24 but in addition transient, then for every $g \in L^1(\mathbb{R}^n;\mathbb{R})$, $g \geq 0$ a.e., the sets $\{x \in \mathbb{R}^n; Gg(x) = \infty\}$ and $\{x \in \mathbb{R}^n; Gg(x) < \infty\}$ are invariant. In addition up to a set of measure zero for all $g \in L^1(\mathbb{R}^n;\mathbb{R})$, $g > 0$ a.e., these sets are independent of g.

Proof: A. In this case we have obviously $G = \tilde{G}$ and the assertion follows from Lemma 3.5.24.

B. There is some $h > 0$ a.e. in $L^1(\mathbb{R}^n;\mathbb{R})$ such that $Gh(x) < \infty$ a.e., i.e. $\lambda^{(n)}(\{x \in \mathbb{R}^n; Gh(x) < \infty\}^c) = 0$ and from Lemma 3.5.22 it follows that $\lambda^{(n)}(\{x \in \mathbb{R}^n; Gg(x) < \infty\}^c) = 0$ for all $g \in L^1(\mathbb{R}^n;\mathbb{R})$, $g \geq 0$ a.e. Hence for all $g \in L^1(\mathbb{R}^n;\mathbb{R})$, $g \geq 0$ a.e., the set $\{x \in \mathbb{R}^n; Gg(x) < \infty\}$ is invariant according to the remark following Definition 3.5.13. The remaining statements are trivial. □

Let $(T_t)_{t\geq 0}$ be an L^p-sub-Markovian semigroup with dual semigroup being sub-Markovian too, and let $A \subset \mathbb{R}^n$ be a $(T_t)_{t\geq 0}$-invariant set. Since $L^p(A) = \chi_A \cdot L^p(\mathbb{R}^n)$ and for $u \in L^p(A)$ we may consider $\chi_A u$ as an extension of u to $\mathbb{R}^n$, we find for $t \geq 0$ that $T_t|_{L^p(A)}u = T_t\chi_A u = \chi_A T_t \chi_A u \in L^p(A)$, i.e. $T_t\chi_A u \in L^p(A)$. Therefore the family $(T_t|_{L^p(A)})_{t\geq 0}$ form an L^p-sub-Markovian semigroup on $L^p(A)$. Further, for $u \in L^p(\mathbb{R}^n;\mathbb{R})$ and $v \in L^{p'}(\mathbb{R}^n;\mathbb{R})$ we have

$$\begin{aligned}\int_A (T_t|_{L^p(A)}u)\chi_A v\,dx &= \int_{\mathbb{R}^n} (\chi_A T_t \chi_A u)\chi_A v\,dv \\ &= \int_{\mathbb{R}^n} (\chi_A u) T_t^* \chi_A v\,dx = \int_{\mathbb{R}^n} \chi_A u T_t^*|_{L^{p'}(A)}(v)\,dx \\ &= \int_A \chi_A u T_t^*|_{L^{p'}(A)}(v)\,dx\end{aligned}$$

i.e. $(T_t|_{L^p(A)})^* = T_t^*|_{L^{p'}(A)}$. Moreover, for $h \in L^1(A;\mathbb{R})$, $h \geq 0$, it follows that $G\chi_A h = \int_0^\infty T_t\chi_A h\,dt = \int_0^\infty T_t|_{L^1(A)}h\,dt = G|_{L^1(A)}h$, i.e. G restricted to the non-negative functions of $L^1(A;\mathbb{R})$ is the potential operator associated with $(T_t|_{L^p(A)})_{t\geq 0}$.

Let $g \in L^1(\mathbb{R}^n;\mathbb{R})$, $g > 0$ a.e., and set $A := \{x \in \mathbb{R}^n; Gg < \infty\}$. From Lemma 3.5.22 it follows that $(T_t|_{L^p(A)})_{t\geq 0}$ is transient implying that $\{x \in \mathbb{R}^n; Gg(x) < \infty\}$ is invariant with respect to $(T_t|_{L^p(A)})_{t\geq 0}$. Moreover, if g_1, g_2 are two strictly positive functions in $L^1(\mathbb{R}^n;\mathbb{R})$ and $A_j := \{x \in \mathbb{R}^n; Gg_j < \infty\}$ then $A_1 \cup A_2$ is invariant too, and $G(g_1 \wedge g_2) < \infty$ on $A_1 \cup A_2$. Thus $(T_t|_{L^p(A_1\cup A_2)})_{t\geq 0}$ is transient implying $Gg_j < \infty$ on $A_1 \cup A_2$ which yields that $A_1 = A_2$. Thus we have proved a Hopf-type decomposition

Lemma 3.5.26. *Let $(T_t)_{t\geq 0}$ be an L^p-sub-Markovian semigroup such that its dual semigroup is sub-Markovian too. Then for every $g \in L^1(\mathbb{R}^n;\mathbb{R})$, $g > 0$ a.e., the set $\{x \in \mathbb{R}^n; Gg(x) < \infty\}$ is invariant and up to a set of measure zero independent of g.*

For an L^p-sub-Markovian semigroup $(T_t)_{t\geq 0}$ with $(T_t^*)_{t\geq 0}$ being sub-Markovian too and for $g \in L^1(\mathbb{R}^n;\mathbb{R})$, $g > 0$ a.e., we decompose

$$\mathbb{R}^n = X_d \cup X_c$$

where $X_d := \{x \in \mathbb{R}^n; Gg(x) < \infty\}$ and $X_c := \{x \in \mathbb{R}^n; Gg(x) = \infty\}$ and call X_d the *dissipative part* and X_c the *conservative part* of $\mathbb{R}^n$ with respect to $(T_t)_{t\geq 0}$. Up to a set of measure zero these two sets are uniquely determined.

Lemma 3.5.27. *Let $f \in L^1(\mathbb{R}^n;\mathbb{R})$, $f \geq 0$ a.e., and set $B := \{x \in X_c; Gf(x) < \infty\}$. Then $f = 0$ and $Gf = 0$ a.e. on B.*

Proof: Let $g \in L^1(\mathbb{R}^n;\mathbb{R})$ be a strictly positive function and $X_c = \{x \in \mathbb{R}^n; Gg(x) = \infty\}$. It follows that B is contained (up to a set of measure zero) in $\{x \in \mathbb{R}^n; \sup\limits_{k\in\mathbb{N}} S_{kh}(g - af)(x) > 0\}$ where S_{kh} is as in Definition 3.5.17 and $h > 0$, $a > 0$ are arbitrary. The Hopf maximal ergodic inequality, Proposition 3.5.21, yields

$$\frac{1}{h}\int_B S_h f\,dx \leq \frac{1}{a}\int_{\mathbb{R}^n} g\,dx$$

implying first $\int_B S_h f\,dx = 0$ and therefore $\int_B Gf\,dx = 0$, hence $Gf = 0$ a.e. on B. An argument similar to that in the proof of Lemma 3.5.22 yields also $f = 0$ on B. $\square$

Proposition 3.5.28. *Let $(T_t)_{t\geq 0}$ be an L^p-sub-Markovian semigroup with dual semigroup being sub-Markovian too.*

A. The semigroup is transient if and only if $X_d = \mathbb{R}^n$ up to a set of measure zero.

B. The semigroup $(T_t)_{t\geq 0}$ is recurrent if and only if $X_c = \mathbb{R}^n$ up to a set of measure zero.

C. If $(T_t)_{t\geq 0}$ is irreducible then we have the dichotomy: $(T_t)_{t\geq 0}$ is either transient or recurrent. In the second case $Gf(x) = \infty$ a.e. for any non-negative measurable function f with $\lambda^{(n)}(x \in \mathbb{R}^n; f(x) > 0\}) > 0$.

Proof: Part A is trivial by Lemma 3.5.22. Suppose that $(T_t)_{t\geq 0}$ is recurent. As in the proof of Lemma 3.5.22 it follows that $f \in L^1(\mathbb{R}^n;\mathbb{R})$, $f \geq 0$, must vanish a.e. on $\{x \in \mathbb{R}^n;\ Gf(x) = 0\}$, thus $\lambda^{(n)}(\{x \in \mathbb{R}^n;\ Gf(x) = 0\}) = 0$ by the recurrence of $(T_t)_{t\geq 0}$ implying $\mathbb{R}^n = X_c$ up to a set of measure zero. Conversely, suppose that $\mathbb{R}^n = X_c$ up to a set of measure zero. Then Lemma 3.5.27 implies the recurrence of $(T_t)_{t\geq 0}$. Finally, if $(T_t)_{t\geq 0}$ is irreducible then $X_c = \mathbb{R}^n$ or $X_c = \varnothing$ up to sets of measure zero, i.e. $(T_t)_{t\geq 0}$ is either recurrent or transient. The last statement in part C follows as the first implication of part B. □

Note that it is possible to prove that if $(T_t)_{t\geq 0}$ is given by $T_t u = \mu_t * u$ with a convolution semigroup $(\mu_t)_{t\geq 0}$ then it is irreducible. Since we will not need this result later on we do not supply its proof, for this we refer to K.-I.Sato [244] for a proof.

We want to relate transience and recurrence to conservativity. As in the case of Feller semigroups, compare Section I.4.8, we may extend each operator T_t of an L^p-sub-Markovian semigroup by taking monotone limits to the space $L^\infty(\mathbb{R}^n;\mathbb{R})$ and this extension is independent of the approximating sequence. Indeed, if $(u_k)_{k\in\mathbb{N}}$, $u_k \in L^p(\mathbb{R}^n;\mathbb{R})$, $u_k \geq 0$, increases to $u \in L^\infty(\mathbb{R}^n;\mathbb{R})$, $u \geq 0$, we first may define $T_t u := \lim_{k\to\infty} T_t u_k$ and for two sequences $(u_k^{(j)})_{k\in\mathbb{N}}$, $u_k^{(j)} \geq 0$, each in $L^p(\mathbb{R}^n;\mathbb{R})$ and increasing to u we have

$$\begin{aligned}\lim_{k\to\infty} T_t u_k^{(1)} &= \lim_{k\to\infty}\lim_{l\to\infty} T_t(u_k^{(1)} \wedge u_l^{(2)}) \\ &= \lim_{l\to\infty}\lim_{k\to\infty} T_t(u_k^{(1)} \wedge u_l^{(2)}) = \lim_{l\to\infty} T_t u_l^{(2)} \quad \text{a.e.}\end{aligned}$$

As usual, for general $u \in L^\infty(\mathbb{R}^n;\mathbb{R})$ we define $T_t u = T_t u^+ - T_t u^-$.

We call an L^p-sub-Markovian semigroup $(T_t)_{t\geq 0}$ *conservative* if $T_t 1 = 1$ a.e. for all $t > 0$. (Whereas in Section I.4.8 it was important to distinguish the extended operators from the original ones, we do not introduce now a separate notation for the extension of T_t to $L^\infty(\mathbb{R}^n;\mathbb{R})$.)

Lemma 3.5.29. *Let $(T_t)_{t\geq 0}$ be an L^p-sub-Markovian semigroup with dual semigroup being sub-Markovian. Then we have $T_t 1 = 1$ a.e. on X_c^*, where X_c^* and X_d^* are defined with respect to $(T_t^*)_{t\geq 0}$. Further if $(T_t^*)_{t\geq 0}$ is recurrent then $(T_t)_{t\geq 0}$ is conservative. However, if $T_t 1 < 1$ a.e., then $(T_t^*)_{t\geq 0}$ is transient.*

Proof: We always have $T_t 1 \leq 1$ a.e. due to the sub-Markovian property. Thus for $f \in L^1(\mathbb{R}^n;\mathbb{R}) \cap L^\infty(\mathbb{R}^n;\mathbb{R})$, $f > 0$ a.e. it follows

$$\begin{aligned}\int_{\mathbb{R}^n} (S_N^* f)(1 - T_t 1)\, dx &= \int_{\mathbb{R}^n} \Big(\int_0^N T_s^* f\, ds\Big)(1 - T_t 1)\, dx \\ &= \int_{\mathbb{R}^n} f \int_0^N T_s(1 - T_t 1)\, ds\, dx \\ &= \int_{\mathbb{R}^n} f\Big(\int_0^t T_s 1\, ds - \int_N^{N+t} T_s 1\, ds\Big)\, dx \\ &\leq \int_{\mathbb{R}^n} f \int_0^t T_s 1\, ds\, dx \leq t \int_{\mathbb{R}^n} f\, dx.\end{aligned}$$

In the limit $N \to \infty$ we arrive at

$$\int_{\mathbb{R}^n} (\tilde{G} f)(1 - T_t 1)\, dx \leq t \int_{\mathbb{R}^n} f\, dx < \infty.$$

Since $\tilde{G} f(x) = \infty$ a.e. on X_c^* we find $T_t 1 = 1$ a.e. on X_c^*, so on $\mathbb{R}^n$ a.e. if $(T_t^*)_{t\geq 0}$ is recurrent which shows conservativeness. If $T_t 1 < 1$ a.e. for all $t > 0$ then $\lambda^{(n)}(X_c^*) = 0$, hence $X_d = \mathbb{R}^n$ up to a set of measure zero implying that $(T_t^*)_{t\geq 0}$ is transient. □

Example 3.5.30. Let $(T_t)_{t\geq 0}$ be a symmetric L^p-sub-Markovian semigroup. Define the new L^p-sub-Markovian semigroup $T_t^{(\lambda)} u := e^{-\lambda t} T_t y$, $\lambda > 0$. Since $T_t 1 \leq 1$ a.e. we find clearly for all $t > 0$ that $T_t^{(\lambda)} 1 = e^{-\lambda t} T_t 1 \leq e^{-\lambda t}$. Hence if $(A, D(A))$ generates a symmetric L^p-sub-Markovian semigroup, then $(A - \lambda, D(A))$, $\lambda > 0$, generates a transient semigroup.

Our next aim is to have comparison criteria for recurrent or transient semigroups. It turns out that for this purpose we have to introduce the notation of an *extended Dirichlet space*. First let us state an auxiliary result.

Lemma 3.5.31. *A. Let $(\mathcal{E}, D(\mathcal{E}))$ be a non-symmetric Dirichlet form in the sense of Definition I.4.7.21.B. Then $(\mathcal{E}^{sym}, D(\mathcal{E}))$ is a symmetric Dirichlet form.*

B. Let $(\mathcal{E}, D(\mathcal{E}))$ be a symmetric Dirichlet form. Then every normal contraction *operates on $(\mathcal{E}, D(\mathcal{E}))$, i.e. if $T : \mathbb{R} \to \mathbb{R}$ such that $T(0) = 0$ and $|T(s) - T(t)| \le |s - t|$ then $T \circ u \in D(\mathcal{E})$ for every $u \in D(\mathcal{E})$ and $\mathcal{E}(T \circ u, T \circ u) \le \mathcal{E}(u, u)$.*

Proof: A. The only non-trivial assertion to prove is the contraction property, i.e. $\mathcal{E}(u^+ \wedge 1, u^+ \wedge 1) \le \mathcal{E}(u, u)$, which, however, follows from adding the two inequalities in (I.4.354).

B. We prove the following statement: If $m \in \mathbb{N}$ and $T : \mathbb{R}^m \to \mathbb{R}$ satisfying

$$|T(x)| \le \sum_{k=1}^{m} |x_k| \quad \text{and} \quad |T(x) - T(y)| \le \sum_{k=1}^{m} |x_k - y_k|$$

for $x, y \in \mathbb{R}^n$, then for all $u_1, \ldots, u_m \in D(\mathcal{E})$ it follows that $T(u_1, \ldots, u_m) \in D(\mathcal{E})$ and

$$\mathcal{E}(T(u_1, \ldots, u_m), T(u_1, \ldots, u_m))^{1/2} \le \sum_{k=1}^{m} \mathcal{E}(u_k, u_k)^{1/2}.$$

Clearly this statement implies that every normal contraction operates on $(\mathcal{E}, D(\mathcal{E}))$ from which it follows that $(\mathcal{E}, D(\mathcal{E}))$ satisfies the contraction property. Thus it remains to show that the contraction property of $(\mathcal{E}, D(\mathcal{E}))$ implies the above statement. We denote the resolvent corresponding to $(\mathcal{E}, D(\mathcal{E}))$ by $(R_\lambda)_{\lambda > 0}$ and by Lemma I.4.7.18.B and D it is sufficient to prove that

$$\begin{aligned} &((\mathrm{id} - \lambda R_\lambda) T(u_1, \ldots, u_m), T(u_1, \ldots, u_m))_0^{1/2} \\ &\quad \le \sum_{k=1}^{m} ((\mathrm{id} - \lambda R_\lambda) u_k, u_k)_0^{1/2} \end{aligned} \tag{3.362}$$

for all $u_k \in L^2(\mathbb{R}^n; \mathbb{R})$ which may be reduced further to be proved only for all $u_k = \sum_{j=1}^{N_k} \tilde{\alpha}_{kj} \chi_{A_{kj}}$ with $N_k \in \mathbb{N}$, $\tilde{\alpha}_{kj} \in \mathbb{R}$ and for each k, $A_{kj_1} \cap A_{kj_2} = \varnothing$ for $j_1 \ne j_2$ where $\lambda^{(n)}(A_{kj}) < \infty$. In fact, changing the sets A_{kj}, $1 \le k \le m$, $1 \le j \le N_k$, we may even assume

$$u_k = \sum_{j=1}^{N} \alpha_{kj} \chi_{A_j}$$

with $\alpha_{kj} \in \mathbb{R}$, $A_j \cap A_i = \emptyset$ for $j \neq i$, and $\lambda^{(n)}(A_j) < \infty$. We set

$$b_{ij} := ((\mathrm{id} - \lambda R_\lambda)\chi_{A_i}, \chi_{A_j})_0$$

and (3.362) becomes

$$\begin{aligned}\Big(\sum_{i,j=1}^{N} T(\alpha_{1i}, \ldots, \alpha_{mi}) T(\alpha_{1j}, \ldots, \alpha_{mj}) b_{ij}\Big)^{1/2} \\ \leq \sum_{k=1}^{m} \Big(\sum_{i,j=1}^{N} \alpha_{ki}\alpha_{kj} b_{ij}\Big)^{1/2}.\end{aligned} \tag{3.363}$$

With $\beta_j = \int_{\mathbb{R}^n} \chi_{A_j}\, dx$ and $a_{ij} := (\lambda R_\lambda \chi_{A_i}, \chi_{A_j})_0$ we find that $b_{ij} = \beta_i \delta_{ij} - a_{ij}$, $1 \leq i, j \leq N$, and therefore by the symmetry of R_λ we have for all $x_1, \ldots, x_N \in \mathbb{R}$

$$\sum_{i,j=1}^{N} b_{ij} x_i x_j = \sum_{i<j} a_{ij}(x_i - x_j)^2 + \sum_{j=1}^{N} m_j x_j^2,$$

where $m_j = \beta_j - \sum_{i=1}^{N} a_{ij}$. Substituting these terms into (3.363) we arrive at

$$\begin{aligned}&\sum_{i<j} a_{ij}(T(\alpha_{1i}, \ldots, \alpha_{mi}) - T(\alpha_{1j}, \ldots, \alpha_{mj}))^2 \\ &\qquad + \sum_{j=1}^{m} \Big(\beta_j - \sum_{i=1}^{N} a_{ij}\Big)(T(\alpha_{1j}, \ldots, \alpha_{mj})^2 \\ &\qquad \leq \sum_{k=1}^{m} \Big(\sum_{i<j} a_{ij}(\alpha_{ki} - \alpha_{kj})^2 + \sum_{j=1}^{m} \Big(\beta_j - \sum_{i=1}^{N} a_{ij}\Big)\alpha_{kj}^2\Big),\end{aligned} \tag{3.364}$$

and taking into account the basic property of T, it follows that (3.363) holds if $a_{ij} \geq 0$ and $m_j \geq 0$. Since λR_λ is sub-Markovian we have indeed $a_{ij} \geq 0$. Further, for $A = \bigcup_{i=1}^{N} A_i$ it follows that

$$\begin{aligned}\sum_{i=1}^{N} a_{ij} &= \lambda \int_{\mathbb{R}^n} \chi_A R_\lambda \chi_{A_j}\, dx = \lambda \int_{\mathbb{R}} (R_\lambda \chi_A)\chi_{A_j}\, dx \\ &\leq \int_{\mathbb{R}^n} \chi_{A_j}\, dx = \beta_j,\end{aligned}$$

implying $m_j \geq 0$ and the lemma is proved. □

Remark 3.5.32. We borrowed the proof of part B from the monograph [204] of Z.-M.Ma and M.Röckner, see also the monograph [35] of N.Bouleau and F.Hirsch.

Let $(T_t)_{t\geq 0}$ be an L^p-sub-Markovian semigroup such that its dual semigroup is sub-Markovian too. It follows in particular that, on $L^p(\mathbb{R}^n;\mathbb{R}) \cap L^2(\mathbb{R}^n;\mathbb{R})$ it coincides with an L^2-sub-Markovian semigroup which we denote again by $(T_t)_{t\geq 0}$ and whose dual semigroup is also sub-Markovian. Considering L^p-sub-Markovian semigroups whose dual semigroup are again sub-Markovian seems to be the correct framework if one wants to work with non-symmetric Dirichlet forms, compare Definition I.4.7.21 and Theorem I.4.7.23. In particular for these Dirichlet forms the sector condition, see Definition I.4.7.12, does hold. For shortness we will call such a sub-Markovian semigroup a *Dirichlet semigroup.* We know that every Dirichlet semigroup is in fact a family of L^p-sub-Markovian semigroups $(T_t^{(p)})_{t\geq 0}$, $1 < p < \infty$, each having a dual semigroup being sub-Markovian too. Furthermore, these sub-Markovian semigroups are analytic.

Theorem 3.5.33. *Assume that $(T_t)_{t\geq 0}$ is a Dirichlet semigroup and denote by $(\mathcal{E}, D(\mathcal{E}))$ the associated non-symmetric Dirichlet form. If $(T_t)_{t\geq 0}$ is transient then there exists a function $g \in L^1(\mathbb{R}^n;\mathbb{R})$ which is a.e. strictly positive and bounded, and moreover it satisfies*

$$\int_{\mathbb{R}^n} |u| g \, dx \leq (\mathcal{E}(u,u))^{1/2} = (\mathcal{E}^{sym}(u,u))^{1/2} \tag{3.365}$$

for all $u \in D(\mathcal{E})$.

Proof: Let $f \in L^1(\mathbb{R}^n;\mathbb{R}) \cap C_\infty(\mathbb{R}^n;\mathbb{R})$ be strictly positive and $(K_k)_{k\in\mathbb{N}}$ a sequence of compact sets $K_k \subset \mathbb{R}^n$ increasing to $\mathbb{R}^n$. We define $A_k := \{x \in K_k;\, Gf(x) \leq k\}$ and without loss of generality we may assume that $\lambda^{(n)}(A_1) > 0$ and we set $c_k := \inf_{x\in A_k} f(x) > 0$. Since $(T_t)_{t\geq 0}$ is transient it follows that $\lambda^{(n)}((\bigcup_{k\in\mathbb{N}} A_k)^c) = 0$. Clearly, we have

$$\int_{A_l} G\chi_{A_k} \, dx \leq \frac{1}{c_k} \int_{A_l} Gf \, dx \leq \frac{l}{c_k} \lambda^{(n)}(A_l) < \infty.$$

The function

$$\tilde{g}(x) := \sum_{k=1}^{\infty} \frac{c_k}{2^k k \lambda^{(n)}(A_k)} \chi_{A_k}(x)$$

is clearly strictly positive a.e. and moreover $\tilde{g}(x) \leq \frac{c_1}{\lambda^{(n)}(A_1)}$ a.e. In addition we have

$$\int_{\mathbb{R}^n} \tilde{g}(x)\,dx = \sum_{k=1}^{\infty} \frac{c_k}{2^k k \lambda^{(n)}(A_k)} \int_{\mathbb{R}^n} \chi_{A_k}(x)\,dx \leq c_1,$$

and by the transience of $(T_t)_{t\geq 0}$ it follows that $G\tilde{g} < \infty$ a.e. Moreover

$$\begin{aligned}\int_{\mathbb{R}^n} \tilde{g} G \tilde{g}\,dx &= \sum_{l=1}^{\infty}\sum_{k=1}^{\infty} \frac{c_l}{2^l l \lambda^{(n)}(A_l)} \frac{c_k}{2^k k \lambda^{(n)}(A_k)} \int_{A_l} G\chi_{A_k}(x)\,dx \\ &\leq \sum_{l=1}^{\infty}\sum_{k=1}^{\infty} \frac{c_l}{2^l 2^k \lambda^{(n)}(A_k)} \leq \frac{c_1}{\lambda^{(n)}(A_1)} < \infty.\end{aligned}$$

Now we define

$$g_1 := \frac{\tilde{g}}{\left(\int_{\mathbb{R}^n} \tilde{g} G \tilde{g}\,dx\right)^{1/2}} \tag{3.366}$$

and find

$$\int_{\mathbb{R}^n} g_1 G g_1\,dx = \frac{1}{\int_{\mathbb{R}^n} \tilde{g} G \tilde{g}\,dx} \int_{\mathbb{R}^n} \tilde{g} G \tilde{g}\,dx = 1.$$

Therefore it follows using first the very properties of the resolvent compare Theorem I.4.7.10, and then the sector condition

$$\begin{aligned}(|u|, g_1)_0 &= \mathcal{E}_\lambda(R_\lambda g_1, |u|) = \mathcal{E}(R_\lambda g_1, |u|) + \lambda(R_\lambda g_1, |u|)_0 \\ &\leq c_0(\mathcal{E}^{sym}(R_\lambda g_1, R_\lambda g_1))^{1/2}(\mathcal{E}^{sym}(|u|, |u|))^{1/2} + (\lambda R_\lambda g_1, |u|)_0.\end{aligned}$$

Now, as $\lambda \to 0$ it follows that $R_\lambda g_1 \uparrow G g_1 < \infty$ implying that $(\lambda R_\lambda g_1, |u|)_0 \to 0$ as $\lambda \to 0$. Further we have

$$\begin{aligned}\mathcal{E}^{sym}(R_\lambda g_1, R_\lambda g_1) &= \mathcal{E}(R_\lambda g_1, R_\lambda g_1) = (R_\lambda g_1, g_1)_0 - \lambda(R_\lambda g_1, R_\lambda g_1)_0 \\ &\leq (R_\lambda g_1, g_1)_0 \leq (G g_1, g_1)_0 \leq 1,\end{aligned}$$

which yields using the fact that normal contractions operate on $\mathcal{E}^{sym}$

$$(|u|, g_1) \leq c_0 (G g_1, g_1)_0^{1/2} (\mathcal{E}^{sym}(|u|, |u|))^{1/2} \leq c_0(\mathcal{E}^{sym}(u,u))^{1/2}.$$

Thus the function $g := \frac{1}{c_0} g_1$ will satisfy

$$(|u|, g)_0 = (|u|, \frac{1}{c_0} g_1)_0 \leq \frac{1}{c_0}(|u|, g_1) \leq (\mathcal{E}^{sym}(u,u))^{1/2}. \quad \square$$

It is important to note that (3.365) already implies the transcience of $(T_t)_{t\geq 0}$, i.e. we have

Corollary 3.5.34. *An L^p-Dirichlet semigroup $(T_t)_{t\geq 0}$ is transient if and only if for the associated Dirichlet form* (3.365) *holds for some $g \in L^1(\mathbb{R}^n; \mathbb{R})$, $g > 0$ a.e., and all $u \in D(\mathcal{E})$.*

Proof: It remains to show that (3.365) implies the transience of $(T_t)_{t\geq 0}$. Suppose that $g \in L^1(\mathbb{R}^n; \mathbb{R}) \cap L^\infty(\mathbb{R}^n; \mathbb{R})$, $g > 0$ a.e., is such that (3.365) holds for all $u \in D(\mathcal{E})$. Since $u := R_\lambda g \in D(\mathcal{E})$, $u \geq 0$, we derive from (3.365)

$$\begin{aligned}\int_{\mathbb{R}^n} (R_\lambda g) g \, dx &\leq (\mathcal{E}(R_\lambda g, R_\lambda g))^{1/2} = (\mathcal{E}_\lambda(R_\lambda g, R_\lambda g) - \lambda(R_\lambda g, R_\lambda g))^{1/2} \\ &\leq (\mathcal{E}_\lambda(R_\lambda g, R_\lambda g))^{1/2} = (R_\lambda g, g)_0^{1/2},\end{aligned}$$

or

$$0 \leq (R_\lambda g, g)_0 \leq (R_\lambda g, g)_0^{1/2}$$

implying $(R_\lambda g, g)_0 \leq 1$, and as $\lambda \to 0$ it follows by monotone convergence that

$$\int_{\mathbb{R}^n} (Gg) g \, dx \leq 1.$$

Since $g > 0$ a.e. we arrive at $Gg < \infty$ a.e., i.e. $(T_t)_{t\geq 0}$ is transient. □

Remark 3.5.35. Note that in (3.365) only the symmetric part of $\mathcal{E}$ enters, i.e. if $(T_t)_{t\geq 0}$ and $(S_t)_{t\geq 0}$ are two Dirichlet semigroups and if the symmetric part of the corresponding Dirichlet forms coincide then (3.365) holds for both semigroups with one and the same function g provided one of them is transient. In particular in this situation $(T_t)_{t\geq 0}$ is transient if and only if $(S_t)_{t\geq 0}$ is transient.

Definition 3.5.36. A. A (non-symmetric) Dirichlet form is called *transient* if and only if the corresponding (L^2-)sub-Markovian semigroup is transient.

B. Let $(T_t)_{t\geq 0}$ be a Dirichlet semigroup and denote by $(\mathcal{E}, D(\mathcal{E}))$ the associated non-symmetric Dirichlet form. By $(T_t^{sym})_{t\geq 0}$ we denote the symmetric L^2-sub-Markovian semigroup associated with the symmetric Dirichlet form $(\mathcal{E}^{sym}, D(\mathcal{E}))$.

According to Remark 3.5.35 the transience of a Dirichlet semigroup depends only on $(T_t^{sym})_{t\geq 0}$. Therefore we will now have a closer look at symmetric L^2-sub-Markovian semigroups and their corresponding Dirichlet forms. More precisely we will have to discuss extended Dirichlet spaces. They arise from the observation that the symmetric part $(\mathcal{E}^{sym}, D(\mathcal{E}))$ of a non-symmetric Dirichlet form in general will not be a scalar product on $D(\mathcal{E})$.

Definition 3.5.37. Let $(\mathcal{E}, D(\mathcal{E}))$ be a symmetric Dirichlet space , $D(\mathcal{E}) \subset L^2(\mathbb{R}^n;\mathbb{R})$. The *extended Dirichlet space* $\mathcal{F}_e$ associated with $(\mathcal{E}, D(\mathcal{E}))$ is the family of all measurable functions $u : \mathbb{R}^n \to \mathbb{R}$, $|u| < \infty$ a.e., such that there is a sequence $(u_k)_{k\in\mathbb{N}}$, $u_k \in D(\mathcal{E})$, which converges a.e. to u and forms a Cauchy sequence with respect to $\mathcal{E}$, i.e. $\mathcal{E}(u_k - u_l, u_k - u_l) \to 0$ as $k, l \to \infty$. The sequence $(u_k)_{k\in\mathbb{N}}$ is called an *approximating sequence* for $u \in \mathcal{F}_e$.

Before we can study $\mathcal{F}_e$ we need some additional results on symmetric Dirichlet forms. For these results we will follow closely the monograph [102] of M.Fukushima, Y.Oshima and M.Takeda.

First we look for a different representation of the form $\mathcal{E}^{(\lambda)}(u,v) = \lambda(u - \lambda R_\lambda u, v)_0$ approximating $\mathcal{E}$.

Lemma 3.5.38. *A. If S is a symmetric sub-Markovian operator on $L^2(\mathbb{R}^n;\mathbb{R})$ then there exists a unique positive symmetric Radon measure σ on $\mathbb{R}^n \times \mathbb{R}^n$ such that for all $u, v \in L^2(\mathbb{R}^n;\mathbb{R})$*

$$(u, Sv)_0 = \iint_{\mathbb{R}^n\times\mathbb{R}^n} u(x)v(x)\sigma(dx, dy) \tag{3.367}$$

holds. In addition we have

$$\sigma(\mathbb{R}^n \times A) \le \lambda^{(n)}(A) \quad \textit{for all Borel sets } A. \tag{3.368}$$

B. For all $u \in L^2(\mathbb{R}^n;\mathbb{R})$ such that $u = v$ a.e. and v is a Borel function it holds

$$\begin{aligned}\mathcal{E}^{(\lambda)}(u,u) = &\frac{1}{2}\lambda \iint_{\mathbb{R}^n\times\mathbb{R}^n} (v(x) - v(y))^2 \sigma_\lambda(dx, dy) \\ &+ \lambda \int_{\mathbb{R}^n} v^2(x)(1 - s_\lambda(x))\, dx,\end{aligned} \tag{3.369}$$

where σ_λ is the measure on $\mathbb{R}^n \times \mathbb{R}^n$ associated with the operator λR_λ accordingly to part A and s_λ is the Radon–Nikodym density of $\sigma_\lambda(\mathbb{R}^n, dy)$ with respect to $\lambda^{(n)}(dy)$ satisfying $0 \le S_\lambda(x) \le 1$.

Proof: A. Since (3.367) clearly implies (3.368) we only have to prove (3.367). For this take $f(x,y) := \sum_{j=1}^N u_j(x)v_j(y)$ with $u_j, v_j \in C_0(\mathbb{R}^n;\mathbb{R})$ and set

$$I(f) := \sum_{j=1}^N (u_j, Sv_j)_0. \tag{3.370}$$

Let us consider I on the space

$$\tilde{C} := \{f \in C_0(\mathbb{R}^n \times \mathbb{R}^n; \mathbb{R});$$
$$f(x,y) = \sum_{j=1}^{N} u_j(x)v_j(y), N \in \mathbb{N} \quad \text{and } u_j, v_j \in C_0(\mathbb{R}^n)\}.$$

Obviously I is a linear form on $\tilde{C}$ and $I(f)$ is independent of the representation of f. Therefore if we can prove that I is a positive functional on $\tilde{C}$, we may extend it as a positive linear functional on $C_0(\mathbb{R}^n \times \mathbb{R}^n; \mathbb{R})$ and the result follows from a variant of F.Riesz' representation theorem, compare Theorem I.2.3.4. So suppose that $f(x,y) \geq 0$ and put $K := \bigcup_{j=1}^{N} \operatorname{supp} u_j$. On K each u_j is uniformly continuous and therefore there exist a finite decomposition $K = \bigcup_{k=1}^{M} E_k$ of K into Borel sets and points $\xi_k \in E_k$ such that with $\tilde{u}_j(x) := \sum_{k=1}^{M} u_j(\xi_k)\chi_{E_k}(x)$ we have $\sup_{x \in K} |u(x) - \tilde{u}_j(x)| < \varepsilon$. It follows that

$$|I(f) - \sum_{j=1}^{N} (\tilde{u}_j, Sv_j)_0| \leq \varepsilon \sum_{j=1}^{N} (\chi_K, |Sv_j|)_0. \tag{3.371}$$

In addition, with $f_{\xi_k}(y) := f(\xi_k, y)$ we have

$$\sum_{j=1}^{N} (\tilde{u}_j, Sv_j)_0 = \sum_{k=1}^{M} (\chi_{E_k}, Sf_{\xi_k})_0 \geq 0,$$

which combined with (3.371) yields $I(f) \geq 0$ and part A is proved. B. Since λR_λ is sub-Markovian there exists a symmetric positive Radon measure σ_λ such that

$$\lambda(u, R_\lambda v)_0 = \lambda \iint_{\mathbb{R}^n \times \mathbb{R}^n} u(x)v(y)\sigma_\lambda(dx, dy). \tag{3.372}$$

Now, for $A \in \mathcal{B}^{(n)}$ we find

$$\sigma_\lambda(\mathbb{R}^n, A) = \iint_{\mathbb{R}^n \times \mathbb{R}^n} \chi_{\mathbb{R}^n}(x)\chi_A(y)\sigma_\lambda(dx, dy)$$

hence $\sigma_\lambda(\mathbb{R}^n, A) = (\chi_A, S1)_0 \leq \lambda^{(n)}(A)$ by the sub-Markov property of S, i.e. $\sigma_\lambda(\mathbb{R}^n, dy)$ is $\lambda^{(n)}(dy)$ continuous implying by the Radon–Nikodym theorem that $\sigma_\lambda(\mathbb{R}^n, dy) = s_\lambda(y)\lambda^{(n)}(dy)$ with a suitable measurable function s_λ which has to satisfy $0 \leq s_\lambda \leq 1$. Therefore we find for $u \in L^2(\mathbb{R}^n;\mathbb{R})$

$$\begin{aligned}\mathcal{E}^{(\lambda)}(u,u) &= \lambda(u - R_\lambda u, u)_0 \\ &= \frac{1}{2}\lambda \iint_{\mathbb{R}^n\times\mathbb{R}^n} (u(x) - u(y)^2 \sigma_\lambda(dx,dy) + \lambda \int_{\mathbb{R}^n} u^2(x)(1 - s_\lambda(x))\,dx,\end{aligned}$$

which implies of course (3.369). □

Further we have

Lemma 3.5.39. *Let $(T_t)_{t\geq 0}$ be a symmetric L^2-sub-Markovian semigroup with corresponding symmetric Dirichlet form $(\mathcal{E}, D(\mathcal{E}))$. For every $g \in L^1(\mathbb{R}^n;\mathbb{R}) \cap L^2(\mathbb{R}^n;\mathbb{R})$, $g \geq 0$ a.e., the equality*

$$\sup_{u\in D(\mathcal{E})} \frac{(|u|,g)_0}{\mathcal{E}(u,u)^{1/2}} = \Big(\int_{\mathbb{R}^n} gGg\,dx\Big)^{1/2} \leq +\infty \tag{3.373}$$

holds where G is the potential operator associated with $(T_t)_{t\geq 0}$.

Proof: If $\sup \frac{(|u|,g)_0}{\mathcal{E}(u,u)^{1/2}} = \infty$, then nothing is to prove by (3.365). As before we denote by S_t the operator $S_t u = \int_0^t T_s u\,ds$. According to Lemma I.4.1.14.A for all $u \in L^2(\mathbb{R}^n;\mathbb{R})$ we have $S_t u \in D(A)$ and since $D(A) \subset D((-A)^{1/2}) = D(\mathcal{E})$ it follows that $S_t u \in D(\mathcal{E})$ for all $u \in L^2(\mathbb{R}^n;\mathbb{R})$. In addition, for $u \in D(A)$ and $v \in D(\mathcal{E})$ it follows that

$$\lim_{t\to 0} \frac{1}{t}(u - T_t u, v)_0 = (-Au, v)_0 = \mathcal{E}(u,v).$$

For $g \in L^1(\mathbb{R}^n;\mathbb{R}) \cap L^2(\mathbb{R}^n;\mathbb{R})$, $g \geq 0$ a.e., we find

$$S_t g - T_s S_t g = -\int_t^{t+s} T_r g\,dr + \int_0^s T_r g\,dr$$

which yields for all $u \in D(\mathcal{E})$

$$\begin{aligned}\lim_{s\to 0} \frac{1}{s}(S_t g - T_s T_t g, u)_0 &= \lim_{s\to 0}\Big(-\frac{1}{s}\int_t^{t+s} T_r g\,dr + \frac{1}{s}\int_0^s T_r g\,dr\Big)_0 \\ &= (g - T_t g, u)_0,\end{aligned}$$

but also

$$\lim_{s\to 0} \frac{1}{s}(S_t g - T_s S_t g, u)_0 = \mathcal{E}(S_t g, u)$$

implying

$$\mathcal{E}(S_t g, u) = (g - T_t g, u)_0. \tag{3.374}$$

Now assume that $\sup_{u \in D(\mathcal{E})} \frac{(|u|, g)_0}{\mathcal{E}(u,u)^{1/2}} = c < \infty$. Taking $u = S_t g \geq 0$ and observing that by (3.374) we have $\mathcal{E}(S_t g, S_t g) \leq (S_t g, g)_0$, we derive the estimate

$$(S_t g, g)_0 \leq c\mathcal{E}(S_t g, S_t g)^{1/2} \leq (S_t g, g)_0^{1/2},$$

or $(S_t g, g)_0^{1/2} \leq c$. Thus

$$\Big(\int_{\mathbb{R}^n} gGg\, dx\Big)^{1/2} = \sup_{t>0}(S_t g, g)_0 \leq c = \sup_{x \in D(\mathcal{E})} \frac{(|u|, g)_0}{\mathcal{E}(u,u)^{1/2}}.$$

Conversely suppose that $\int_{\mathbb{R}^n} gGg\, dx < \infty$. Since $(T_s g, g)_0 = (T_{s/2} g, T_{s/2} g)_0 = \|T_{s/2} g\|_0^2$ and $\|T_{s/2} g\|_0^2$ is non-increasing as $s \to \infty$ due to the contraction property of $(T_t)_{t \geq 0}$ it follows that $(T_s g, g)_0 \to 0$ as $s \to \infty$ because $\int_0^\infty (T_s g, g)_0\, ds = \int_{\mathbb{R}^n} gGg\, dx < \infty$. Now (3.374) yields

$$\begin{aligned}(|u|, g)_0 &= \mathcal{E}(|u|, S_t g) + (|u|, T_t g)_0 \\ &\leq \mathcal{E}(S_t g, S_t g)^{1/2} \mathcal{E}(u,u)^{1/2} + (u,u)_0^{1/2} (T_t g, T_t g)_0^{1/2} \\ &\leq (S_t g, g)_0^{1/2} \mathcal{E}(u,u)^{1/2} + \|u\|_0 (T_{2t} g, g)_0^{1/2},\end{aligned}$$

where we used as before $\mathcal{E}(S_t g, S_t g) \leq (S_t g, g)_0$ which follows also from (3.374). Thus for $t \to \infty$ we arrive at

$$(|u|, g)_0 \leq \Big(\int_{\mathbb{R}^n} gGg\, dx\Big)^{1/2} \mathcal{E}(u,u)^{1/2}$$

implying

$$\sup_{u \in D(\mathcal{E})} \frac{(|u|, g)_0}{\mathcal{E}(u,u)^{1/2}} \leq \Big(\int_{\mathbb{R}^n} gGg\, dx\Big)^{1/2}$$

proving the lemma. □

Remark 3.5.40. For a (symmetric) transient Dirichlet semigroup estimate (3.373) sharpens Theorem 3.5.33.

Lemma 3.5.41. *Let $(\mathcal{E}, D(\mathcal{E}))$ be a symmetric Dirichlet form with associated resolvent $(R_\lambda)_{\lambda>0}$ and semigroup $(T_t)_{t\geq 0}$. Then $\mathcal{E}^{(\lambda)}(u,u) = \lambda(u - \lambda R_\lambda u, u)_0$ is non-decreasing as λ tends to infinity and $\lim_{\lambda \to \infty} \mathcal{E}^{(\lambda)}(u,u) = \mathcal{E}(u,u)$. Further $\frac{1}{t}(u - T_u, u)_0$ is non-decreasing as t tends to 0 and $\lim_{t \to 0} \frac{1}{t}(u - T_t u, u)_0 = \mathcal{E}(u,u)$.*

Proof: Using the spectral projection $(E_\mu)_\mu$ associated with the generator of $(\mathcal{E}, D(\mathcal{E}))$ we find

$$\mathcal{E}^{(\lambda)}(u,u) = \int_0^\infty \frac{\lambda\mu}{\lambda+\mu} d(E_\mu u, u)$$

and

$$\frac{1}{t}(u - T_t u, u)_0 = \int_0^\infty \frac{1-e^{-t\mu}}{t} d(E_\mu u, u)$$

proving the lemma. □

Theorem 3.5.42. *Let $(\mathcal{E}, D(\mathcal{E}))$, $D(\mathcal{E}) \subset L^2(\mathbb{R}^n;\mathbb{R})$, be a symmetric Dirichlet form with associated sub-Markovian semigroup $(T_t)_{t\geq 0}$.*

A. For every $u \in \mathcal{F}_e$ and approximating sequence $(u_k)_{k\in\mathbb{N}}$ for u the limit $\lim\limits_{k\in\infty} \mathcal{E}(u_k, u_k)$ exists and is independent of the choice of the approximating sequence. We will denote this limit once again by $\mathcal{E}(u,u)$.

B. If $u \in \mathcal{F}_e$ and $u = v$ a.e. for a Borel function v then

$$\frac{\lambda}{2}\iint_{\mathbb{R}^n\times\mathbb{R}^n} (v(x)-v(y))^2 \sigma_\lambda(dx,dy) + \lambda\int_{\mathbb{R}^n} v(x)^2(1-s_\lambda(x))\,dx$$

increases to $\mathcal{E}(u,u)$ as λ tends to ∞.

C. It holds $D(\mathcal{E}) = \mathcal{F}_e \cap L^2(\mathbb{R}^n;\mathbb{R})$.

Proof: First note that there is no problem to extend the right-hand side of (3.369) by monotone convergence to all Borel functions. Since $\mathcal{E}^{1/2}$ is a seminorm the triangle inequality yields

$$|\mathcal{E}(u_k,u_k)^{1/2} - \mathcal{E}(u_l,u_l)^{1/2}| \leq \mathcal{E}(u_k-u_l, u_k-u_l)^{1/2}$$

implying that $\lim\limits_{k\to\infty} \mathcal{E}(u_k,u_k)$ exists. Now let $\tilde{u}_k$ and $\tilde{u}$ be some Borel modification of u_k and u respectively satisfying $\lim\limits_{k\to\infty} \tilde{u}_k(x) = \tilde{u}(x)$ for all $x \in \mathbb{R}^n$. Fatou's lemma yields

$$\begin{aligned}\mathcal{E}^{(\lambda)}(\tilde{u}_k - \tilde{u}, \tilde{u}_k - \tilde{u}) &\leq \liminf_{l\to\infty} \mathcal{E}^{(\lambda)}(\tilde{u}_k - \tilde{u}_l, \tilde{u}_k - \tilde{u}_l) \\ &\leq \lim_{l\to\infty} \mathcal{E}(u_k - u_l, u_k - u_l),\end{aligned}$$

and since $(u_k)_{k\in\mathbb{N}}$ is an $\mathcal{E}$-Cauchy sequence it follows that

$$\lim_{k\to\infty} \mathcal{E}^{(\lambda)}(\tilde{u}_k, \tilde{u}_k) = \mathcal{E}^{(\lambda)}(\tilde{u}, \tilde{u})$$

and $\mathcal{E}^{(\lambda)}(\tilde{u},\tilde{u})$ is non-decreasing with λ by Lemma 3.5.41. Moreover it follows that

$$\begin{aligned}&\left|\mathcal{E}(u,u)^{1/2}-\mathcal{E}^{(\lambda)}(\tilde{u},\tilde{u})^{1/2}\right|\\&\quad\le\left|\mathcal{E}(u,u)^{1/2}-\mathcal{E}(u_k,u_k)^{1/2}\right|+\left|\mathcal{E}(u_k,u_k)^{1/2}-\mathcal{E}^{(\lambda)}(\tilde{u}_k,\tilde{u}_k)^{1/2}\right|\\&\quad+\mathcal{E}^{(\lambda)}(\tilde{u}_k-\tilde{u},\tilde{u}_k-\tilde{u})^{1/2}\end{aligned}$$

implying that $\lim\limits_{\lambda\to\infty}\mathcal{E}^{(\lambda)}(\tilde{u},\tilde{u})=\mathcal{E}(u,u)$, which gives first part B and further, if $u=0$ a.e. then $\mathcal{E}^{(\lambda)}(\tilde{u},\tilde{u})=0$, hence $\mathcal{E}(u,u)=0$ implying, with the previous remark that $\lim\limits_{k\to\infty}\mathcal{E}(u_k,u_k)$ exists, also part A. Part C is clear from part B, compare also Lemma I.4.7.18. □

Remark 3.5.43. Suppose that $(T_t)_{t\ge0}$ is a general Dirichlet semigroup, i.e. not necessarily symmetric, with associates Dirichlet form $(\mathcal{E},D(\mathcal{E}))$. We may construct $\mathcal{F}_e^{sym}$ as before by using $(T_t^{sym})_{t\ge0}$ or $\mathcal{E}^{sym}$. Then the sector condition implies that we may extend $\mathcal{E}$ to $\mathcal{F}_e$ ($=\mathcal{F}_e^{sym}$ the extended Dirichlet space associated with $\mathcal{E}^{sym}$) by $\mathcal{E}(u,v):=\lim\limits_{k\to\infty}\mathcal{E}(u_k,v_k)$ with approximating sequences $(u_k)_{k\in\mathbb{N}}$ and $(v_k)_{k\in\mathbb{N}}$ for u and v, respectively.

Let $(\mathcal{E},D(\mathcal{E}))$ be a symmetric Dirichlet form with corresponding sub-Markovian semigroup $(T_t)_{t\ge0}$. Since $(T_t)_{t\ge0}$ is analytic, compare Stein's theorem, Theorem I.4.2.12, $D(\mathcal{E})$ is invariant under T_t, i.e. $T_tD(\mathcal{E})\subset D(\mathcal{E})=D((-A)^{1/2})$ for all $t\ge0$ and in addition for all $u\in D(\mathcal{E})$

$$\begin{aligned}\mathcal{E}(T_tu-u,T_tu-u)&=\|(-A)^{1/2}(T_tu-u)\|_0^2\\&=\|T_t((-A)^{1/2}u)-(-A)^{1/2}u\|_0^2,\end{aligned}$$

implying

$$\lim_{t\to0}\mathcal{E}(T_tu-u,T_tu-u)=0. \tag{3.375}$$

Moreover, for $u\in D(\mathcal{E})$

$$\mathcal{E}(T_tu,T_tu)=\|(-A)^{1/2}T_tu\|_0^2=\|T_t(-A)^{1/2}u\|_0^2\le\|(-A)^{1/2}u\|_0^2$$

i.e.

$$\mathcal{E}(T_tu,T_tu)\le\mathcal{E}(u,u). \tag{3.376}$$

Lemma 3.5.44. *Both* (3.375) *and* (3.376) *extend to* $\mathcal{F}_e$.

Proof: Take an approximating sequence $(u_k)_{k\in\mathbb{N}}$, $u_k \in D(\mathcal{E})$, for $u \in \mathcal{F}_e$. From (3.375) it follows that $(T_t u_k)_{k\in\mathbb{N}}$ is a Cauchy sequence with respect to $\mathcal{E}$. Now, for $\omega \in L^2(\mathbb{R}^n;\mathbb{R})$ the contraction property of $(T_t)_{t\geq 0}$ implies $(T_t\omega, T_t\omega)_0 \leq (T_{t/2}\omega, T_{t/2}\omega)_0 \leq (\omega, T_t\omega)$ and therefore it follows that

$$\begin{aligned}\frac{1}{t}&\|(u_k - u_l) - T_t(u_k - u_l)\|_0^2\\ &= \frac{1}{t}((u_k - u_l) - T_t(u_k - u_l), u_k - u_l)_0\\ &\quad - \frac{1}{t}((u_k - u_l) - T_t(u_k - u_l), T_t(u_k - u_l))_0\\ &\leq \frac{1}{t}((u_k - u_l) - T_t(u_k - u_l), u_k - u_l)_0 \leq \mathcal{E}(u_k - u_l, u_k - u_l),\end{aligned}$$

where in the last step we used the second part of Lemma 3.5.41. It follows that $(u_k - T_t u_k)_{k\in\mathbb{N}}$ is a Cauchy sequence in $L^2(\mathbb{R}^n;\mathbb{R})$. Thus a subsequence of this sequence converges a.e. and since $(u_k)_{k\in\mathbb{N}}$ converges a.e. to u it follows that a subsequence $(T_t u_{k_l})_{l\in\mathbb{N}}$ converges a.e. to some function $v \in L^2(\mathbb{R}^n;\mathbb{R})$ and this limit is independent from the choice of the approximating sequence. If we define $T_t u := v$ then $T_t u \in \mathcal{F}_e$, $(T_t u_{k_l})_{l\in\mathbb{N}}$ is an approximating sequence for $T_t u$ and $u - T_u \in \mathcal{F}_e \cap L^2(\mathbb{R}^n;\mathbb{R}) = D(\mathcal{E})$. Since

$$\begin{aligned}\mathcal{E}(T_t u - u, T_t u - u)^{1/2} &\leq \mathcal{E}(T_t u_k - u_k, T_t u_k - u_k)^{1/2}\\ &+ \mathcal{E}(T_t(u - u_k), T_t(u - u_k))^{1/2} + \mathcal{E}(u - u_k, u - u_k)^{1/2}\end{aligned}$$

we find if first $t \to 0$ and then $k \to \infty$ that (3.375) holds and (3.376) follows by an analogous argument. □

Now we may start to relate $\mathcal{F}_e$ to the transience of $(T_t)_{t\geq 0}$.

Proposition 3.5.45. *If the symmetric Dirichlet form* $(\mathcal{E}, D(\mathcal{E}))$ *is transient then its extended Dirichlet space* $\mathcal{F}_e$ *is complete with respect to the scalar product* $\mathcal{E}$.

Proof: For $u \in \mathcal{F}_e$ and an approximating sequence $(u_k)_{k\in\mathbb{N}}$ we have $\mathcal{E}(u - u_k, u - u_k) = \lim_{l\to\infty} \mathcal{E}(u_l - u_k, u_l - u_k)$ implying that $(u_k)_{k\in\mathbb{N}}$ converges with respect to $\mathcal{E}$ to u. Now let $(\mathcal{E}, D(\mathcal{E}))$ be transient and $(u_k)_{k\in\mathbb{N}}$ be a Cauchy sequence in $\mathcal{F}_e$. From the above remark follows the existence of a sequence $(v_k)_{k\in\mathbb{N}}$, $v_k \in D(\mathcal{E})$ and $(v_k)_{k\in\mathbb{N}}$ is a Cauchy sequence with respect to

$\mathcal{E}$ such that $\lim_{k\to\infty} \mathcal{E}(u_k - v_k, u_k - v_k) = 0$, but now Theorem 3.5.33 yields that $(u_k - v_k)_{k\in\mathbb{N}}$ is a Cauchy sequence in the weighted L^1-space $L^1(\mathbb{R}^n; g\lambda^{(n)})$. Therefore a subsequence $(v_{k_l})_{l\in\mathbb{N}}$ converges a.e. to a function u. Thus $u \in \mathcal{F}_e$ and furthermore

$$\begin{aligned}\mathcal{E}(u_k - u, u_k - u)^{1/2} &\le \mathcal{E}(u_k - u_{k_l}, u_k - u_{k_l})^{1/2} \\ &+ \mathcal{E}(u_{k_l} - v_{k_l}, u_{k_l} - v_{k_l})^{1/2} + \mathcal{E}(v_{k_l} - u, v_{k_l} - u)^{1/2}.\end{aligned}$$

If we let first l tend to $+\infty$ and then k to $+\infty$ it follows that $(u_k)_{k\in\mathbb{N}}$ converges with respect to $\mathcal{E}$ to u. □

Now we are able to characterize the extended Dirichlet space of a transient Dirichlet space.

Theorem 3.5.46. *Let $(\mathcal{E}, D(\mathcal{E}))$ be a symmetric transient Dirichlet form on $L^2(\mathbb{R}^n; \mathbb{R})$. Then for its extended Dirichlet space $\mathcal{F}_e$ with scalar product $\mathcal{E}$ we have*

i) $\mathcal{F}_e$ is a Hilbert space with inner product $\mathcal{E}$;

ii) there exists a bounded strictly positive function $g \in L^1(\mathbb{R}^n; \mathbb{R})$ such that

$$\int_{\mathbb{R}^n} |u(x)|g(x)\,dx \le \mathcal{E}(u,u)^{1/2}$$

holds for all $u \in \mathcal{F}_e$;

iii) $\mathcal{F}_e \cap L^2(\mathbb{R}^n; \mathbb{R})$ is dense in $L^2(\mathbb{R}^n; \mathbb{R})$ and $(\mathcal{F}_e, \mathcal{E})$;

iv) every normal contraction acts on $(\mathcal{F}_e, \mathcal{E})$, i.e. if T is a normal contraction then $Tu \in \mathcal{F}_e$ for all $u \in \mathcal{F}_e$ and $\mathcal{E}(Tu, Tu) \le \mathcal{E}(u,u)$.

In addition $\mathcal{F}_e \cap L^2(\mathbb{R}^n; \mathbb{R}) = D(\mathcal{E})$. Conversely, suppose that for a pair $(\mathcal{F}_e, \mathcal{E})$ i)-iv) are fulfilled. Then $(\mathcal{F}_e, \mathcal{E})$ is the extended Dirichlet space of the transient symmetric Dirichlet form $(\mathcal{E}, \mathcal{F}_e \cap L^2(\mathbb{R}^n; \mathbb{R}))$.

Proof: If $(\mathcal{E}, D(\mathcal{E}))$ is a symmetric transient Dirichlet form and $\mathcal{F}_e$ its extended Dirichlet space we have already proved i)-iii). To see iv) take $u \in \mathcal{F}_e$ and an approximating sequence $(u_k)_{k\in\mathbb{N}}$. Further let T be a normal contraction and define $\omega_k := (Tu \wedge |u_k|)^+ \in L^2(\mathbb{R}^n; \mathbb{R})$. As usual let us denote a Borel modification of a function $v \in L^2(\mathbb{R}^n; \mathbb{R})$ by $\tilde{v}$. It follows that

$$\begin{aligned}\mathcal{E}^{(\lambda)}(\omega_k, \omega_k) &\le \mathcal{E}^{(\lambda)}(\widetilde{Tu}, \widetilde{Tu}) + \mathcal{E}^{(\lambda)}(u_k, u_k) \\ &\le \mathcal{E}^{(\lambda)}(\tilde{u}, \tilde{u}) + \mathcal{E}^{(\lambda)}(u_k, u_k) \le \mathcal{E}(u,u) + \mathcal{E}(u_k, u_k).\end{aligned}$$

Since $\omega \in L^2(\mathbb{R}^n; \mathbb{R})$ it follows that $\omega_k \in D(\mathcal{E})$ and since $\mathcal{E}(u_k, u_k)$ converges to $\mathcal{E}(u,u)$ it follows that $\mathcal{E}(\omega_k, \omega_k) \le c$ uniformly in k. Thus by the Banach–Saks theorem, Theorem I.2.7.2, the arithmetic mean of some subsequence of $(\omega_k)_{k\in\mathbb{N}}$ converges with respect to $\mathcal{E}$ to some $\omega \in \mathcal{F}_e$. Since $(Tu)^+ = \lim_{k\to\infty} \omega_k = \omega \in \mathcal{F}_e$, i.e. $(Tu)^+ \in \mathcal{F}_e$, and analogously $(Tu)^- \in \mathcal{F}_e$ we find first $Tu \in \mathcal{F}_e$ and then the estimate $\mathcal{E}(Tu, Tu) \leq \mathcal{E}(u, u)$ follows from the extension of (3.369) to $\mathcal{F}_e$, see Theorem 3.5.42. Suppose now that i)-iv) hold. Then it is clear that $(\mathcal{E}, \mathcal{F}_e \cap L^2(\mathbb{R}^n; \mathbb{R}))$ is a symmetric Dirichlet form on $L^2(\mathbb{R}^n; \mathbb{R})$ and ii) implies the transience of this Dirichlet form. Further it is clear that the extended Dirichlet space associated with $(\mathcal{E}, \mathcal{F}_e \cap L^2(\mathbb{R}^n; \mathbb{R}))$ must be $\mathcal{F}_e$. □

Definition 3.5.47. A pair $(\mathcal{F}_e, \mathcal{E})$ satisfying i)-iv) of Theorem 3.5.46 is called a *transient extended Dirichlet space.*

We need

Lemma 3.5.48. *Let $(T_t)_{t\geq 0}$ be a symmetric transient sub-Markovian semigroup with generator A. Then $\lim_{t\to\infty} T_t u = 0$ for all $u \in L^2(\mathbb{R}^n; \mathbb{R})$.*

Proof: Suppose that there is some $u_0 \in L^2(\mathbb{R}^n; \mathbb{R})$ such that $\lim_{t\to\infty} T_t u_0 \neq 0$. Using the spectral projections associated with the selfadjoint operator A, the spectral theorem yields

$$\lim_{t\to\infty} \int_0^\infty e^{-t\mu} dE_\mu u_0 \neq 0$$

which is only possible if $T_t u_0 = u_0$ for all t. But for any $u \in L^1(\mathbb{R}^n; \mathbb{R}) \cap L^2(\mathbb{R}^n; \mathbb{R})$ it follows from the transience of the semigroup that

$$\lim_{t\to\infty} \int_t^{t+r} T_s u \, ds = 0, \quad r > 0,$$

and since $L^1(\mathbb{R}^n; \mathbb{R}) \cap L^2(\mathbb{R}^n; \mathbb{R})$ is dense in $L^2(\mathbb{R}^n; \mathbb{R})$ such a $u_0 \in L^2(\mathbb{R}^n; \mathbb{R})$ can not exists proving the lemma. □

Most important is now

Theorem 3.5.49. *There is a one-to-one correspondence of all transient extended Dirichlet spaces $(\mathcal{F}_e, \mathcal{E})$ and all symmetric transient L^2-sub-Markovian semigroups $(T_t)_{t\geq 0}$ on $L^2(\mathbb{R}^n; \mathbb{R})$: For any non-negative measurable function f such that $\int_{\mathbb{R}^n} fGf \, dx < \infty$ we have*

$$Gf \in \mathcal{F}_e \quad \text{and } \mathcal{E}(Gf, v) = (f, v)_0 \text{ for all } v \in \mathcal{F}_e. \tag{3.377}$$

Proof: Let $(T_t)_{t\geq 0}$ be a symmetric transient L^2-sub-Markovian semigroup with corresponding extended Dirichlet space $(\mathcal{F}_e, \mathcal{E})$. We have to show that for every non-negative measurable function f such that $\int_{\mathbb{R}^n} fGf\,dx < \infty$ we have $Gf \in \mathcal{F}_e$ and $\mathcal{E}(Gf, v) = (f, v)_0$ for all $v \in \mathcal{F}_e$, and that $(\mathcal{F}_e, \mathcal{E})$ is uniquely determined by $(T_t)_{t\geq 0}$. For $f \in L^2(\mathbb{R}^n; \mathbb{R})$, $f \geq 0$ a.e. and $\int_{\mathbb{R}^n} fGf\,dx < \infty$ it follows from (3.374) that for $t > t' > 0$

$$\begin{aligned}\mathcal{E}(S_t f - S_{t'} f, S_t f - S_{t'} f) &= (T_{t'} f - T_t f, S_t f - S_{t'} f)_0 \\ &= \int_{\mathbb{R}^n} f \cdot \Big(\int_{2t'}^{t+t'} T_s f\,ds - \int_{t+t'}^{2t} T_s f\,ds \Big)\,dx \\ &\leq \int_{\mathbb{R}^n} f \int_{t+t'}^{2t} T_s f\,ds\,dx.\end{aligned}$$

Since $\int_{\mathbb{R}^n} f \int_{t+t'}^{2t} T_s f\,ds\,dx \to 0$ as $t, t' \to 0$ it follows that $(S_k f)_{k\in\mathbb{N}}$ is a Cauchy sequence with respect to $\mathcal{E}$ which converges a.e. to Gf, i.e. $Gf \in \mathcal{F}_e$. Now, (3.374), i.e. $\mathcal{E}(S_t g, u) = (g - T_t g, u)_0$, and Lemma 3.5.48 yield first for all $v \in L^2(\mathbb{R}^n; \mathbb{R}) \cap \mathcal{F}_e$ that

$$\mathcal{E}(Gf, v) = (f, v)_0$$

and now the density of $\mathcal{F}_e \cap L^2(\mathbb{R}^n; \mathbb{R})$ in $\mathcal{F}_e$ and $L^2(\mathbb{R}^n; \mathbb{R})$ implies (3.377).

Next if f is any measurable function such that $\int_{\mathbb{R}^n} fGf\,dx < \infty$ then we choose $g \in L^1(\mathbb{R}^n; \mathbb{R}) \cap L^2(\mathbb{R}^n; \mathbb{R})$, $g > 0$ a.e., and consider the sequence $f_k := f \wedge (kg) \in L^2(\mathbb{R}^n; \mathbb{R})$. For f_k we may apply (3.377) to find

$$\begin{aligned}\mathcal{E}(Gf_k - Gf_l, Gf_k - Gf_l) &= (f_k, Gf_k - Gf_l)_0 + (f_l, Gf_k - Gf_l)_0 \\ &\leq \int_{\mathbb{R}^n} f_k G f_k\,dx - \int_{\mathbb{R}^n} f_l G f_l\,dx.\end{aligned}$$

Thus for $k, l \to \infty$ it follows that $\mathcal{E}(Gf_k - Gf_l, Gf_k - Gf_l) \to 0$ implying first that $Gf \in \mathcal{F}_e$ and then we may pass to the limit in $\mathcal{E}(Gf_k, v) = (f_k, v)_0$ to extend (3.377) to all measurable $f, f \geq 0$ a.e., such that $\int_{\mathbb{R}^n} fGf\,dx < \infty$. Finally, since for $(T_t)_{t\geq 0}$ given, the set of all Gf, $f \geq 0$ and measurable such that $\int_{\mathbb{R}^n} fGf\,dx < \infty$, is sufficiently rich to approximate elements in $\mathcal{F}_e$ (after decomposing into positive and negative part), it follows that $(T_t)_{t\geq 0}$ determines uniquely the extended Dirichlet space.

Conversely, let a transient extended Dirichlet space $(\mathcal{F}_e, \mathcal{E})$ be given. We have to show that (3.377) determines $(T_t)_{t\geq 0}$ uniquely. Since $Gf =$

$\lim_{N\to\infty} R_{1/N}f$ a.e. for $f \in L^1(\mathbb{R}^n;\mathbb{R}) \cap L^2(\mathbb{R}^n;\mathbb{R})$, $f \geq 0$ a.e., we may extend the resolvent equation with $R_0 := G$ by

$$Gf = R_0 f = R_\lambda f + \lambda G R_\lambda f.$$

If in addition $\int_{\mathbb{R}^n} fGf\,dx < \infty$ then

$$\begin{aligned}\int_{\mathbb{R}^n} (R_\lambda f) G R_\lambda f\,dx &\leq \frac{1}{\lambda}\int_{\mathbb{R}^n} R_\lambda f G f\,dx \\ &\leq \frac{1}{\lambda}\int_{\mathbb{R}^n} f G R_\lambda f\,dx \leq \frac{1}{\lambda^2}\int_{\mathbb{R}^n} fGf\,dx < \infty.\end{aligned}$$

It follows from (3.377) that $GR_\lambda f \in \mathcal{F}_e$ and $\mathcal{E}(GR_\lambda f, v) = (R_\lambda f, v)_0$ for all $v \in \mathcal{F}_e$. By the extended resolvent equation we obtain that $R_\lambda f \in \mathcal{F}_e$ and further

$$\mathcal{E}(R_\lambda f, v) = \mathcal{E}(Gf, v) - \lambda\mathcal{E}(GR_\lambda f, v) = (f, v)_0 - \lambda(R_\lambda f, v)_0$$

for all $v \in \mathcal{F}_e$. Therefore we may deduce $R_\lambda f \in \mathcal{F} = \mathcal{F}_e \cap L^2(\mathbb{R}^n;\mathbb{R})$ and

$$\mathcal{E}_\lambda(R_\lambda f, v) = (f, v)_0 \quad \text{for all } v \in D(\mathcal{E}).$$

This relation holds for all $f = h \cdot g$, g being a reference function in Theorem 3.5.33 and h any bounded function. By a density argument it follows that $(R_\lambda)_{\lambda\geq 0}$ is uniquely determined by $(\mathcal{F}_e, \mathcal{E})$ proving the theorem. □

It is now straightforward to extend basic properties of symmetric Dirichlet forms to transient extended Dirichlet spaces. In particular we have

$$u, v \in \mathcal{F}_e \quad \text{imply } u \vee v, u \wedge v, u \wedge 1 \in \mathcal{F}_e; \tag{3.378}$$

$$\begin{aligned}&u \in \mathcal{F}_e \text{ and } u_k := ((-k) \vee u) \wedge k \text{ imply } u_k \in \mathcal{F}_e \\ &\qquad \text{and } \mathcal{E}(u_k - u, u_k - u) \to 0 \text{ as } k \to \infty;\end{aligned} \tag{3.379}$$

$$\begin{aligned}&u \in \mathcal{F}_e \text{ and } u_\varepsilon := u - ((-\varepsilon) \vee u) \wedge \varepsilon \text{ for } \varepsilon > 0 \text{ imply } u_\varepsilon \in \mathcal{F}_e \\ &\qquad \text{and } \mathcal{E}(u_\varepsilon - u, u_\varepsilon - u) \to 0 \text{ as } \varepsilon \to 0.\end{aligned} \tag{3.380}$$

Further, if $\varphi : \mathbb{R} \to \mathbb{R}$ is such that $\varphi(0) = 0$, $|\varphi(t) - \varphi(s)| \leq |t - s|$, then $u \in \mathcal{F}_e$ implies $\varphi(u) \in \mathcal{F}_e$ and in addition

$$\begin{aligned}&u_k, u \in \mathcal{F}_e \text{ such that } \mathcal{E}(u_k - u, u_k - u) \to 0 \text{ implies} \\ &\mathcal{E}(\varphi(u_k) - \varphi(u), v) \to 0 \text{ for all } v \in \mathcal{F}_e.\end{aligned} \tag{3.381}$$

Theorem 3.5.50. *Let $(\mathcal{F}_e, \mathcal{E})$ be a transient extended Dirichlet space with corresponding semigroup $(T_t)_{t\geq 0}$. A necessary and sufficient condition that $\mathcal{F}_e \subset L^1_{loc}(\mathbb{R}^n; \mathbb{R})$ and that for any compact set $K \subset \mathbb{R}^n$*

$$\int_K |u(x)|\, dx \leq c_K \mathcal{E}(u,u)^{1/2} \tag{3.382}$$

holds for all $u \in \mathcal{F}_e$ is

$$\int_{\mathbb{R}^n} fGf\, dx = \int_0^\infty (T_t f, f)_0\, dt < \infty \tag{3.383}$$

for all $f \in C_0(\mathbb{R}^n; \mathbb{R}^n)$, $f \geq 0$.

Proof: We only have to combine Lemma 3.5.39 and Theorem 3.5.49. □

Now we are in a position to discuss further examples.

Theorem 3.5.51. *Let $(\mu_t)_{t\geq 0}$ be a symmetric convolution semigroup with corresponding continuous negative definite function $\psi : \mathbb{R}^n \to \mathbb{R}$ and symmetric Dirichlet form $(\mathcal{E}^\psi, D(\mathcal{E}^\psi)) = (\mathcal{E}^\psi, H^{\psi,1}(\mathbb{R}^n; \mathbb{R}))$. Further denote the associated semigroup by $(T_t^\psi)_{t\geq 0}$. The following assertions are equivalent:*
i) $(T_t^\psi)_{t\geq 0}$ is transient;
ii) for any compact set $K \subset \mathbb{R}^n$ it holds

$$\mathcal{K}(K) := \int_0^\infty \mu_t(K)\, dt < \infty; \tag{3.384}$$

iii) for all $u \in C_0(\mathbb{R}^n; \mathbb{R})$, $u \geq 0$, we have

$$\int_0^\infty (T_t^\psi u, u)_0\, dt < \infty; \tag{3.385}$$

iv) the function $\frac{1}{\psi}$ belongs to $L^1_{loc}(\mathbb{R}^n; \mathbb{R})$.

Proof: If ii) holds then

$$\int_0^\infty T_t^\psi \chi_K(x)\, dt = \mathcal{K}(K-x) < \infty$$

for every compact set $K \subset \mathbb{R}^n$ and almost all $x \in \mathbb{R}^n$ which implies

$$\int_0^\infty \int_{\mathbb{R}^n} (T_t^\psi u) u\, dx\, dt < \infty$$

for every $u \in C_0(\mathbb{R}^n;\mathbb{R})$, $u \geq 0$, i.e. ii) implies iii). Of course iii) implies i) as well as i) implies ii). Now assume that iii) holds. Let $K \subset \mathbb{R}^n$ be compact and $u \in C_0(\mathbb{R}^n;\mathbb{R})$, $u \geq 0$, such that $|\hat{u}(\xi)| \geq 1$ for all $\xi \in K$ (clearly such a function u exists). It follows that

$$\begin{aligned}\int_K \frac{d\xi}{\psi(\xi)} &\leq \int_{\mathbb{R}^n} \frac{|\hat{u}(\xi)|^2}{\psi(\xi)}\,d\xi = \int_0^\infty \int_{\mathbb{R}^n} (T_t^\psi u)^\wedge(\xi)\overline{\hat{u}(\xi)}\,d\xi \\ &= \int_0^\infty \int_{\mathbb{R}^n} (T_t^\psi u)(x)u(x)\,dx\,dt < \infty,\end{aligned}$$

where we used $\int_0^\infty e^{-t\psi(\xi)}dt = \frac{1}{\psi(\xi)}$, the fact that $(T_t^\psi u)^\wedge(\xi) = e^{-\psi(\xi)t}\hat{u}(\xi)$ and Plancherel's theorem. Thus iii) implies iv). On the other hand with u as before and $v(\xi) := (2\pi)^{-n/2}(u * u)(\xi)$. We find first that $\operatorname{supp} v \subset \operatorname{supp} u + \operatorname{supp} u \subset K'$, where $K' \subset \mathbb{R}^n$ is a compact set. Further it follows that

$$\begin{aligned}\mathcal{K}(K) &\leq \lim_{T\to\infty} \int_0^T \left\{\int_{\mathbb{R}^n} |\hat{u}(\xi)|^2 \mu_t(d\xi)\right\} dt \\ &= \lim_{T\to\infty} \int_0^T \left\{\int_{\mathbb{R}^n} \hat{v}(\xi)\mu_t(d\xi)\right\} dt \\ &= (2\pi)^{-n/2} \lim_{T\to\infty} \int_0^T \left\{\int_{K'} v(\xi)\hat{\mu}_t(d\xi)\right\} dt \\ &= (2\pi)^{-n/2} \int_{K'} \frac{v(\xi)}{\psi(\xi)}\,d\xi < \infty,\end{aligned}$$

where we used in the last step the non-negativity of v and $\hat{\mu}_t$, hence iv) implies ii) and the theorem is proved. □

Corollary 3.5.52. *Let $\psi : \mathbb{R}^n \to \mathbb{C}$ be a continuous negative definite function satisfying*

$$|\operatorname{Im}\psi(\xi)| \leq c \operatorname{Re}\psi(\xi). \tag{3.386}$$

If $\frac{1}{\operatorname{Re}\psi} \in L^1_{loc}(\mathbb{R}^n)$ then the semigroup $(T_t^\psi)_{t\geq 0}$ associated with ψ, compare Example I.4.1.5, is transient.

Proof: We know from Example I.4.7.32 that $(T_t^\psi)_{t\geq 0}$ corresponds to a non-symmetric Dirichlet form if and only if (3.386) holds. Hence $(T_t^\psi)_{t\geq 0}$ is a Dirichlet semigroup which is transient if and only if $(T_t^{\psi,sym})_{t\geq 0}$ is transient. But $(T_t^{\psi,sym})_{t\geq 0} = (T_t^{\operatorname{Re}\psi})_{t\geq 0}$ and the Corollary follows. □

Remark 3.5.53. Since (3.386) implies that

$$\frac{1}{(1+c^2)\operatorname{Re}\psi} \leq \operatorname{Re}\frac{1}{\psi} \leq \frac{1}{\operatorname{Re}\psi}$$

it follows that $(T_t^\psi)_{t\geq 0}$ is transient if and only if $\operatorname{Re}\frac{1}{\psi} \in L^1_{loc}(\mathbb{R}^n;\mathbb{R})$

Example 3.5.54. Consider the Poisson semigroup on $\mathbb{R}^n$ given by $\mu_t = \sum_{k=0}^\infty e^{-t}\frac{t^k}{k!}\varepsilon_{hk}$, $h \in \mathbb{R}^n$. According to Example I.3.6.20 it corresponds to the continuous negative definite function $\psi(\xi) = (1-e^{-ih\xi}) = (1-\cos h\xi)+i\sin h\xi$. Since

$$\operatorname{Re}\frac{1}{\psi(\xi)} = \frac{1-\cos h\xi}{(1-\cos h\xi)^2 + (\sin h\xi)^2} = \frac{1}{2}$$

it follows that the operator semigroup associated to $(\mu_t)_{t\geq 0}$ is transient. Since for $0 \leq \xi\cdot h \leq \frac{\pi}{2}$ we have $\sin h\xi \geq \frac{2}{\pi}h\cdot\xi$ and $1-\cos h\xi \leq \frac{\sqrt{2}}{2\pi^2}(h\cdot\xi)^2$ it follows that for $0 < \xi\cdot h \leq \frac{\pi}{2}$

$$\frac{\operatorname{Im}\psi(\xi)}{\operatorname{Re}\psi(\xi)} = \frac{\sin h\xi}{1-\cos h\xi} \geq \frac{4\pi}{\sqrt{2}}\frac{1}{h\cdot\xi}.$$

Hence we can not fulfill (3.386) showing that this is not a necessary condition for the transience of $(T_t^\psi)_{t\geq 0}$.

Example 3.5.55. Let $\psi : \mathbb{R}^n \to \mathbb{R}$ be a continuous negative definite function such that $\frac{1}{\psi} \in L^1_{loc}(\mathbb{R}^n;\mathbb{R})$. The symmetric Dirichlet space $(\mathcal{E}, D(\mathcal{E}))$ associated with ψ is given according to Example I.4.7.28 by

$$D(\mathcal{E}) = H^{\psi,1}(\mathbb{R}^n;\mathbb{R})$$

and

$$\mathcal{E}(u,v) = \int_{\mathbb{R}^n} \psi(\xi)\hat{u}(\xi)\overline{\hat{u}(\xi)}\,d\xi.$$

The corresponding extended Dirichlet space $\mathcal{F}_e$ is now

$$\mathcal{F}_e = H^{\psi,1}_{(e)}(\mathbb{R}^n;\mathbb{R}) := \{u \in S'(\mathbb{R}^n;\mathbb{R}), u \in L^1_{loc}(\mathbb{R}^n;\mathbb{R}) \text{ and } \hat{u}\psi^{1/2} \in L^2(\mathbb{R}^2)\}.$$

Note that $\hat{u}\psi^{1/2} \in L^2(\mathbb{R}^2)$ if and only if $\hat{u}$ belongs to the weighted L^2-space $L^2(\mathbb{R}^n;\psi\lambda^{(n)})$. Clearly the extended form is given once again by

$$\mathcal{E}(u,v) = \int_{\mathbb{R}^n} \psi(\xi)\hat{u}(\xi)\overline{\hat{u}(\xi)}\,d\xi.$$

In general we do not know whether $H^{\psi,1}_{(e)}(\mathbb{R}^n;\mathbb{R}) \subset L^p(\mathbb{R}^n;\mathbb{R})$ for some $p \geq 1$. The following discussion of the space $H^s_{(e)}(\mathbb{R}^n;\mathbb{R})$, $0 < s < 1$, shows the difficulties we may encounter.

Example 3.5.56. Consider the semigroup generated by $-(-\Delta)^\alpha$, $0 < \alpha < 1$, i.e.

$$T_t^{2\alpha}u(x) = (2\pi)^{-n/2}\int_{\mathbb{R}^n} e^{ix\cdot\xi}e^{-t|\xi|^{2\alpha}}\,\hat{u}(\xi)\,d\xi$$

with corresponding continuous negative definite function $\psi_{2\alpha} : \mathbb{R}^n \to \mathbb{R}$, $\xi \mapsto \psi_{2\alpha}(\xi) = |\xi|^\alpha$, $0 < \alpha < 1$. Since

$$\int_{B_R(0)} \frac{1}{|\xi|^{2\alpha}}\,d\xi = c_n \int_0^R r^{n-1-2\alpha}\,dr$$

it follows that $\frac{1}{\psi_{2\alpha}} \in L^1_{loc}(\mathbb{R}^n;\mathbb{R})$ if and only if $2\alpha < n$, i.e. the semigroup $(T_t^{(2\alpha)})_{t\geq 0}$ is transient and if $n = 2$ only the Brownian semigroup is not transient. Since the Gauss kernel is strictly positive it follows that the Brownian semigroup in $\mathbb{R}^2$ is recurrent. For $n = 1$ it follows first by the strict positivity of the Gauss kernel that all of the semigroups $(T_t^{(2\alpha)})_{t\geq 0}$ are irreducible since they are obtained from the Brownian semigroup by subordination with respect to the one-sided stable convolution semigroup of order α, $0 < \alpha < 1$, compare Example 3.5.15 , thus for $0 < \alpha < \frac{1}{2}$ these semigroups are transient whereas for $\frac{1}{2} \leq \alpha \leq 1$ they are recurrent.

Now, suppose $2\alpha < n$. We will characterize the extended Dirichlet space associated with $(T_t^{2\alpha})_{t\geq 0}$ in terms of Riesz potentials. First note that for $2\alpha < n$ and $u \in S(\mathbb{R}^n;\mathbb{R})$, $u \geq 0$, it follows that

$$\begin{aligned} G^{2\alpha}u(x) &= \int_0^\infty T_t^{2\alpha}u(x)\,dt = (2\pi)^{-n/2}\int_0^\infty \int_{\mathbb{R}^n} e^{ix\cdot\xi}e^{-t|\xi|^{2\alpha}}\,\hat{u}(\xi)\,d\xi\,dt \\ &= (2\pi)^{-n/2}\int_{\mathbb{R}^n} e^{ix\cdot\xi}\int_0^\infty e^{-t|\xi|^{2\alpha}}\,dt\,\hat{u}(\xi)\,d\xi \\ &= (2\pi)^{-n/2}\int_{\mathbb{R}^n} e^{ix\cdot\xi}\frac{1}{|\xi|^{2\alpha}}\hat{u}(\xi)\,d\xi, \end{aligned}$$

or by the convolution theorem

$$G^{2\alpha}u = R^{2\alpha} * u$$

where $R^{2\alpha}(x) = c_{2\alpha,n}|\xi|^{-n+2\alpha}$ is the *Riesz kernel* of order 2α. In the following we will write $R^\beta u$ for $R^\beta * u$. Note that from Theorem I.3.3.6.C it is clear that

$R^{2\alpha}(x)$ must be homogeneous of degree $-n+2\alpha$. Thus we find for $H^\alpha_{(e)}(\mathbb{R}^n;\mathbb{R})$, the extended Dirichlet space associated with $(T_t^{2\alpha})_{t\geq 0}$, or equivalently with $H^\alpha(\mathbb{R}^n;\mathbb{R})$,

$$H^\alpha_{(e)}(\mathbb{R}^n;\mathbb{R}) := \{u \in L^1_{loc}(\mathbb{R}^n;\mathbb{R}); \quad u = R^\alpha f \quad \text{for some } f \in L^2(\mathbb{R}^n;\mathbb{R}),$$

and of course, the corresponding form is given by

$$\mathcal{E}(u,v) = \int_{\mathbb{R}^n} \hat{u}(\xi)\overline{\hat{v}(\xi)}|\xi|^{2\alpha}\,d\xi = \int_{\mathbb{R}^n} R^\alpha f(x)\cdot R^\alpha g(x)\,dx$$

with $u = R^\alpha f$ and $v = R^\alpha g$.

From our general theory we do only know that $H^\alpha_{(e)}(\mathbb{R}^n;\mathbb{R})$ is a subspace of $L^1_{loc}(\mathbb{R}^n;\mathbb{R})$. But the homogeneity of the kernel R^α allows us to prove a variant of Sobolev's embedding theorem to find $H^\alpha_{(e)}(\mathbb{R}^n;\mathbb{R}) \subset L^p(\mathbb{R}^n;\mathbb{R})$, $p = \frac{2n}{n-2\alpha}$.

Before proving Sobolev's theorem let us follow a consideration given in E.M.Stein [256]. We want to prove

$$\|R^\alpha f\|_{L^p} \leq c\|f\|_{L^q} \tag{3.387}$$

for some p and q. For $\rho > 0$ we find with $f_\rho(x) = f(\rho x)$ that

$$\|f_\rho\|_{L^q} = \rho^{-n/q}\|f\|_{L^q}$$

and

$$\|R^\alpha f_\rho\|_{L^p} = \rho^{-\alpha}\|(R^\alpha f)_\rho\|_{L^p} = \rho^{-\alpha-\frac{n}{p}}\|R^\alpha f\|_{L^p}.$$

Thus (3.387) would imply

$$\rho^{-\alpha-\frac{n}{p}}\|R^\alpha f\|_{L^p} \leq c\rho^{-n/q}\|f\|_{L^q}$$

for all suitable f and $\rho > 0$. Thus we shall expect that $\alpha + \frac{n}{p} = \frac{n}{q}$, or $\frac{1}{p} = \frac{1}{q} - \frac{\alpha}{n}$. In particular, for $q = 2$ we should have $p = \frac{2n}{n-2\alpha}$.

Theorem 3.5.57 (Sobolev's inequality). *Let $1 < q < p < \infty$, $0 < \alpha < n$ and $\frac{1}{p} = \frac{1}{q} - \frac{\alpha}{n} > 0$. Then for all $u \in \mathcal{S}(\mathbb{R}^n;\mathbb{R})$ (or $u \in L^q(\mathbb{R}^n;\mathbb{R})$)*

$$\|R^\alpha u\|_{L^p} \leq c_0\|u\|_{L^q} \tag{3.388}$$

holds.

Proof: We use a variant of the *Marcinkiewicz interpolation theorem*, compare Theorem I.2.8.9: If T is as in Theorem I.2.8.9 and $1 \leq q_i < p_i \leq \infty$, $i = 0, 1$, $q_0 < q_1$, $p_0 \neq p_1$, then, if T is of weak type (q_i, p_i) $i = 0, 1$, compare Definition I.2.8.8, then it satisfies $\|Tu\|_{L^p} \leq \|u\|_{L^q}$ for $\frac{1}{q} = \frac{1-\theta}{q_0} + \frac{\theta}{q_1}$ and $\frac{1}{p} = \frac{1-\theta}{p_0} + \frac{\theta}{p_1}$, $0 < \theta < 1$.

Without loss of generality we may assume $u \geq 0$ a.e. We decompose $R^\alpha u$ for $r > 0$ according to

$$\begin{aligned} R^\alpha u(x) &= \int_{B_{2r}(x)} |x-y|^{\alpha-n} u(y)\, dy + \int_{B^c_{2r}(x)} |x-y|^{\alpha-n} u(y)\, dy \\ &= u_1(x) + u_2(x). \end{aligned}$$

Since $\alpha q < n$, Hölder's inequality yields with $q' = \frac{q}{q-1}$ that

$$\begin{aligned} u_2(x) &\leq \Big(\int_{B^c_{2r}(x)} |x-y|^{(\alpha-n)q'}\, dy \Big)^{1/q'} \cdot \Big(\int_{B^c_{2r}(x)} |u(y)|^q\, dy \Big)^{1/q} \\ &\leq c_1 r^{((\alpha-n)q'+n)/q'} \|u\|_{L^q}. \end{aligned}$$

For $\mu > 0$ take $r > 0$ such that

$$c_1 r^{((\alpha-n)q'+n)/q'} \|u\|_{L^q} = \mu$$

which implies that

$$\begin{aligned} \lambda^{(n)}(\{x \in \mathbb{R}^n; R^\alpha u(x) > 2\mu\}) &\leq \lambda^{(n)}(\{x \in \mathbb{R}^n; u_1(x) > \mu\}) \\ &\leq \int_{\mathbb{R}^n} \Big(\frac{u_1(x)}{\mu} \Big)^q dx \leq c_2 \Big(\frac{r^\alpha}{\mu} \|u\|_{L^q} \Big)^q \\ &= c_2 \Big(\frac{\|f\|_{L^q}}{\mu} \Big)^p. \end{aligned}$$

Thus R^α is of weak-type (q, p) and the same calculation yields that it is always of weak-type $(1, p)$. Thus an applications of the Marcinkiewicz interpolation theorem yields the result. □

Remark 3.5.58. The proof of Theorem 3.5.57 is essentially that given in E.M.Stein [256] but we used much the presentation in Y.Mizuta [219].

Corollary 3.5.59. *For* $2\alpha < n$ *we have* $H^\alpha_{(e)}(\mathbb{R}^n; \mathbb{R}) \hookrightarrow L^p(\mathbb{R}^n; \mathbb{R})$, $p = \frac{2n}{n-2\alpha}$, *and the estimate*

$$\|u\|_{L^p} \leq c(\mathcal{E}(u,u))^{1/2} \tag{3.389}$$

holds for all $u \in H^\alpha_{(e)}(\mathbb{R}^n; \mathbb{R})$.

Proof: We just have to apply Sobolev's type inequality to $u \in R^\alpha f$, note the equality $\hat{f}(\xi) = |\xi|^\alpha \hat{u}(\xi)$. $\square$

Corollary 3.5.60. *Let $\psi : \mathbb{R}^n \to \mathbb{R}$ be a continuous negative definite function such that $c_0|\xi|^{2\alpha} \le \psi(\xi)$ holds for some $c_0 > 0$, $0 < \alpha < 1$ and all $\xi \in \mathbb{R}^n$. If $2\alpha < n$ then the semigroup $(T_t^\psi)_{t\ge 0}$ corresponding to ψ is transient and the extended Dirichlet space $H_{(e)}^{\psi,1}(\mathbb{R}^n;\mathbb{R})$ is continuously embedded into $L^p(\mathbb{R}^n;\mathbb{R})$, $p = \frac{2n}{n-2\alpha}$.*

Proof: First we observe that

$$\int_{B_R(0)} \frac{1}{\psi(\xi)}\, d\xi \le \frac{1}{c_0} \int_{B_R(0)} \frac{1}{|\xi|^{2\alpha}}\, d\xi < \infty,$$

and further we have

$$\|u\|_{L^p}^2 \le c \int_{\mathbb{R}^n} |\xi|^{2\alpha} |\hat{u}(\xi)|^2\, d\xi \le \tilde{c} \int_{\mathbb{R}^n} \psi(\xi) |\hat{u}(\xi)|^2\, d\xi,$$

i.e.

$$\|u\|_{L^p} \le \tilde{c}(\mathcal{E}^\psi(u,u))^{1/2},$$

where $\mathcal{E}^\psi(u,v) = \int_{\mathbb{R}^n} \psi(\xi)\hat{u}(\xi)\overline{\hat{v}(\xi)}\, d\xi$ is the scalar product in the extended Dirichlet space $H_{(e)}^{\psi,1}(\mathbb{R}^n;\mathbb{R})$. $\square$

The idea behind the theory of pseudo-differential operators and its applications to stochastic processes is to reduce properties of operators (or stochastic processes) to properties of the corresponding symbol. Often we will use comparison of a given symbol $q(x,\xi)$ with a fixed continuous negative definite function $\psi(\xi)$ in order to derive results for $-q(x,D)$ (and the corresponding stochastic process) analogously to those for $-\psi(D)$ (and the associated Lévy process). We want to apply this idea to the study of the transience and recurrence of a semigroup for this we need additional preparations.

Recall that $X_c := \{x \in \mathbb{R}^n; Gg(x) = \infty\}$ where G denotes the potential operator associated with $(\mathcal{E}; D(\mathcal{E}))$.

Lemma 3.5.61. *Let $(\mathcal{E}, D(\mathcal{E}))$ be a symmetric Dirichlet form on $L^2(\mathbb{R}^n;\mathbb{R})$ with corresponding sub-Markovian semigroups $(T_t)_{t\ge 0}$. Further suppose that $g \in L^1(\mathbb{R}^n;\mathbb{R}) \cap L^\infty(\mathbb{R}^n;\mathbb{R})$ such that $g|_{X_c} > 0$. Then the semigroup $(T_t^g)_{t\ge 0}$ associated with the symmetric Dirichlet form $(\mathcal{E}^g, D(\mathcal{E}))$ defined by*

$$\mathcal{E}^g(u,v) := \mathcal{E}(u,v) + \int_{\mathbb{R}^n} u(x)v(x)g(x)\, dx \tag{3.390}$$

is transient. Moreover, $G^g g \le 1$ a.e. on $\mathbb{R}^n$ and, on X_c we have $G^g g = 1$ where G^g is the potential operator associated with $(\mathcal{E}^g, D(\mathcal{E}))$.

Proof: We denote by $(R_\lambda^g)_{\lambda>0}$ the resolvent corresponding to $(\mathcal{E}^g, D(\mathcal{E}))$ and by $(R_\lambda)_{\lambda>0}$ the resolvent corresponding to $(\mathcal{E}, D(\mathcal{E}))$. Since

$$\mathcal{E}_\lambda(R_\lambda^g f, u) = \mathcal{E}_\lambda^g(R_\lambda^g f, u) - \int_{\mathbb{R}^n} ((R_\lambda^g f)u)g\,dx = (f - gR_\lambda^g f, u)_0, \tag{3.391}$$

we find

$$R_\lambda^g f = R_\lambda(f - gR_\lambda^g f). \tag{3.392}$$

In particular we have for $f \in L^1(\mathbb{R}^n; \mathbb{R})$, $f \ge 0$ a.e.,

$$R_\lambda^g f \le R_\lambda f, \quad \lambda > 0.$$

It follows now that every $(T_t)_{t\ge0}$-invariant set (which is by symmetry strongly $(T_t)_{t\ge0}$-invariant) is also $(T_t^g)_{t\ge0}$-invariant. In fact, for $A \subset \mathbb{R}^n$ being $(T_t)_{t\ge0}$-invariant we have

$$R_\lambda^g \chi_A u \le R_\lambda \chi_A u = \chi_A R_\lambda u$$

implying invariance, i.e. $R_\lambda^g \chi_A u = \chi_A R_\lambda^g \chi_A u$, and by symmetry the strong invariance follows. Moreover, for X_d defined as before by $X_d := \{x \in \mathbb{R}^n; Gg(x) < \infty\}$ it follows that $X_d \subset X_d^g := \{x \in \mathbb{R}^n; G^g f(x) < \infty\}$ except for a set of measure zero. Since

$$\begin{aligned}(R_\lambda^g g, gR_\lambda^g g)_0 &\le \mathcal{E}_\lambda^g(R_\lambda^g g, R_\lambda^g g) = (g, R_\lambda^g g)_0 \\ &\le \Big(\int_{\mathbb{R}^n} (R_\lambda^g g)^2 g\,dx\Big)^{1/2}\Big(\int_{\mathbb{R}^n} g\,dx\Big)^{1/2}\end{aligned}$$

we find that

$$\int_{\mathbb{R}^n} (R_\lambda^g g)^2 g\,dx \le \int_{\mathbb{R}^n} g\,dx$$

and monotone convergence for $\lambda \downarrow 0$ shows that a.e. on X_c we have $G^g < \infty$. Now for any $h \in L^1(\mathbb{R}^n; \mathbb{R})$, $h > 0$, we consider the function $f = h\chi_{X_d} + g\chi_{X_c} \in L^1(\mathbb{R}^n; \mathbb{R})$ which is strictly positive a.e. It follows that

$$G^g f = G^g(h\chi_{X_d} + g\chi_{X_c}) = \chi_{X_d} G^g h + \chi_{X_c} G^g g < \infty$$

implying by Lemma 3.5.22 that $(T_t^g)_{t\geq 0}$ is transient. Next we prove $G^g g \leq 1$ a.e. on $\mathbb{R}^n$. Since for $\varepsilon > 0$

$$\begin{aligned}\mathcal{E}(R_\varepsilon^g(\varepsilon f+fg),v) &+ \int_{\mathbb{R}^n} (R_\varepsilon^g(\varepsilon f + fg))v(g+\varepsilon)\,dx \\ &= \mathcal{E}^g(R_\varepsilon^g(\varepsilon f + fg),v) + \varepsilon \int_{\mathbb{R}^n} R_\varepsilon^g(\varepsilon f + fg)v\,dx \\ &= (\varepsilon f + fg, v)_0 = \int_{\mathbb{R}^n} (fv)(g+\varepsilon)\,dx,\end{aligned}$$

we may consider $R_\varepsilon^g((\varepsilon + g)f)$ as the resolvent at 1 of the Dirichlet form $(\mathcal{E}, D(\mathcal{E}))$ with reference L^2-space $L^2(\mathbb{R}^n; (g+\varepsilon)\lambda^{(n)})$. By sub-Markovianity we have $R_\varepsilon^g((\varepsilon+g)f) \leq 1$ a.e. for any $f \in L^2(\mathbb{R}^n;\mathbb{R})$ such that $0 \leq f \leq 1$ a.e. Now we first let ε decrease to 0 and then letting f increase pointwise to 1, we obtain $G^g g \leq 1$ a.e. on $\mathbb{R}^n$. Finally observe that (3.392) implies $G(g(1-G^g)) = G^g g < \infty$ a.e. Since $X_c \subset \{x \in \mathbb{R}^n; g(x) > 0\}$ we find with Lemma 3.5.27 that $g(1 - G^g g) = 0$ a.e. and now $g > 0$ a.e. on X_c yields that $G^g g = 1$ a.e. on X_c. □

Corollary 3.5.62. *A. Let $g \in L^1(\mathbb{R}^n;\mathbb{R}) \cap L^\infty(\mathbb{R}^n;\mathbb{R})$ be such that $g > 0$ a.e. on X_c and $g = 0$ on X_d. Then the function $u_k := R_{1/k}^g \in D(\mathcal{E})$ satisfies*

$$\lim_{k\to\infty} u_k(x) = \chi_{X_c}(x) \quad \text{a.e. and} \lim_{k\to\infty} \mathcal{E}(u_k,u_k) = 0. \tag{3.393}$$

B. We have $\chi_{X_c} \in \mathcal{F}_e$ and $\mathcal{E}(\chi_{X_c}, \chi_{X_c}) = 0$.

Now we may characterize transience and recurrence by using the extended Dirichlet space.

Theorem 3.5.63. *Let $(\mathcal{E}, D(\mathcal{E}))$ be a symmetric Dirichlet form with associated semigroup $(T_t)_{t\geq 0}$. The following conditions are equivalent:*
i) $(T_t)_{t\geq 0}$ is transient;
ii) if $u \in \mathcal{F}_e$ and $\mathcal{E}(u,u) = 0$ then $u = 0$ a.e.;
iii) $\mathcal{F}_e$ is a (real) Hilbert space with inner product $\mathcal{E}$.

Proof: We know already that the transience of $(T_t)_{t\geq 0}$ implies iii), compare Theorem 3.5.49, and from iii) it follows obviously that ii) holds. But ii) implies by Corollary 3.5.62.B that $\chi_{X_c} = 0$ a.e. i.e. the transience of $(T_t)_{t\geq 0}$. □

Theorem 3.5.64. *Let $(\mathcal{E}, D(\mathcal{E}))$ be a symmetric Dirichlet form with associated semigroup $(T_t)_{t\geq 0}$. The following conditions are equivalent:*

i) $(T_t)_{t\geq 0}$ is recurrent;

ii) there is a sequence $(u_k)_{k\in\mathbb{N}}$, $u_k \in D(\mathcal{E})$, such that $\lim_{k\to\infty} u_k = 1$ a.e. and $\lim_{k\to\infty} \mathcal{E}(u_k, u_k) = 0$;

iii) the constant 1 belongs to $\mathcal{F}_e$ and $\mathcal{E}(1,1) = 0$.

Proof: The equivalence of ii) and iii) is clear. Further, Corollary 3.5.62.B guarantees that i) implies iii). We prove that ii) implies i). Suppose that $(T_t)_{t\geq 0}$ is not recurrent, i.e. $\lambda^{(n)}(X_d) > 0$, and that ii) holds. By passing to unit contractions we may assume that $0 \leq u_k \leq 1$ a.e. Define $D := \{u|_{X_d}; u \in D(\mathcal{E})\}$ and $\mathcal{E}_D(u|_{X_d}, v|_{X_d}) := \mathcal{E}(\chi_{X_d}u, \chi_{X_d}v)$. From Theorem 3.5.11 it follows that $(\mathcal{E}_D, D)$ is a symmetric Dirichlet form on $L^2(X_d, \mathbb{R})$ and the associated semigroup is given by $(S_t)_{t\geq 0}$, $S_t(u|_{X_d}) := T_t(u\chi_{X_d})$ for $u \in D(\mathcal{E})$. Clearly $(S_t)_{t\geq 0}$ is sub-Markovian and transient on $L^2(X_d;\mathbb{R})$. Hence by Lemma 3.5.22 there exists $g \in L^1(X_d;\mathbb{R})$, $g > 0$, such that

$$\int_{X_d} u_k g\,dx \leq (\mathcal{E}(\chi_{X_d}u_k, \chi_{X_d}u_k))^{1/2} \leq (\mathcal{E}(u_k,u_k))^{1/2},$$

where we used for the last inequality once again Theorem 3.5.11. Thus, as $k \to \infty$ it follows that $\int_{X_d} g\,dx = 0$ which is of course a contradiction, hence the semigroup $(T_t)_{t\geq 0}$ is recurrent. □

Now we have arrived at a comparison theorem for transient and recurrent Dirichlet forms.

Theorem 3.5.65. *Let $(\mathcal{E}^{(i)}, D(\mathcal{E}^{(i)}))$, $i = 1,2$, be two symmetric Dirichlet forms on $L^2(\mathbb{R}^n;\mathbb{R})$ with associated semigroup $(T_t^i)_{t\geq 0}$, $i = 1,2$. Suppose further that*

$$D(\mathcal{E}^{(2)}) \subset D(\mathcal{E}^{(1)}) \quad \text{and } \mathcal{E}^{(1)}(u,u) \leq c\,\mathcal{E}^{(2)}(u,u) \text{ for all } u \in D(\mathcal{E}^{(1)}). \tag{3.394}$$

If $(T_t^1)_{t\geq 0}$ is transient, then $(T_t^2)_{t\geq 0}$ is transient too, and if $(T_t^2)_{t\geq 0}$ is recurrent, then $(T_t^1)_{t\geq 0}$ is recurrent too.

Proof: If $(T_t^1)_{t\geq 0}$ is transient there exists $g \in L^1(\mathbb{R}^n;\mathbb{R})$, $g > 0$ a.e., such that

$$\int_{\mathbb{R}^n} |u| g\,dx \leq (\mathcal{E}^{(1)}(u,u))^{1/2}$$

implying

$$\int_{\mathbb{R}^n} |u| \frac{g}{\sqrt{c}}\, dx \le \frac{1}{\sqrt{c}} (\mathcal{E}^{(1)}(u,u))^{1/2} \le \mathcal{E}^{(2)}(u,u),$$

i.e. the transience of $(T_t^2)_{t\ge 0}$. If however $(T_t^2)_{t\ge 0}$ is recurrent then $1 \in D(\mathcal{E}^{(2)})_e \subset D(E^{(1)})_e$ and

$$\mathcal{E}^{(1)}(1,1) \le c\,\mathcal{E}^{(1)}(1,1) = 0$$

implying by Theorem 3.5.64 the recurrence property of $(T_t^1)_{t\ge 0}$. □

Finally we give a characterization of the conservativeness of a symmetric L^2-sub-Markovian semigroup which is due to Y.Oshima [227].

Theorem 3.5.66. *Let $(T_t)_{t\ge 0}$ be a symmetric L^2-sub-Markovian semigroup with associated Dirichlet form $(\mathcal{E}, D(\mathcal{E}))$. Then the following conditions are equivalent:*

i) $(T_t)_{t\ge 0}$ is conservative, i.e. $T_t 1 = 1$;

ii) there exists a sequence $(u_k)_{k\in\mathbb{N}}$, $u_k \in D(\mathcal{E})$, such that $0 \le u_k \le 1$ a.e., $lim_{k\to\infty} u_k = 1$ a.e., and $\lim\limits_{k\to\infty} \mathcal{E}(u_k, v) = 0$ for all $v \in \mathcal{F}_e$;

iii) there exists a sequence $(u_k)_{k\in\mathbb{N}}$, $u_k \in D(\mathcal{E})$, satisfying $0 \le u_k \le 1$ a.e., $\lim\limits_{k\to\infty} u_k = 1$ a.e., and for some $v = R_\lambda f$, $\lambda > 0$ and $f \in L^1(\mathbb{R}^n;\mathbb{R}) \cap L^2(\mathbb{R}^n;\mathbb{R})$, $f > 0$, a.e., it follows that $\lim\limits_{k\to\infty} \mathcal{E}(u_k, v) = 0$.

Proof: Suppose that $(T_t)_{t\ge 0}$ is conservative and that $(f_k)_{k\in\mathbb{N}}$, $f_k \in L^2(\mathbb{R}^n;\mathbb{R})$ is a sequence satisfying $0 \le f_k \le f_{k+1} \le 1$ a.e. and $f_k \uparrow 1$ a.e. For $u_k := R_1 f_k$ we find $u_k \in D(\mathcal{E})$, $u_k = \int_0^\infty e^{-t} T_t f_k\, dt \uparrow 1$ a.e. since $T_t f_k \uparrow 1$, and therefore for any $v \in D(\mathcal{E}) \cap L^1(\mathbb{R}^n;\mathbb{R})$ it follows (using $\lambda R_\lambda 1 = 1$ which holds for a conservative semigroup) that

$$\lim_{k\to\infty} \mathcal{E}(u_k, v) = \lim_{k\to\infty} (f_k - R_1 f_k, v)_0 = 0$$

proving that i) implies ii) and it is obvious that ii) implies iii). Now suppose that $\lim\limits_{k\to\infty} \mathcal{E}(u_k, R_\lambda f) = 0$ holds for some $\lambda > 0$ and $f \in L^1(\mathbb{R}^n;\mathbb{R}) \cap L^2(\mathbb{R}^n;\mathbb{R})$, $f > 0$, and $u_k \in D(\mathcal{E})$, $0 \le u_k \le 1$ a.e. and $\lim\limits_{k\to\infty} \mathcal{E}(u_k, u_k) = 0$. Passing to the limit $k \to \infty$ in $\mathcal{E}(u_k, R_\lambda f) = (u_k - \lambda R_\lambda u_k, f)_0$ we get $(1 - \lambda R_\lambda 1, f) = 0$, or $\lambda R_\lambda 1 = 1$ a.e. From the resolvent equation it follows that $\mu R_\mu 1 = 1$ a.e. for all $\mu > 0$ implying the conservativeness of $(T_t)_{t\ge 0}$. □

Now we are going to discuss these results in the light of several examples of sub-Markovian semigroups generated by (non-local) pseudo-differential operators with negative definite symbols. For examples of (local) second order differential operators we refer to M.Fukushima, Y.Oshima and M.Takeda [102].

Example 3.5.67. In Chapter 2 we encountered many pseudo-differential operators $q(x, D)$ with a symbol $q(x, \xi)$ being comparable to some continuous negative definite function $\psi : \mathbb{R}^n \to \mathbb{R}$. In addition we often could prove a Gårding inequality

$$B(u, u) \geq \kappa_0 \|u\|^2_{H^{\psi,1}} - \lambda_0 \|u\|^2_0 \tag{3.395}$$

for all $u \in H^{\psi,1}(\mathbb{R}^n; \mathbb{R})$ with some $\kappa_0 > 0$ and $\lambda_0 > 0$. Furthermore $(B_\lambda, H^{\psi,1}(\mathbb{R}^n; \mathbb{R}))$ often was a non-symmetric Dirichlet form for $\lambda \geq \lambda_0$. We may rewrite (3.395) as

$$B(u, u) \geq \kappa_0 \int_{\mathbb{R}^n} \psi(\xi) |\hat{u}(\xi)|^2 \, d\xi - (\lambda_0 - \kappa_0) \|u\|^2_0. \tag{3.396}$$

Suppose that $\frac{1}{\psi} \in L^1_{loc}(\mathbb{R}^n; \mathbb{R})$, $\kappa_0 \geq \lambda_0$ and that B is symmetric. Then it follows that $(B, H^{\psi,1}(\mathbb{R}^n; \mathbb{R}))$ is a transient Dirichlet form. Of course, $B_\lambda(u, u)$, $\lambda > \lambda_0$ is always transient since the translation invariant Dirichlet form $\int_{\mathbb{R}^n} (1 + \psi(\xi)) |\hat{u}(\xi)|^2 \, d\xi$ is always transient.

Next we continue Example I.4.7.31.

Example 3.5.68. For $1 \leq j \leq n$ let $\psi_j^2 : \mathbb{R} \to \mathbb{R}$ be continuous negative definite functions and set $\psi^2(\xi) = \sum_{j=1}^n \psi_j^2(\xi_j)$. Further let $b_j : \mathbb{R}^{n-1} \to \mathbb{R}$ be functions independent of x_j belonging to $L^\infty(\mathbb{R}^{n-1}; \mathbb{R})$ and being bounded from below by $\lambda_0 > 0$, i.e. $b_j(x) \geq \lambda_0$ for all $x \in \mathbb{R}^{n-1}$, $1 \leq j \leq n$. Then

$$\mathcal{E}(u, v) = \sum_{j=1}^n (b_j(\cdot) \psi_j(D_j) u(\cdot), \psi_j(D_j) v(\cdot))_0 \tag{3.397}$$

is a symmetric Dirichlet form with domain $H^{\psi^2,1}(\mathbb{R}^n; \mathbb{R})$ satisfying

$$\mathcal{E}(u, u) \geq \lambda_0 \sum_{j=1}^n \|\psi_j(D_j) u\|^2_0.$$

Further the transience of the Dirichlet form

$$\mathcal{E}^{\psi^2}(u, v) = \sum_{j=1}^n \int_{\mathbb{R}^n} \psi_j^2(\xi_j) \hat{u}(\xi) \overline{\hat{v}(\xi)} \, d\xi$$

will imply the transience of $\mathcal{E}$. It is clear that $C_0^\infty(\mathbb{R}^n;\mathbb{R})$ is a dense subset of $D(A)$, the domain of the generator of $\mathcal{E}$, and on $C_0^\infty(\mathbb{R}^n;\mathbb{R})$ we have

$$Au(x) = -q(x,D)u(x) = -(2\pi)^{-n/2}\int_{\mathbb{R}^n} e^{ix\xi}\sum_{j=1}^n b_j(\tilde{x}_j)\psi_j^2(\xi_j)\hat{u}(\xi)\,d\xi \tag{3.398}$$

where $\tilde{x}_j = (x_1,\dots,x_{j-1},x_{j+1},\dots,x_n)$ and we consider b_j as an element in $L^\infty(\mathbb{R}^n;\mathbb{R})$ with $\frac{\partial b_j}{\partial x_j} = 0$.

We will now construct a special Dirichlet form belonging to the class considered in Example 3.5.69 and we will apply to this example Y.Oshima's conservativeness criterion.

This example is taken from our joint paper [132] with W.Hoh.

Example 3.5.69. Let $\mathcal{E}$ be the Dirichlet form (3.397) with generator $-q(x,D)$ given by (3.398). Suppose in addition that for all j, $1\le j\le n$, we have $\psi_j^2(0)=0$. Then the semigroup $(T_t)_{t\ge 0}$ associated with $\mathcal{E}$ is conservative.

Proof: Let $\varphi\in C_0^\infty(\mathbb{R}^n;\mathbb{R})$ such that $0\le\varphi\le 1$, $\operatorname{supp}\varphi\subset B_1(0)$ and $\varphi|_{B_{1/2}(0)}=1$. Further let $\varphi_k(x) := \varphi(\frac{x}{k})$ for any $k\in\mathbb{N}$. The sequence $(\varphi_k)_{k\in\mathbb{N}}$ is uniformly bounded and for any $x\in\mathbb{R}^n$ we have $\lim_{k\to\infty}\varphi_k(x)=1$. Now we claim that

$$\|q(x,D)\varphi_k\|_{L^\infty}\to 0 \quad \text{as } k\to\infty \tag{3.399}$$

holds. For this note that

$$\begin{aligned}
\sup_{x\in\mathbb{R}^n}|q(x,D)\varphi_k(x)| &= \sup_{x\in\mathbb{R}^n}(2\pi)^{-n/2}\Big|\int_{\mathbb{R}^n} e^{ix\xi}\sum_{j=1}^n b_j(\tilde{x}_j)\psi_j^2(\xi_j)\hat{\varphi}_k(\xi)\,d\xi\Big|\\
&\le (2\pi)^{-n/2}c\int_{\mathbb{R}^n}\sum_{j=1}^n\psi_j^2(\xi_j)|\hat{\varphi}_k(\xi)|\,d\xi\\
&= c'\int_{\mathbb{R}^n}\psi^2(\xi)|\hat{\varphi}(k\xi)|\,d\xi\\
&= c'\int_{|\xi|\le\frac{1}{\sqrt{k}}}\psi^2(\xi)k^n|\hat{\varphi}(k\xi)|\,d\xi + c'\int_{|\xi|\ge\frac{1}{\sqrt{k}}}\psi^2(\xi)k^n|\hat{\varphi}(k\xi)|\,d\xi.
\end{aligned}$$

It follows that

$$\begin{aligned}\int_{|\xi|\leq\frac{1}{\sqrt{k}}}\psi^2(\xi)k^n|\hat{\varphi}(k\xi)|\,d\xi &\leq \Big(\sup_{|\xi|\leq\frac{1}{\sqrt{k}}}\psi^2(\xi)\Big)\int_{\mathbb{R}^n}k^n|\hat{\varphi}(k\xi)|\,d\xi\\ &= \Big(\sup_{|\xi|\leq\frac{1}{\sqrt{k}}}\psi^2(\xi)\Big)\int_{\mathbb{R}^n}|\hat{\varphi}(k\xi)|\,d\xi.\end{aligned}$$

Since $\sup_{|\xi|\leq\frac{1}{\sqrt{k}}}\psi^2(\xi)\to 0$ as $k\to\infty$ and the integral of $|\hat{\varphi}|$ is finite, we get

$$\int_{|\xi|\leq\frac{1}{\sqrt{k}}}\psi^2(\xi)k^n|\hat{\varphi}(k\xi)|\,d\xi\to 0\quad\text{as } k\to\infty. \tag{3.400}$$

The function $\hat{\varphi}$ belongs to $S(\mathbb{R}^n)$, hence we have $|\hat{\varphi}(\xi)|\leq c|\xi|^{-(n+3)}$ for all $\xi\neq 0$. Thus we get using that $\psi^2(\xi)\leq\tilde{c}(1+|\xi|^2)$ for any continuous negative definite function that

$$\begin{aligned}&\int_{|\xi|>\frac{1}{\sqrt{k}}}\psi^2(\xi)k^n|\hat{\varphi}(k\xi)|\,d\xi\\ &\qquad\leq c\int_{\frac{1}{\sqrt{k}}\leq|\xi|\leq 1}k^n|\hat{\varphi}(k\xi)|\,d\xi + c\int_{|\xi|>1}|\xi|^2k^n|\hat{\varphi}(k\xi)|\,d\xi\\ &\qquad\leq c\frac{1}{k^3}\int_{\frac{1}{\sqrt{k}}\leq|\xi|\leq 1}|\xi|^{-(n+3)}\,d\xi + c\frac{1}{k^3}\int_{|\xi|>1}|\xi|^{-(n+1)}\,d\xi\\ &\qquad= c\frac{1}{k^3}(k^{3/2}-1)+c\frac{1}{k^3},\end{aligned}$$

which gives

$$\int_{|\xi|>\frac{1}{\sqrt{k}}}\psi^2(\xi)k^n|\hat{\varphi}(k\xi)|\,d\xi\to 0\quad\text{as } k\to\infty. \tag{3.401}$$

From (3.400) and (3.401) we get (3.399). Since $C_0^\infty(\mathbb{R}^n;\mathbb{R})\subset H^{\psi^2,1}(\mathbb{R}^n;\mathbb{R})$ and $\varphi_k\in C_0^\infty(\mathbb{R}^n)$, φ_k lies in the domain of the Dirichlet form $\mathcal{E}$ and $0\leq\varphi_k\leq 1$ as well as $\lim_{k\to\infty}\varphi_k=1$. It remains to show that for all $v\in H^{\psi^2,1}(\mathbb{R}^n;\mathbb{R})\cap L^1(\mathbb{R}^n;\mathbb{R})$ it follows that $\mathcal{E}(\varphi_k,v)\to 0$ as $k\to\infty$, then Theorem 3.5.66 will imply the conservativeness of $(T_t)_{t\geq 0}$. Since $\varphi_k\in C_0^\infty(\mathbb{R}^n;\mathbb{R})$ we find

$$\mathcal{E}(\varphi_k,v)=(q(x,D)\varphi_k,v)_0$$

and using (3.399) we get

$$|\mathcal{E}(\varphi_k, v)| = |(q(x,D)\varphi_k, v)_0| \le \|q(x,D)\varphi_k\|_{L^\infty}\|v\|_{L^1} \to 0$$

for $k \to \infty$, which proves the conservativeness of $(T_t)_{t\ge 0}$. □

Remark 3.5.70. A. The proof of Example 3.5.69 uses much ideas from the thesis of W.Hoh [121] and in volume III we will come back to these techniques when dealing with the martingale problem for pseudo-differential operators.

B. Example 3.5.69 fits very well to the observations that a Feller semigroup generated by a pseudo-differential operator with symbol $q(x,\xi)$ is often conservative if $q(x,0) = 0$ for all $x \in \mathbb{R}^n$. For a detailed discussion of this point we refer to the paper [249] of R.Schilling.

Let $L(x,D) = \sum_{k,l=1}^n \frac{\partial}{\partial x_k}(a_{kl}(x)\frac{\partial}{\partial x_l}\cdot)$ be a second order elliptic differential operator satisfying all assumptions in order that the Aronson estimates (2.442) hold, i.e. the operator extends to a generator of a symmetric sub-Markovian semigroup $(T_t)_{t\ge 0}$ having a symmetric density thus $T_t u(x) = \int_{\mathbb{R}^n} \Gamma(t,x,y)u(y)\,dy$ and

$$\kappa_1 \Gamma^{\gamma_1}(t,x,y) \le \Gamma(t,x,y) \le \kappa_2 \Gamma^{\gamma_2}(t,x,y) \tag{3.402}$$

holds where Γ^{γ_j} is given by

$$\Gamma^{\gamma_j}(t,x,y) = \frac{1}{(4\pi\gamma_j t)^{n/2})} e^{-\frac{|x-y|^2}{4\gamma_j t}}.$$

Now let $f : (0,\infty) \to \mathbb{R}$ be a Bernstein function and $(\eta_t)_{t\ge 0}$ the associated convolution semigroup with $\operatorname{supp}\eta_t \subset [0,\infty)$. For the subordinate semigroup $(T_t^f)_{t\ge 0}$ we have, compare Proposition 2.9.16

$$T_t^f u(x) = \int_{\mathbb{R}^n}\int_0^\infty \Gamma(s,x,y)\eta_t(ds)u(y)\,dy = \int_{\mathbb{R}^n} \Gamma^f(t,x,y)u(y)\,dy$$

and the estimates

$$\kappa_1 \Gamma^{\gamma_1,f}(t,x,y) \le \Gamma^f(t,x,y) \le \kappa_2 \Gamma^{\gamma_2,f}(t,x,y)$$

with

$$\Gamma^{\gamma_j,f}(t,x,y) = \int_0^\infty \Gamma^{\gamma_j}(s,x,y)\eta_t(ds).$$

We set

$$T_t^{\gamma_j,f}u(x) = \int_{\mathbb{R}^n} \Gamma^{\gamma_j,f}(t,x,y)u(y)\,dy.$$

If $n \geq 3$ then we know that $(T_t^{\gamma_j,f})_{t\geq 0}$, $j = 1,2$, is transient hence it follows for $g \in L^1(\mathbb{R}^n;\mathbb{R})$, $g > 0$ a.e., that

$$\int_0^\infty T_t^f g(x)\,dt \leq \kappa_2 \int_0^\infty \Gamma_t^{\gamma_2,f} g(x)\,dt < +\infty \quad \text{a.e.,}$$

i.e. $(T_t^f)_{t\geq 0}$ will be transient too.

For $n = 2$ and f satisfying $f(s) \geq c_0 s^\alpha$, $\alpha < 1$, for $0 < s \leq s_0$ it follows that $(T_t^{\gamma_j,f})_{t\geq 0}$ is also transient. Indeed, the corresponding continuous negative definite function is given by $f(\gamma_j|\xi|^2) \geq c_0\gamma_j^\alpha|\xi|^{2\alpha}$ and $\frac{1}{f(\gamma_j|\xi|^2)} \leq \frac{1}{c_0\gamma_j^\alpha|\xi|^{2\alpha}}$ for $0 \leq |\xi| \leq \frac{s_0}{c_0\gamma_j}$ implying that $(T_t^{\gamma_j,f})_{t\geq 0}$ is transient, hence $(T_t^f)_{t\geq 0}$ is transient by the same argument as before. In case $n = 1$ the assumption $f(s) \geq c_0 s^\alpha$, $\alpha < \frac{1}{2}$, for $0 < s \leq s_0$, implies that $(T_t^{\gamma_j,f})_{t\geq 0}$, hence $(T_t^f)_{t\geq 0}$ is transient. However for $f(s) \leq c_0 s^\alpha$, $\alpha > \frac{1}{2}$ and $0 < s \leq s_0$, it follows that $f(\gamma_1|\xi|^2) \leq c_1\gamma_1^\alpha|\xi|^\alpha$ or $\frac{1}{c_1\gamma_1^\alpha|\xi|^{2\alpha}} \leq \frac{1}{f(\gamma_1|\xi|^2)}$ for $0 \leq |\xi| \leq \frac{s_0}{c_1\gamma_1}$, and therefore $(T_t^{\gamma_j,f})_{t\geq 0}$ is recurrent, implying the recurrence of $(T_t^f)_{t\geq 0}$ by

$$\int_0^\infty T_t^{\gamma_1,f} g(x)\,dt \leq \int_0^\infty T_t^f g(x)\,dt.$$

The considerations just made give rise some new problems, for example

1. Investigate whether subordinating a transient sub-Markovian semigroup will always lead to a transient semigroup.
2. Find estimates corresponding $T_t\chi_A(x)$ with $\mu_t(x - A)$ where $(\mu_t)_{t\geq 0}$ is a suitable convolution semigroup.

It turns out that a satisfactory answer to some of these questions is related to a different problem related to the potential operator which will lead us to K.Yosida's theory of abstract potential operator, see [284]–[288], and the commentary of S.Watanabe in [279]. For an L^p-sub-Markovian semigroup we have defined its potential operator by

$$Gf := \int_0^\infty T_t f\,dt = \lim_{N\to\infty} \int_0^N T_t f\,dt, \quad f \geq 0, \tag{3.403}$$

and clearly the definition extends to Feller semigroups. Further for those $f \in L^p(\mathbb{R}^n;\mathbb{R})$ (or $f \in C_\infty(\mathbb{R}^n;\mathbb{R})$), $f \geq 0$ a.e., for which $Gf \in D(A)$, A being the generator of $(T_t)_{t\geq 0}$, we should expect $AGf = GAf = -f$, i.e. $-A^{-1} = G$. But so far (3.403) gives not a linear operator mapping from $L^p(\mathbb{R}^n;\mathbb{R})$ into $D(A)$ (in the case of Feller semigroups replace $L^p(\mathbb{R}^n;\mathbb{R})$ by $C_\infty(\mathbb{R}^n;\mathbb{R})$). Thus we may ask when there is a linear operator coinciding with G for "good" functions. This was exactly K.Yosida's approach to define abstract potential operators. Many of his results can be formulated for arbitrary strongly continuous contraction semigroups on some Banach space $(X, \|\cdot\|_X)$. We will follow in our discussion K.Yosida [288] as well as Chr.Berg and G.Forst [26].

Let $(X, \|\cdot\|_X)$ be a Banach space and $(T_t)_{t\geq 0}$ be a strongly continuous contraction semigroup on X with generator $(A, D(A))$. Since now we have not necessarily an order structure on X we define the potential operator associated with $(T_t)_{t\geq 0}$ by

$$D(G) := \Big\{u \in X;\ \lim_{N\to\infty} \int_0^N T_t u\, dt \quad \text{exists in } X\Big\}. \tag{3.404}$$

and

$$G := \int_0^\infty T_t u\, dt := \lim_{N\to\infty} \int_0^N T_t u\, dt. \tag{3.405}$$

First we aim to extend some results of Lemma I.4.1.14 to G.

Lemma 3.5.71. *Let $(T_t)_{t\geq 0}$ be a strongly continuous contraction semigroup on $(X, \|\cdot\|_X)$ with generator $(A, D(A))$ and potential operator $(G, D(G))$. The domain $D(G)$ of G is invariant under T_t for all $t > 0$ and for all $u \in D(G)$ it follows that $T_tGu = GT_tu$. Moreover, for $u \in D(G)$ we have*

$$T_tGu - Gu = -\int_0^t T_t u\, ds. \tag{3.406}$$

Proof: For $t > 0$ and $u \in D(G)$ we find

$$\int_0^N T_s(T_t u)\, ds = T_t \int_0^N T_s u\, ds, \qquad N \in \mathbb{N}.$$

As N tends to infinity $T_t(\int_0^N T_s u\, ds) \to T_tGu$, implying first $T_tD(G) \subset D(G)$ and then $T_tGu = GT_tu$. Now, a straightforward calculation for $u \in D(G)$

yields

$$\begin{aligned} T_t Gu &= \lim_{N\to\infty} \int_0^N T_s T_t u \, ds = \lim_{N\to\infty} \int_0^N T_{s+t} u \, ds \\ &= \lim_{N\to\infty} \int_t^{N+t} T_s u \, ds = \lim_{N\to\infty} \Big(\int_0^{N+t} T_s u \, ds - \int_0^t T_s u \, ds \Big) \\ &= Gu - \int_0^t T_s u \, ds \end{aligned}$$

proving (3.406). □

Lemma 3.5.72. *Let $(T_t)_{t\geq 0}$, $(A, D(A))$ and $(G, D(G))$ be as in Lemma 3.5.71. Then it follows that $D(G) \subset \overline{R(G)}$, and*

$$\lim_{t\to 0} \|T_t u\| = 0 \quad \textit{for} \quad u \in \overline{R(G)}. \tag{3.407}$$

Moreover, $R(G) \subset D(A)$ and

$$AGu = -u \quad \textit{for } u \in D(G), \tag{3.408}$$

implying in particular the injectivity of G.

Proof: Using (3.406) we find for $u \in D(G)$

$$G\Big(\frac{1}{t}(u - T_t u)\Big) = \frac{1}{t} \int_0^t T_s u \, ds$$

which yields as $t \to 0$ that u is the limit in $(X, \|\cdot\|_X)$ of elements in $R(G)$, i.e. $D(G) = \overline{R(G)}$. Next observe that $\{u \in X;\ \lim_{t\to\infty} T_t u = 0\}$ is closed which follows from the estimate $\|T_t u\|_X \leq \|u - u_\nu\|_X + \|T_t u_\nu\|_X$. Further, (3.406) implies for $u \in D(G)$ that $\lim_{t\geq\infty} T_t(Gu) = 0$, hence $\overline{R(G)} \subset \{u \in X;\ \lim_{t\to\infty} T_t u = 0\}$, i.e. (3.407) is proved. Finally, again using (3.406) it follows for $u \in D(G)$ and $t > 0$

$$\frac{1}{t}(T_t(Gu) - Gu) = -\frac{1}{t} \int_0^t T_s u \, ds$$

implying as $t \to 0$ that $Gu \in D(A)$ and $AGu = -u$ for $u \in D(G)$. □

Lemma 3.5.73. *In the situation of Lemma* 3.5.71 *we have for* $u \in D(G)$ *and* $N > 0$

$$\int_0^N T_s u\, ds \in D(G). \tag{3.409}$$

Proof: For $t > 0$ it follows that

$$\int_0^t T_r \Big(\int_0^N T_s u\ ds \Big) dr = \int_0^t \Big(\int_0^N T_{r+s} u\, ds \Big)\, dr$$
$$= \int_0^N \Big(\int_s^{s+t} T_r u\, dr \Big) ds = \int_0^N \Big(\int_0^{s+t} T_r u\, dr - \int_0^s T_r u\, dr \Big) ds.$$

Now, given $\varepsilon > 0$ choose $t_0 > 0$ such that

$$\Big\| Gu - \int_0^t T_s u\, ds \Big\|_X < \varepsilon \quad \text{for } t \geq t_0$$

and observe that for $t \geq t_0$

$$\Big\| \int_0^N \Big(\int_0^{s+t} T_r u\, dr \Big) ds - NGu \Big\|_X$$
$$= \Big\| \int_t^{N+t} \Big(\int_0^s T_r u\, dr - Gu \Big) ds \Big\|_X \leq N\varepsilon,$$

implying that $\lim\limits_{t \to 0} \int_0^t T_r \Big(\int_0^N T_s u\ ds \Big) dr \in D(G)$. $\square$

The next result, due to Chr.Berg [25], gives conditions in order that (3.407) does hold for all $u \in X$.

Theorem 3.5.74. *Let* $(T_t)_{t \geq 0}$ *be a strongly continuous contraction semigroup on* $(X, \|\cdot\|_X)$ *with generator* $(A, D(A))$ *and potential operator* $(G, D(G))$. *Then the assertions*

i) $\overline{D(G)} = X$;
ii) $\overline{R(G)} = X$;
iii) $\lim\limits_{t \to \infty} \|T_t u\| = 0$ *for all* $u \in X$

are equivalent. If any (hence all) of these conditions hold then G *is a densely defined closed operator on* X, A *is injective and satisfies*

$$G = -A^{-1} \quad \textit{and} \quad A = -G^{-1}. \tag{3.410}$$

Proof: We know already from Lemma 3.5.72 that i) implies ii) and that ii) implies iii). Now suppose that iii) holds. By Lemma I.4.1.14 we have for $u \in D(A)$ that

$$\int_0^t T_s Au \, ds = T_t u - u$$

and therefore iii) implies that

$$\lim_{t \to \infty} \int_0^t T_s Au \, ds = -u, \tag{3.411}$$

i.e. $Au \in D(G)$ and $GAu = -u$. Thus A is injective and maps $D(A)$ into $D(G)$ which together with Lemma 3.5.72 yields (3.410). The closedness of A implies now that of G and from (3.411) we deduce by Lemma 3.5.73 that $D(A) \subset \overline{D(G)}$. Since $D(A)$ is dense in $(X, \|\cdot\|_X)$ we find $\overline{D(G)} = X$. □

We can now introduce K.Yosida's abstract potential operator.

Definition 3.5.75. Let $(T_t)_{t \geq 0}$ be a strongly continuous contraction semigroup on a Banach space $(X, \|\cdot\|_X)$ with generator $(A, D(A))$, potential operator $(G, D(G))$ and resolvent $(R_\lambda)_{\lambda > 0}$. The *abstract potential operator* or *zero order resolvent* $(R_0, D(R_0))$ associated with $(T_t)_{t \geq 0}$ is the operator with domain

$$D(R_0) := \{u \in X; \lim_{\lambda \to 0} R_\lambda u \quad \text{exists}\} \tag{3.412}$$

and

$$R_0 u := \lim_{\lambda \to 0} R_\lambda u, u \in D(R_0). \tag{3.413}$$

First let us prove some results for R_0 analogous to those for G.

Lemma 3.5.76. *Let $(T_t)_{t \geq 0}$ be as in Definition 3.5.75.*
A. For all $\lambda > 0$ we have $R_\lambda(D(R_0)) \subset D(R_0)$.
B. If $\lambda > 0$ and $u \in D(R_0)$ then $R_\lambda(R_0 u) = R_0(R_\lambda u)$.
C. The resolvent equation extends in the sense that

$$R_0 u = R_\lambda u + \lambda R_\lambda (R_0 u) \tag{3.414}$$

holds for $\lambda > 0$ and $u \in D(R_0)$.

D. It holds $D(R_0) \subset \overline{R(R_0)}$.

E. For $u \in \overline{R(R_0)}$ *it follows that*

$$\lim_{\lambda \to 0} \lambda R_\lambda u = 0 \tag{3.415}$$

F. It holds $R(R_0) \subset D(A)$ *and* $A(R_0 u) = -u$ *for* $u \in D(R_0)$ *implying that* R_0 *is one-to-one.*

Proof: The assertions A and B clearly follow from $R_\lambda R_\mu = R_\mu R_\lambda$, $\lambda, \mu > 0$, and (3.414) is an obvious consequence of the resolvent equation. Further, the strong continuity of $(\lambda R_\lambda)_{\lambda>0}$ implies C, and (3.414) that

$$R_0(\lambda(u - \lambda R_\lambda u)) = \lambda N_\lambda u \to 0,$$

i.e. $D(R_0) \subset \overline{R(R_0)}$ and D is proved. Again as in the proof of Lemma 3.5.72 it is easy to see that $\{u \in X; \lim_{\lambda \to 0} \lambda R_\lambda u = 0\}$ is closed and using C we find for $u \in D(R_0)$

$$\lim_{\lambda \to 0} \lambda R_\lambda(R_0 u) = R_0 u - \lim_{\lambda \to 0} R_\lambda = 0,$$

or $\overline{R(R_0)} \subset \{u \in X; \lim_{\lambda \to 0} \lambda R_\lambda u = 0\}$ and hence E is shown. Note that $R_\lambda = (\lambda \,\mathrm{id} - A)^{-1}$. From this it follows that

$$A(R_\lambda u) = -u + \lambda R_\lambda u \quad \text{for } u \in X,$$

and for $u \in D(R_0)$ it follows now

$$\lim_{\lambda \to 0} A R_\lambda u = -u.$$

But the operator $(A, D(A))$ is closed implying that $\lim_{\lambda \to 0} R_\lambda u = R_0 u \in D(A)$ and $A(R_0 u) = -u$ and the lemma is proved. □

Corollary 3.5.77. *The abstract potential operator* R_0 *satisfies*

$$\|\lambda R_0 u\|_X \le \|u + \lambda R_0 u\|_X, \quad \lambda > 0, \tag{3.416}$$

for all $u \in D(R_0)$.

Proof: The extended resolvent equation yields $\lambda R_0 u = \lambda R_\lambda(u + \lambda R_0 u)$. implying (3.416) since $\|\lambda R_\lambda\|_X \le 1$. □

Now we may prove K.Yosida result, see [288].

Theorem 3.5.78. *The following conditions are equivalent:*

i) $D(R_0)$ *is dense in* X*;*

ii) $R(R_0)$ *is dense in* X*;*

iii) for all $u \in X$ *it follows that* $\lim_{\lambda \to 0} \lambda R_\lambda u = 0$.

If one, hence all of, i)-iii) hold then R_0 *is a densely defined closed operator on* $(X, \|\cdot\|_X)$ *and*

$$R_0 = -A^{-1} \quad \textit{and} \quad A = -R_0^{-1} \tag{3.417}$$

hold.

Proof: From Lemma 3.5.76 we deduce immediately that i) implies ii) and ii) implies iii). Now assume that iii) holds. For $u \in D(A)$ we find

$$AR_\lambda u = R_\lambda A u = -u + \lambda R_\lambda u, \quad \lambda > 0,$$

and it follows that $\lim_{\lambda \to 0} R_\lambda(Au) = -u$, i.e. $Au \in D(R_0)$ and $R_0(Au) = -u$, thus A is one-to-one from $D(A)$ into $D(R_0)$ which yields by Lemma 3.5.76 that (3.417) holds, in particular it follows that R_0 is closed. Moreover, for $u \in D(A)$ we find $\lim_{\lambda \to 0} A(R_\lambda u) = -u$, thus $D(A) \subset \overline{D(R_0)}$ since $A(R_\lambda u) \in D(R_0)$. But $D(A) \subset X$ is dense, hence $D(R_0)$ is dense in X, i.e. iii) implies i) and the theorem is proved. □

Finally we may relate the potential operator G and the abstract potential operator or zero order resolvent R_0 to each other.

Proposition 3.5.79. *The abstract potential operator* $(R_0, D(R_0))$ *is an extension of the potential operator* $(G, D(G))$*. However, if* $\lim_{t \to \infty} T_t u = 0$ *for all* $u \in X$ *then they coincide, i.e.* $G = R_0$.

Proof: For $u \in D(G)$ we have to show that $u \in D(R_0)$ and $R_0 u = Gu$. For $\lambda > 0$ and $t > 0$ we find for $u \in D(G)$ from an integration by parts

$$\int_0^t e^{-\lambda s} T_s u \, ds = e^{-\lambda s} \int_0^s T_r u \, dr \Big|_0^t + \int_0^t \lambda e^{-\lambda s} \Big(\int_0^s T_r u \, dr \Big) ds$$

and for $t \to \infty$ we get

$$R_\lambda u = \int_0^\infty \lambda e^{-\lambda s} \Big(\int_0^s T_r u \, dr \Big) ds = \int_0^\infty e^{-r} \Big(\int_0^{r/\lambda} T_s u \, ds \Big) dr.$$

Now passing to the limit $\lambda \to 0$ we arrive at

$$\lim_{\lambda\to 0} R_\lambda u = Gu$$

proving that $(R_0, D(R_0))$ extends $(G, D(G))$.

By Theorem 3.5.74, $\lim_{t\to\infty} T_t u = 0$ for all $u \in X$ implies that $D(G)$ is dense in X, hence $\overline{D(G)} \subset \overline{D(R_0)} = X$, and therefore Theorem 3.5.78 gives the second assertion of the proposition. □

Corollary 3.5.80. *Let $(T_t)_{t\geq 0}$ be a symmetric transient sub-Markovian semigroup on $L^2(\mathbb{R}^n;\mathbb{R})$ with generator $(A, D(A))$. Then the corresponding potential operator and the abstract potential operator as L^2-operator do coincide and satisfy $A = -G^{-1}$ as well as $G = -A^{-1}$.*

Proof: We just have to combine Lemma 3.5.48 with Proposition 3.5.79. □

Let us return to our investigation on the transience of sub-Markovian semigroups. We will now discuss the case of translation invariant sub-Markovian semigroups, in particular subordinated semigroups which originate from convolution semigroups. We will follow partly the presentation of Chr.Berg and G.Forst [26].

Let $(\mu_t)_{t\geq 0}$ be a convolution semigroup on $\mathbb{R}^n$. The resolvent of $(\mu_t)_{t\geq 0}$ is then defined by

$$\rho_\lambda = \int_0^\infty e^{-\lambda t}\mu_t \, dt$$

with the interpretation

$$\begin{aligned}\rho_\lambda(u) = \langle \rho_\lambda, u\rangle = \int_{\mathbb{R}^n} u(x)\rho_\lambda(dx) &= \int_0^\infty e^{-\lambda t}\int_{\mathbb{R}^n} u(x)\mu_t(dx)\, dt \\ &= \int_0^\infty e^{-\lambda t}\mu_t(dx)\, dt\end{aligned}$$

for all $u \in C_0(\mathbb{R}^n;\mathbb{R})$, $u \geq 0$, compare Remark I.3.9.11. We call $(\mu_t)_{t\geq 0}$ a *transient convolution semigroup* if

$$\lim_{\lambda\to 0} \rho_\lambda(u) < \infty \quad \text{for all} \quad u \in C_0(\mathbb{R}^n;\mathbb{R}), \quad u \geq 0,$$

and in this case we may define the kernel κ

$$\kappa(u) := \lim_{\lambda\to 0} \rho_\lambda(u), \quad u \in C_0(\mathbb{R}^n;\mathbb{R}), \quad u \geq 0.$$

The kernel κ is called the *potential kernel* associated with the convolution semigroup $(\mu_t)_{t\geq 0}$.

Since for $y \in \mathbb{R}^n$ the function $u_y(\cdot) = u(\cdot - y) \in C_0(\mathbb{R}^n)$ if $u \in C_0(\mathbb{R}^n)$, we find that $(\mu_t)_{t\geq 0}$ is transient if and only if the associated operator semigroup $(T_t^\psi)_{t\geq 0}$ defined by $T_t u(x) = \int_{\mathbb{R}^n} u(x-y)\mu_t(dy)$, $\hat{\mu}_t(\xi) = e^{-t\psi(\xi)}$, is transient.

Next let $(\eta_t)_{t\geq 0}$ be a convolution semigroup on $\mathbb{R}$ with $\operatorname{supp}\nu_t \subset [0,\infty)$ and associated Bernstein function f. The following result is taken from Chr.Berg and G.Frost [26].

Proposition 3.5.81. *Let $(\eta_t)_{t\geq 0}$ and f be as above. Either $f(x) = 0$ for all $x \in (0,\infty)$ implying $\eta_t = \varepsilon_0$ for all $t > 0$, or $f(x) > 0$ for all $x \in (0,\infty)$. Then, in the latter case, $(\eta_t)_{t\geq 0}$ is transient and its potential kernel τ is characterized by*

$$\mathcal{L}(\tau) = \frac{1}{f}. \tag{3.418}$$

Proof: Since f is increasing and concave it follows that $f(x) = 0$ for all x if it is 0 for some $x > 0$. Now, if $f(x) > 0$ for all $x > 0$ then

$$\int_0^\infty \mathcal{L}(\eta_t)(x)\,dt = \int_0^\infty e^{-tf(x)}\,dt = \frac{1}{f(x)}, \tag{3.419}$$

where we used the defining identity $\mathcal{L}(\eta_t)(x) = e^{-tf(x)}$. Given $\varphi \in C_0(\mathbb{R}^n;\mathbb{R})$, $\varphi \geq 0$, we find $c > 0$ such that for $s \geq 0$ we have $\varphi(s) \leq ce^{-s}$. Therefore we have

$$\int_0^\infty \int_{\mathbb{R}} \varphi(x)\eta_t(dx)\,dt \leq c\int_0^\infty (\mathcal{L}(\eta_t))(1)\,dt = \frac{c}{f(1)} < \infty,$$

which implies that the transience of $(\eta_t)_{t\geq 0}$ and then by (3.419) that $\mathcal{L}(\tau) = \frac{1}{f}$. $\square$

In the following we will assume that $(\eta_t)_{t\geq 0}$ is associated with f, $f(x) > 0$, and denote its potential kernel always by τ. Since τ is a positive measure we may define for $g \in L^1((0,\infty);\mathbb{R})$, $g \geq 0$ a.e.,

$$\int_0^\infty g(s)\tau(ds) := \lim_{k\to\infty} \int_0^\infty g_k(s)\tau(ds),$$

where $g_k \geq 0$ is a sequence in $C_0(\mathbb{R};\mathbb{R})$ increasing to g. Thus if $(T_t)_{t\geq 0}$ is a transient Dirichlet semigroup on $L^2(\mathbb{R}^n;\mathbb{R})$ and $h \in S(\mathbb{R}^n;\mathbb{R})$, $h > 0$, then we find

$$\int_0^\infty T_t^f h\,dt = \int_0^\infty \Big(\int_0^\infty T_t h\,\eta_t(ds)\Big)\,dt = \int_0^\infty (T_s h)(\tau(ds)), \tag{3.420}$$

i.e. we have

Lemma 3.5.82. *Let $(T_t)_{t\geq 0}$ be a transient Dirichlet semigroup and $(\eta_t)_{t\geq 0}$ a convolution semigroup supported in $[0,\infty)$ and associated Bernstein function f, $f(r) > 0$ for all $r > 0$. Denote the potential kernel associated with $(\eta_t)_{t\geq 0}$ by τ. If for some $h \in S(\mathbb{R}^n;\mathbb{R})$, $h > 0$, it follows that*

$$\int_0^\infty (T_s h)(x)\tau(ds) < \infty \tag{3.421}$$

for all $x \in \mathbb{R}^n$, then $(T_t^f)_{t\geq 0}$ is transient.

To our knowledge there are no general results assuring (3.421), but if $T_t u(x) = \int_{\mathbb{R}^n} u(x-y)\mu_t(dy)$, then there is a complete satisfactory theory due to F.Hirsch [115] and M.Itô [146] and we will discuss their results immediately following Chr.Berg and G.Frost [26]. However, having in mind the Aronson estimate (2.442), then we may formulate the following corollary.

Corollary 3.5.83. *Let $(T_t)_{t\geq 0}$ be a transient Dirichlet semigroup and $(\eta_t)_{t\geq 0}$ a convolution semigroup supported on $[0,\infty)$ associated with the Bernstein function f, $f(r) > 0$ for $r > 0$. If there is a convolution semigroup $(\mu_t)_{t\geq 0}$ such that with some non-negative constant c_0 we have*

$$T_t\chi_A(x) \leq c_0\mu_t(x-A) \tag{3.422}$$

for all bounded Borel sets, then $(T_t^f)_{t\geq 0}$ is transient if $\int_0^\infty \mu_s(h)\tau(ds) < \infty$ for some $h \in S(\mathbb{R}^n;\mathbb{R})$, $h > 0$.

Remark 3.5.84. A. If $(T_t)_{t\geq 0}$ is generated by $-q(x,\xi)$ and $q(x,\xi) \sim \psi(\xi)$, where ψ is a continuous negative definite function it would be natural to seek for some $(\mu_t)_{t\geq 0}$ such that $\hat{\mu}_t(\xi) = e^{-tc_0\psi(c_1\xi)}$ and (3.422) holds.

B. Clearly we do not need a convolution semigroup in (3.422), any estimate of the type $T_t h(x) \leq g_t(x)$ such that $\int_0^\infty g_s(x)\tau(ds) < \infty$ is sufficient. Hence we are longing for estimates for the kernel $p_t(x,A) = T_t\chi_A(x)$, this will be the theme of the next section.

As stated before our aim now is to investigate $\int_0^\infty \mu_s\tau(ds)$ for a convolution semigroup $(\mu_t)_{t\geq 0}$ and τ as in Proposition 3.5.81.

As we know many properties of $(\mu_t)_{t\geq 0}$ are characterized by properties of its associated continuous negative definite function $\psi : \mathbb{R}^n \to \mathbb{C}$, and the set $\Gamma^\psi := \{\xi \in \mathbb{R}^n; \psi(\xi) = 0\}$ as well we $\Gamma^{\operatorname{Re}\psi} : \{\xi \in \mathbb{R}^n; \operatorname{Re}\psi(\xi) = 0\}$ are of importance when considering questions like conservativeness, characterization of domains or dealing with transience. In the rest of this section unless it is stated otherwise let us assume that $\Gamma^\psi = \Gamma^{\operatorname{Re}\psi} \subset \{0\}$, i.e. the periodicity group of $(\mu_t)_{t\geq 0}$ is either empty or trivial. (In fact we prove this condition only to avoid a discussion of periodicity which is never of relevance when having an assumption as (2.302).) It is known from Lemma I.3.6.27 that each of these sets form a closed subgroup of $(\mathbb{R}^n, +)$. For a set $A \subset \mathbb{R}^n$ we denote for a moment by

$$A^\perp := \{\xi \in (\mathbb{R}^n)^*; e^{ix\cdot\xi} = 1 \quad \text{for all } x \in A\}$$

its group-theoretical orthogonal complement.

Note that under the general assumptions made above we have

$$G_0 = (\Gamma^\psi)^\perp = \mathbb{R}^n,$$

where $G_0 \subset \mathbb{R}^n$ is the smallest closed subset of $(\mathbb{R}^n, +)$ with the property that $\bigcup_{t>0}(\operatorname{supp}\mu_t) \subset G_0$, compare Chr.Berg and G.Forst [26].

For the following the reader should recall that measures are distributions of order zero, hence for $u \in C_0(\mathbb{R}^n)$ we may write $\langle \mu, f\rangle$ instead $\int_{\mathbb{R}^n} f(x)\mu(dx)$.

The following result is again taken from Chr.Berg and G.Frost [26].

Theorem 3.5.85. *Let $(T_t)_{t\geq 0}$ be the L^2-sub-Markovian semigroup associated with the convolution semigroup $(\mu_t)_{t\geq 0}$ and denote the corresponding continuous negative definite function by ψ.*

A. The abstract potential operator $(R_0, D(R_0))$ is densely defined if and only if $\psi|_K \neq 0$ a.e. for any compact set $K \subset \mathbb{R}^n$.

B. The potential operator $(G, D(G))$ is densely defined if and only if $(\operatorname{Re}\psi)|_K \neq 0$ for all compact sets $K \subset \mathbb{R}^n$.

Proof: We denote by $(R_\lambda)_{\lambda>0}$ the resolvent of $(T_t)_{t\geq 0}$ and by $(\rho_\lambda)_{\lambda>0}$ the resolvent of $(\mu_t)_{t\geq 0}$. If $\psi(0) > 0$ then

$$\frac{|(T_t u, v)_0|}{\|u\|_0 \|v\|_0} \leq e^{-t\psi(0)}$$

implying that $\|T_t\|_{L^2 - L^2} \leq e^{-t\psi(0)}$. Thus for $u \in L^2(\mathbb{R}^n; \mathbb{R})$ we find

$$\int_0^\infty \|T_t u\|_{L^2}\, dt \leq \frac{\|u\|_{L^2}}{\psi(0)}$$

which implies that $\int_0^\infty T_t u\,dt$ exists in $L^2(\mathbb{R}^n;\mathbb{R})$, thus

$$\lim_{N\to\infty}\int_0^N T_t u\,dt = \int_0^\infty T_t u\,dt$$

and

$$\lim_{\lambda\to 0} R_\lambda u = \int_0^\infty T_t u\,dt$$

do exists in $L^2(\mathbb{R}^n;\mathbb{R})$. Thus in this case R_0 and G are defined for all $u \in L^2(\mathbb{R}^n;\mathbb{R})$ and in fact they coincide as operators being bounded with operator norm less or equal to $\frac{1}{\psi(0)}$. Now suppose that $\psi(0) = 0$. For $u \in C_0(\mathbb{R}^n;\mathbb{R})$ it follows that

$$F^{-1}(\lambda R_\lambda(F^{-1}u)) = \frac{\lambda}{\lambda+\psi(\cdot)}u(\cdot), \qquad \lambda > 0.$$

As $\lambda \to 0$ it follows that $\frac{\lambda}{\lambda+\psi}u \to 0$ a.e. and since $\left|\frac{\lambda}{\lambda+\psi}u\right| \le |u|$ the dominated convergence theorem yields now

$$\lim_{\lambda\to 0}\|F^{-1}(\lambda R_\lambda(F^{-1}u))\|_{L^2} = \lim_{\lambda\to 0}\int_{\mathbb{R}^n}\left|\frac{\lambda}{\lambda+\psi(\xi)}u(\xi)\right|^2 d\xi = 0.$$

Since $S(\mathbb{R}^n;\mathbb{R}) \subset \{F^{-1}u; u \in C_0(\mathbb{R}^n;\mathbb{R})\}$ is dense in $L^2(\mathbb{R}^n;\mathbb{R})$ and $\{u \in L^2(\mathbb{R}^n;\mathbb{R});\ \lim\limits_{\lambda\to\infty}\lambda R_{\lambda^n} = 0\}$ is closed it follows by Theorem 3.5.78.iii) that R_0 is densely defined. If there is a bounded set A, $\lambda^{(n)}(A) > 0$ such that $\psi|_A = 0$, then Γ^ψ must be an open subgroup of $(\mathbb{R}^n,+)$ and $(\Gamma^\psi)^\perp$ must be a compact subgroup of $(\mathbb{R}^n,+)$ since in any locally compact Abelian group the orthogonal complement of any open subgroup is compact, compare H.Reiter and J.Stegeman [235], Proposition 4.2.21. Hence the measure $(\lambda\rho_\lambda) * (\lambda S\rho_\lambda)$, $Sx = -x$, are concentrated on $(\Gamma^\psi)^\perp$. Now, with $u \in C_0(\mathbb{R}^n;\mathbb{R})$, $u \ge 0$, and $u * Su \ge 1$ on the compact set $(\Gamma^\psi)^\perp$ we find for all $\lambda > 0$

$$1 < \langle \lambda\rho_\lambda * \lambda S\rho, u * Su\rangle = \int_{\mathbb{R}^n}|\lambda\rho_\lambda * u|^2\,dx = \|\lambda R_\lambda u\|_{L^2}^2$$

and from Theorem 3.5.78 we deduce that $D(R_0)$ is not dense in $L^2(\mathbb{R}^n;\mathbb{R})$ and part A is proved. Now, $\operatorname{Re}\psi$ is a continuous negative definite function too and therefore $\Gamma^{\operatorname{Re}\psi}$ is again a closed subgroup of $\mathbb{R}^n$. For $u \in C_0(\mathbb{R}^n)$ we find

$$F^{-1}(T_t(F^{-1}u)) = e^{-t\psi}u, \quad t > 0,$$

and if $\operatorname{Re}\psi|_K \neq 0$ a.e. it follows as above that

$$\lim_{t\to\infty} \|T_t F^{-1}u\|_{L^2}^2 = \lim_{t\to\infty} \int_{\mathbb{R}^n} e^{-2t\operatorname{Re}\psi(\xi)} |u(\xi)|^2\, d\xi = 0$$

which by Theorem 3.5.74 yields that G is densely defined. Conversely, if $(\operatorname{Re}\psi)|_A \neq 0$, A bounded and $\lambda^{(n)}(A) > 0$, we may argue as before that $(\Gamma^{\operatorname{Re}\psi})^\perp$ is a compact subgroup on $\mathbb{R}^n$ and with $u \in C_0(\mathbb{R}^n;\mathbb{R})$, $u \geq 0$, such that $u * Su \geq 1$ on $(\Gamma^{\operatorname{Re}\psi})^\perp$ we find for all $t > 0$

$$1 < \langle \mu_t * S\mu_t, u * Su\rangle = \int_{\mathbb{R}^n} |\mu_t * u|^2(x)\, dx = \|T_t u\|_{L^2}^2,$$

and by Theorem 3.5.74 deduce that $D(G)$ is not dense in $L^2(\mathbb{R}^n;\mathbb{R})$. □

Corollary 3.5.86. *Let $(T_t)_{t\geq 0}$ be an L^2-sub-Markovian semigroup associated with a convolution semigroup $(\mu_t)_{t\geq 0}$. Its abstract potential operator is densely defined except when $\mu_t = \varepsilon_0$ for all $t > 0$, whereas its potential operator is densely defined whenever $(\mu_t)_{t\geq 0}$ is not a translation semigroup.*

Proof: If R_0 is not densely defined then Γ^ψ is an open subgroup of $(\mathbb{R}^n.+)$ as it follows from the arguments in the proof of Theorem 3.5.85. Hence $(\Gamma^\psi)^\perp \subset \mathbb{R}^n$ must be compact, hence $(\Gamma^\psi)^\perp = \{0\}$ being the only compact subgroup of $(\mathbb{R}^n, +)$, thus $\Gamma^\psi = \mathbb{R}^n$, implying that $\mu_t = \varepsilon_0$ for all $t > 0$. If G is not densely defined then $\Gamma^{\operatorname{Re}\psi}$ is an open subgroup of $(\mathbb{R}^n, +)$ and as before we conclude that $\Gamma^{\operatorname{Re}\psi} = \mathbb{R}^n$ hence $\operatorname{Re}\psi(\xi) = 0$ for all $\xi \in \mathbb{R}^n$, i.e. $\psi(\xi) = il(\xi)$ and, by Example I.3.6.19, $l : \mathbb{R}^n \to \mathbb{R}$ must be linear, i.e. $\psi(\xi) = ib \cdot \xi$ for some $b \in \mathbb{R}^n$ and μ_t is the translation semigroup $\mu_t = \varepsilon_{bt}$. □

Remark 3.5.87. If $(T_t)_{t\geq 0}$ is an L^2-sub-Markovian semigroup associated with a convolution semigroup $(\mu_t)_{t\geq 0}$ then whenever $(\mu_t)_{t\geq 0}$ is not of translation type $(\varepsilon_{bt})_{t\geq 0}$, $b \in \mathbb{R}^n$, then the abstract potential operator and the potential operator related to $(T_t)_{t\geq 0}$ coincide and they are densely defined.

Remark 3.5.88. A careful inspection of the proof of Theorem 3.5.85 shows that we do not really use much L^2-norms, but the representation of all operators using the Fourier transform. From the early examples in Section I.4.1 we may deduce that the assertions of Theorem 3.5.85 indeed do hold for all L^p-sub-Markovian semigroups $(T_t^{(p)})_{t\geq 0}$, $2 < p < \infty$, and for $p = \infty$ we have to take the Feller semigroup associated with $(\mu_t)_{t\geq 0}$. In addition, since $\overline{\psi}$ is also a continuous negative definite function we may use the standard duality argument to find that Theorem 3.5.85 holds for all the L^p-sub-Markovian semigroups, $1 < p < \infty$, associated with $(\mu)_{t\geq 0}$. In view of our transient-recurrent observation (p-independence) this should not be surprising.

To proceed further we need

Lemma 3.5.89. *Let $(T_t)_{t\geq 0}$ be a Feller semigroup. Then the following conditions are equivalent*
i) $C_0(\mathbb{R}^n;\mathbb{R}) \subset D(G)$;
ii) $C_0(\mathbb{R}^n;\mathbb{R}) \subset D(R_0)$.

Proof: Since by Proposition 3.5.79 $(R_0, D(R_0))$ is an extension of $(G, D(G))$ it is obvious that i) implies ii). Suppose that $C_0(\mathbb{R}^n;\mathbb{R}) \subset D(R_0)$ and take $u \in C_0(\mathbb{R}^n;\mathbb{R})$, $u \geq 0$. Since $(T_t)_{t\geq 0}$ is a Feller semigroup it follows that $R_0 u \in C_\infty(\mathbb{R}^n;\mathbb{R})$ and $R_0 u \geq 0$. Moreover, for every $x \in \mathbb{R}^n$ we have

$$R_0 u(x) = \lim_{\lambda\to 0} R_\lambda u(x) = \int_0^\infty T_t u(x)\, dt$$

implying

$$\lim_{N\to\infty} \int_0^N T_t u(x)\, dt = (R_0 u)(x) \quad \text{for } x \in \mathbb{R}^n. \tag{3.423}$$

We claim that the limit in (3.423) is even uniform. For $\varepsilon > 0$ given, take $K \subset \mathbb{R}^n$ compact such that

$$0 \leq R_0 u(x) \leq \varepsilon \quad \text{in } K^c. \tag{3.424}$$

From Dini's theorem, compare Theorem I.2.1.6, it follows that the convergence in (3.423) is uniform on K, i.e. there exists N_0 such that for all $N \geq N_0$

$$\left| \int_0^N T_t u(x)\, dt - R_0 u(x) \right| \leq \varepsilon \quad \text{for all } x \in K,$$

and together with (3.424) the lemma follows. □

We want to emphasize that in Lemma 3.5.89 we do not assume $(T_t)_{t\geq 0}$ to be associated with a convolution semigroup, i.e. to be translation invariant.

Denote by τ_x, $x \in \mathbb{R}^n$, the right translation $\tau_x : \mathbb{R}^n \to \mathbb{R}^n$, $y \mapsto y + x$.

Definition 3.5.90. A. A Borel measure μ on $\mathbb{R}^n$ is called *shift-bounded* if the set $\{\tau_x\mu; x \in \mathbb{R}^n\}$ is vaguely bounded.

B. A Borel measure μ is said to *vanish at infinity* if $\mu * u \in C_\infty(\mathbb{R}^n; \mathbb{R})$ for all $u \in C_0(\mathbb{R}^n; \mathbb{R})$.

C. A Feller semigroup $(T_t)_{t\geq 0}$ is said to be *integrable* if one of the equivalent conditions of Lemma 3.5.89 holds; a convolution semigroup $(\mu_t)_{t\geq 0}$ is called *integrable* if its associated Feller semigroup is integrable.

Recall that, by definition, a family M of positive Borel measures is vaguely bounded if

$$\sup_{\mu\in M}\left|\int_{\mathbb{R}^n} u(x)\mu(dx)\right| < \infty \quad \text{for all } u \in C_0(\mathbb{R}^n; \mathbb{R}). \tag{3.425}$$

Thus if μ is a shift-bounded positive measure and $K \subset \mathbb{R}^n$ is compact then $x \mapsto \mu(x + K)$ is bounded on $\mathbb{R}^n$, which follows from (3.425) by taking $u \in C_0(\mathbb{R}^n; \mathbb{R})$, $u|_K = 1$. Conversely, if the function $x \mapsto \mu(x + K)$ is for all compact sets $K \subset \mathbb{R}^n$ bounded, then μ is shift-bounded which follows when taking $\operatorname{supp} u \subset K$.

Now let $(\mu)_{t\geq 0}$ be a convolution semigroup on $\mathbb{R}^n$ with associated continuous negative definite function $\psi : \mathbb{R}^n \to \mathbb{C}$, resolvent $(\rho_\lambda)_{\lambda>0}$ and potential kernel κ.

Lemma 3.5.91. *Let $(\mu_t)_{t\geq 0}$ be a transient convolution semigroup on $\mathbb{R}^n$ with potential kernel κ. Then the measure $\frac{1}{2}(\kappa + S\kappa)$, $Sx = -x$, is positive definite and shift bounded.*

Proof: Recall that a measure is positive definite if the corresponding distribution is positive definite, compare Definition I.3.5.8. For the resolvent $(\rho_\lambda)_{\lambda>0}$ we find

$$\frac{1}{2}(\rho_\lambda + S\rho_\lambda)^\wedge(\xi) = \operatorname{Re}\frac{1}{\lambda + \psi(\xi)} = \frac{\lambda + \operatorname{Re}\psi(\xi)}{|\lambda + \psi(\xi)|^2} \geq 0,$$

implying

$$\langle \frac{1}{2}(\rho_\lambda + S\rho_\lambda), u * \overline{Su}\rangle = \int_{\mathbb{R}^n} \frac{\lambda + \operatorname{Re}\psi(\xi)}{|\lambda + \psi(\xi)|^2}|\hat{u}(\xi)|^2\, d\xi \geq 0,$$

hence $\frac{1}{2}(\rho_\lambda + S\rho_\lambda)$ is positive definite. Next we prove that every positive definite measure is shift-bounded. Indeed, given $u \in C_0(\mathbb{R}^n; \mathbb{R})$, $u \geq 0$, we find $g \in C_0(\mathbb{R}^n; \mathbb{R})$ such that $f \leq g * Sg$ implying that

$$\mu * u \leq \mu * g * Sg.$$

Since $\mu * g * Sg$ is a positive definite function it is bounded. Hence μ is shift-bounded. As we already proved, $\frac{1}{2}(\kappa + S\kappa)$ is positive definite, the lemma follows. □

Now we may relate the transience of a convolution semigroup to its integrability.

Proposition 3.5.92. *A convolution semigroup $(\mu)_{t\geq 0}$ is integrable if and only if it is transient and its potential kernel κ vanishes at infinity.*

Proof: Let $(T_t)_{t\geq 0}$ be the Feller semigroup associated with $(\mu_t)_{t\geq}$. If $(\mu_t)_{t\geq 0}$ is integrable it follows that the positivity preserving translation invariant operator $R_0|_{C_0(\mathbb{R}^n;\mathbb{R})}$ is a convolution operator, i.e.

$$R_0 u(x) = (\kappa * u)(x)$$

for all $u \in C_0(\mathbb{R}^n;\mathbb{R})$ and $x \in \mathbb{R}^n$. The Feller property of $(T_t)_{t\geq 0}$ implies that $R_0 u \in C_\infty(\mathbb{R}^n;\mathbb{R})$, i.e. κ vanishes at infinity. Moreover for the resolvent $(\rho_\lambda)_{\lambda>0}$ and for $u \in C_0(\mathbb{R}^n;\mathbb{R})$, $u \geq 0$, it follows that

$$\begin{aligned}\lim_{\lambda\to 0}\langle \rho_\lambda, u\rangle &= \lim_{\lambda\to 0} R_\lambda(Su)(0) = R_0(Su)(0)\\ &= (\kappa * Su)(0) = \langle \kappa, u\rangle,\end{aligned}$$

implying that $(\mu)_{t\geq 0}$ is transient with potential kernel κ (vanishing at infinity). Conversely, suppose that $(\mu)_{t\geq 0}$ is transient and has a potential kernel κ vanishing at infinity. For $u \in C_0(\mathbb{R}^n;\mathbb{R})$, $u \geq 0$, we have

$$\lim_{\lambda\to 0} R_\lambda u(x) = \lim_{\lambda\to 0}\langle \rho_\lambda, \tau_x(Su)\rangle = \langle \kappa, \tau_x(Su)\rangle = (\kappa * u)(x)$$

as a pointwise, increasing limit. But by assumption $\kappa * u$ belongs to $C_\infty(\mathbb{R}^n;\mathbb{R})$ implying with the arguments in the proof of Lemma 3.5.89 that

$$\lim_{\lambda\to 0} R_\lambda u = \kappa * u \quad \text{uniformly on } \mathbb{R}^n,$$

i.e. $u \in D(R_0)$ and $R_0 u = \kappa * u$. Thus $C_0(\mathbb{R}^n;\mathbb{R}) \subset D(R_0)$ and consequently $(\mu_t)_{t\geq 0}$ is integrable. □

Finally we return to the question when a subordinate convolution semigroup is transient.

Proposition 3.5.93. *Let $(\mu_t)_{t\geq 0}$ be a convolution semigroup on $\mathbb{R}^n$ and $(\eta_t)_{t\geq 0}$ a convolution semigroup supported on $[0,\infty)$ associated with the Bernstein function f. The subordinate semigroup $(\mu_t^f)_{t\geq 0}$ is transient if and only if as a vague limit*

$$\kappa_\tau := \int_0^\infty \mu_s \tau(ds) \tag{3.426}$$

exists where τ is the potential kernel associated with $(\eta_t)_{t\geq 0}$.

Proof: From Proposition 3.5.92 we deduce for $u \in C_0(\mathbb{R}^n;\mathbb{R})$, $u \geq 0$, that

$$\int_0^\infty \langle \mu_t^f, u\rangle dt = \int_0^\infty \Big(\int_0^\infty \langle \mu_s, u\rangle \eta_t(ds)\Big) dt = \int_0^\infty \langle \mu_s, u\rangle \tau(ds)$$

implying the proposition. □

Lemma 3.5.94. *Let $(\mu)_{t\geq 0}$ be a transient convolution semigroup with potential kernel κ. For every shift-bounded measure τ on $\mathbb{R}$ with* $\operatorname{supp}\tau \subset [0,\infty)$ *the vague integral*

$$\kappa_\tau := \int_0^\infty \mu_t \tau(dt)$$

exists. The measure κ_τ is shift-bounded and in addition it vanishes at infinity if κ does.

Proof: From the comments following Definition 3.5.90 we know that $c := \sup_{k\geq 0} \tau([k,k+1]) < \infty$. Define

$$\sigma := \int_0^1 \mu_t \, dt$$

as a vague integral which gives clearly a positive bounded measure on $\mathbb{R}^n$ and $\sigma(\mathbb{R}^n) > 0$. Now given $h \in C_0(\mathbb{R}^n;\mathbb{R})$, $h \geq 0$, there exists $g \in C_0(\mathbb{R}^n;\mathbb{R})$, $g \geq 0$, such that $h \leq \sigma * g$. Indeed this follows from the fact that for every $K \subset \mathbb{R}^n$ compact there exists $f \in C_0(\mathbb{R}^n;\mathbb{R})$ such that $\mu * f \leq 1$ on K. For $k \geq 0$, $s \in [k,k+1]$ and $x \in \mathbb{R}^n$ we find further

$$\begin{aligned}(\mu_s * h)(x) \leq (\mu_s * \sigma * g)(x) &= \int_0^1 (\mu_s * \mu_t * g)(x)\, dt \\ &= \int_s^{s+1} (\mu_t * g)(x)\, dt \leq \int_k^{k+2} (\mu_t * g)(x)\, dt,\end{aligned}$$

and therefore

$$\int_k^{k+1}(\mu_s * h)(x)\tau(ds) \le \tau([k,k+1]) \max_{s\in[k,k+1]}(\mu_s * h)(x)$$
$$\le c\int_k^{k+2}(\mu_t * g)(x)\,dt.$$

Thus we find

$$\begin{aligned}\int_0^\infty(\mu_t * h)(x)\,\tau(ds) &= \sum_{k=0}^\infty\int_k^{k+1}(\mu_t * h)(x)\,\tau(ds)\\ &\le c\sum_{k=0}^\infty\int_k^{k+2}(\mu_t * h)(x)\,dt\\ &\le 2c\int_0^\infty(\mu_t * h)(x)\,dt = 2c(\kappa * g)(x).\end{aligned}$$

Taking $x=0$ we arrive for every $h\in C_0(\mathbb{R}^n;\mathbb{R})$, $h\ge 0$, at

$$\int_0^\infty\langle\mu_t, Sh\rangle\tau(ds) \le 2c\langle\kappa, Sg\rangle < \infty,$$

showing that the vague integral exists and that $\kappa * h \le 2c\,\kappa * g$. By Lemma 3.5.91 the measure κ is shift-bounded implying that κ_τ vanishes at infinity if κ vanishes at infinity. □

Finally we arrive at the result already announced and which is due to F.Hirsch and M.Itô.

Theorem 3.5.95. *Let $(\mu)_{t\ge 0}$ be a transient convolution semigroup with potential kernel κ and let $(\eta_t)_{t\ge 0}$ be a convolution semigroup supported on $[0,\infty)$ with associated Bernstein function f, $f(x)>0$ for $x>0$, and potential kernel τ. Then the subordinate semigroup $(\mu_t^f)_{t\ge 0}$ is transient and has the potential kernel*

$$\kappa = \int_0^\infty \mu_t\tau(dt). \tag{3.427}$$

If $(\mu_t)_{t\ge 0}$ is integrable then $(\mu_t^f)_{t\ge 0}$ in integrable.

Proof: From Lemma 3.5.91 we know that τ is shift-bounded and $\operatorname{supp}\tau\subset[0,\infty)$. Therefore by Lemma 3.5.94 the integral (3.427) exists and by Proposition 3.5.93 it follows that $(\mu_t^f)_{t\ge 0}$ is transient with κ_τ being its potential kernel. Further Proposition 3.5.92 and Proposition 3.5.93 yield the integrability of $(\mu_t^f)_{t\geq 0}$ if $(\mu_t)_{t\geq 0}$ is integrable. $\square$

Let us recollect where we are by now in terms of examples. We have a complete characterization for the transience or recurrence of translation invariant Dirichlet semigroups. Further we can deduce the transience of semigroups of the type $T_t^\lambda = e^{-\lambda t}T_t$, $\lambda \geq \lambda_0$, when $(T_t)_{t\geq 0}$ is generated by a symmetric pseudo-differential operator $-q(x,D)$ satisfying a Gårding inequality

$$(q(x,D)u,u)_0 \geq c_0\|u\|^2_{\psi,1} - \lambda_0\|u\|_0^2.$$

We know special examples related to symbols of the form $\sum_{j=1}^n b_j(\tilde{x}_j)\psi_j^2(\xi_j)$, $\tilde{x}_j = (x_1,\ldots,x_{j-1},x_{j+1},\ldots,x_n)$. But apart of these special cases we are not aware of general theorems of the type: conditions on $q(x,\xi)$ imply $(T_t)_{t\geq 0}$ is transient or recurrent.

However we have quite satisfying conditions on $q(x,y)$ in order that the Feller semigroup $(T_t)_{t\geq 0}$ generated by $-q(x,D)$ is conservative, namely we need to have in particular $q(x,0) = 0$, compare R.Schilling [248] and [249].

Such a general theory does in fact not exists by now, however more can be proved when we take more care on the operator T_t. We will use Lemma 3.5.82 to explain this remark. Before doing so we should make a comment on the deep work by F.Hirsch upon which our discussion is based. As Lemma 3.5.82 is stated, the potential kernel τ, not the convolution semigroup $(\eta_t)_{t\geq 0}$ is clearly the central object, i.e. one may ask for kernels operating on semigroups or their potential operators. This is essentially the theme of F.Hirsch's papers [115]–[118], see also the discussion in Chr.Berg and G.Forst [26]. We cut this discussion short by posing the following question:

Are there potential kernels τ related to convolution semigroups supported on $[0,\infty)$ with associated Bernstein function f such that estimates of some operator norms of $(T_t)_{t\geq 0}$ will imply the transience of $(T_t^f)_{t\geq 0}$?

This question is already a bridge to the next section. Let us make our point clear by an example.

Example 3.5.96. Since $\mathcal{L}\big(\frac{r^{\alpha-1}}{\Gamma(\alpha)}\big)(s) = \frac{1}{s^\alpha}$, $0 < \alpha < 1$, it follows from Proposition 3.5.81 that

$$\tau(ds) := \chi_{(0,\infty)}(s)\frac{1}{\Gamma(\alpha)}s^{\alpha-1}\lambda^{(1)}(ds) \tag{3.428}$$

is the potential kernel associated with the one-sided stable convolution semigroup $(\eta_t^\alpha)_{t\geq 0}$ of order α. Let $(T_t)_{t\geq 0}$ be a transient Dirichlet semigroup on

$L^2(\mathbb{R}^n;\mathbb{R})$ and for some $h \in S(\mathbb{R}^n;\mathbb{R})$, $h > 0$, consider

$$\int_0^\infty (T_s h)(x)\,\tau(ds) = c\int_0^\infty (T_s h)(x) s^{\alpha-1}\,ds.$$

Suppose that in addition T_t is an L^∞-contraction and maps also $L^1(\mathbb{R}^n;\mathbb{R})$ boundedly into $L^\infty(\mathbb{R}^n;\mathbb{R})$ with some operator norm estimate $\|T_t\|_{L^1-L^\infty} \leq g(t)$. We find now

$$\begin{aligned}\int_0^\infty T_s h(x) s^{\alpha-1}\,ds &= \int_0^1 (T_s h)(x)s^{\alpha-1}\,ds + \int_1^\infty (T_s h)(x) s^{\alpha-1}\,ds\\ &\leq \|h\|_\infty \int_0^1 s^{\alpha-1}\,ds + \|h\|_{L^1}\int_1^\infty g(s)s^{\alpha-1}\,ds\\ &= \frac{\|h\|_\infty}{\alpha} + \|h\|_{L^1}\int_1^\infty g(s)s^{\alpha-1}\,ds.\end{aligned}$$

Thus whenever the integral $\int_1^\infty g(s)s^{\alpha-1}\,ds$ is finite then $(T_t^{f_\alpha})_{t\geq 0}$, $f_\alpha(r) = r^\alpha$, is transient.

In [117] F.Hirsch gave much more examples of potential kernels $\tau(ds)$ having a density with respect to $\lambda^{(1)}(ds)$ and being associated to convolution semigroups supported on $[0,\infty)$.

Definition 3.5.97. A function $k : (0,\infty) \to \mathbb{R}$ not being identically zero is called *logarithmically convex* if $k(x) > 0$ for all $x > 0$ and $\log k$ is convex.

Theorem 3.5.98 (F.Hirsch). *Let $k : (0,\infty) \to \mathbb{R}$ be decreasing and logarithmically convex. Then there exists a Bernstein function f, $f(x) > 0$ for $x > 0$, such that $\chi_{(0,\infty)}(s)k(s)\lambda^{(1)}(ds)$ is the potential kernel of the convolution semigroup supported on $[0,\infty)$ and associated with f.*

For a proof of Theorem 3.5.98 we refer to F.Hirsch [118] or [26].

Note that every Stieltjes transform in the sense of Definition I.3.8.17 (with $a = 0$ in (I.4.220)) is a decreasing logarithmically convex function.

3.6 Nash-Type and Sobolev-Type Inequalities — a Short Outline

In our treatise we do not handle the analysis of local generators of L^p-sub-Markovian semigroups or Dirichlet forms, and the related diffusion processes.

Our interest lies in non-local operators which generate jump-processes. Therefore we do not make any attempt to discuss "The geometry of Markov Generators", a headline we borrowed from M.Ledoux [192]. For this highly interesting and rapidly developing topic relating analysis, geometry and stochastics we refer beside to Ledoux's paper to N.Varopoulos, L.Saloff-Coste and T.Coulhon [277], the survey of A.Grigoryan [110], the most recent book of L.Saloff-Coste [241] and the references given therein. Of interest is also the survey [262] of K.-Th.Sturm and the recent paper [108] of V.Gol'dshtein and M.Troyanov. (It would be interesting to know whether the approach of A.Parmeggiani [231], see also [230] and the short, but very informative survey of C.Fefferman [89], may provide in some situation a suitable setting for handling the operators we are interested in.)

In this section we want to handle only one very special topic, namely how to obtain bounds for some operator norms of T_t, $(T_t)_{t\geq 0}$ being a symmetric sub-Markovian semigroup, by using estimates for the associated Dirichlet form. Estimates for the resolvent associated with a Dirichlet form were first derived by M.Fukushima [98] using lower L^p-bounds, $p \geq 2$, for the Dirichlet form by taking up ideas of G.Stampacchia's work [254] on second order elliptic differential operators. His ideas had been taken up by M.Tomisaki [264], in [150] and more recently by M.Kassmann [170] and [171]. It seems that N.Varopoulos [276] was the first who used consequently Nash-type and Sobolev-type inequalities to get bounds for the norms $\|T_t\|_{L^p - L^q}$ for symmetric sub-Markovian semigroups $(T_t)_{t\geq 0}$.

The following fundamental result is taken from N.Varopoulos et.al., [277].

Theorem 3.6.1. *Let $(T_t)_{t\geq 0}$ be a symmetric conservative sub-Markovian semigroup on $L^2(\mathbb{R}^n;\mathbb{R})$ with corresponding regular Dirichlet form $(\mathcal{E}, D(\mathcal{E}))$. Further let $p > 2$ and $N := \frac{2p}{p-2} > 2$. The following estimates are equivalent*

$$\|u\|_{L^p}^2 \leq c\mathcal{E}(u,u) \quad \textit{for all } u \in D(\mathcal{E}); \tag{3.429}$$

$$\|u\|_{L^2}^{2+4/N} \leq c\mathcal{E}(u,u)\|u\|_{L^1}^{4/N} \quad \textit{for all } u \in D(\mathcal{E}) \cap L^1(\mathbb{R}^n); \tag{3.430}$$

$$\|T_t\|_{L^1 - L^\infty} \leq c' t^{-N/2} \quad \textit{for all } t > 0, \tag{3.431}$$

where $\|B\|_{X-Y}$ denotes the operator norm of $B : X \to Y$.

Note that (3.429) is a *Sobolev-type inequality* and (3.430) is a *Nash-type inequality*.

An application of the Dunford–Pettis theorem, Theorem I.2.6.20, to (3.431) yields that T_t has a kernel representation, i.e.

$$T_t u(x) = \int_{\mathbb{R}^n} u(y) K_t(x,y)\, dy, \quad K_t(x,y) \geq 0, \tag{3.432}$$

and

$$\operatorname*{ess\,sup}_{x,y\in\mathbb{R}^n} K_t(x,y) \leq c_1 t^{-N/2}. \tag{3.433}$$

Clearly, when we have instead of (3.429) a *Gårding-type inequality*

$$\|u\|_{L^p}^2 \leq c_2 \mathcal{E}_\lambda(u,u) = c_2(\mathcal{E}(u,u) + \lambda(u,u)_0), \quad \lambda > 0, \tag{3.434}$$

we obtain instead of (3.431)

$$\|T_t\|_{L^1-L^\infty} \leq c_3(\lambda) \frac{e^{\lambda t}}{t^{N/2}} \tag{3.435}$$

and

$$\operatorname*{ess\,sup}_{x,y\in\mathbb{R}^n} K_t(x,y) \leq c_4(\lambda) \frac{e^{\lambda t}}{t^{N/2}}. \tag{3.436}$$

Obviously, (3.431) and (3.435) (or (3.433) and (3.436)) are *diagonal estimates* (for $K_t(x,y)$). Now, in the case of diffusion semigroups, i.e. sub-Markovian semigroups with a local generator, from (3.431) or (3.435) one may deduce using *Davies' method*, see E.B.Davies [61] or [62], to derive *off-diagonal estimates* for $K_t(x,y)$. In [44] E.Carlen, S.Kusuoka and D.Stroock gave a certain extension of Davies' method to non-local generators, i.e. non-diffusion semigroups. But to our best knowledge nobody was able to solve in a non-trivial concrete situation of interest to us the final optimization problem. Thus in case for non-local generators we have so far only diagonal estimates to our disposal and they follow almost easily from a well documented theory. For this reason we do not provide proofs for these results here.

A straightforward application of Theorem 3.6.1 to symmetric L^2-sub-Markovian semigroups generated by a pseudo-differential operator $-q(x,D)$ is possible when using (3.434). For a lot of examples in Chapter 2 we have proved the *Gårding inequality*

$$\|u\|_{H^{\psi,1}}^2 \leq c_5(B(u,u) + \lambda_0 \|u\|_{L^2}^2), \quad \lambda_0 \geq 0, \tag{3.437}$$

for some continuous negative definite function $\psi : \mathbb{R}^n \to \mathbb{R}$, and B given on $C_0^\infty(\mathbb{R}^n;\mathbb{R})$ by

$$B(u,v) = (q(x,D)u,v)_0. \tag{3.438}$$

Thus if $H^{\psi,1}(\mathbb{R}^n;\mathbb{R})$ is continuously embedded into $L^p(\mathbb{R}^n;\mathbb{R})$, $p>2$, i.e. if we have for some $p>2$ the estimate

$$\|u\|_{L^p} \le c\|u\|_{H^{\psi,1}}, \tag{3.439}$$

we arrive at (3.434), and Varopoulos' theory is at our disposal. However, minimal growth conditions on ψ, i.e.

$$\psi(\xi) \ge c_0|\xi|^{r_0} \quad \text{for } |\xi| \text{ large, } c_0>0,\ r_0\in(0,2], \tag{3.440}$$

immediately implies (3.439), compare Theorem I.3.10.12. It seems that this is the best we can at the moment get from the general theory. However, there is a fine comparison result due to M.Tomisaki [265] and interesting examples can be provided. We state first Tomisaki's result in the formulation of [132].

Theorem 3.6.2 (M.Tomisaki). *Let $(\mathcal{E}^{(1)}, D(\mathcal{E}^{(1)}))$ and $(\mathcal{E}^{(2)}, D(\mathcal{E}^{(2)}))$ be two regular, symmetric Dirichlet forms on $L^2(\mathbb{R}^n;\mathbb{R})$ and denote by $(T_t^{(1)})_{t\ge0}$ and $(T_t^{(2)})_{t\ge0}$, respectively, the corresponding L^2-sub-Markovian semigroups. Assume that $D(\mathcal{E}^{(2)}) \subset D(\mathcal{E}^{(1)})$ and that for $c_0>0$ we have*

$$\mathcal{E}^{(1)}(u,u) \le c_0\mathcal{E}^{(2)}(u,u) \tag{3.441}$$

for all $u\in D(\mathcal{E}^{(2)})$. Moreover, suppose that

$$\|T_t^{(1)}\|_{L^1-L^\infty} \le g(t) \tag{3.442}$$

for all $t>0$ with a right continuous nonincreasing function g such that

$$H(s) := \int_s^\infty \frac{G(t)}{g(G(t))}\,dt < \infty \tag{3.443}$$

for all $s>0$, where G is the left continuous inverse of g. Suppose further that $(T_t^{(2)})_{t\ge0}$ is conservative, i.e. $T_t^{(2)}1 = 1$ a.e. for all $t\ge0$. Then we have

$$\|T_t^{(2)}\|_{L^1-L^\infty} \le 2h\Big(\frac{t}{2c_0}\Big), \tag{3.444}$$

where h is the inverse function of H.

Let us discuss this result first with $g(t) = \gamma_0 t^{-\kappa}$, $t > 0$, with some $\kappa > 0$. The function g has the inverse function

$$G(t) = \gamma_0^{1/\kappa} t^{-1/\kappa}$$

and the function H becomes

$$H(t) = K\gamma_0^{1/\kappa} t^{-1/\kappa},$$

leading to

$$h(t) = \kappa^\kappa \gamma_0 t^{-\kappa} = \kappa^\kappa g(t).$$

Thus we have the two corollaries:

Corollary 3.6.3. *In the situation of Theorem 3.6.2 suppose that for some $\kappa > 0$ it holds*

$$\|T_t^{(1)}\|_{L^1-L^\infty} \leq \gamma_0 t^{-\kappa} \quad \textit{for all } t > 0.$$

Then it follows that

$$\|T_t^{(2)}\|_{L^1-L^\infty} \leq 2\Big(\frac{\kappa}{2c_0}\Big)^\kappa \gamma_0 t^{-\kappa} \tag{3.445}$$

holds for all $t > 0$.

Corollary 3.6.4. *In the situation of Theorem 3.6.2 suppose that for some $\kappa > 0$ we have*

$$\frac{1}{\gamma_0} t^{-\kappa} \leq \|T_t^{(1)}\|_{L^1-L^\infty} \leq \gamma_0 t^{-\kappa} \tag{3.446}$$

for all $t > 0$. Then it follows that

$$\|T_t^{(2)}\|_{L^1-L^\infty} \leq 2\Big(\frac{\kappa}{2c_0}\Big)^\kappa \gamma_0^2 \|T_t^{(1)}\|_{L^1-L^\infty} \tag{3.447}$$

holds for all $t > 0$.

Clearly, if we have in addition a lower bound for $\|T_t^{(2)}\|_{L^1-L^\infty}$ by the function $t \mapsto \gamma_1 t^{-\kappa}$, $\gamma_1 > 0$, we also get the estimates

$$\eta_1 \|T_t^{(1)}\|_{L^1-L^\infty} \leq \|T_t^{(2)}\|_{L^1-L^\infty} \leq \eta_2 \|T_t^{(1)}\|_{L^1-L^\infty}, \tag{3.448}$$

with suitable constants $\eta_1, \eta_2 > 0$, provided $(T_t^{(1)})_{t\geq 0}$ is also a conservative semigroup.

We want to apply these result to a special example which fits to the class introduced with W.Hoh in [131], compare Example I.4.7.31. Our example is taken from the joint paper [132] with W.Hoh. We need some preparations.

Lemma 3.6.5. *Assume $f : \mathbb{R}_+ \to \mathbb{R}_+$ is an increasing function such that $f(x) > 0$ for all $x > 0$. Moreover, assume that the Laplace transform $\mathcal{L}\,f(t) = \int_0^\infty e^{-tx} f(x)\,dx$ exists for all $t > 0$. Then there is a continuous negative definite function $\psi : \mathbb{R} \to \mathbb{R}_+$, $\psi(0) = 0$, such that the corresponding semigroup $(T_t)_{t\geq 0}$ of operators associated with the convolution semigroup $(\nu_t)_{t\geq 0}$ of measures given by*

$$\hat{\nu}_t(\xi) = (2\pi)^{-1/2} e^{-t\psi(\xi)} \tag{3.449}$$

satisfies

$$\|T_t\|_{L^1-L^\infty} = \mathcal{L}(f)(t), \quad t > 0. \tag{3.450}$$

Proof: Let $\tilde{F}(x) = \int_0^x f(s)\,ds$ for all $x \geq 0$. Then $\tilde{F}$ is continuous, increasing and convex on $[0,\infty)$. Furthermore, $\tilde{F}(0) = 0$ and $\lim_{x\to\infty} \tilde{F}(x) = \infty$ and therefore $\tilde{F}$ maps $[0,\infty)$ bijectively onto itself. Define

$$\psi(\xi) = \tilde{F}^{-1}\Big(\frac{|\xi|}{\pi}\Big), \quad \xi \in \mathbb{R}. \tag{3.451}$$

Then ψ is continuous, even and restricted to $[0,\infty)$ it is increasing and concave, hence by Polya's theorem, Theorem I.3.5.22 combined with Corollary I.3.6.10, it follows that ψ is a continuous negative definite function. Moreover,

$$\mathcal{L}\,f(t) = \int_0^\infty e^{-tx} f(x)\,dx = \frac{1}{\pi}\int_0^\infty e^{-t\tilde{F}^{-1}(\xi/\pi)}\,d\xi = \frac{1}{2\pi}\int_{\mathbb{R}} e^{-t\psi(\xi)}\,d\xi. \tag{3.452}$$

Thus, $\xi \mapsto e^{-t\psi(\xi)}$ is integrable for all $t > 0$ and

$$\|T_t\|_{L^1-L^\infty} = \sup_{x\in\mathbb{R}} \frac{1}{2\pi}\int_{\mathbb{R}} e^{ix\cdot\xi} e^{-t\psi(\xi)}\,d\xi = \frac{1}{2\pi}\int_{\mathbb{R}} e^{-t\psi(\xi)}\,d\xi = \mathcal{L}\,f(t). \quad \square$$

Now, for $\rho > 0$ consider the function $f_\rho : \mathbb{R}_+ \to \mathbb{R}_+$ given by

$$f_\rho(x) = \sum_{k=0}^\infty \frac{x^{\rho^k}}{k!\Gamma(\rho k + 1)} \tag{3.453}$$

which converges for all $x \geq 0$ and for the Laplace transform of f_ρ we find

$$\mathcal{L}\,f_\rho(t) = \sum_{k=0}^\infty \frac{1}{k!}\mathcal{L}\Big(\frac{x^{\rho^k}}{\Gamma(\rho k+1)}\Big)(t) = \sum_{k=0}^\infty \frac{1}{k!} t^{-\rho k - 1} = \frac{1}{t} e^{t^{-\rho}}. \tag{3.454}$$

Denote by $\psi_\rho : \mathbb{R} \to \mathbb{R}_+$ the continuous negative definite function associated with f_ρ by Lemma 3.6.5. Suppose $\rho_1, \ldots, \rho_n > 0$ are given. Then a Dirichlet form is given on $L^2(\mathbb{R}^n; \mathbb{R})$ by

$$\mathcal{E}(u,v) = \int_{\mathbb{R}^n} \sum_{j=1}^{n} \psi_{\rho_j}(\xi_j)\hat{u}(\xi)\overline{\hat{v}(\xi)}\, d\xi, \tag{3.455}$$

and

$$D(\mathcal{E}) = H^{\psi_\rho,1}(\mathbb{R}^n;\mathbb{R}), \quad \psi_\rho(\xi) = \sum_{j=1}^{n} \psi_{\rho_j}(\xi_j). \tag{3.456}$$

Since $\psi_{\rho_j}(0) = 0$ the corresponding semigroup is conservative.

Now, let $\psi_j : \mathbb{R} \to \mathbb{R}$, $j = 1, \ldots, n$, be continuous negative definite functions satisfying

$$\psi_j(\xi_j) \geq \psi_{\rho_j}(\xi_j) \quad \text{for all } \xi_j \in \mathbb{R},\ \psi_j(0) = 0. \tag{3.457}$$

It is clear that the Dirichlet form $(\mathcal{E}^\psi, D(\mathcal{E}^\psi))$ given by

$$\mathcal{E}^\psi(u,v) = \int_{\mathbb{R}^n} \sum_{j=1}^{n} \psi_j(\xi_j)\hat{u}(\xi)\overline{\hat{v}(\xi)}\, d\xi, \tag{3.458}$$

and

$$D(\mathcal{E}^\psi) = H^{\psi,1}(\mathbb{R}^n;\mathbb{R}), \quad \psi(\xi) = \sum_{j=1}^{n} \psi_j(\xi_j), \tag{3.459}$$

give rise to a semigroup $(T_t^\psi)_{t\geq 0}$ we may apply Theorem 3.6.2 too. Hence we find

$$\begin{aligned}
\|T_t^\psi\|_{L^1-L^\infty} &= \prod_{j=1}^{n} (2\pi)^{-1} \int_{\mathbb{R}} e^{-t\psi_j(\xi_j)}\, d\xi_j \\
&\leq \prod_{j=1}^{n} (2\pi)^{-1} \int_{\mathbb{R}} e^{-t\psi_{\rho_j}(\xi_j)}\, d\xi_j \\
&= \prod_{j=1}^{n} \mathcal{L} f_{\rho_j}(t) = \prod_{j=1}^{n} \left(\frac{1}{t} e^{t^{-\nu_j}}\right),
\end{aligned}$$

i.e.

$$\|T_t^\psi\| \leq \frac{1}{t} e^{\sum_{j=1}^{n} t^{-\nu_j}}.$$

Of course, it is now straightforward to produce an example with variable coefficients: take the Lévy–Khinchin, i.e. the Beurling–Deny representation of $\mathcal{E}^{\psi}$ and insert bounded coefficients bounded uniformly from below by a strictly positive constant.

Bibliography

[1] Adams, D. R., and L. I. Hedberg, *Function spaces and potential theory.* Grundlehren der mathematischen Wissenschafte, Vol. 314, Springer Verlag, Berlin 1996.

[2] Agmon, S., A. Douglis and L. Nirenberg, *Estimates near the boundary for solutions of elliptic partial differential equations satisfying general boundary conditions.* I. Comm. Pure Appl. Math. **12** (1959), 623–727.

[3] Agmon, S., A. Douglis and L. Nirenberg, *Estimates near the boundary for solutions of elliptic partial differential equations satisfying general boundary conditions.* II. Comm. Pure Appl. Math. **17** (1964), 35–92.

[4] Alberverio, S., and Z.-M. Ma, *A note on quasicontinuous kernels representing quasi-linear positive maps.* Forum Math. **3** (1991), 389–400.

[5] Amann, H., *Linear and quasilinear parabolic problems.* Vol. I : Abstract linear theory. Monographs in Mathematics, Vol. 89, Birkhäuser Verlag, Basel 1995.

[6] Arendt, W., Ch. Batty, M. Hieber and F. Neubrander, *Vector-valued Laplace transforms and Cauchy problems.* Monographs in Mathematics, Vol. 96, Birkhäuser Verlag, Basel 2001.

[7] Aronson, D., *Bounds for the fundamental solution of a parabolic equation.* Bull. Amer. Math. Soc. **73** (1967), 890–896.

[8] Bakry, D., *Transformations de Riesz pour les semigroupes symétriques.* In: Sém. de Probabilités XIX, Lecture Notes in Mathematics, Vol. 1123, Springer Verlag, Berlin 1985, 130–174.

[9] Bakry, D., *Inégalités de Sobolev faibles: un critère* Γ_2. In: Sém. de Probabilités XXV, Lecture Notes in Mathematics, Vol. 1485, Springer Verlag, Berlin 1991, 234–261.

[10] Bakry, D., *L' hypercontractivité et son utilisation en théorie des semigroupes.* In: Ecole d'Eté de Probabilités de St-Flour. 1992, Lecture Notes in Mathematics, Vol. 1581, Springer Verlag, Berlin 1994, 1–114.

[11] Bakry, D., and M. Ledoux, *Sobolev inequalities and Myer's diameter theorem for an abstract Markov generator.* Duke Math. J. **85** (1996), 253–270.

[12] Bakry, D., and Z. Qian, *Some new results on eigenvectors via dimension, diameter, and Ricci curvature.* Adv. Math. **155** (2000), 98–153.

[13] Baldus, F., *$S(M,g)$-pseudo-differential calculus with spectral invariance on $\mathbf{R}^n$ and manifolds for Banach function spaces.* Dissertation Johannes Gutenberg-Universität Mainz, Logos Verlag, Berlin 2001.

[14] Baldus, F., *Application of the Weyl–Hörmander calculus to generators of Feller semigroups.* Math. Nachr. (to appear)

[15] Barndorff-Nielsen, O. E., and S. Z. Levendorskiĭ, *Feller processes of normal inverse Gaussian type.* Quantitative Finance **1** (2001), 318–331.

[16] Bass, R., *Uniqueness in law for jump Markov processes.* Probab. Th. Rel. Fields **79** (1988), 271–287.

[17] Bass, R., *Probabilistic techniques in analysis.* Probability and its Applications. Springer Verlag, New York 1995.

[18] Bass, R., *On Aronson's upper bound for heat kernels.* Preprint 2002.

[19] Bauer, H., *Probability theory.* de Gruyter Studies in Mathematics, Vol. 23, Walter de Gruyter Verlag, Berlin 1996.

[20] Bauer, H., *Measure and integration theory.* de Gruyter Studies in Mathematics, Vol. 26, Walter de Gruyter Verlag, Berlin 2001.

[21] Beals, R., *A general calculus of pseudo-differential operators.* Duke Math. J. **42** (1975), 1–42.

[22] Beals, R., and C. Fefferman, *Spatially inhomogeneous pseudo-differential Operators.* I. Comm. Pure Appl. Math. **27** (1974), 1–24.

[23] Benedeck, A., and R. Panzone, *The space L^p with mixed norms.* Duke Math. J. **28** (1961), 301–324.

[24] Bennett, C., and R. Sharpley, *Interpolation of operators.* Pure and Applied Mathematics, Vol. 129, Academic Press, Boston 1988.

[25] Berg, Chr., *On the potential operators associated with a semigroup.* Studia Math. **51** (1974), 109–111.

[26] Berg, Chr., and G. Forst., *Potential theory on locally compact Abelian groups.* Ergebnisse der Mathematik und ihrer Grenzgebiete (Ser.II), Vol. 87, Springer Verlag, Berlin 1975.

[27] Bertoin, J., *Lévy processes.* Cambridge Tracts in Mathematics, Vol.121, Cambridge University Press, Cambridge 1996.

[28] Bertoin, J., *Subordination: Examples and Applications.* In: Ecole d'Eté de Probabilités de St-Flour 1997, Lecture Notes in Mathematics, Vol. 1717, Springer Verlag, Berlin 1999, 1–91.

[29] Bisgaard, T. M., and Z. Sasvari, *Characteristic functions and moment sequences. Positive definiteness in probability.* Nova Science Publisher, Huntington NY 2000. (Addition to the bibliography of volume I)

[30] Bliedtner, J., and W. Hansen., *Potential theory.* Universitext, Springer Verlag, Berlin 1986.

[31] Blumenthal, R., and R. Getoor, *Markov processes and potential theory.* Pure and Applied Mathematics, Vol. 29, Academic Press, New York 1968.

[32] Bochner, S., *Harmonic analysis and the theory of probability.* California Monographs in Mathematical Science, University of California Press, Berkeley CA 1955.

[33] Bogachev, V. I., P. Lescot and M. Röckner, *The martingale problem for pseudo-differential operators on infinite-dimensional spaces.* Nagoya Math. J. **153** (1999), 101–118.

[34] Bogachev, V. I., M. Röckner and B. Schmuland, *Generalized Mehler semigroups and applications.* Probab. Th. Rel. Fields **105** (1996), 193–225.

[35] Bouleau, N., and F. Hirsch, *Dirichlet forms and analysis on Wiener spaces.* de Gruyter Studies in Mathematics, Vol. 14, Walter de Gruyter Verlag, Berlin 1991.

[36] Brenner, P., V. Thomée and L. Wahlbin, *Besov spaces and applications to difference methods for initial value problems.* Lecture Notes in Mathematics, Vol. 434, Springer Verlag, Berlin 1975.

[37] Browder, F. E., *On the spectral theory of elliptic differential operators.* Math. Anal. **142** (1961), 22–130.

[38] Browder, F. E., *Nonlinear elliptic boundary value problems.* Bull. Amer. Math. Soc. **69** (1963), 863–874.

[39] Burckel, R., *Vorlesungen über analytische Mengen und Borelsche Abbildungen.* Mathematisches Institut Universität Erlangen-Nürnberg, Erlangen 1987.

[40] Calderón, A., *Uniqueness in the Cauchy problem for partial differential equations.* Amer. J. of Math. **80** (1958), 16–36.

[41] Calderón, A. P., and R. Vaillancourt, *On the boundedness of pseudo-differential operators.* J. Math. Soc. Japan **23** (1972), 374–378.

[42] Calderón, A., and A. Zygmund, *Algebras of certain singular integrals.* Amer. J. of Math. **79** (1957), 801–821.

[43] Caps, O., *Linear and quasilinear evolution equations in scales of Banach spaces.* Dissertation Johannes Gutenberg-Universität Mainz, 2000.

[44] Carlen, E., S. Kusuoka and D. W. Stroock, *Upper bounds for symmetric Markov transition functions.* Ann. Inst. Henri Poincaré, Probabilités et Statistiques, Sup. au n° **23:2** (1987), 245–287.

[45] Carleson, L., *Selected problems on exceptional sets.* Van Nostrand Mathematical Studies, Vol. 13, D. van Nostrand Comp., Princton NJ 1967.

[46] Carasso, A.S., and T. Kato, *On subordinated holomorphic semigroups.* Trans. Amer. Math. Soc. **327** (1991), 867–877.

[47] Chernikov, V., and S. Vodop'yanov, *Sobolev spaces and hypoelliptic equations I.* Siberian Advances in Math. **6.3** (1996), 27–67.

[48] Choquet, G., *Theory of capacities.* Ann. Inst. Fourier (Grenoble) **5** (1953/54), 131–295.

[49] Choquet, G., *Lectures on analysis, Vol.I. Integration and topological vector spaces.* Mathematical Lecture Notes Series, Vol. 24, W.A. Benjamin, Inc., Reading M.A. 1969.

[50] Choquet, G., *Lectures on analysis, Vol. II. Representation theory.* Mathematical Lecture Notes Series, Vol. 25, W.A. Benjamin, Inc., Reading M.A. 1969.

[51] Choquet, G., *Lectures on analysis, Vol.III. Infinite dimensional measures and problem solutions.* Mathematical Lecture Notes Series, Vol. 26, W.A. Benjamin, Inc., Reading M.A. 1969.

[52] Chung, K. L., *Lectures from Markov processes to Brownian motion.* Grundlehren der mathematischen Wissenschaften, Vol. 249, Springer Verlag, New York 1982.

[53] Clarkson, J. A., *Uniformly convex spaces.* Trans. Amer. Math. Soc. **40** (1936), 396–414.

[54] Coifman, R. R., and G. Weiss, *Transference methods in analysis.* CBMS Regional Conference Series in Mathematics, Vol. 31, Amer. Math. Soc., Providence RI 1977.

[55] Courrège, Ph., *Sur la forme intégro-différentielle des opérateurs de C_K^∞ dans C satisfaisant au principe du maximum.* In: Sém. Théorie du Potentiel 1965/66. Exposé 2, 38 pp.

[56] Cowling, M. G., *On Littlewood–Paley–Stein theory.* Suppl. Rend. Circolo Mat. Palermo (Ser. II) **1** (1981), 21–55.

[57] Cowling, M. G., *Harmonic analysis on semigroups.* Ann. Math. **117** (1983), 267–283.

[58] Danilov, V. G., and S. M. Frolovitchev, *Exact asymptotics of the density of the transition probability for discontinuous Markov processes.* Math. Nachr. **215** (2000), 55–90.

[59] Dappa, H., and W. Trebels, *On L^1–criteria for quasi-radial Fourier multipliers with application to some anisotropic function spaces.* Analysis Math. **9** (1983), 275–289.

[60] Davies, E.B., *One-parameter semigroups.* London Mathematical Society Monographs, Vol. 15, Academic Press, London 1980.

[61] Davies, E. B., *Explicit constants for Gaussian upper bounds on heat kernels.* Amer. J. of Math. **109** (1987), 319–934.

[62] Davies, E.B., *Heat kernels and spectral theory.* Cambridge Tracts in Mathematics, Vol. 92, Cambridge University Press, Cambridge 1989.

[63] Dellacherie, C., B. Maisonneuve et P.-A. Meyer, *Probabilités et potentiel, Chapitres XVII à XXIV: Processus de Markov (fin). Compléments de calcul stochastique.* Publication de l'Institut de Mathématique de l'Université de Strasbourg, Vol. 20, Edition Hermann, Paris 1992.

[64] Dellacherie, C., et P.-A. Meyer, *Probabilités et potentiel, Chapitres I á IV.* Publications de l'Institut de Mathématique de l'Université de Strasbourg, Vol. 15, Edition Hermann, Paris 1975.

[65] Dellacherie, C., et P.-A. Meyer, *Probabilités et potentiel, Chapitres V á VIII: Théorie des martingales.* Publications de l'Institut de Mathématique de l'Université de Strasbourg, Vol. 17, Edition Hermann, Paris 1980.

[66] Dellacherie, C., et P.-A. Meyer, *Probabilités et potentiel, Chapitres IX á XI: Théorie discrète du potentiel.* Publicaitons de l'Institut de Mathématique de l'Université de Strasbourg, Vol. 18, Edition Hermann, Paris 1983.

[67] Dellacherie, C., et P.-A. Meyer, *Probabilités et potentiel, Chapitres XII á XVI: Théorie du potentiel associée à une résolvante. Théorie des processus de Markov.* Publicaitons de l'Institut de Mathématique de l'Université de Strasbourg, Vol. 19, Edition Hermann, Paris 1987.

[68] Demuth, M., and J. van Casteren, *Stochastic spectral theory for self-adjoint Feller operators.* Probability and its Applications, Birkäuser Verlag, Basel 2000.

[69] Doob, J. L., *Regularity properties of certain families of chance variables.* Trans. Amer. Math. Soc. **47** (1940), 455–486.

[70] Doob, J. L., *Stochastic processes.* John Wiley and Sons Publishing Comp., New York 1953.

[71] Doob, J. L., *Classical potential theory and its probabilistic counterpart.* Grundlehren der mathematischen Wissenschaften, Vol. 262, Springer Verlag, Berlin 1984.

[72] Dudley, R. M., *Real analysis and probability.* The Wadsworth & Brooks Mathematical Series, Wadsworth and Brooks/Cole, Pacific Grove C.A. 1989.

[73] Dunford, N., and J. T. Schwartz, *Convergence almost everywhere of operator averages.* J. Rational Mech. Anal. **5** (1956), 129–178.

[74] Dunford, N., and J.T. Schwartz, *Linear Operators, Part I. General theory.* Pure and Applied Mathematics, Vol. 7, John Wiley — Interscience Publisher, New York 1963.

[75] Dunford, N., and J.T. Schwartz, *Linear Operators, Part II. Spectral theory — Self adjoint operators in Hilbert space.* Pure and Applied Mathematics, Vol. 7, John Wiley — Interscience Publisher, New York 1963.

[76] Dunford, N., and J.T. Schwartz, *Linear Operators, Part III. Spectral operators.* Pure and Applied Mathematics Vol. 7, John Wiley — Interscience Publisher, New York 1971.

[77] Dynkin, E. B., *Markov processes, Vol. I.* Grundlehren der mathematischen Wissenschaften, Vol. 121, Springer Verlag, Berlin, 1965.

[78] Dynkin, E. B., *Markov processes, Vol. II.* Grundlehren der mathematischen Wissenschaften, Vol. 122, Springer Verlag, Berlin, 1965.

[79] Eberle, A. *Uniqueness and non-uniqueness of semigroups generated by singular diffusion operators.* Lecture Notes in Mathematics, Vol. 1718, Springer Verlag, Berlin 1999.

[80] Edmunds, D., and W. D. Evans, *Spectral theory and differential operators.* Oxford Mathematical Monographs, Clarendon Press, Oxford 1987.

[81] Egorov, Yu., and V. Kondratiev, *On spectral theory of elliptic operators.* Operator Theory — Advances and Applications, Vol. 89, Birkhäuser Verlag, Basel 1996.

[82] Engel, K.J., and R. Nagel, *One-parameter semigroups for linear evolution equations.* Graduate Texts in Mathematics, Vol. 194, Springer Verlag, New York 2000.

[83] Ethier, S.N., and Th.G. Kurtz, *Markov processes — characterization and convergence.* Wiley Series in Probability and Mathematical Statistics, John Wiley & Sons, New York 1986.

[84] Fabes, E., and D. Stroock, *A new proof of Moser's parabolic Harnack inequality using the old ideas of Nash.* Arch. Rat. Mech. Anal. **96** (1986), 327–338.

[85] Faradzheva, M. R., *The R-identity of second-order elliptic operators with continuous coefficients.* Differential Equations **35** (1999), 1709–1715.

[86] Farkas, W., N. Jacob and R.L. Schilling, *Feller semigroups, L^p-sub-Markovian semigroups, and applications to pseudo-differential operators with negative definite symbols.* Forum Math. **13** (2001), 51–90.

[87] Farkas, W., N. Jacob and R. Schilling, *Function spaces related to continuous negative definite functions: Ψ-Bessel potential spaces.* Dissertationes Mathematicae **CCCXCIII** (2001), 1–62.

[88] Farkas, W., and H.-G. Leopold, *Characterisations of function spaces of generalized smoothness.* Preprint (2001).

[89] Fefferman, C., *Symplectic subunit balls and algebraic functions.* In: Christ, M., C. Kenig and C. Sadosky (eds.), Harmonic Analysis and Partial Differential Equations. Essays in Honor of Alberto P. Calderón. The University of Chicago Press, Chicago 1999, 199–205.

[90] Fefferman, C., and D. H. Phong, *Subelliptic eigenvalue problems* In: W. Beckner et al. (eds.), Conference on Harmonic Analysis in Honor of Antoni Zygmund, Wadsworth, Pacific Grove 1983, 590–606.

[91] Fefferman, C., and A. Sanchez-Calle, *Fundamental solutions for second order elliptic operators.* Ann. Math. **124** (1986), 247–272.

[92] Feyel, D., and A. de La Pradelle, *Capacités gaussiennes.* Ann. Inst. Fourier **41** (1991), 49–76.

[93] Frehse, J., *Capacity methods in the theory of partial differential equations.* Jahresber. Deutsche Math. Verein. **84** (1982), 1–44.

[94] Friedman, A., *Partial differential equations of parabolic type.* Prentice-Hall Inc., Englewood Cliffs NJ 1964.

[95] Friedman, A., *Partial Differential Equations.* Robert E. Krieger Publishing Company, Huntington NY, 1976

[96] Fuhrman, M., and M. Röckner, *Generalized Mehler semigroups: The non-Gaussian case.* Potential Analysis **12** (2000), 1–47.

[97] Fukushima, M., *Dirichlet spaces and strong Markov processes.* Trans. Amer. Math. Soc. **162** (1971), 185–224.

[98] Fukushima, M., *On an L^p-estimate of resolvents of Markov processes.* Publ. R. I. M. S. Kyoto Univ. **13** (1977), 191–202.

[99] Fukushima, M., *Dirichlet forms and Markov processes.* North Holland Math. Library, Vol. 23, North Holland Publishing Company, Amsterdam 1980.

[100] Fukushima, M., *Two topics related to Dirichlet forms: quasi everywhere convergence and additive functionals.* In: Fabes, E., et al. (eds.), Dirichlet Forms, Lecture Notes in Mathematics, Vol. 1563, Springer Verlag, Berlin 1993, 21–53.

[101] Fukushima, M. and H. Kaneko, *On (r,p)-capacities for general Markovian semigroups.* In: Albeverio, S., (ed.), Infinite Dimensional Analysis and Stochastic Processes, Research Notes in Mathematics, Vol. 124, Pitman Publishing, Boston M.A. 1985, 41–47.

[102] Fukushima, M., Y. Oshima and M. Takeda, *Dirichlet forms and symmetric Markov processes.* de Gruyter Studies in Mathematics, Vol. 19, Walter de Gruyter Verlag, Berlin 1994.

[103] Gagliardo, E., *Proprietá di alcune classi di funzioni in piu variabili.* Richerche Mat. **7** (1958), 102–137.

[104] Garroni, M. G., and J. L. Menaldi, *Green functions for second order parabolic integro-differential problems.* Research Notes in Mathematics, Vol. 275, Longman Scientific & Technical, Harlow, Essex 1992.

[105] Gianazza, U., and V. Vesperi, *Analytic semigroups generated by square Hörmander operators.* Rend. Istit. Mat. Univ. Trieste Suppl. Vol. 28 (1997), 201–220.

[106] Gilbarg, D., and N. Trudinger, *Elliptic partial differential equations of second order.* Grundlehren der mathematischen Wissenschaften, Vol. 224, Springer Verlag, Berlin 1977.

[107] Goldstein, J., A., *Semigroups of linear operators and applications.* Oxford Mathematical Monographs, Oxford University Press, Oxford 1985.

[108] Gol'dsthein, V., and M. Troyanov, *Axiomatic theory of Sobolev spaces.* Expo. Math. **19** (2001), 289–336.

[109] Gramsch, B., *Relative Inversion in der Störungstheorie von Operatoren und Ψ-Algebren.* Math. Ann. **269** (1984), 27–71.

[110] Grigor'yan, A., *Analytic and geometric background of recurrence and non-explosion of the Brownian motion on Riemannian manifolds.* Bull. Amer. Math. Soc. (2) **36** (1999), 135–249.

[111] Grubb, G., *Functional calculus of pseudo-differential boundary problems.* Progress in Mathematics, Vol. 65, Birkhäuser Verlag, Basel 1986.

[112] Grubb, G., *Parabolic pseudo-differential boundary problems and applications.* In: Cattabriga, L., and L. Rodino (eds.), Microlocal Analysis and Applications, Lecture Notes in Mathematics, Vol. 1495, Springer Verlag, Berlin 1991.

[113] Gruschin, V. V., *On a class of hypoelliptic operators.* Math. USSR Sbornik **12** (1970), 458–476.

[114] Gundy, R., *A decomposition for L^1 bounded martingales.* Ann. Math. Statist. **39** (1968), 134–138.

[115] Hirsch, F., *Sur une généralisation d'un théorème de M. Itô.* C. R. Acad. Sci. Paris **271** (1970), 1236–1238.

[116] Hirsch, F., *Intégrales de résolvantes et calcul symbolique.* Ann. Inst. Fourier (Grenoble) **22.4** (1972), 239–264.

[117] Hirsch, F., *Transformation de Stieltjes et fonctions opérant sur les potentiels abstraits.* In Faraut, J., (ed.), Théorie du Potentiel et Analyse Harmonique, Lecture Notes in Mathematics, Vol. 404, Springer Verlag, Berlin 1974, 149–163.

[118] Hirsch, F., *Familles d'opérateurs potentials.* Ann. Inst. Fourier (Grenoble) **25.3** (1975), 263–288.

[119] Hirsch, F., *Lipschitz functions and fractional Sobolev spaces.* Potential Analysis **11** (1999), 415–429.

[120] Hirsch, M., and S.Smale, *Differential equations, dynamical systems, and linear algebra.* Pure and Applied Mathematics, Vol. **60**, Academic Press, New York 1974.

[121] Hoh, W., *Das Martingalproblem für eine Klasse von Pseudodifferentialoperatoren.* Dissertation Universität Erlangen-Nürnberg, Erlangen 1992.

[122] Hoh, W., *Some commutator estimates for pseudo differential operators with negative definite functions as symbol.* Integr. Equat. Oper. Th., **17** (1993), 46–53.

[123] Hoh, W., *The martingale problem for a class of pseudo differential operators.* Math. Ann. **300** (1994), 121–147.

[124] Hoh, W., *Pseudo differential operators with negative definite symbols and the maringale problem.* Stoch. and Stoch. Rep. **55** (1995), 225–252.

[125] Hoh, W., *Feller semigroups generated by pseudo differential operators.* In: Ma, Z.-M., M.Röckner and J.A. Yan (eds.), Intern. Conf. Dirichlet Forms and Stoch. Processes, Walter de Gruyter Verlag, Berlin 1995, 199–206.

[126] Hoh, W., *Un calcul symbolique pour des opérateurs pseudo-différentiels engendrant des semi-groupes de Feller.* C. R. Acad. Sci. Paris **325** (1997), 1121 1124.

[127] Hoh, W., *A symbolic calculus for pseudo differential operators generating Feller semigroups.* Osaka J. Math. **35** (1998) 798–820.

[128] Hoh, W., *Pseudo differential operators generating Markov processes.* Habilitationsschrift, Universität Bielefeld, Bielefeld 1998.

[129] Hoh, W., *Pseudo differential operators with negative definite symbols of variable order.* Rev. Mat. Iberoam. **16** (2000), 219–241.

[130] Hoh, W., *On pertubations of pseudo differential operators with negative definite symbols.* Applied Anal. Optimization **45** (2002). (in press)

[131] Hoh, W., and N. Jacob, *Some Dirichlet forms generated by pseudo differential operators.* Bull. Sc. Math. **116** (1992) 383–398.

[132] Hoh, W., and N. Jacob, *Upper bounds and conservativeness for semigroups associated with a class of Dirichlet forms generated by pseudo differential operators.* Forum Math. **8** (1996), 107–120.

[133] Hoh, W., and N. Jacob, *Towards an L^p-potential theory for sub-Markovian semigroups: Variational inequalities and balayage theory.* Preprint 2002.

[134] Hoh, W., N. Jacob and A.G.Tokarev, (work in preparation).

[135] Hörmander, L., *On the general theory of partial differential operators.* Acta Math. **94** (1955), 161–248.

[136] Hörmander, L., *Estimates for translation invariant operators in L^p spaces.* Acta Math. **104** (1960), 93–140.

[137] Hörmander, L., *Pseudo-differential operators.* Comm. Pure Appl. Math. **18** (1965), 501–517.

[138] Hörmander, L., *Pseudo-differential operators and hypoelliptic equations.* In: Calderón, A. P. (ed.), Singular Integrals, Proc. Symp. Pure Mathematics, Vol. X, Amer. Math. Soc., Providence RI 1967.

[139] Hörmander, L., *Hypoelliptic second order differential equations.* Acta Math. **119** (1967), 147–171.

[140] Hörmander, L., *The Weyl calculus of pseudo-differential operators.* Commun. Pure Appl. Math. **32** (1979), 359–443.

[141] Hörmander, L., *The analysis of linear partial differential operators III.* Grundlehren der mathematischen Wissenschaften, Vol. 274, Springer Verlag, Berlin 1985.

[142] Hörmander, L., *On Sobolev spaces associated with some Lie algebras.* In: Okya, Y., et al. (eds.) Current Topics in Partial Differential Equations. Kinokuniya, Tokyo 1986.

[143] Ikeda, N., and S. Watanabe, *Stochastic differential equations and diffusion processes.* North-Holland Mathematical Library, Vol. 24, North-Holland Publishing Company, Amsterdam, and Kodansha Ltd., Tokyo, 2nd ed., 1985.

[144] Il'in, A. M., A. Kalašnikov and O. Oleinik, *Second-order linear equations of parabolic type.* Russian Math. Surveys **17.3** (1962), 1–143.

[145] Itô, K., and H. P. McKean, *Diffusion processes and their sample paths.* Grundlehren der mathematischen Wissenschaften, Vol. 125, Springer Verlag, Berlin 1974.

[146] Itô, M., *Sur une famille sous-ordennée au nogau de convolution de Hunt donné.* Nagoya Math. J. **41** (1973), 45–56.

[147] Itô, S., *Diffusion equations.* Translations of Mathematical Monographs, Vol. 114, Amer. Math. Soc., Providence RI 1992.

[148] Iwasaki, C., *The fundamental solution for pseudo differential operators of parabolic type.* Osaka J. Math. **14** (1977), 569–592.

[149] Jacob, N., *Dirichlet forms and pseudo differential operators.* Exop. Math. **6** (1988), 313–351.

[150] Jacob, N., *L^∞-bounds for solutions of pseudo differential boundary problems in Dirichlet spaces.* Expo. Math. **6** (1988), 363–371.

[151] Jacob, N., *A Gårding inequality for certain anisotropic pseudo differential operators with non-smooth symbos.* Osaka J. Math. **26** (1989), 857–879.

[152] Jacob, N., *Commutator estimates for pseudodifferential operators with negative definite functions as symbol.* Forum Math. **2** (1990), 155–162.

[153] Jacob, N., *Feller semigroups, Dirichlet forms, and pseudo differential operators.* Forum Math. **4** (1992), 433 446.

[154] Jacob, N., *Further pseudo differential operators generating Feller semigroups and Dirichlet forms.* Rev. Mat. Iberoam. **9** (1993), 373–407.

[155] Jacob, N., *A class of Feller semigroups generated by pseudo-differential operators.* Math. Z. **215** (1994), 151–166.

[156] Jacob, N., *Non-local (semi-) Dirichlet forms generated by pseudo differential operators.* In Ma, Z.-M., M. Röckner and J. A. Yan (ed.), Intern. Conf. Dirichlet Forms and Stoch. Processes, Walter de Gruyter Verlag, Berlin 1995, 223–233.

[157] Jacob, N., *Lineare partielle Differentialgleichungen.* Akademie Verlag, Berlin 1995.

[158] Jacob, N., *Pseudo-Differential Operators and Markov Processes.* Mathematical Research Notes, Vol. 94, Akademie Verlag, Berlin 1996.

[159] Jacob, N., *Characteristic functions and symbols in the theory of Feller processes.* Potential Analysis **8** (1998), 61–68.

[160] Jacob, N., *Pseudo-differential operators and Markov processes. Vol. 1: Fourier analysis and semigroups.* Imperial College Press, London 2001.

[161] Jacob, N., *Pseudo-differential operators and Markov processes. Vol. 3: Markov processes and applications.* (in preparation).

[162] Jacob, N., and H.-G. Leopold, *Pseudo-differential operators with variable order of differentiation generating Feller semigroups.* Integr. Equat. Oper. Th. **17** (1993), 544–553.

[163] Jacob, N., and R.L. Schilling, *Subordination in the sense of S.Bochner. — An approach through pseudo differential operators.* Math. Nachr. **178** (1996), 199–231.

[164] Jacob, N., and R. Schilling, *Estimates for Feller semigroups generated by pseudo differential operators.* In: Rakosnik, J. (ed.), Function Spaces, Differential Operators and Nonlinear Analysis. Prometheus Publishing House, Praha 1996, 27–49.

[165] Jacob, N., and R.L. Schilling, *An analytic proof of the Lévy–Khinchin formula on* $\mathbf{R}^n$. Publ. Math. Debrecen **53** (1998), 69–89.

[166] Jacob, N., and R. Schilling, *Towards an* L^p*-potential theory for sub-Markovian semigroups: Kernels and capacities.* Preprint 2002.

[167] Jacobs, K., *Measure and integral.* Probability and Mathematical Statistics, Academic Press, New York 1978.

[168] Jacod, J., *Calcul stochastique et problèms de martingales.* Lecture Notes in Mathematics, Vol. 714, Springer Verlag, Berlin 1979.

[169] Jacod, J., and A. N. Shiryaev, *Limit theorems for stochastic processes.* Grundlehren der mathematischen Wissenschaften, Vol. 288, Springer Verlag, Berlin 1987.

[170] Kassmann, M., *Harnack-Ungleichungen für nichtlokale Differentialoperatoren und Dirichlet-Formen.* Bonner Mathematische Schriften, Vol. 336, Bonn 2001.

[171] Kassmann, M., *On regularity for Beurling–Deny type Dirichlet forms.* (submitted)

[172] Kazumi, T., and I. Shigekawa, *Measures of finite (r,p)-energy and potentials on a separable metric space.* In: Séminaire de Probabilités XXVI, Lecture Notes in Mathematics, Vol. 1526, Springer Verlag, Berlin 1992, 415–444.

[173] Kikuchi, K., and A. Negoro, *Pseudo differential operators and spaces of variable order of differentiation,* Rep. Fac. Liberal Arts, Shizuoka University **31** (1995), 19–27.

[174] Kikuchi, K., and A. Negoro, *On Markov processes generated by pseudo differential operators of variable order.* Osaka J. Math. **34** (1997), 319–335.

[175] Kinderlehrer, D., and G. Stampacchia., *An introduction to variational inequalities and their applications.* Pure and Applied Mathematics, Vol. 88, Academic Press, New York 1980.

[176] Kochubei, A. N., *Parabolic pseudodifferential equations, hypersingular integrals and Markov processes.* Math. USSR Izvestija **33** (1989), 233–259.

[177] Kohn, J. J., *Pseudo-differential operators and hypoellipticity.* In: Spencer, D. C. (ed.), Proc. Symp. Pure Math., Vol. 23, Amer. Math. Soc., Providence RI 1973, 61–69.

[178] Kohn, J. J., *Subelliptic estimates.* In: Weiss, G., and S. Wainger (eds.), Harmonic Analysis in Euclidean Spaces, Proc. Symp. Pure Math., Vol 35.2, Amer. Math. Soc., Providence RI 1979, 143–152.

[179] Kohn, J. J., and Nirenberg, *An algebra of pseudo-differential operators.* Comm. Pure Appl. Math. **18** (1965), 265–305.

[180] Kolokoltsov, V. N., *Symmetric stable laws and stable-like jump-diffusions.* Proc. London Math. Soc. (3) **80** (2000), 725–768.

[181] Kolokoltsov, V. N., *Semiclassical analysis for diffusions and stochastic processes.* Lecture Notes in Mathematics, Vol. 1724, Springer Verlag, Berlin 2000.

[182] Komatsu, T., *Markov processes associated with certain integro-differential operators.* Osaka J. Math. **10** (1973), 271–303.

[183] Komatsu, T., *Pseudo-differential operators and Markov processes.* J. Math. Soc. Japan **36** (1984), 387–418.

[184] Komatsu, T., *On the martingale problem for generators of stable processes with perturbations.* Osaka J. Math. **21** (1984), 113–132.

[185] Komatsu, T., *Continuity estimates for solutions of parabolic equations associated with jump type Dirichlet forms.* Osaka J. Math. **25** (1988), 697–728.

[186] Komatsu, T., *Uniform estimates for fundamental solutions associated with non-local Dirichlet forms.* Osaka J. Math. **32** (1995), 833–860.

[187] Komatsu, T., *On stable-like processes.* In: S.Watanabe et al. (eds.), 7th Japan–Russian Symposium on Probability Theory and Mathematical Statistics. World Scientific Press, Singapore 1996.

[188] Krengel, U., *Ergodic theorems.* de Gruyter Studies in Mathematics, Vol. 6, Walter de Gruyter Verlag, Berlin 1985.

[189] Kumano-go, H., *Pseudo-differential operators.* MIT Press, Cambridge M. A. 1974.

[190] Landkof, N. S., *Foundations of modern potential theory.* Grundlehren der mathematischen Wissenschaften, Vol. 180, Springer Verlag, Berlin 1972.

[191] Lax, P., *On Cauchy's problem for hyperbolic equations and the differentialbility of solutions of elliptic equations.* Comm. Pure Appl. Math. **8** (1955), 615–633.

[192] Ledoux, M., *The geometry of Markov diffusion generators.* Ann. Fac. Sci. Toulouse **9** (2000), 305–366.

[193] LeJan, Y., *Balayage et formes de Dirichlet.* Z. Wahrscheinlichkeitstheorie verw. Gebiete **37** (1977), 297–319.

[194] LeJan, Y., *Mesures associées à une forme de Dirichlet.* Applications. Bull. Soc. Math. France, **106** (1978), 61–112.

[195] Leopold, H.-G., *Pseudodifferentialoperatoren und Functionenräume variabler Glattheit.* Dissertation B, Friedrich-Schiller-Universität Jena, Jena 1987.

[196] Leopold, H.-G., *On function spaces of variable order of differentiation.* Forum Math. **3** (1991), 69–82.

[197] Lescot, T., and M. Röckner, *Generators of Mehler-type semigroups as pseudo-differential operators.* Infinite Dimensional Analysis and Quantum Probability. (to appear)

[198] Levi, E. E., *Sulle equazioni lineari totalmente ellittiche alle derivate parziali.* Rend. del. Circ. Mat. Palermo **24** (1907), 275–317.

[199] Long, R., *Martingale spaces and inequalities.* Peking University Press-Vieweg Publishing, Beijing-Wiesbaden 1993.

[200] Louhivaara, I., and C. Simader, *Fredholmsche verallgemeinerte Dirichletprobleme für koerzitive lineare partielle Differentialgleichungen.* In: Proc. of the Rolf Nevanlinna symposium on complex analysis, Silivri. Publ. Math. Research Inst. Istanbul, Vol. 7, Istanbul 1978, 47–57.

[201] Louhivaara, I., and C. Simader, *Über koerzitive lineare partielle Differentialoperatoren: Fredholmsche verallgemeinerte Dirichletprobleme und deren Klasseneinteilung.* In: Boboljubov, N. N., et al. (eds.), Complex Analysis and its Applications. A collection of papers dedicated to I.N.Vekua on his 70th birthday. Izdat. Nanka Moscow 1978, 342–345.

[202] Lunardi, A., *Analytic semigroups and optimal regularity in parabolic problems.* Progress in Nonlinear Differential Equations and their Applications, Vol.16, Birkhäuser Verlag, Basel 1995.

[203] Ma, Z.-M., L. Overbeck and M. Röckner, *Markov Processes associated with semi-Dirichlet forms.* Osaka J. Math. **32** (1995), 97–119.

[204] Ma, Z.-M., and M. Röckner, *An introduction to the theory of (non-symmetric) Dirichlet forms.* Universitext, Springer Verlag, Berlin 1992.

[205] Malliavin, P., *Stochastic calculus of variation and hypoelliptic operators.* In Itô, K., (ed.), Proc. Intern. Symp. on Stochastic Differential Equations. Kinokuniya, Tokyo 1978, 195-263.

[206] Malliavin, P., *Parametrix trajectorielle.* C. R. Acad. Sa. Paris **281** (1975), 241–245.

[207] Malliavin, P., *Stochastic Analysis.* Grundlehren der mathematischen Wissenschaften, Vol. 313, Springer Verlag, Berlin 1997.

[208] Malliavin, P., Private communication. Bielefeld 2001.

[209] Martinez Carracedo, C., and M. Sanz Alix, *The theory of fractional powers of operators.* North-Holland Mathematics Studies, Vol. 187, Elsevier, Amsterdam 2001.

[210] Maslov, V. P., and I. A. Shishmarev, *On the behavior of one-parameter operator semigroups* e^{tA} *as* $t \to \infty$. Uspehi Mat. Nauk **29** (1973), 177. (in Russian)

[211] Maz'ya, V. G., and V. P. Havin, *Nonlinear potential theory.* Russian Math. Surveys **27** (1972), 71–148.

[212] Maz'ya, V. G., and T. Shaposhnikova, *An elementary proof of the Brezis and Mironescu theorem on the composition operator in fractional Sobolev spaces.* J. Evolution Equations **2** (2002), 113–125.

[213] Meyer, P.-A., *Démonstration probabiliste de certaines inégalités de Littlewood–Paley.* Exposé II: L'opérateur carrée du champ. In: Sém. de Probabilités X, Lecture Notes in Mathematics, Vol. 511, Springer Verlag, Berlin 1976, 142–161.

[214] Meyers, N., and J. Serrin, $H = W$. Proc. Math. Acad. Sci. U.S.A. **51** (1964), 1055–1056.

[215] Mikulevicius, R. and H. Pragarauskus, *On the Cauchy problem for certain integro-differential operators in Sobolev and Hölder spaces.* Liet. Matem. Rink **32** (1992), 299–331.

[216] Mikulevicius, R. and H. Pragarauskus, *On the martingale problem associated with non degenerate Lévy operators.* Lithuanian Math. J. **32** (1992), 297–311.

[217] Mikulevicius, R. and B. L. Rozovskii, *Martingale problems for stochastic PDE's.* In: Carmona, R. A. and B. Rozovskii (eds.), Mathematical Surveys and Monographs, Vol. **64**, Amer. Math. Soc., Providence RI 1999, 243–325.

[218] Minty, G., *On a monotonicity method for the solution of non linear equations in Banach spaces.* Proc. Nat. Acad. Sci. U.S.A. **50** (1963), 1038–1041.

[219] Mizuta, Y., *Potential theory in Euclidean spaces.* Gakuto International Series, Mathematical Sciences and Applications, Vol. 6, Gakkōtosho, Tokyo 1996.

[220] Nagel, A., E.M. Stein and S. Wainger, *Balls and metrics defined by vector fields I: Basic properties.* Acta Math. **155** (1985), 103–147.

[221] Nash, J., *Continuity of solutions of parabolic and elliptic equations.* Amer. J. of Math. **80** (1958), 931–954.

[222] Negoro, A., *Stable-like processes: Construction of the transition density and behavior of sample path near $t = 0$.* Osaka J. Math. **31** (1994), 189–214.

[223] Negoro, A., and M. Tsuchiya, *Stochastic processes and semigroups associated with degenerate Lévy generating operators.* Stoch. and Stoch. Rep. **26** (1989), 29–61.

[224] Nirenberg, L., *On elliptic partial differential equations.* Ann. Scuola Norm. Sup. Pisa **13** (1959), 115–161.

[225] Oleinik, O., and E. Radkevic, *Second order equations with non negative characteristic form.* Amer. Math. Soc. and Plenum Press, Providence RI and New York 1973.

[226] Oshima, Y., *Dirichlet spaces.* Lecture Notes Universität Erlangen-Nürnberg, Erlangen 1988.

[227] Oshima, Y., *On conservativeness and recurrence criteria for Markov processes.* Potential Analysis **1** (1992), 115–131.

[228] Oshima, Y., *A short introduction to the general theory of Dirichlet forms.* Lecture Notes Universität Erlangen-Nürnberg, Erlangen 1994.

[229] Ouhabaz, E. M., *Propriétés d'ordre et de contractivité des semi-groupes avec applications auc opératurs elliptiques* Thèse de l'Université de Franche-Comté, 1992.

[230] Parenti, C., and A. Parmeggiani, *Some remarks on almost-positivity of ψdo's.* Boll. U. M. I. (8) **1-B** (1998), 187–215.

[231] Parmeggiani, A., *Subunit balls for symbols of pseudo differential operators.* Adv. Math. **131** (1997), 357–452.

[232] Pazy, A., *Semigroups of linear operators and applications to partial differential equations.* Applied Mathematical Sciences, Vol. 44, Springer Verlag, New York 1983.

[233] Podolynny, S. I., and N. I. Portenko, *On multidimensional stable processes with locally unbounded drift.* Random Oper. and Stoch. Equ. **3** (1995), 113–124.

[234] Portenko, N. I., *Some perturbations of drift-type for symmetric stable processes.* Random Oper. and Stoch. Equ. **2** (1994), 211–224.

[235] Reiter, H., and J.Stegeman, *Classical harmonic analysis and locally compact groups.* London Mathematical Society Monographs, New Series, Vol. 22, Clarendon Press, Oxford 2000.

[236] Rogers, C. A., *Hausdorff Measures.* Cambridge Mathematical Library. Cambridge University Press, Cambridge 1998.

[237] Rota, G. C., *An "Alternierende Verfahren" for general positive operators.* Bull. Amer. Math. Soc. **68** (1962), 95–102.

[238] Rothschild, L., and E.M. Stein, *Hypoelliptic differential operators and nilpotent groups.* Acta Math. **137** (1976), 247–320.

[239] Rubin, B., *Fractional integrals and potentials.* Monographs and Surveys in Pure and Appl. Mathematics, Vol. 82, Longman Harlow (Essex) 1996.

[240] Rudin, W., *Functional analysis.* McGraw-Hill Series in Higher Mathematics. McGraw-Hill Book Company, New York 1973.

[241] Saloff-Coste, L., *Aspects of Sobolev-type inequalities.* London Math. Soc. Lecture Notes Series, Vol. 289, Cambridge University Press, Cambridge 2002.

[242] Saloff-Coste, L., and D. Stroock, *Opérateurs uniformément sous-elliptiques sur les groupes de Lie.* J. Funct. Anal. **98** (1991), 97–121.

[243] Samko, S., A. Kilbas and O. Marichev, *Fractional integrals and derivatives.* Theory and Applications. Gordon and Breach, Amsterdam 1993.

[244] Sato, K., *Lévy processes and infinitely divisible distributions.* Cambridge Studies in Advanced Mathematics, Vol. 68, Cambridge University Press, Cambridge 1999.

[245] Schechter, M., *Modern methods in partial differential equations.* McGraw-Hill, Inc. New York 1977.

[246] Schilling, R. L., *On the domain of the generator of a subordinate semigroup.* In Kral, J., et al. (eds.), Proc Conf. on Potential Theory, Walter de Gruyter Verlag, Berlin 1996, 449–462.

[247] Schilling, R. L., *Subordination in the sense of Bochner and a related functional calculus.* J. Austral. Math. Soc. (Ser.A) **64** (1998), 368–396.

[248] Schilling, R. L., *Conservativeness and extensions of Feller semigroups.* Positivity **2** (1998), 239–256.

[249] Schilling, R. L., *Conservativeness of semigroups generated by pseudo differential operators.* Potential Analysis **9** (1998), 91–104.

[250] Schilling, R. L., *Dirichlet operators and the positive maximum principle.* Integr. Equ. Oper. Theory **41** (2001), 74–92.

[251] Silverstein, M. L., *Symmetric Markov processes.* Lecture Notes in Mathematics, Vol. 426, Springer Verlag, Berlin 1974.

[252] Simader, C., *On Dirichlet's boundary value problem.* Lecture Notes in Mathematics, Vol. 268, Springer Verlag, Berlin 1972.

[253] Sogge, Chr. D., *Fourier integrals in classical analysis.* Cambridge Tracts in Mathematics, Vol. 105, Cambridge University Press, Cambridge 1993.

[254] Stampacchia, G., *Le problème de Dirichlet pour les équations elliptiques du second ordre à coefficients discontinus.* Ann. Inst. Fourier (Grenoble) **15** (1965), 189–258.

[255] Stein, E. M., *Topics in Harmonic Analysis (Related to the Littlewood–Paley Theory).* Annals of Mathematics Studies, Vol. 63, Princeton University Press, Princeton NJ 1970.

[256] Stein, E. M., *Singular integrals and differentiability properties of functions.* Princeton Mathematical Series, Vol. 30, Princeton University Press, Princeton NJ 1970.

[257] Stroock, D. W., *Diffusion processes associated with Lévy generators.* Y. Wahrscheinlichkeitstheorie verw. Gebiete **32** (1975), 209–244.

[258] Stroock, D. W., *Diffusion semigroups corresponding to uniformly elliptic divergence form operators.* In: Séminaire de Probabilités XXII. Lecture Notes in Mathematics, Vol. 1321, Springer Verlag, Berlin 1988, 316–347.

[259] Stroock, D. W., *Probability theory — An analytic view.* Cambridge University Press, Cambridge 1993.

[260] Sturm, K.-Th., *Störung von Hunt-Prozessen durch signierte additive Funktionale.* Dissertation Universität Erlangen-Nürnberg, Erlangen 1989.

[261] Sturm, K.-Th., *Gauge Theorems for resolvents with applications to Markov processes.* Probab. Th. Rel. Fields **89** (1991), 387–406.

[262] Sturm, K.-Th., *Diffusion processes and heat kernels on metric spaces.* Ann. Prob. **26** (1998), 1–55.

[263] Tanabe, H., *Equations of evolution.* Monographs and Studies in Mathematics, Vol. 6, Pitman Publishing, London 1979.

[264] Tomisaki, M., *Some estimates for solutions of equations related to non-local Dirichlet forms.* Rep. Fac. Sci. Eng. Saga Univ. **6** (1978), 1–7.

[265] Tomisaki, M., *Comparison theorems on Dirichlet norms and their applications.* Forum Math. **2** (1990), 277–295.

[266] Trèves, F., *Introduction to pseudo differential and Fourier integral operators, Vol. 1.* The University Series in Mathematics. Plenum Press, New York 1980.

[267] Triebel, H., *Interpolation theory, function spaces, differential operators.* North Holland Mathematical Library, Vol. 18, North-Holland Publishing Company, Amsterdam 1978.

[268] Triebel, H., *Theory function spaces.* Monographs in Mathematics, Vol. 78, Birkhäuser Verlag, Basel 1983.

[269] Triebel, H., *Theory function spaces II.* Monographs in Mathematics, Vol. 84, Birkhäuser Verlag, Basel 1992.

[270] Triebel, H., *Truncations of functions.* Forum Math. **12** (2000), 371–356.

[271] Triebel, H., *The structure of functions.* Monographs in Mathematics, Vol. 97, Birkhäuser Verlag, Basel 2001.

[272] Tsuchiya, M., *Lévy measure with generalized polar decomposition and the associated SDE with jumps.* Stoch. and Stoch. Rep. **38** (1992), 95–117.

[273] Tsutsumi, C., *The fundamental solution for a degenerate parabolic pseudo-differential operator.* Proc. Japan Acad. **50** (1974), 11–15.

[274] Unterberger, A., and J. Bokobza, *Les opérateurs pseudo différentiels d'ordre variable.* C. R. Acad. Sci. Paris **261** (1965), 2271–2273.

[275] Ushakov, N. G., *Selected topics in characteristic functions.* Modern Probability and Statistics, VSP, Utrecht 1999. (Addition to the bibliography of volume I)

[276] Varopoulos, N. Th., *Hardy–Littlewood theory for semigroups.* J. Funct. Anal. **63** (1985), 240–260.

[277] Varopoulos, N. Th., L. Saloff-Coste and T. Coulhon, *Analysis and geometry on groups.* Cambridge Tracts in Mathematics, Vol. 100, Cambridge University Press, Cambridge 1992.

[278] Voigt, J., *One-parameter semigroups acting simultaneously on different L_p spaces.* Bull. Soc. Royale Sci. Liège **61** (1992), 465–470.

[279] Watanabe, S., *Comments on K. Yosida's papers on potential theory.* In: K. Itô (ed.), Kôsuku Yosida "Collected Papers", Springer Verlag, Tokyo 1992, 563-564.

[280] Westphal, U., *Ein Kalkül für gebrochene Potenzen infinitesimaler Erzeuger von Halbgruppen und Gruppen von Operatoren. Teil I: Halbgruppenerzeuger.* Compositio Math. **22** (1970), 67–103.

[281] Westphal, U., *Ein Kalkül für gebrochene Potenzen infinitesimaler Erzeuger von Halbgruppen und Gruppen von Operatoren. Teil II: Gruppenerzeuger.* Compositio Math. **22** (1970), 104–136.

[282] Xu, Ch.-J., *Subelliptic variational problems.* Bull. Soc. Math. France. **118** (1990), 147–169.

[283] Xu, Ch.-J., *Dirichlet problems for the quasilinear second order subelliptic equations.* Acta Math. Sinica (N. S.) **12** (1996), 18–32.

[284] Yosida, K., *The existence of the potential operator associated with an equi-continuous semi-group of class (C_0).* Studia Math. **31** (1958), 531–533.

[285] Yosida, K., *On the existence and a characterization of abstract potential operators.* In: 3ième Colloq. sur l'Analyse Fonctionnelle Liège 1970, Vander, Louvain 1971, 129–136.

[286] Yosida, K., *On the pre-closedness of Hunt's potential operators and its applications.* In: Yosida, K., (ed.), Proc. Intern. Conf. Functional Analysis and Related Topics, University of Tokyo Press, Tokyo 1970, 324–331.

[287] Yosida, K., *Abstract potential operators on Hilbert spaces.* Publ. R. I. M. S. **8** (1972), 201–205.

[288] Yosida, K., *Functional analysis.* 4th ed. Grundlehren der mathematischen Wissenschaften, Vol. 123, Springer Verlag, Berlin 1974.

[289] Zeidler, E., *Nonlinear functional analysis and its applications II/B: Non linear monotone operators.* Springer Verlag, New York 1990.

[290] Ziemer, W., *Weakly differentiable functions.* Graduate Texts in Mathematics, Vol. 120, Springer Verlag, New York 1989.

[291] Zolotarev, V., *One-dimensional stable distributions.* Translations of Mathematical Monographs, Vol. 65, Amer. Math. Soc., Providence RI 1986.

[292] Zygmund, A., *Trigonometric Series.* 2nd ed. Vol. I & II combined. Cambridge Mathematical Library, Cambridge University Press, Cambridge 1959/1988.

Author Index

Subject Index